STUDIES IN GEOMETRY

A SERIES OF BOOKS IN MATHEMATICS

Editors: R. A. Rosenbaum
G. Philip Johnson

STUDIES IN GEOMETRY

Leonard M. Blumenthal
University of Missouri

Karl Menger
Illinois Institute of Technology

W. H. FREEMAN AND COMPANY
San Francisco

Printed in the United States of America
Library of Congress Catalog Number: 74–75624
International Standard Book Number: 0–7167–0437–4

1 2 3 4 5 6 7 8 9 0

PREFACE

PARTS 1 AND 3

The "algebraic" concept of a *lattice* and the "geometric" notion of a *metric space* are two important devices by means of which vast areas of mathematics are unified. Although the history of mathematics suggests that the great mathematicians of the past achieved their classic results by obeying the classical injunction *divide et impera*, modern investigators have made many of their most notable contributions by adopting the very different principle of *conjunge et impera* (unify and rule). The process of unification by abstraction had its initial successes in the study of abstract groups (the last of the great concepts of mathematics, according to Oswald Spengler!), which began with Kronecker and Frobenius during the last quarter of the nineteenth century, and in the development of abstract spaces, initiated by Fréchet in the first decade of this century.

What we call unification by abstraction was clearly what E. H. Moore had in mind when, in his New Haven Colloquium lectures of 1906, he stated: "The existence of analogies between central features of various theories implies the existence of a general theory which underlies the particular theories and unifies them with respect to those central features."

Nowadays, those "central features" appear as abstract, primitive notions whose fundamental properties are given in a set of postulates. This procedure is followed explicitly in Part 1, and is implicit in Part 3, since the theory developed there is part of the distance geometry of a general metric space. Thus, unification by abstraction is the *leitmotiv* of these two studies. Parts 1 and 3 embody results that, for the most part, are due to me and some of my students, and which were published in research journals over the nearly two decades that elapsed since the appearance of my *Theory and Applications of Distance Geometry*, Clarendon Press, Oxford, 1953. Much of the contents of these parts finds detailed exposition here for the first time.

Part 1 contains more than enough material for a semester's course in some geometrical aspects of lattice theory. In addition to dealing with the geometry of metric lattices, I develop both the distance geometry and the topology of Boolean metric spaces, a theory due entirely to me and two of my students, David O. Ellis and C. J. Penning.

Part 3 comprises a course in two important topics of the distance geometry of metric spaces. It contains a detailed metric characterization of Banach and euclidean spaces and a purely metric study of the differential geometry of curves and (to a lesser extent) surfaces. The study of curves found in Part 3 could be combined with Part 4 to form a course in metric and topological curve theory in metric spaces.

Studies in Geometry is intended for graduate students of mathematics and their teachers—at all stages of their careers. Exercises (some of which are better called problems) abound; these are intended to illustrate the theory presented in the text, but many actually extend the theory.

June 1969   LEONARD M. BLUMENTHAL

PARTS 2 AND 4

Parts 2 and 4 of this book are studies in the geometry of projective and related spaces and in curve theory, respectively. Projective geometry is presented as an algebra of two operations, corresponding to the joining and intersecting of flats. This theory, developed in the author's Vienna Colloquium (1928–1936) and only recently perfected by the work of Dr. Helen Skala, has not heretofore been published in its entirety.

The essential source of the algebra of (projective) geometry, which overlaps with lattice theory, is one dual pair of formulas concerning two operations. In conjunction with simple assumptions about the existence of a universe, a vacuum and a certain configuration, these *Projectivity Laws* yield the entire theory of finite-dimensional projective spaces. In particular, one can introduce a part relation and define a point as "that which has no part"—a definition that is the initial sentence in Euclid's *Elements*, but which had lain fallow until I incorporated it in the algebra of geometry 40 years ago. Similarly, one introduces lines, planes, and so on as elements bearing special relations to the operations—not as sets of points.

Part 2 also contains a special treatment of two- and three-dimensional projective, affine, and related spaces, which include a self-dual fragment of the affine plane, convex plane regions, and the hyperbolic plane of Bolyai and Lobachevsky. Their noneuclidean geometry is here developed exclusively in terms of joins of points and meets of lines. Thus, in Hilbert's terminology, this is a development on the basis of axioms on alignment alone, without the traditional axioms on order, parallelism, congruence, and perpendicularity. In euclidean geometry, on the other hand, assumptions about concepts other than join and meet are indispensable. This situation does not seem to bear out a remark repeatedly made by Poincaré: that among the geometries that might be used in describing physical space, euclidean geometry would always be distinguished by its greater simplicity. Developed by students in the author's Notre Dame Colloquium (1937–1946), hyperbolic geometry in the terms of joining and intersecting, has not previously been published in book form.

New applications of fragments of affine and hyperbolic planes are herein made to the kinematics of Galileo and of Einstein and Minkowski. The kinematics of a 1-dimensional space are derived from assumptions about the undefined concepts of *event* and *motion*, just as the geometries are developed as postulational theories in terms of *points* and *lines*.

Part 4 deals with the concepts of *curve* and *ramification*. Everyone associates intuitive ideas with these terms. But when mathematicians tried to make these ideas precise, they came up with so many divergent proposals that in 1914, in the first edition of his famous *Mengenlehre*, Hausdorff expressed doubts about the possibility of finding a satisfactory definition. Less than 10 years later, however, the problem was solved, and the definitions yielded an extensive theory of curves and ramification, which was essentially completed in 1932. (The entire theory, plus the related work of American and Polish mathematicians, was presented in my book *Kurventheorie*, reprinted by the Chelsea Press, The Bronx, N.Y., 1967.) During the past two decades, however, curve theory has been somewhat neglected. This is regrettable, because the extension of the ramification concept to surfaces still presents open problems, as do applications of curve theory to real functions. Results of curve theory have recently been utilized in unexpected characterizations of analytic and algebraic complex functions. Moreover, a theorem on graphs, suggested by general curve-theoretical studies, is now widely used in electrical engineering.

Parts 2 and 4 thus introduce the student to particular geometric theories of two altogether different types. One is based on assumptions in terms of undefined concepts; the other, on definitions in terms of sets. After absorbing these theories, students and beginning teachers of mathematics might be better equipped to understand a great number of other theories.

There is another group that could profit from the study of the book— students of philosophy. Not only philosophers of science, but every philosopher should understand the unifying aspects of some modern mathematical theories, the details of some particular postulational theory, and the procedure of explicating concepts, which are nowhere more clearly demonstrated than in curve theory. Philosophers who doubt this should remember the famous words that Plato wrote on the entrance gate to his academy: *Let no one unacquainted with geometry enter here.*

June 1969   KARL MENGER

CONTENTS

Part 2   PROJECTIVE AND RELATED STRUCTURES

Part 3   METRIC GEOMETRY

10 DIFFERENTIAL GEOMETRY OF METRIC ARCS   319

11 METRIZATIONS OF SURFACE CURVATURE   362

Part 4   CURVE THEORY

12 CLASSICAL DEFINITIONS OF CURVES   391

STUDIES IN GEOMETRY

LATTICE GEOMETRIES

CHAPTER 1

Preliminary Notions
Algebraic and Geometric Structures
Spaces and Geometries

1.1 INTRODUCTORY REMARKS

The source of the power and fecundity of modern mathematics lies in its formulation of abstract notions and in its employment of the postulational procedure that the introduction of such notions naturally suggests. Whereas the talent of mathematicians of the past manifested itself in part by their ability to define fruitful concepts, the outstanding mathematicians of today are distinguished rather by knowing what *not* to define; that is, their talent is for detecting the basic features that are common to several mathematical disciplines and developing a more general theory in which those features appear as undefined notions.

Abstraction results in unification. By not specifying, or by voluntarily ignoring, the nature of the elements of a set, a theory is obtained that applies to sets of numbers as well as to sets of functions or sets of chairs. Of course, those properties that a set has by virtue of the nature of its elements are not within the scope of a theory of abstract sets. As properties of specific sets, they are lost by abstraction (generalized out of existence), but it has proved of great value to have at hand a theory that yields important information about every one of a numerous class of things, even though the theory by its very nature is inadequate to investigate exhaustively any one member of the class.

In a similar way, geometries and theories of spaces are unified by the process of abstraction applied to the nature of their elements and to certain fundamental relations (for example, " part of ") which are postulated as basic in a unified theory. In this way, various classes of abstract *spaces* are obtained with geometries depending upon those properties (metric, projective, topological, and so on) that are selected for investigation.

Since our purpose is to investigate certain abstract spaces and their geometries, and since such spaces are always defined over an abstract set, we begin our study by presenting some elementary principles of set theory, with which the reader is already, perhaps, acquainted. But first some remarks concerning the notion of a set are in order.

EXERCISES

1. What do you think Euclid's point of view was regarding the basic notions of geometry?

2. In regard to the basic notions, how does a modern presentation of euclidean geometry differ from Euclid's *Elements*?

3. What significant development in the nineteenth century led to the modern view of geometry?

1.2   THE SET CONCEPT

If the reader tries to express his intuitive notion of a set without using a synonymous term such as class, collection, or aggregate, he may begin to realize that the simplicity of this concept, of which so many exemplifications lie close at hand, is more apparent than real. A closer examination would reveal that his intuition concerning the use of the word cannot be trusted, because it would involve him in contradictions. Many such contradictions arising from an unrestricted notion of set have been formulated. One of the first of these so-called *logical antinomies*, due to the British logician Bertrand Russell (1872–1970), may be stated in the following way.

Call a set *ordinary* in case it is not one of its own elements. Probably most of the sets that the reader can think of are ordinary sets; for example, the set of even natural numbers, the set of all elephants, and many others. But the set of all things that are *not* even natural numbers is surely not an even natural number, and hence is one of its own elements. Let us call a set *extraordinary* if it is a member of itself.

It seems clear that each set is either ordinary or extraordinary, and no set is both ordinary and extraordinary. Let Ω denote the set of all ordinary sets and apply to Ω the preceding remark. Then the set Ω is either an ordinary set or an extraordinary set, but not both.

But if Ω is an ordinary set, it is a member of the set of all ordinary sets; that is, Ω is a member of itself, and hence Ω is an extraordinary set. If the set Ω is an extraordinary set, however, then it is a member of itself and hence it is an ordinary set, since all and only such sets are members of Ω.

If this contradiction is to be avoided, the notion of a set must be restricted in some manner; that is, "set" must be given a *technical* meaning. To avoid the Russell antinomy, for example, one should agree that a collection of things, one of which is the collection itself, does not constitute a *set*. The resolutions of other antinomies introduce additional restrictions, but apparently all of the known antinomies can be resolved in terms of a set concept rich enough for the purposes of mathematics.

In this book we shall not attempt to formulate a definition of set, preferring to regard the notion (along with that of a real number, for example) as one of the basic primitive concepts with which we deal. Also, if a set has elements, we shall feel free to select one of them, and if any family of such sets be given, we shall often select an element from each of the sets, and consider the set of all elements so selected. All of the operations on sets performed in this book are admissible on the basis of postulational set theory; the operation just described, for example, is justified by the so-called *Axiom of Choice* introduced in 1904 by the German mathematician E. Zermelo (1871–1953). The reader should keep in mind that if S denotes a set, then (1) there is no other alternative for an entity than to belong or not to belong to S, and (2) there is no other alternative for each pair of entities of S than to be distinct or not.

If the criterion of membership is not satisfied by any entity, it is convenient to agree that the criterion determines a set with no elements, called a *null* or *empty* set. With regard to the definition of equality for sets given in the next section, each two null sets are equal and so set theory does not distinguish one null set from another (for example, the set of living centaurs is the same as the set of natural numbers that are both even and odd). Thus we speak of *the* null set instead of *a* null set, and denote it by $\varnothing$.

EXERCISES

1. What contradiction arises if each set that consists of just one element is identified with that element?

2. Let P denote the set of pairs $(1, 2)$, $(3, 4)$, ..., $(m, m + 1)$, ... What is the justification of this assertion: There exists a set Q of natural numbers that has exactly one element in common with each of the pairs of numbers in the set P?

3. Consider the set of all infinite sequences of real numbers, and arrange this set into sets by putting in the same set all (and only) those sequences that differ only in the order of their terms. What justifies the assertion that there exists a set containing one and only one sequence from each of those sets of sequences?

1.3   SET ALGEBRA

If x denotes a member of a set S, we write $x \in S$; $x \notin S$ signifies that the entity denoted by x is not a member of S. A set B is a *subset* of S if and only if each member of B is a member of S (that is, if and only if $x \in B$ implies $x \in S$). We symbolize this relation between the sets B and S by writing either $B \subset S$ or $S \supset B$. The reader should note that (1) $\varnothing \subset S$ for every set S, and (2) the relations $B \subset S$ and $S \subset B$ might both be valid.

DEFINITION   Two sets A, B are *equal* (written $A = B$) if and only if $A \subset B$ and $B \subset A$.

In case $A \subset B$ but A, B are not equal (written $A \neq B$), then A is called a *proper* subset of B.

Associated with a given set S is the family of all sets that are subsets of S. This family is frequently denoted by 2^S. Observe that $\varnothing \in 2^S$ and $S \in 2^S$. The important operations of *sum* (union, join), *product* (intersection, meet), and *complementation* are now defined for the elements of 2^S.

DEFINITION   Suppose $A \in 2^S$ and $B \in 2^S$. The unique subset of S consisting of all those elements of S that are members of A or of B is denoted by $A \cup B$, and called the *sum* of A and B. The unique subset of S consisting of all those elements of S that are members of A and also members of B is denoted by $A \cap B$, and called the *product* of A and B. The unique subset of S consisting of all those elements of S that are *not* members of A is denoted by $C(A)$ or *com A* and called the *complement* of A.

The following elementary properties of set sum and set product are easily established: If A, B, $C \in 2^S$, then

(i) $\qquad A \cup B \in 2^S$, $\qquad\qquad\qquad A \cap B \in 2^S$,

(ii) $(A \cup B) \cup C = A \cup (B \cup C)$, $\quad (A \cap B) \cap C = A \cap (B \cap C)$,

(iii) $\qquad A \cup B = B \cup A$, $\qquad\qquad A \cap B = B \cap A$,

(iv) $\qquad A \cup A = A$, $\qquad\qquad\quad A \cap A = A$,

(v) $\qquad A \cup B = A \qquad$ if and only if $\qquad A \cap B = B$.

Two sets A, B are *mutually exclusive* or *disjoint* provided $A \cap B = \emptyset$.

Property (i) states that the set 2^S is *closed* with respect to the two binary operations "sum" and "product," each of which, according to properties (ii), (iii), (iv), is associative, commutative, and idempotent. Property (v) establishes a connection between these two otherwise independent operations.

EXERCISES

1. *The De Morgan Formulas.* Prove that if A, $B \in 2^S$,
   (a) $\text{com}(A \cup B) = \text{com } A \cap \text{com } B$,
   (b) $\text{com}(A \cap B) = \text{com } A \cup \text{com } B$.

2. *Distributivity Formulas.* Prove that if A, B, $C \in 2^S$,
   (a) $A \cap (B \cup C) = (A \cap B) \cup (A \cap C)$,
   (b) $A \cup (B \cap C) = (A \cup B) \cap (A \cup C)$.

3. Extend the definitions of set sum and set product to any family $\mathfrak{F}$ of subsets of S and denote the resulting sets by $\bigcup_{A \in \mathfrak{F}} A$ and $\bigcap_{A \in \mathfrak{F}} A$, respectively. Do the De Morgan and Distributivity Formulas remain valid?

1.4 ALBEGRAIC STRUCTURE. LATTICES

We have remarked in a previous section that unification of diverse mathematical disciplines is achieved by abstraction, and we are now in position to illustrate this by an important example.

Two binary operations, set sum and set product, have been defined for each two elements of the set Σ consisting of all of the subsets of a given abstract set S, and these two operations have the properties (i) to (v) of Section 1.3. There are many interesting mathematical systems that exhibit a similar structure. Let us consider a few examples (in which we denote the analogues of set sum and set product by $\oplus$ and $\cdot$, respectively).

Example 1.   In the set N of natural numbers, let a $\oplus$ b denote the least common multiple of the numbers a, b, and let the greatest common divisor of a, b be denoted by $a \cdot b$. These two binary operations in N obviously have properties (i), (iii), (iv), and (v), and the reader may readily show that they possess property (ii) also.

Example 2.   In the set R of real numbers. let $a \oplus b = \max\{a, b\}$, and $a \cdot b = \min\{a, b\}$. Then all five of the properties (i) to (v) are obvious in this context.

Example 3.   Consider the set S of all points, lines, and planes of a three-dimensional euclidean (or affine) space $\mathbf{E}_3$, together with the null set and the $\mathbf{E}_3$ itself. If a, $b \in S$ denote by $a \oplus b$ that unique element of S *with smallest dimension* that contains both of the elements a, b, and let $a \cdot b$ denote the set product of a, b. It is verified at once that this structure has all the properties (i)–(v).

Example 4.   In Example 3, substitute "three-dimensional *projective* space $\mathbf{P}_3$" for "three-dimensional *euclidean* (or affine) space $\mathbf{E}_3$". (For reasons that will be clear later, it is desirable to separate Examples 3 and 4.)

Example 5.   Let P denote the set of all propositions. Symbolize by $p \oplus q$ the logical disjunction of the propositions p, q (that is, $p \oplus q$ denotes the proposition "p and/or q") and by $p \cdot q$ the logical conjunction of p, q (that is, $p \cdot q$ denotes the proposition "p and q"). Interpreting the equality sign to signify "has the same truth value," it is easily seen that properties (i) to (v) are all valid in this system.

Example 6.   Let C denote the set of all real, continuous functions defined in the closed interval $[0, 1]$. If $f(x)$, $g(x) \in C$, define $h(x) = \max\{f(x), g(x)\}$ and $k(x) = \min\{f(x), g(x)\}$, $x \in [0, 1]$. Writing $f \oplus g = h$ and $f \cdot g = k$, the reader should verify that all of the properties (i) to (v) persist in this new environment.

These examples, chosen from very different parts of mathematics (and many more that might be given) indicate that an abstract formulation of properties (i) to (v) forms the core or skeleton of many diverse mathematical structures, and they suggest that the study of those structures might be unified by studying the abstract system obtained by that abstraction. In this manner, all properties that are common to all of those structures can be derived without any reference to the particular nature of any one of them, and the applications of such properties to a special structure is obtained merely by reversing the process of abstraction. Of course, the development of the abstract system (the common core), by its very nature, will not yield properties that are peculiar to specific systems. Those properties are reserved for special investigation. The utility of the unification by abstraction procedure must be determined in each instance by a careful assessment of the results that follow. If only trivialities are obtainable, "too much territory" has been covered by the abstraction.

The abstraction that we have been leading up to was formulated in the recent past (1928) by K. Menger and is intensively studied today under the name of *Lattice Theory*.

DEFINITION   A set L in which two associative, commutative, idempotent operations, "sum" or "join" (denoted by $+$) and "product" or "meet" (denoted by $\cdot$ or by juxtaposition) are defined such that $a \cdot b = a$ if and only if $a + b = b$, for each two elements a, b of the set, forms a *lattice* $\mathscr{L}$ with respect to those two operations. We write $\mathscr{L} = \{L; +, \cdot\}$.

Hence the set of all subsets of a given set is a lattice with respect to set sum and set product, and the six structures defined in Examples 1 through 6 are additional examples of lattices. Thus, a lattice arises from an abstract set by imposing upon it an *algebraic structure in conformity with the requirements of the above definition*.

EXERCISES

1. Are the properties of a lattice as given in the definition dependent (that is, is any one of them a consequence of the others)?

2. Let Γ denote the set of all subgroups of an arbitrary group G. If for $\alpha, \beta \in \Gamma$, $\alpha \cdot \beta$ denotes the set product $\alpha \cap \beta$, and $\alpha + \beta$ stands for the smallest *subgroup* of G that contains each of the subgroups α, β, show that the system $\{\Gamma; \cdot, +\}$ is a lattice.

3. Let Σ denote the set of all k-dimensional subsimplices of a simplex S of euclidean three-dimensional space, $k = -1, 0, 1, 2, 3$ (the null set is the only subsimplex of dimension -1). If for $\alpha, \beta \in \Sigma$, $\alpha \cdot \beta$ denotes the subsimplex of greatest dimension contained in each of the subsimplices α, β, and $\alpha + \beta$ denotes the subsimplex of smallest dimension containing both α and β, show that $\{\Gamma; \cdot, +\}$ is a lattice.

1.5   LATTICES AS PARTIALLY ORDERED SETS

We have seen how the notion of a lattice may arise from the set 2^S of all sub-sets of a set S by abstractions applied to the elements of 2^S and to the binary operations of sum and product that are performed on its elements.

Since the elements of 2^S are subsets of a set S, each two elements A, B of 2^S are such that (1) either $A \subset B$ or $A \not\subset B$ (where $\not\subset$ symbolizes "is not a subset of"), (2) $A \subset A$, (3) if $A \subset B$ and $B \subset A$, then $A = B$, and (4) if $A \subset B$ and $B \subset C$, then $A \subset C$. Thus there is *defined* in the set 2^S a binary relation which is *reflexive*, *asymmetric*, and *transitive*. We obtain the notion of an (abstract) *partially ordered set* by abstracting the nature of the specific relation $\subset$, and that of the elements of 2^S.

DEFINITION   If a reflexive, asymmetric, transitive binary relation is de-fined in a set, then the set is said to be *partially ordered* with respect to that relation.

The symbol $\prec$ is used to denote such a relation. The relation $x \prec y$ may be read, "x precedes y"—for lack of a better (neutral) term. Define $x \succ y$ ("x follows y") to mean $y \prec x$.

The qualification "partially" arises from the lack of any requirement that, for each two elements x, y of the set, either $x \prec y$ or $y \prec x$; that is, it is *not* necessary that each two elements of a partially ordered set be *comparable* with respect to the partial ordering relation $\prec$. If each two elements of such a set *are* comparable, then the qualification is omitted, and the set is said to be *ordered*.

The reader may observe that in the set 2^S there is a connection between the partial ordering relation $\subset$ and the set product (or sum); for it is clear that if $A, B \in 2^S$, then $A \subset B$ if and only if $A \cap B = A$ (or equivalently, $A \cup B = B$).

Now in any lattice $\mathscr{L} = \{L; +, \cdot\}$ a partial ordering $\prec$ may be defined in L by writing $a \prec b$ if and only if $a \cdot b = a$ (or equivalently, $a + b = b$), for $a, b \in L$. For, since $a \cdot b \in L$ when $a, b \in L$, the binary relation $\prec$ is then defined in L, while the necessary reflexive, asymmetric, and transitive properties of the binary relation follow at once from the idempotent, commutative, and associative properties of the lattice product.

Examining again the set 2^S, we observe that for each two of its elements A, B it contains unique elements, temporarily denoted by $[A, B]$ and $\{A, B\}$, such that ($*$) $[A, B] \subset A$, $[A, B] \subset B$, while if $X \in 2^S$ such that $X \subset A$ and $X \subset B$, then $X \subset [A, B]$; and ($**$) $A \subset \{A, B\}$, $B \subset \{A, B\}$, while if $Y \in 2^S$ such that $A \subset Y$ and $B \subset Y$, then $\{A, B\} \subset Y$.

The reader will recognize that $[A, B]$ and $\{A, B\}$ are the set product $A \cap B$ and set sum $A \cup B$, respectively. Not every partially ordered set contains, for each two of its elements, a, b, the analogous elements $[a, b]$ (a *greatest lower bound* of a, b) and $\{a, b\}$ (a *least upper bound* of a, b). Let the reader give examples to verify this.

Now consider a partially ordered set $\{L; \prec\}$, which contains for each two of its elements a, b a greatest lower bound $[a, b]$ and a least upper bound $\{a, b\}$. *We show that the system* $\{L; a \cdot b = [a, b], a + b = \{a, b\}\}$ *is a lattice.*

1. *The set L is closed with respect to each of the operations* $[a, b]$ *and* $\{a, b\}$.

2. *Each of the operations* $[a, b], \{a, b\}$ *is associative.* Since $[[a, b], c] \prec [a, b]$, it is clear that $[[a, b], c]$ precedes *each* of the elements a, b, c, and so does the element $[a, [b, c]]$. It follows that each of the elements $[a, [b, c]]$, $[[a, b], c]$ precedes the other, and hence the two elements are equal. The associativity of the operation $\{a, b\}$ is similarly established.

3. *Each of the operations* $[a, b], \{a, b\}$ *is commutative.* The proof is clear.

4. *Each of the operations is idempotent.* Clearly $[a, a] \prec a$, and from $a \prec a$ follows $a \prec [a, a]$; that is, $[a, a] = a$. In a similar manner, $\{a, a\} = a$.

5. $[a, b] = a$ *if and only if* $\{a, b\} = b$. If $[a, b] = a$, then $a \prec b$, and so $\{a, b\} \prec b$. Since by definition, $b \prec \{a, b\}$, the equality $\{a, b\} = b$ is established. Conversely, if $\{a, b\} = b$, then again $a \prec b$ and consequently $a \prec [a, b]$, which, since $[a, b] \prec a$, gives the desired equality.

If in the lattice $\{L; [a, b], \{a, b\}\}$ we define a binary relation $\prec *$ by $a \prec * b = [a, b] = a$, then $a \prec * b$ obviously implies $a \prec b$. Conversely, $a \prec b$ gives, as above, $[a, b] = a$, and so $a \prec * b$. Hence $\prec *$ is a partial ordering of L and the two partially ordered sets $\{L; \prec\}, \{L; \prec*\}$ are isomorphic (that is, abstractly the same.)

We conclude from the foregoing that any lattice may be considered either (i) purely algebraically, as a double composition system $\{L; \cdot, +\}$, or (ii) as a partially ordered set $\{L, \prec\}$ which contains, for each two of its elements, a greatest lower bound and least upper bound. The equivalence of the two points of view is provided by the equivalence: $a \prec b$ if and only if $a \cdot b = a$.

EXERCISES

1. Can any partial ordering relation defined in a set S be interpreted as the set-inclusion relation in a family of subsets of S?

2. In each of the examples of Section 1.4 find the corresponding partial ordering relation.

3. A subset of a euclidean space is called *convex* if it contains the line segment joining any two of its points. Does the set of all convex subsets of the plane form a lattice when the partial ordering relation is set-inclusion?

1.6 SPECIAL KINDS OF LATTICES.
BOOLEAN ALGEBRA,
MODULAR AND NORMED LATTICES

In showing how the notion of a lattice may be obtained by abstraction from the structure possessed by the set 2^S of all subsets of a set S, we selected just five properties of that structure as the basis for abstraction. But 2^S has, with respect to the binary operations of sum and product (or equivalently, with respect to the binary partial ordering relation $\subset$), many properties that are *not* logical consequences of those five properties, and consequently such properties are not possessed by all lattices. Among the most important of these properties are the following.

1. 2^S contains both a *first* element $\varnothing$ (that is, an element that *precedes* each of its elements) and a *last* element S (that is, an element that *follows* each of its elements). Such elements are unique whenever they exist. But the lattice given by Example 2, Section 1.4, does *not* have either a first or a last element.

2. 2^S is *complemented*; that is, 2^S contains for each of its elements A an element $C(A)$ such that $A \cap C(A) = \varnothing$ and $A \cup C(A) = S$. The element $C(A)$ of 2^S is that subset of S that consists of all elements of S that are not members of the subset A of S; that is $C(A)$ is the set complement of A. A lattice without first and last elements is surely not complemented, but even the presence of such elements does not, of course, ensure that the lattice is complemented.

3. 2^S is *distributive*; that is, if A, B, $C \in 2^S$, then $(A \cup B) \cap C = (A \cap C) \cup (B \cap C)$ and $(A \cap B) \cup C = (A \cup C) \cap (B \cup C)$ (see Exercise 2, Section 1.3).

Distributive properties are not possessed by the lattice of Example 4, Section 1.4, for if a, b are two intersecting lines, each skew to a line c, but whose plane is parallel to line c, then $(a \oplus b) \cdot c \neq a \cdot c \oplus b \cdot c$, since the left member is a point and the right member is the null element. Parentheses are unnecessary in the right member since the reader will not fail to interpret it correctly without them.

DEFINITION.   A lattice that is complemented and distributive is a *Boolean algebra*.

Boolean algebras were first studied, in a very special case, by the British logician George Boole [1815–1864].

THEOREM 1.1.   *The set of all subsets of a set is a Boolean algebra with respect to the operations of set sum and set product.*

In 1900 the German mathematician R. Dedekind [1831–1916] investigated a structure that is an example of a lattice $\mathscr{L}$ that is *not* distributive, but in which the following *weakened* form of the distributive property is valid:

If a, b, $c \in \mathscr{L}$ and $a \prec c$, then $(a + b) \cdot c = a + b \cdot c$.

Such lattices form the important class of so called *Dedekind* or *modular* lattices. The lattice of all linear subspaces of a projective space belongs to this class (we have seen above that this lattice is *not* distributive).

We remark that the lattice of Example 3, Section 1.4, is not modular, for if b, c are two (distinct) mutually parallel lines and a is a point of line c, then $a \prec c$, but $(a \oplus b) \cdot c \neq a \oplus b \cdot c$, since $(a \oplus b) \cdot c = c$, $a \oplus b \cdot c = a$, and $c \neq a$.

An important subclass of the class of modular lattices is the class of normed lattices introduced by K. Menger in 1928. A lattice is called *normed* provided that to each of its elements x there corresponds a nonnegative real number $\|x\|$, the norm of x, such that (i) if $x \prec y$ and $x \neq y$, then $\|x\| < \|y\|$, and (ii) for each two elements x, y of the lattice, $\|x + y\| + \|x \cdot y\| = \|x\| + \|y\|$.

The lattice of Example 4, Section 1.4, may be normed by defining $\|x\| = 1 + \text{dimension } x$, while that of Example 3, Section 1.4, cannot be normed (we shall prove later that every normed lattice is modular).

EXERCISES

1. Prove that a lattice has at most one first element and at most one last element.

2. Let L denote the set of all ordered pairs of real numbers. If $a = (a_1, a_2)$ and $b = (b_1, b_2)$ are elements of L, define $a \prec b$ if $a_1 \leq b_1$ and $a_2 \leq b_2$. Show that $\mathscr{L} = \{L; \prec\}$ is a distributive lattice. Clearly $\mathscr{L}$ has neither a first nor a last element. The set of all ordered pairs of *nonnegative* numbers, partially ordered as above, is a normed, distributive lattice with a first element (the element $(0,0)$) but no last element, where $\|a\| = a_1 + a_2$.

3. If the set L of Exercise 2 is partially ordered (lexicographically) by defining $a\,b \prec$ whenever (i) $a_1 < b_1$ or (ii) if $a_1 = b_1$, then $a_2 \leqq b_2$, what special kind of lattice $\mathscr{L} = \{L; \prec\}$ results?

1.7   GEOMETRIC STRUCTURE. SPACES

Using the operations of set sum and set product as guides, we have obtained the notion of an (abstract) lattice—a structure that is exemplified by many diverse mathematical objects, which admit, therefore, of a simultaneous, unified development.

There is another means of imposing a structure upon an abstract set that yields a concept whose role as a unifier is even more important than that of an abstract lattice. This concept is that of an abstract space.

It is likely that even the most mathematically naive person who uses the words "set" and "space" would not fail to assign different meanings to those terms; for, intuitively, a space is expected to possess a kind of arrangement or order that is not required of a set. The necessity of a *structure* in order for a set to qualify as a space may be rooted in the feeling that a notion of "proximity" (in some sense not necessarily quantitative) is inherent in our concept of a space. Thus a space differs from the mere set of its elements by possessing a structure which in some way (however vague) gives expression to that notion. Let us examine a few ways of defining such a structure.

Example 1.   *Fréchet* **V**-*space*.   This very general method of establishing a structure in an abstract set S is due to the French mathematician Maurice Fréchet [1878–   ]. It assumes merely that with each element x of the set S there is associated a nonempty class $\{V_x\}$ of subsets V_x of S. Each member of $\{V_x\}$ is called a neighborhood of x. [The French word for neighborhood is *voisinage*—hence the name **V**-space.] A neighborhood V_x does not necessarily contain the element x, though it turns out that for the (*topological*) purposes for which the space was defined, we may just as well suppose that it does.

The notion of proximity embodied in this very "loose" structure can be expressed by saying that an element x of S is "near" a subset E of S provided that every neighborhood V_x of x contains an element of E. Thus the proximity of an element of the space to a subset is relative to the choice of neighborhoods of that element. This is as it should be, since the structure of the whole set (that is, the space itself) is determined by the selection of the set of neighborhoods.

Example 2.   *Fréchet L-Space.*   Let $\mathfrak{S}$ denote the family of all infinite sequences of elements of an abstract set S and let $\mathscr{L}$ denote any relation such that for each $\sigma \in \mathfrak{S}$ and each $p \in S$ either $p\mathscr{L}\sigma$ or $(p\mathscr{L}\sigma)'$, where the prime denotes negation. The relation $\mathscr{L}$ yields a Fréchet $\mathscr{L}$-space whenever the relation $\mathscr{L}$ conforms to the following three requirements:

(i)   If $p_m = p (m = 1, 2, \ldots)$, then $p\mathscr{L}\sigma$, where $\sigma = \{p_m\}$,

(ii)   if $p, q \in S$ and $\sigma \in \mathfrak{S}$ such that $p\mathscr{L}\sigma$ and $q\mathscr{L}\sigma$, then $p = q$,

(iii)   if $p\mathscr{L}\sigma$, $\sigma = \{p_m\}$, and $m_1 < m_2 < \cdots$ is any monotone increasing infinite sequence of natural numbers, then $p\mathscr{L}\sigma*$, where $\sigma* = \{p_{m_i}\}$. The sequence $\sigma*$ is called a *subsequence* of the sequence σ.

An element p of S is "near" a subset E of S provided that there exists an infinite sequence $\sigma = \{p_m\}$, $p_m \in E$ $(m = 1, 2, \ldots)$ such that $p\mathscr{L}\sigma$.

Example 3. *Kuratowski closure space.*   The following method of establishing a structure in an abstract set S is associated with the Polish mathematician Casimir Kuratowski [1896–   ]. Let f denote any *mapping* of the set 2^S of all subsets of S into itself; that is, to each subset E of S there corresponds exactly one subset $f(E)$ of S. The subset $f(E)$ corresponding to E is called the *closure* of E, provided the mapping f has the following properties:

(i)   if $X, Y \in 2^S$, then $f(X \cup Y) = f(X) \cup f(Y)$,

(ii) $X \in 2^S$ implies $f(f(X)) = f(X)$, and

(iii)   If $X = \varnothing$ or if X contains just one element, then $f(X) = X$.

An element p of S is "near" a subset E of S provided $p \in f(E)$. The closure $f(X)$ of X is frequently denoted by $\overline{X}$ or cl X.

Example 4.   *Topological and Hausdorff topological spaces.*   A topological space, whose rich theory has been extensively developed, arises from a Fréchet V-space by assuming that the neighborhoods of such a space have the following four properties:

(i)   each neighborhood V_x of an element x of S contains that element,

(ii)   if U_x, V_x are neighborhoods of x, there exists a neighborhood W_x such that $W_x \subset U_x \cap V_x$,

(iii)   if $x, y \in S$, $x \neq y$, there exists a neighborhood V_x such that $y \notin V_x$,

(iv)   if $x \in V_y$, there exists a V_x such that $V_x \subset V_y$.

A still richer theory is obtained by strengthening property (iii) to assume (iii)*: if $x, y \in S$, $x \neq y$, neighborhoods V_x, V_y exist such that $V_x \cap V_y = \varnothing$. Such a topological space is called a Hausdorff topological space, after the German mathematician F. Hausdorff [1868–1942], who formulated the definition of the space in his famous Grundzüge der Mengenlehre (1914).

Example 5. Semimetric and metric spaces. A direct, quantitative measure of proximity is introduced in an abstract set S by associating with each ordered pair x, y of its elements (called "points") a *nonnegative real number*, denoted by xy and called the "distance" from x to y. If the association of numbers with ordered pairs of points is such that (i) $xy = 0$ if and only if $x = y$, and (ii) $xy = yx$, the resulting space is called *semimetric*. It is called *metric* provided, in addition to (i), (ii), the association fulfills a third requirement: (iii) if x, y, $z \in S$, the sum of any two of the numbers xy, yz, xz equals or exceeds the remaining one. The distance function in a metric space is called the *metric* of the space. Referring to the inequality in (iii) as the *triangle inequality*, we see that a metric space is a semimetric space that satisfies the triangle inequality. This very important class of spaces was introduced and studied by Fréchet during the first decade of this century.

We shall see later that if we suitably define "neighborhood," "closure," and "limit of an infinite sequence of elements," then each metric space belongs to each of the other four classes of spaces defined in this section.

EXERCISES

1. Show that the closure operator (Example 3) is *monotone* (that is, if $X \subset Y$, then $\bar{X} \subset \bar{Y}$), that $X \in 2^S$ implies $X \subset \bar{X}$, and $\bar{S} = S$.

2. In a Kuratowski closure space **K** define a subset X of **K** to be a neighborhood of an element x of **K** if and only if (i) $x \in X$ and (ii) $cl(com\ X) = com\ X$. Prove that in terms of this definition of neighborhood, **K** is a topological space.

3. Let R denote the set of all real numbers and associate with each pair, x, y of its elements the real number $xy = |x - y|^{\alpha}$. Show that the resulting space is semimetric, but not metric, for $\alpha > 1$, and that it is metric for $0 < \alpha \leq 1$. What is the nature of the space when $\alpha < 0$?

4. Let S denote any abstract set, and let $\mathfrak{M}$ denote the family of all semimetric spaces with element-set S. If $\mathbf{M}_1$, $\mathbf{M}_2 \in \mathfrak{M}$, put $\mathbf{M}_1 \prec \mathbf{M}_2$ provided p, $q \in S$ implies $pq_{\mathbf{M}_1} \leq pq_{\mathbf{M}_2}$, where $pq_{\mathbf{M}_i}$ denotes the distance of p, q in the space $\mathbf{M}_i$ ($i = 1, 2$). Is the system $\{\mathfrak{M}; \prec\}$ a lattice?

5. Consider $\{\mathfrak{M}; \prec\}$ when $\mathfrak{M}$ denotes the set of all metric spaces with element-set S, and $\prec$ is defined as in Exercise 4. Is it a lattice?

6. If p, q, r are points of a semimetric space, put $a = pq$, $b = qr$, $c = pr$. Show that p, q, r satisfy the triangle inequality (and hence form a metric triple) if and

only if the determinant

$$D(p, q, r) = \begin{vmatrix} 0 & 1 & 1 & 1 \\ 1 & 0 & a^2 & b^2 \\ 1 & a^2 & 0 & c^2 \\ 1 & b^2 & c^2 & 0 \end{vmatrix} \leq 0.$$

1.8 EXAMPLES OF METRIC SPACES

In Section 1.4 we have called the reader's attention to a few of the many special structures whose theories (up to a point) are subsumed in the general theory of lattices. This unification of algebraic structures by means of abstraction is paralleled by a similar unification of the theories of many different kinds of spaces. We list a few of the many spaces, each one of which is a *metric space* (and hence a member of the other classes of spaces defined in the preceding section) whose theories are (partially) subsumed in that of a general metric space.

Example 1. The n-dimensional euclidean space E_n. The point set of this space is the set of all ordered n-tuples $(x_1, x_2, \ldots, x_n)$ of real numbers. The set is given a structure by defining the distance xy of each two of its points by setting

$$xy = \left[\sum_{i=1}^{n} (x_i - y_i)^2 \right]^{1/2}.$$

Putting $x = y$ provided $x_i = y_i$ $(i = 1, 2, \ldots, n)$, it is clear that $xy \geq 0$, $xy = 0$ if and only if $x = y$, and $xy = yx$. It remains to prove the triangle inequality.

If $x = (x_1, x_2, \ldots, x_n)$, $y = (y_1, y_2, \ldots, y_n)$, $z = (z_1, z_2, \ldots, z_n)$ are any three points, set $a_i = x_i - y_i$, $b_i = y_i - z_i$ $(i = 1, 2, \ldots, n)$. The obvious inequality

$$0 \leq \tfrac{1}{2} \sum_{i, j=1}^{n} (a_i b_j - a_j b_i)^2 = \sum_{i, j=1}^{n} (a_i^2 b_j^2 - a_i b_j a_j b_i)$$

yields

$$\left(\sum_{i=1}^{n} a_i b_i \right) \cdot \left(\sum_{j=1}^{n} a_j b_j \right) \leq \sum_{i, j=1}^{n} a_i^2 b_j^2 = \left(\sum_{i=1}^{n} a_i^2 \right) \cdot \left(\sum_{j=1}^{n} b_j^2 \right).$$

That is,

$$(*) \qquad \left(\sum_{i=1}^{n} a_i b_i\right)^2 \leq \left(\sum_{i=1}^{n} a_i^2\right) \cdot \left(\sum_{i=1}^{n} b_i^2\right).$$

This is the well-known Cauchy inequality, which it is useful to write also in the form

$$\begin{vmatrix} \Sigma a_i^2 & \Sigma a_i b_i \\ \Sigma b_i a_i & \Sigma b_i^2 \end{vmatrix} \geqq 0,$$

all summations going from 1 to n.

From $(*)$ follows

$$\Sigma a_i b_i \leq |\Sigma a_i b_i| \leq (\Sigma a_i^2 \cdot \Sigma b_i^2)^{1/2},$$

$$\Sigma a_i^2 + 2\Sigma a_i b_i + \Sigma b_i^2 \leq \Sigma a_i^2 + 2(\Sigma a_i^2 \cdot \Sigma b_i^2)^{1/2} + \Sigma b_i^2,$$

$$[\Sigma(a_i + b_i)^2]^{1/2} \leqq (\Sigma a_i^2)^{1/2} + (\Sigma b_i^2)^{1/2};$$

and hence $xz \leqq xy + yz$.

Example 2.   The n-dimensional Minkowski space $\mathbf{M}_n$.   We consider first the class of two-dimensional spaces $\mathbf{M}_2(\Gamma)$, each member of which is determined by the choice of a closed, convex curve Γ of the euclidean plane $\mathbf{E}_2$ which is symmetric about the origin. The curve Γ is a simple closed curve which, together with its interior, is a convex subset of the plane (that is, the set contains with each two of its points the unique segment determined by them), and each line of the plane through the origin o (that is, each *diameter* of Γ) intersects Γ in exactly two points p, p' such that $op = op'$. The space $\mathbf{M}_2(\Gamma)$ (the Minkowski plane with *indicatrix* Γ) is obtained by defining the Minkowski distance $m(x, y)$ of two points x, y of the point set of the $\mathbf{E}_2$ in the following way (Fig. 1).

If $x \neq y$, $m(x, y) = 2(xy/d_{xy})$, where xy is the euclidean distance of x, y, and d_{xy} is the length of that chord of Γ contained in the diameter of Γ that is parallel to (or coincident with) the line of $\mathbf{E}_2$ determined by the points x, y; if $x = y$, we put $m(x, y) = 0$. Clearly $m(x, y)$ is a nonnegative real number that vanishes if and only if $x = y$ and $m(x, y) = m(y, x)$.

To prove the validity of the triangle inequality, we note that if x, y, z are points of a euclidean line, the Minkowski distances $m(x, y)$, $m(y, z)$ $m(x, z)$ are proportional to the euclidean distances xy, yz, xz, respectively, and hence satisfy the triangle inequality. Note that euclidean lines are also Minkowski lines.

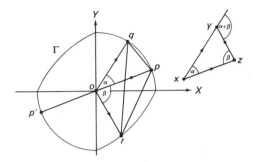

Figure 1

If, now, x, y, z are noncollinear, let p, q, r denote those points of Γ such that the three vectors $\overrightarrow{op}, \overrightarrow{oq}, \overrightarrow{or}$ make angles of zero degrees with the three vectors $\overrightarrow{xz}, \overrightarrow{xy}, \overrightarrow{yz}$, respectively. It is easily verified that vector $\overrightarrow{op}$ lies within the angle made by the two vectors $\overrightarrow{oq}$ and $\overrightarrow{or}$, and, since Γ is convex, p is either a point of the segment with end points q, r, or p lies *outside* the triangle with vertices o, q, r. In either case

$$(*) \qquad A(o, p, q) + A(o, p, r) \geqq A(o, q, r),$$

where $A(s, t, u)$ denotes the area of the triangle whose vertices are s, t, u.

Putting $\alpha = \sphericalangle[\overrightarrow{op}, \overrightarrow{oq}] = \sphericalangle[\overrightarrow{xy}, \overrightarrow{xz}]$ and $\beta = \sphericalangle[\overrightarrow{op}, \overrightarrow{or}] = \sphericalangle[\overrightarrow{xz}, \overrightarrow{yz}]$, elementary trigonometry gives the following equalities:

$$A(o, p, q) = \tfrac{1}{2}op \cdot oq \sin \alpha = \Delta/[m(x, y)\, m(x, z)],$$

$$A(o, p, r) = \tfrac{1}{2}op \cdot or \sin \beta = \Delta/[m(x, z)\, m(y, z)],$$

$$A(o, q, r) = \tfrac{1}{2}oq \cdot or \sin(\alpha + \beta) = \Delta/[m(x, y)\, m(y, z)],$$

since $op = xz/m(x, z)$, $oq = xy/m(x, y)$, $or = yz/m(y, z)$,

where Δ denotes the area of the triangle with vertices x, y, z. Substitution in the inequality $(*)$ yields the desired inequality,

$$m(x, y) + m(y, z) \geqq m(x, z),$$

and so each Minkowski plane is a metric space.

To consider the n-dimensional Minkowski space $\mathbf{M}_n$, let Σ denote any closed, $(n-1)$-dimensional "surface" of the $\mathbf{E}_n$ that is cut by each two-dimensional plane through the origin in a curve Γ *of the kind considered above.* Then each line through o cuts Σ in exactly two points, which are equidistant from o, and the distance $m(x, y)$ of points x, y is defined by $m(x, y) = 2(xy/d_{xy})$ if $x \neq y$; $m(x, y) = 0$ if $x = y$, where, again, xy is the euclidean distance of points x, y and d_{xy} is the length of that chord of Σ contained in the line through o that is parallel to (or coincident with) the line determined by x, y. The space $\mathbf{M}_n(\Sigma)$ so defined is clearly semimetric.

The validity of the triangle inequality for three collinear points is immediate. If x, y, z are noncollinear points, let $\pi(x', y', z')$ denote the unique plane through o that is parallel to (or coincident with) the plane determined by x, y, z, where x', y', z' denote the orthogonal projections of x, y, z respectively, on that plane. Putting $\Gamma = \Sigma \cap \pi(x', y', z')$, then the mutual distances in $\mathbf{M}_n(\Sigma)$ of the points x, y, z, are the same as the mutual distances in $\mathbf{M}_2(\Gamma)$ of x', y', z'. Since $\mathbf{M}_2(\Gamma)$ has been shown to be a metric space, it follows that $\mathbf{M}_n(\Sigma)$ is also a metric space.

We observe that if Σ is the unit sphere of $\mathbf{E}_n$, then $\mathbf{M}_n(\Sigma) = \mathbf{E}_n$, and so the class of Minkowski spaces contains the class of euclidean spaces.

Example 3.　*Normed linear space.*　This important class of spaces, that contains the class of Minkowski spaces as a proper subclass, was defined about forty years ago by three mathematicians working independently of one another; namely, Hans Hahn (1879–1934) of Austria, Stefen Banach [1892–1945) of Poland, and the American mathematician Norbert Wiener [1894–1964].

Let S denote any nonempty set that contains with each of its elements x and each real number λ a unique element $\lambda \cdot x$ (called the *scalar multiple* of x by λ), and with each two of its elements x, y, a unique element $x + y$ (the sum of x and y). The system $\{S; \cdot, +\}$ is a *linear* or *vector* space provided the operations of scalar multiplication and sum have the following properties:

(i) $x + y = y + x$

(ii) $x + (y + z) = (x + y) + z$,

(iii) $x + y = x + z$ implies $y = z$,

(iv) $\lambda \cdot (x + y) = \lambda \cdot x + \lambda \cdot y$,

(v) $(\lambda_1 + \lambda_2) \cdot x = \lambda_1 \cdot x + \lambda_2 \cdot x$　　$(\lambda_1, \lambda_2$ real),

(vi) $\lambda_1 \cdot (\lambda_2 \cdot x) = (\lambda_1 \lambda_2) \cdot x$　　$(\lambda_1, \lambda_2$ real),

(vii) $1 \cdot x = x$.

Remark.　The linear space $\{S; \cdot, +\}$ contains a *unique* element θ such that $x + \theta = x$ for every element x of the space. For if a is any arbitrarily selected element of S, the space contains an element $0 \cdot a$ (where 0 denotes the real number zero). Clearly $a + 0 \cdot a = (1 + 0) \cdot a = 1 \cdot a = a$, and if $x \in \{S; \cdot, +\}$, let $z = x + 0 \cdot a$. Then $z + a = (x + 0 \cdot a) + a = x + (a + 0 \cdot a)$, and so $z + a = x + a$. Hence $z = x$ by property iii; that is $x + 0 \cdot a = x$, for every element x of the space. Similarly, if $b \in S$, we obtain $0 \cdot b = 0 \cdot b + 0 \cdot a = 0 \cdot a + 0 \cdot b = 0 \cdot a$. Writing $\theta = 0 \cdot a, a \in S$, the remark is proved, since $x + \theta = x + \theta_1$ implies $\theta_1 = \theta$.

A linear space is normed provided there is associated with each of its elements x a nonnegative real number, called its *norm* and denoted by $\|x\|$, such that

(viii) $\|x\| = 0$ if and only if $x = \theta$,

(ix) $\|\lambda \cdot x\| = |\lambda| \cdot \|x\|$ (λ real),

(x) $\|x + y\| \leq \|x\| + \|y\|$.

A metric space arises by attaching to each ordered pair of elements x, y of the normed, linear space $\{S; \cdot, +, \| \|\}$, the nonnegative real number $\|x - y\| = \|x + (-1) \cdot y\|$ as distance. For $\|x - y\| = 0$ if and only if $x - y = \theta$ (that is, if and only if $[x + (-1)y] + y = \theta + y = y$, $x + 0 \cdot y = y$, $x = y$), and $\|x - y\| = \|(-1)(y - x)\| = |-1| \cdot \|y - x\| = \|y - x\|$. Hence the distance function is positive, definite, and symmetric. Since, moreover, by property (x),

$$xz = \|x - y + y - z\| \leq \|x - y\| + \|y - z\| = xy + yz,$$

the triangle inequality is satisfied, and the space is metric.

EXERCISES

1. Let $\mathscr{L} = \{L; +, \cdot\}$ denote any normed lattice and put

   $$d(x, y) = \|x + y\| - \|x \cdot y\|,$$

   for each pair x, $y \in L$. Taking $d(x, y)$ as the distance of x, y, show that the space so obtained is metric.

2. If $\mathfrak{C}$ denotes the family of all real functions of a real variable that are continuous in the interval $I = [0, 1]$, let distance $fg = \max_{t \in I} |f(t) - g(t)|, f, g \in \mathfrak{C}$. Prove that the resulting space is metric.

3. Consider the set of all ordered n-tuples $x = (x_1, x_2, \ldots, x_n)$ of real numbers, and define distance $xy = [\sum_{i=1}^{n} |x - y|^p]^{1/p}$, where p denotes a (fixed) real number, $p \geq 1$. Is the resulting space metric? What is the nature of the space when $0 < p < 1$?

4. If each two distinct points of a semimetric space has distance at least 1 and at most 2, is the space metric?

5. In the euclidean space E_n (Example 1), "distort" the euclidean distance ab of points a, b by putting distance $d(a, b) = \varphi(ab)$, where φ is a monotone increasing, concave function (that is, no arc of the curve $y = \varphi(x)$ determined by two points of the curve lies below the chord joining the two points) with $\varphi(0) = 0$. Is the space with "distorted" distance metric?

6. Consider the set of all infinite sequences $\{x_n\}$ $(n = 1, 2, \ldots)$ of real numbers such that the infinite series $x_1^2 + x_2^2 + \cdots + x_n^2 + \cdots$ is convergent. If $x = \{x_n\}$, $y = \{y_n\}$, define distance $xy = [\sum_{i=1}^{\infty} (x_i - y_i)^2]^{1/2}$ and show that the resulting (Hilbert) space is metric.

7. Let J denote any simple closed (Jordan) curve of the euclidean plane. If the distance $d(x, y)$ of points x, y in the interior of J is defined by

$$d(x, y) = \log \left[\left(\max_{p \in J} \frac{px}{py} \right) \middle/ \left(\min_{p \in J} \frac{px}{py} \right) \right],$$

where px, py denote euclidean distances, prove that the resulting space is metric. Identify this space when J is a circle.

1.9   DEFINITIONS OF SOME BASIC CONCEPTS

This section is devoted to defining some of the notions that are basic for the developments that follow. In each definition, the concept being defined is printed in italic, and the word "Definition," which usually prefaces a definition, is sometimes omitted. Whenever "provided" appears in a definition, it is used in the sense of "if and only if". Other concepts will be defined as they are needed.

A set is *countable* provided there exists a one-to-one correspondence between its elements and the elements of a subset of the set of natural numbers. Infinite countable sets are called *denumerable*.

Remark.   The null set is countable.

An element p of a metric space $\mathbf{M}$ is a limit of the infinite sequence $\{p_n\}$ of elements of $\mathbf{M}$ (symbolized by $\lim_{n \to \infty} p_n = p$, or $\{p_n\} \to p$) if and only if the limit of the infinite sequence $\{pp_n\}$ of nonnegative real numbers is zero ($\lim_{n \to \infty} pp_n = 0$).

If $E \subset \mathbf{M}$, $E^* \subset \mathbf{M}^*$ ($\mathbf{M}$, $\mathbf{M}^*$ metric spaces), a *mapping of E into E^** is defined by associating with each element p of E exactly one element p^* of E^*.

Writing $p^* = f(p)$, we call p^* the *image* of p by the mapping f, and put $f(E) = [q \in E^* | q = f(p), p \in E]$. The subset $f(E)$ of E^* is the *image* or *transform* of E by f. If $T^* \subset E^*$ the subset $\varphi(T^*) = [p \in E | f(p) \in T^*]$ of E is called the *f-model* of T^*. In case $\varphi(T^*) \neq \varnothing$, for every $T^* \subset E^*$, then f is said to map E *onto* E^*; in case the f-model of each one-element subset $\{q\}$ of E^* is a one-element subset $\{p\}$ of E, a mapping of E^* onto E is obtained (symbolized by $p = \varphi(q)$, or $p = f^{-1}(q)$) called the mapping *inverse* to f. The mapping f has *domain* E and *range* $f(E)$, and the mapping f^{-1} has domain $f(E)$ and range E (whenever it exists). A mapping is frequently called a *transformation*. Note that $\mathbf{E} \subset \mathbf{M}$ indicates that distance in $\mathbf{E}$ is the same as in $\mathbf{M}$.,

Remark.   A mapping f of E onto $f(E)$ has an inverse if and only if the association of $p(p \in E)$ with $f(p)$, $(f(p) \in f(E))$ is one-to-one.

A mapping f of E onto $f(E)$, $(E, f(E)$ subsets of metric spaces) is *continuous at* p_0, $(p_0 \in E)$, if and only if for every infinite sequence $\{p_n\}$ of element of E with $\lim_{n \to \infty} p_n = p_0$, the infinite sequence $\{f(p_n)\}$ of elements of $f(E)$ has limit $f(p_0)$. A mapping f is *continuous in* E provided it is continuous at every element of E, and it is *bicontinuous* provided its inverse exists and is continuous in $f(E)$. Bicontinuous mappings are called *homeomorphisms*, and two sets are *homeomorphic* provided one of the sets is a bicontinuous image of the other.

Remark.   Every homeomorphism establishes a one-to-one correspondence between the elements of its domain and the elements of its range.

A mapping f with domain E and range $f(E)$, $(E, f(E)$ subsets of metric spaces) is a *congruence* or an *isometric* mapping provided it is distance-preserving (that is, if $p, q \in E$, then $pq = f(p)f(q)$). The sets E, $f(E)$ are then said to be *congruent* or *isometric*, symbolized by $E \approx f(E)$. If f is a congruent mapping of a finite set $p_1, p_2, \ldots, p_n$ onto the finite set $p'_1, p'_2, \ldots, p'_n$, and $p'_i = f(p_i)$, we write

$$p_1, p_2, \ldots, p_n \approx p'_1, p'_2, \ldots, p'_n.$$

Remark.   A congruence $E \approx f(E)$ $(E, f(E)$ subsets of metric spaces) establishes a one-to-one correspondence between the elements of E and those of $f(E)$ (Exercise 6.)

A point p of a metric space $\mathbf{M}$ is an *accumulation* point of a subset E of M provided each $\varepsilon > 0$ implies the existence of a point q of E such that $0 < pq < \varepsilon$. The set of all accumulation points of E, called the *derived* set of E, is denoted by E'. A subset E of $\mathbf{M}$ is *closed* provided $E' \subset E$, and a subset of $\mathbf{M}$ is *open* provided its complement is closed.

Following Fréchet, we call a subset E of a metric space $\mathbf{M}$ *compact* provided the derived set of each infinite subset of E is nonnull.

Remark.   Each finite metric subset is compact, and if E is compact, then so are all of its subsets.

A metric subset that is both closed and compact is called a *compactum*. Since every space is closed, every compact metric space is a compactum.

If E, F are subsets of a metric space $\mathbf{M}$, the set F is *dense in* E provided $E \subset F \cup F'$.

A metric space $\mathbf{M}$ is called *separable* if and only if it contains a countable subset that is dense in $\mathbf{M}$.

EXERCISES

1. Show that with respect to the definition of limit of an infinite sequence of elements of a metric space $\mathbf{M}$, given in this section, $\mathbf{M}$ is a Fréchet L-space, and hence an infinite sequence of points of a metric space has at most one limit.

2. If $\{p_n\}$, $\{q_n\}$ are infinite sequences of points of a metric space $\mathbf{M}$ and

$$\lim_{n \to \infty} p_n = r = \lim_{n \to \infty} q_n, \quad r \in \mathbf{M},$$

   show that the "staggered" sequence $p_1, q_1, p_2, q_2, \ldots, p_n, q_n, \ldots$ has limit r.

3. If (p_n) is an infinite sequence of points of a metric space $\mathbf{M}$ and

$$\lim_{n \to \infty} p_n \neq p, \quad p \in \mathbf{M},$$

   show that an infinite subsequence $\{p_{n_k}\}$ of $\{p_n\}$ exists such that no subsequence of $\{p_{n_k}\}$ has limit p.

4. Give an example of a mapping f of a subset E of the euclidean plane $\mathbf{E}_2$ onto a subset $f(E)$ of $\mathbf{E}_2$ that is one-to-one and continuous in E, but whose inverse mapping is *not* continuous on $f(E)$.

5. Prove that if an infinite sequence $\{p_n\}$ of points of a metric space $\mathbf{M}$ has limit p, $p \in \mathbf{M}$, then $\{p_n\}$ is a Cauchy sequence; that is, $\varepsilon > 0$ implies the existence of a natural number N such that if $i, j > N$, then $p_i p_j < \varepsilon$.

6. Show that a congruent mapping is (a) one-to-one and (b) a homeomorphism. Give an example of a homeomorphism that is not a congruence.

7. Let A, B be two subsets of a metric space such that $B = f(A)$, where f is one-to-one and continuous in A, and $A = g(B)$, where g is one-to-one and continuous in B. Are A and B necessarily homeomorphic?

1.10   WHAT IS A GEOMETRY?

The question posed in the heading of this section is metamathematical rather than mathematical, and consequently mathematicians of unquestioned competence may (and, indeed, do) differ in the answers they give to it. Even among those mathematicians called geometers there is no generally accepted definition of the term. It has been observed that the abstract, postulational method that has permeated nearly all parts of modern mathematics makes it difficult, if not meaningless, to mark with precision the boundary of that mathematical domain which should be called *geometry*. To some, geometry is not so much a subject as it is a point of view—a way of looking at a subject— so that geometry is the mathematics that a geometer does! To others, geometry is a *language* that provides a very useful and suggestive means of discussing almost every part of mathematics (just as, in former days, French was the language of diplomacy); and there are, doubtless, some mathematicians who find such a query without any real significance and who, consequently, will distain to vouchsafe any answer at all to it.

But the readers of a book entitled *Studies in Geometry* might not unreasonably expect to be provided with a modern definition of geometry, and this section is devoted to meeting that eventual expectation.

In a famous paper, *Vergleichende Betrachtungen über neuere geometrische Forschungen*, published at Erlangen in 1872 (which soon became known as the *Erlanger Program*) the German mathematician Felix Klein [1849–1925] formulated his frequently quoted definition of a geometry. According to Klein, a geometry is a system $\{S; \mathfrak{G}\}$ where S denotes a space and $\mathfrak{G}$ denotes a *group of mappings (transformations) of* S *onto itself*. Using the word *figure* to denote any subset of the element-set of S (with the same structure that converted the element set of S into the *space* S), the Klein geometry $\{S; \mathfrak{G}\}$ studies those and only those properties of a figure that are *invariant* under every transformation of the group $\mathfrak{G}$. Thus, a property $\mathcal{P}$ of a figure F is an object of study in the Klein geometry $\{S; \mathfrak{G}\}$ if and only if $\mathscr{P}$ is a property of each figure $g(F)$ for every $g \in \mathfrak{G}$, where $g(F)$ denotes the figure into which F is mapped by the transformation g of S onto itself.

Now it is clear that an equivalence relation is established in the class of all figures of S by defining two figures F and H to be equivalent (in symbols, $F \sim H$) provided there exists an element g of $\mathfrak{G}$ such that $H = g(F)$. For since $\mathfrak{G}$ contains an identity element, it follows that $F \sim F$ (see Exercise 6), and since $\mathfrak{G}$ contains the inverse of each of its elements, the relation $F \sim H$ implies that $H \sim F$. Finally, if A, B, C are figures such that $A \sim B$ and $B \sim C$, then $B = g_1(A)$, $C = g_2(B)$, $(g_1, g_2 \in \mathfrak{G})$, and so $C = g_2 g_1(A) = g_3(A)$, where g_3 is that element of $\mathfrak{G}$ that is the group product of g_1, g_2 in the order indicated. Hence $A \sim C$, and so the transitive property of the equivalence relation is established.

Hence a Klein geometry $\{S; \mathfrak{G}\}$ studies those and only those properties that a figure has in common with each figure equivalent to it, where equivalence of figures is *defined* by the *special convention adopted above*.

At the time that it was given, Klein's definition was considered very general. It not only embraced all of the known geometries of his day, but it greatly extended the applicability of the term. Thus, for example, Klein wrote in 1912, "What the modern physicists call 'relativity theory' is the theory of the invariants of a four-dimensional space-time continuum (Minkowski's world) with respect to a given group of collineations (the Lorentz group), and hence is a geometry."

An interesting Klein geometry over a space S arises when the group of transformations of S onto itself consists of just one element—the identity transformation! In this case, (1) each figure is equivalent *only* to itself, and consequently (2) *every property of every figure is invariant*. So every property of every figure of S is a fit object of study in this geometry, which is, therefore, simply the *theory of the space S*. Thus space theory is subsumed under the notion of a Klein geometry. It is that geometry over the space whose group is the most trivial of all groups!

An objection to Klein's definition of a geometry is that it is not geometrical! The definition relates a geometry to a group (an algebraic concept) of transformations which are usually given by formulas expressing the *coordinates* (another nongeometrical concept) of a point in terms of the coordinates of the transformed point, or vice versa.

In seeking the geometric essence of Klein's definition, one is led to conclude that the essential element in forming a geometry is the *choice of equivalent figures*, and not the particular convention by means of which the choice is established. This observation suggests the following definition.

DEFINITION(∗)   A *geometry* $\mathfrak{G}$ *over a set* Σ is a system $\{\Sigma; \mathscr{E}\}$, where $\mathscr{E}$ denotes an equivalence relation defined in the set of all subsets (figures) of Σ. The geometry $\{\Sigma; \mathscr{E}\}$ studies those and only those properties of a figure F that the figure has in common with all figures equivalent to F; these are the invariant properties.

It is clear that every Klein geometry is also a geometry in the sense of the above definition. But the new definition admits important subjects that mathematicians would unanimously call geometries, but which do not qualify as such under Klein's definition. Two such cases are the following.

I. Distance geometry over an arbitrary metric space **M**

It is easily seen that an equivalence relation is established in the set of all figures of **M** by defining two figures to be equivalent if and only if they are congruent. This yields a geometry in the sense of definition (∗)—*the distance geometry over the metric space* **M**—which, in general, is not a geometry in the Klein sense since (1) in important instances (for example, when **M** is Hilbert space, or any Minkowski space (not euclidean) or any elliptic space) congruences between figures are not necessarily extendible to a congruence of the space with itself, and (2) the set of congruences (of figures) do *not* form a group.

   (Parenthetically, we note that the distance geometry as defined above, over a *euclidean space of given dimension*, is a Klein geometry only when reflections are added to the "rigid motions.")

II. Topology over an arbitrary metric space **M**

An equivalence relation is established in the set of all figures of **M** by defining two figures to be equivalent if and only if they are homeomorphic. Again we obtain a geometry (*topology*) in the sense of Definition (∗), but not necessarily in the sense of Klein, for reasons analogous to those discussed in the second paragraph of I: (1) homeomorphisms between two figures of **M** are not necessarily extendible to a homeomorphism of **M** with itself, and (2) homeomorphisms between figures of **M** do not, in general, form a group.

   The reader may have observed that Definition (∗) of a geometry is in terms of a *set* Σ rather than a space **S**. This widens the applicability of the notion. We may, for example, consider the theory of cardinal numbers as a geometry over a set Σ, by defining two figures (subsets) of Σ to be equivalent if and only if there exists a one-to-one correspondence between their elements. But in most of the important geometries, the convention that establishes equivalence of figures makes use of the *structure* by virtue of which a set becomes the element-set of a space. Hence one speaks more often of a geometry over a *space* than over a *set*. Topology and distance geometry are two quite different geometries that may be defined over the same (metric) space **M**. They arise from different definitions of equivalent figures. It happens that two figures that are equivalent in the sense of distance geometry (congruent or metrically equivalent figures) are also equivalent in the topological sense, but not conversely. This justifies calling distance geometry over **M** a subgeometry of topology over **M**. The first geometry has "more" invariants to study than the second, since every topological invariant is a metric invariant, but not conversely.

EXERCISES

1. Define five important geometries over the euclidean plane E_2 by defining five different equivalence relations in the set of all subsets of E_2.

2. If a geometry over a set Σ is defined by calling each two subsets of S equivalent, what properties of figures (subsets) are to be studied?

3. Let S denote an abstract set whose elements are called "points," and let T denote an abstract set whose elements are (for suggestiveness) called "transformations." Assume that $X \subset S$ implies the existence of an element t of T such that the pair $\{X; t\}$, also written $t(X)$, is a subset (figure) of S, and that for each $t \in T$, a figure X of S exists such that $t(X)$ is a figure. If a figure X is defined to be *equivalent* to a figure Y if and only if an element t of T exists such that $Y = t(X)$, what algebraic properties of T follow from the reflexive, symmetric, and transitive properties of the equivalence relation?

   For each figure X, how are the subsets T_X, T_Y of T related, where $T_X = [t \in T | t(X) \subset S]$ and $T_Y = [t \in T | t(Y) \subset S]$, with $Y = t(X)$? Does T_X contain an identity element?

4. If the set T of "transformations" of Exercise 3 is such that for *every* $t \in T$ and *every* figure X of S, $t(X)$ is a figure, and the relation $Y = t(X)$ is an equivalence relation (X, Y subsets of S), show that T is a *semigroup* (that is, T contains for each ordered pair t_1, t_2 of its elements, a "product" $t_1 t_2$, and the product operation is associative). What can be said concerning unit and inverse elements?

5. If two figures of a metric space **M** are called equivalent provided there exists a homeomorphism of **M** with itself that maps one figure onto the other, show that the resulting geometry is a Klein geometry. This geometry is classically called topology over **M**.

6. If G denotes a group of mappings of a set Σ *onto* itself, show that the identity element of G is the identity mapping of Σ. Is this necessarily the case if each element of G maps Σ into itself?

1.11   SOME PROPERTIES OF METRIC SPACES

It is not our purpose to present here a systematic account of either Distance Geometry or Topology over a metric space. But in order that this book may be reasonably complete in itself, we shall establish in this and in other sections some standard theorems of both subjects that are needed later.

A real function $f(p, q)$ of point-pairs p, q of a metric space $\mathbf{M}$ is continuous at $p_0, q_0, (p_0, q_0 \in \mathbf{M})$, provided for each two infinite sequences $\{p_n\}, \{q_n\}$, of points of $\mathbf{M}$, $\lim_{n \to \infty} p_n = p_0$, $\lim_{n \to \infty} q_n = q_0$ imply $\lim_{n \to \infty} p_n q_n = p_0 q_0$.

THEOREM 1.2.   *The metric of a metric space is continuous at each point-pair p_0, q_0 of the space.*

Proof. If $\{p_n\}, \{q_n\}$ are any two sequences of points of the space, with $\lim_{n \to \infty} p_n = p_0, \lim_{n \to \infty} q_n = q_0$, then

$$|p_0 q_0 - p_n q_n| = |p_0 q_0 - p_0 q_n + p_0 q_n - p_n q_n| \le |p_0 q_0 - p_0 q_n| + |p_0 q_n - p_n q_n|.$$

The triangle inequality applied to the triples p_0, q_0, q_n and p_0, p_n, q_n gives $|p_0 q_0 - p_0 q_n| \le q_0 q_n$ and $|p_0 q_n - p_n q_n| \le p_0 p_n$. Hence

$$0 \le |p_0 q_0 - p_n q_n| \le p_0 p_n + q_0 q_n,$$

and consequently

$$\lim_{n \to \infty} |p_0 q_0 - p_n q_n| = 0.$$

That is, $\lim_{n \to \infty} p_n q_n = p_0 q_0$.

The next theorem establishes a very useful relation between the notions of accumulation element of a set and the limit of an infinite sequence of elements.

THEOREM 1.3   *If p is an accumulation element of a subset E of a metric space $\mathbf{M}$, then there exists an infinite sequence $\{p_n\}$ of elements of E such that $p_i \ne p_j$ $(i, j = 1, 2, \ldots, i \ne j)$, $p_i \ne p$ $(i = 1, 2, \ldots)$, and $\lim_{n \to \infty} p_n = p$.*

Proof. Let p_1 denote any element of E with $0 < pp_1 < 1$, $p_2 \in E$ with $0 < pp_2 < \min[\frac{1}{2}, pp_1]$, ..., $p_n \in E$ with $0 < pp_n < \min[1/n, pp_{n-1}]$. By induction, an infinite sequence $p_1, p_2, \ldots, p_n, \ldots$ of points of E are obtained which clearly satisfies the demands of the theorem.

COROLLARY   *Each infinite sequence $\{p_n\}$ of elements of a closed and compact subset E of a metric space contains an infinite subsequence $\{p_{i_n}\}$ such that $\lim_{n \to \infty} p_{i_n} = p \, (p \in E)$.*

The reader should provide a proof of this frequently used corollary.

DEFINITION   If $p \in M$ and $\rho > 0$, the set $U(p; \rho)$ of all points q of M such that $pq < \rho$ is the *spherical neighborhood of p with radius ρ.*

LEMMA 1.1   *If $p \in M$ and $\rho > 0$, the set $U(p; \rho)$ is open.*

Proof. The complementary set, com $U(p; \rho)$ consists of all points q of **M** such that $pq \geqq \rho$. If a is an accumulation point of that set then, by the preceding theorem, there exists an infinite sequence $\{q_n\}$, $n = 1, 2, \ldots$, of its points with $\lim_{n \to \infty} q_n = a$. Since $q_n \in \operatorname{com} U(p; \rho)$ then $pq_n \geqq \rho, n = 1, 2, \ldots$, and hence $\lim_{n \to \infty} pq_n \geqq \rho$. By Theorem 1.2, $\lim_{n \to \infty} pq_n = pa$; consequently $pa \geqq \rho$ and $a \in \operatorname{com} U(p; \rho)$. Thus com $U(p; \rho)$ is closed, and so $U(p; \rho)$ is open.

THEOREM 1.4   *A metric space **M** is a Hausdorff topological space for which a neighborhood of a point is any open set containing the point.*

Proof. Referring to Section 1.7 for the definition of a Hausdorff topological space, it is trivial to observe that condition (i) is satisfied.

To establish condition (ii) it suffices to show that the product $U \cap V$ of any two open sets U, V is open. Now $C(U \cap V) = C(U) \cup C(V)$, and if we suppose $p \in \mathbf{M}$ and $p \notin C'(U)$, then $\rho_0 > 0$ exists such that $U(p; \rho_0) \cap C(U) = \varnothing$. Assuming $p \in C'(U \cap V)$, then each $\rho > 0$ implies the existence of a point q of $C(U \cap V)$ such that $0 < pq < \rho$. Taking $\rho < \rho_0$, then $q \in C(V)$ and so $p \in C'(V)$. It follows that $C(U \cap V)$ is closed and so $U \cap V$ is open.

Let x, y denote distinct elements of **M**. Since by Lemma 1.1, sets $U(p; \rho)$ are open, condition (iii)* is established by showing the existence of sets $U(x; \rho)$, $U(y; \rho)$ such that $U(x; \rho) \cap U(y; \rho) = \varnothing$.

Supposing that no such positive ρ exists, then for each natural number n a point r_n exists such that $xr_n < 1/n$ and $yr_n < 1/n$. Hence $\lim_{n \to \infty} r_n = x$ and $\lim_{n \to \infty} r_n = y$ and so, by continuity of the metric, $xy = \lim_{n \to \infty} r_n r_n = 0$, which contradicts the choice of x, y.

From our definition of neighborhood, condition (iv) is obviously satisfied, and the proof is complete.

THEOREM 1.5 *A subset E of a metric space* **M** *is open if and only if each point p of E is the center of a spherical neighborhood that is contained in E.*

Proof. If E is open and contains a point p such that for every $\rho > 0$, $U(p; \rho) \cap C(E) \neq \varnothing$, then clearly $p \in C'(E)$ and hence, since $C(E)$ is closed, $p \in C(E)$, which is impossible.

Conversely, if for each point p of E there is a $\rho(p) > 0$ such that $U(p; \rho(p)) \subset E$, then no accumulation point of $C(E)$ is a point of E. Consequently $C(E)$ is closed and E is open.

COROLLARY *If $E \subset$ **M**, then $p \in E'$ if and only if every open set containing p contains an element of E distinct from p.*

The proof follows from Lemma 1.1 and the preceding theorem.

THEOREM 1.6 *A compact, nonnull, subset E of* **M** *contains for every $\varepsilon > 0$ a finite subset such that every point of E has distance less than ε from at least one point of that finite subset.*

Proof.   If we assume the contrary, then there exists an $\varepsilon > 0$ and an infinite sequence $p_1, p_2, \ldots, p_n, \ldots$ of points of E such that $p_i p_j \geqq \varepsilon$ ($i, j = 1, 2, \ldots, i \neq j$). The point-set of this sequence is an infinite subset of E, *without* an accumulation element, which violates the compactness of E.

COROLLARY *A compact subset of a metric space is bounded (that is, there exists a constant C such that each pair of points of the subset has distance less than C).*

THEOREM 1.7 *Each compact subset E of a metric space contains a countable subset F such that $E \subset F \cup F'$.*

Proof.   For each positive integer k, a finite subset $P^k = \{p_1^k, p_2^k, \ldots, p_{n_k}^k\}$ of E exists such that each point of E has distance less than $1/k$ from at least one point of P^k (Theorem 1.6). If F denotes the union of the sets P^k ($k = 1, 2, \ldots$) and $p \in E$, then for every positive integer m at least one of the potnis of F has distance from p less than $1/m$. It follows that $p \in F \cup F'$.

COROLLARY.   *A compact metric space is separable.*

EXERCISES

1. Let $\mathscr{F}$ denote any family of closed subsets of a metric space M. Show that the product of all members of the family is a closed set. What can you say about the sum of all of the members of any family of open subsets of M?

2. If X, Y are subsets of a metric space, prove that $(X \cup Y)' = X' \cup Y'$.

3. Show that the derived set of any finite or null subset of a metric space is null.

4. Prove that the operator $f(X) = X \cup X'$, defined for each subset X of a metric space M, has the properties (i), (ii), (iii) of the Kuratowski closure function, and hence each metric space is a Kuratowski closure space with cl $X = X \cup X'$. In metric spaces $X \cup X'$ is denoted by cl$X$.

5. Show that if X is any subset of a metric space, its derived set X' is closed.

6. If X is any subset of a metric space, prove that cl X is the product of all closed subsets of the space that contain X, and hence is the "smallest" closed set that contains X. Show that X is a closed set if and only if $X = $ cl X.

7. Prove that a point p of a metric space is an accumulation point of a subset E of the space if and only if every open subset of the space that contains p contains infinitely many points of E.

8. Let $\mathbf{M}$ denote a metric space. If $p \in \mathbf{M}$ and $\{p_n\}$ is an infinite sequence of its points, prove that $p = \lim_{n \to \infty} p_n$ if and only if there corresponds to each open subset U of $\mathbf{M}$ that contains p, a natural number N such that $p_n \in U$ whenever $n > N$.

9. Does the Axiom of Choice (Section 1.2) enter in the proof of Theorem 1.3?

1.12   ADDITIONAL PROPERTIES OF METRIC SPACE

If $\mathbf{M}$ is a separable metric space and P is a countable subset that is dense in $\mathbf{M}$, the family $\mathfrak{S}$ of all spherical neighborhoods $U(p; 1/n)$, for $p \in P$ and $n = 1, 2, 3, \ldots$ is a countable family of open (Lemma 1.1) subsets of $\mathbf{M}$.

This family $\mathfrak{S}$ plays an important role in the theory of separable metric spaces, as may be seen, for example, in the proof of the following theorem.

THEOREM 1.8 *Every separable metric space* $\mathbf{M}$ *is hereditarily separable; that is, if* $X \subset \mathbf{M}$, *then* $X \subset \mathrm{cl}\ Y$, *where* Y *is a countable subset of* X.

Proof. If P denotes a countable subset of $\mathbf{M}$ that is dense in $\mathbf{M}$, we obtain set Y by selecting one element from each of the sets $X \cap U(p; 1/n)$, $p \in P$ ($n = 1, 2, 3, \ldots$), which is not null *Axiom of Choice*). Clearly, Y is a countable subset of X; we proceed to show that $X \subset Y \cup Y'$.

If $x \in X$ and U is any open subset of $\mathbf{M}$ that contains x, there exists a spherical neighborhood $U(x; \rho)$, with center x and radius ρ, that is contained in U (Theorem 11.5). Let n be any natural number such that $1/n < \rho/2$. Since $\mathbf{M} = P \cup P'$, there exists $p \in P$ such that $px < 1/n$. Now the set $U(p; 1/n)$ is a member of the family $\mathfrak{S}$ with the element x in common with X (since $px < 1/n$) and so $U(p; 1/n)$ contains an element y of set Y. Since $xy \leqq px + py < 1/n + 1/n < \rho$, then $y \in U(x; \rho)$. But $U(x; \rho) \subset U$, and consequently each open set containing x, $(x \in X)$, contains an element of set Y. It follows (Corollary, Theorem 1.5) that $x \in Y \cup Y'$, and so $X \subset \mathrm{cl}\ Y$.

Remark. In the proof of the preceding theorem, we showed that if x is any element of any open set U, there is a member $U(p; 1/n)$ of the family $\mathfrak{S}$ that contains x and is contained in U. This implies at once the following important property of separable metric spaces.

THEOREM 1.9 *Each nonnull open subset of a separable metric space is the sum of sets belonging to the countable family* $\mathfrak{S}$ *of subsets of the space.*

The family $\mathfrak{S}$ is said to form a *basis* for the space, and Theorem 1.9 is frequently stated: *every separable metric space has a countable basis.*

DEFINITION A subset F of a subset E of a metric space is *closed in* E provided $E \cap F' \subset F$. A subset G of E is open in E provided the set $E - G$ is closed in E, where $E - G = E \cap \mathrm{com}\ G$.

Remark. A closed set is closed in any set containing it, and the product of any set E with a closed set is closed in E.

THEOREM 1.10 *A mapping f of a subset E of a metric space $\mathbf{M}$ onto a subset E^* of a metric space $\mathbf{M}^*$ is continuous in E if and only if the f-model T of each subset T^* of E^* that is closed in E^* is closed in E.*

Proof. First suppose that f is continuous in E, and let T^* be any subset of $E^* = f(E)$ which is closed in E^*. We show that its model

$$T = [p \in E \,|\, f(p) \in T^*]$$

is closed in E.

If $p_0 \in E \cap T'$, let $p_1, p_2, \ldots, p_n, \ldots$ be an infinite sequence of elements of T with $\lim_{n \to \infty} p_n = p_0$, and $p_i \ne p_0$, $i = 1, 2, \ldots$ (Theorem 1.3). Since f is continuous in E, the infinite sequence $f(p_1), f(p_2), \ldots, f(p_n), \ldots$ of elements of T^* has the element $f(p_0)$ as limit.

Now if $f(p_0) \notin T^*$, then $f(p_n) \ne f(p_0)$, $n = 1, 2, \ldots$, and it follows that $f(p_0)$ is an accumulation element of T^*, and hence $f(p_0) \in T^*$ (since $f(p_0) \in E^*$ and T^* is closed in E^*). This contradication yields $f(p_0) \in T^*$; consequently $p_0 \in T$, and so T is closed in E.

Conversely, assume that f is a mapping of E onto $E^* = f(E)$, and each subset T^* of E^* that is closed in E^* has its model T closed in E. We show that f is continuous at each element p_0 of E.

Let $p_1, p_2, \ldots p_n, \ldots$ be any infinite sequence of elements of E with $\lim_{n \to \infty} p_n = p_0$. To show that $\lim_{n \to \infty} f(p_n) = f(p_0)$, let V denote any open subset of $\mathbf{M}^*$ containing $f(p_0)$. Since com V is closed, the product $T^* = E^* \cap$ com V is closed in E^*, and hence the model $T = [p \in E \,|\, f(p) \in T^*]$ is closed in E. Now $p_0 \notin T$ (since $f(p_0) \notin T^*$) and since $p_0 \in E$, and T is closed in E, then $p_0 \notin T'$. Hence there exists an open subset U of $\mathbf{M}$ that contains p_0 and does not contain any point of T (Exercise 7, Section 1.10).

Since $\lim_{n \to \infty} p_n = p_0$, there exists a natural number N such that $p_n \in U$ for every $n > N$ (Exercise 8, Section 1.10). Hence $n > N$ implies $p_n \notin T$, and consequently $f(p_n) \notin T^* = E^* \cap$ com V. Since $f(p_n) \in E^*$ for every index n, it follows that for $n > N$, $f(p_n) \in V$, and we may conclude that $\lim_{n \to \infty} f(p_n) = f(p_0)$.

DEFINITIONS   Two subsets A, B of a metric space are *separated* provided $A \ne \varnothing$, $B \ne \varnothing$, $A \cap B = A \cap B' = A' \cap B = \varnothing$.

A subset E of a metric space is *connected* provided E is not the sum of two separated sets.

THEOREM 1.11   *A subset E of a metric space is connected if and only if E is not the sum of two nonnull, mutually exclusive subsets, each of which is closed in E.*

Proof.   Suppose $E = A \cup B$, where $A \ne \varnothing$, $B \ne \varnothing$, $A \cap B = \varnothing$, and $A \supset E \cap A'$, $B \supset E \cap B'$. Then $A = A \cap E$, $B = B \cap E$, and

$$B \cap A' = (B \cap E) \cap A' = B \cap (E \cap A') \subset B \cap A = \varnothing,$$

$$A \cap B' = (A \cap E) \cap B' = A \cap (E \cap B') \subset A \cap B = \varnothing.$$

Hence, A, B are separated sets, and so E is not connected.

Conversely, if E is not connected, sets A, B exist such that $E = A \cup B$, $A \neq \varnothing$, $B \neq \varnothing$, $A \cap B = A' \cap B = A \cap B' = \varnothing$. Then

$$A' \cap E = A' \cap (A \cup B) = (A' \cap A) \cup (A' \cap B) = A' \cap A \subset A,$$

$$B' \cap E = B' \cap (A \cup B) = (B' \cap A) \cup (B' \cap B) = B' \cap B \subset B,$$

and consequently A, B are each closed in E.

THEOREM 1.12 *If E denotes a connected subset of a metric space, and f is a continuous mapping of E onto a subset $f(E)$ of a metric space, then $f(E)$ is connected.*

Proof. If $f(E)$ is not connected, then subsets A^*, B^* of $f(E)$ exist such that $f(E) = A^* \cup B^*$, $A^* \neq \varnothing$, $B^* \neq \varnothing$, $A^* \cap B^* = \varnothing$, and each of the sets A^*, B^* is closed in $f(E)$ (Theorem 1.11).

Let A_1, B_1 denote the models of A^*, B^*, respectively, in E. Then $A_1 \neq \varnothing$, $B_1 \neq \varnothing$, $A_1 \cap B_1 = \varnothing$, and $E = A_1 \cup B_1$. Moreover, since f is continuous in E, it follows by Theorem 1.10 that A_1 and B_1 are each closed in E. But this implies that E is not connected (Theorem 1.11) in contradiction with the hypothesis of the theorem.

Examples are easily found to show that a continuous image of a closed set is not necessarily closed, and a continuous image of a compact set need not be compact. Such examples give point to the following well-known result.

THEOREM 1.13 *Let E and $f(E)$ be subsets of metric spaces. If E is closed and compact, and f is continuous in E, then $f(E)$ is closed and compact.*

Proof. We show first that $f(E)$ is compact. If Q denotes any infinite subset of $f(E)$, then Q contains an infinite sequence $q_1, q_2, \ldots, q_n, \ldots$ of pairwise distinct points. (Why?) Now $q_n \in f(E)$ implies the existence of p_n ($p_n \in E$), such that $q_n = f(p_n)$, $n = 1, 2, \ldots$, and since $q_i \neq q_j$ ($i \neq j$), then $p_i \neq p_j$ ($i \neq j$). Hence the set of points $P = \{p_1, p_2, \ldots, p_n, \ldots\}$ is an infinite subset of E, and since E is compact, that set has an accumulation point p_0 which (since E is closed) belongs to E. Hence $q_0 = f(p_0)$ is a point of $f(E)$.

It follows from Theorem 1.3 that P contains an infinite sequence of points, each distinct from p_0, with limit p_0. Such a sequence may be written

$$p_{n_1}, p_{n_2}, \ldots, p_{n_k}, \ldots$$

with $n_1 < n_2 < \cdots < n_k < \cdots$, and we have $\lim_{k \to \infty} p_{n_k} = p_0$. Since f is continuous at p_0, the $\lim_{k \to \infty} q_{n_k} = q_0$, where $q_{n_k} = f(p_{n_k})$, $k = 1, 2, \ldots$. The points of the sequence $\{q_{n_k}\}$ being pairwise distinct, its limit q_0 is an accumulation point of the set $\{q_{n_1}, q_{n_2}, \ldots, q_{n_k}, \ldots\}$, and since that set is a subset of Q, then $q_0 \in Q'$. Hence each infinite subset of $f(E)$ has a nonnull derived set, and so $f(E)$ is compact.

Now if $q_0 \in f'(E)$ and $q_1, q_2, \ldots, q_n, \ldots$ is an infinite sequence of pairwise distinct points of $f(E)$, each distinct from q_0, with $\lim_{n \to \infty} q_n = q_0$, let p_1, $p_2, \ldots, p_n, \ldots$ denote an infinite sequence of points of E, where p_k is an arbitrarily selected point in the model of q_k, $k = 1, 2, \ldots$. Then $p_i \neq p_j$ ($i \neq j$), the set $\{p_1, p_2, \ldots, p_n, \ldots\}$ is an infinite subset of E, and has an accumulation element p_0 in E. Then, as in the first part of the proof, we have $\lim_{k \to \infty} p_{n_k} = p_0$, $\lim_{k \to \infty} q_{n_k} = \lim_{k \to \infty} f(p_{n_k}) = f(p_0)$. But $\lim_{n \to \infty} q_n = q_0$, and since the sequence $\{q_{n_k}\}$ is a subsequence of the sequence $\{q_n\}$, then $f(p_0) = \lim_{k \to \infty} q_{n_k} = \lim_{n \to \infty} q_n = q_0$ (Exercise 1, Section 1.9). Hence $q_0 \in f(E)$, and so $f(E)$ is closed.

DEFINITION   A subset of a metric space is a *continuum* provided it is closed, compact, connected, and contains more than one point.

THEOREM 1.14   *Let E and $f(E)$ be subsets of metric spaces. If E is a continuum and f is continuous in E, then $f(E)$ is a continuum if it contains at least two points.*

Proof.   The theorem is an immediate consequence of Theorems 1.12 and 1.13.

COROLLARY   *Let E and $f(E)$ be subsets of metric spaces. If E is a continuum and f is one-to-one and continuous in E, then $f(E)$ is a continuum.*

Proof.   Since E is a continuum, E contains at least two points, and since f is one-to-one in E, $f(E)$ contains at least two points. Consequently, $f(E)$ is a continuum, by the preceding theorem.

Let E denote the subset of the euclidean (coordinate) plane consisting of all points with coordinates (x, y) for $y = 1$, $0 \leq x \leq 1$, x rational, and $y = 2$, $0 < x < 1$, x irrational. A one-to-one and continuous mapping f of E onto the (closed) unit interval $I = [0, 1]$ is defined by associating with $(x, y) \in E$ the point $(x, 0)$ of I. (The mapping f is the (vertical) projection of E onto I.) The reader can easily show that the inverse function is *totally discontinuous* in I; that is, f^{-1} is not continuous at any point of I. Thus a one-to-one and continuous mapping f of a subset E of a metric space onto a subset $f(E)$ of a metric space may not have a continuous inverse (and hence may not be a homeomorphism). The following theorem shows that this cannot happen if E is closed and compact. (Note that, as a subset of the plane, the set E of the example just given is compact but *not* closed; if E is considered by itself as a metric space, it is, indeed, closed—but then it is not compact.)

THEOREM 1.15.   *Let E and $f(E)$ be subsets of metric spaces. If E is closed and compact, and f is one-to-one and continuous in E, then the inverse f^{-1} of f is continuous in $f(E)$.*

Proof.   If f^{-1} is not continuous at a point q_0 of $f(E)$, then an infinite sequence $q_1, q_2, \ldots, q_n, \ldots$ of points of $f(E)$ exists such that $\lim_{n \to \infty} q_n = q_0$, *without* $\lim_{n \to \infty} f^{-1}(q_n) = f^{-1}(q_0)$ holding.

Now since $\lim_{n \to \infty} q_n = q_0$, at most one element of the sequence $\{q_n\}$ can occur infinitely often, and if one does do so, that element must be q_0. In that case, an infinite subsequence $\{q_{n_i}\}$ of $\{q_n\}$ exists in which q_0 does not occur, for in the contrary event, a natural number N exists such that $q_n = q_0$ for $n > N$. But then $f^{-1}(q_n) = f^{-1}(q_0)$ for $n > N$, and so $\lim_{n \to \infty} f^{-1}(q_n) = f^{-1}(q_0)$, in violation of the above property of the sequence $\{q_n\}$. The subsequence $\{q_{n_i}\}$ also has limit q_0, and since each of its elements recurs just finitely often, it contains a subsequence of pairwise distinct elements, each distinct from q_0, with limit q_0. Hence, if f^{-1} is not continuous at q_0 ($q_0 \in f(E)$), there exists an infinite sequence $t_1, t_2, \ldots, t_n, \ldots$ of pairwise distinct elements of $f(E)$, each distinct from q_0, such that $\lim_{n \to \infty} t_n = q_0$, without $\lim_{n \to \infty} f^{-1}(t_n) = f^{-1}(q_0)$ holding (note that $\lim_{n \to \infty} f^{-1}(t_n)$ might not even exist).

Writing $p_n = f^{-1}(t_n)$, $n = 1, 2, \ldots$, the infinite sequence $p_1, p_2, \ldots, p_n, \ldots$ of points of E has its elements pairwise distinct (and different from $p_0 = f^{-1}(q_0)$), and hence forms an infinite subset of E. Since $\lim_{n \to \infty} p_n = p_0$ does not hold, there exists an open set U containing p_0 and an infinite subsequence $\{p_{n_k}\}$ of $\{p_n\}$ with $p_{n_k} \notin U$ ($k = 1, 2, \ldots$). Putting $p_{n_k} = s_k$ ($k = 1, 2, \ldots$), the set $\{s_1, s_2, \ldots\}$ is an infinite subset of E, and since E is closed and compact, there exists an infinite subsequence $\{s_{n_k}\}$ and an element s_0 of E such that $\lim_{k \to \infty} s_{n_k} = s_0$.

From the continuity of f in E, it follows that $\lim_{k \to \infty} f(s_{n_k}) = f(s_0)$. But $f(s_{n_k}) = q_{n_k}$ $(k = 1, 2, \ldots)$, and since $\{q_{n_k}\}$ is a subsequence of $\{q_n\}$, $\lim_{k \to \infty} q_{n_k} = q_0$. Hence $f(s_0) = q_0 = f(p_0)$, and since f is one-to-one in E, then $p_0 = s_0$. But this is impossible, since $p_0 \in U$ and none of the elements s_{n_k} of the sequence $\{s_{n_k}\}$ belongs to U.

That contradiction establishes the theorem.

COROLLARY   *Let E and $f(E)$ be subsets of a metric space. If f is one-to-one and continuous in E, and E is closed and compact, then f is a homeomorphism.*

DEFINITION   Let E denote a subset and p a point of a metric space **M**. A point f_p of E is a *foot* of p on E provided $pf_p \leqq px$ for every point x of E.

THEOREM 1.16   *If E is a closed and compact subset of a metric space **M** and $p \in$ **M**, there exists a foot of p on E.*

Proof.   Since the numbers px, $x \in E$, are nonnegative, there exists a greatest lower bound d of those numbers. Hence for each natural number n, a point p_n of E exists such that $d \leqq pp_n < d + 1/n$, and a point q of E exists with $\lim_{n \to \infty} p_{i_n} = q$, where $\{p_{i_n}\}$ is an infinite subsequence of $\{p_n\}$. (Why?)

Evidently

$$d = \lim_{n \to \infty} pp_{i_n} = pq$$

(why?) and q is a foot of p on E.

Remark.   Each subsequence of $\{p_n\}$ that has a limit defines a foot of p on E and consequently p might have uncountably many feet on E.

EXERCISES

1. If $\mathfrak{F}$ is any family of open subsets of a separable metric space, show that a countable subfamily of $\mathfrak{F}$ exists, the sum of the members of which is the same as the sum of the members of the family $\mathfrak{F}$.

Hint: With each member of the family $\mathfrak{S}$ that is contained in a member of the family $\mathfrak{F}$, associate one such member of $\mathfrak{F}$. Consider the sum of those members of $\mathfrak{S}$, taking into account Theorem 1.9 and the Remark preceding it.

2. *Lindelöf property.* If a subset E of a separable metric space is contained in the sum of the members of a family $\mathfrak{F}$ of open subsets of the space, prove that E is contained in the sum of countably many members of that family.

3. *Cantor property.* Let $F_1 \supset F_2 \supset \cdots \supset F_n \supset \cdots$ denote any infinite sequence of monotone decreasing, nonnull, closed subsets of a metric space. If at least one of the subsets of the sequence is compact, show that the product $\bigcap_{i=1}^{\infty} F_i \neq \varnothing$.

Hint: Consider a set $P = \{p_1, p_2, \ldots p_n, \ldots\}$, where $p_i \in F_i$ $(i = 1, 2, \ldots)$, and examine the two possibilities: (a) P is a finite set; (b) P is an infinite set.

4. Give examples to show that for metric spaces the Cantor property is not valid if any other part of the hypothesis is suppressed.

5. *Borel property.* Let F denote a closed and compact subset of a metric space. If $\{U_n\}$, $n = 1, 2, \ldots$, is any infinite sequence of open subsets of the space such that

$$F \subset U_1 \cup U_2 \cup \cdots \cup U_n \cup \cdots,$$

show there exists a natural number N such that

$$F \subset U_1 \cup U_2 \cup \cdots \cup U_N.$$

Hint: Consider the "partial sums" $S_n = U_1 \cup U_2 \cup \cdots \cup U_n$, and the sets $F_n - F \cap \text{com } S_n$ $(n = 1, 2, \ldots)$. Show that if $F_n \neq \varnothing$ $(n = 1, 2, \ldots)$, the Cantor property applied to $F_1 \supset F_2 \supset \cdots$ yields a contradiction. Hence there exists a natural number N for which $F_N = \varnothing$, and the desired result follows.

6. *Borel-Lebesgue property.* If a closed and compact subset E of a separable metric space is contained in the sum of the members of a family $\mathfrak{F}$ of open subsets of the space, prove that E is contained in the sum of a finite number of the members of $\mathfrak{F}$.

REFERENCES

Section 1.2   For an axiomatic study of set theory, see P. Bernays, *Axiomatic Set Theory* (with a historical introduction by A. A. Fraenkel), Studies in Logic and the Foundations of Mathematics, North-Holland Publishing Co., Amsterdam, 1958. The book contains an excellent bibliography.

E. Zermelo, "Beweis, dass jede Menge wohlgeordnet werden kann," Math. Ann., **59**: 514–516 (1904); see also *Math. Ann.*, **65**: 107–128 (1908).

Section 1.4   Dedekind investigated lattices (especially modular lattices) under the name *Dualgruppe* in his article "Uber die von drei Moduln erzeute Dualgruppe," Math. Ann., **53**: 371–403 (1900). Menger introduced lattices (especially normed lattices) in "Axiomatik der endlichen Mengen und der elementargeometrischen Verknüpfungsbeziehungen," Jber. Deutsch. Math. Verein., **37**:309–325 (1928).

The use of the word *lattice* in this context seems to be due to Garrett Birkhoff, whose "Lattice Theory," Amer. Math. Soc. Coll. Pub., vol. 25, 1940, was the first book devoted to the subject.

Section 1.6   In a sense, the study of Boolean algebra marked the beginning of lattice theory; see G. Boole, *The Mathematical Analysis of Logic*, Cambridge, 1847, and his more famous work, *An Investigation into the Laws of Thought*, London. 1854.

Richard Dedekind investigated modular lattices under the name Dualgruppe; see his article, "Über die von drei Moduln erzeugte Dualgruppe," *Math. Ann.*, **53**: 371–403 (1900).

Normed lattices were introduced by K. Menger in the reference cited above.

Section 1.7   For historicial references to semimetric and metric spaces, see L. M. Blumenthal, *Distance Geometries: A Study of the Development of Abstract Metrics*, University of Missouri Studies, Vol. 13 (1938), No. 2.

Section 1.8   Additional examples of metric spaces are given in L. M. Blumenthal, *Theory and Applications of Distance Geometry*, The Clarendon Press, Oxford, 1953. In future references this book will be referred to by the symbol DG.

CHAPTER 2

Geometric Aspects of Lattice Theory

In the preceding chapter a lattice has been defined (1) algebraically, as a certain double composition system $\mathscr{L} = \{L; \cdot, +\}$, and also (2) as a partially ordered set $\{L; \prec\}$ that contains with each two of its elements their greatest lower and least upper bounds. In this chapter we present a brief development of lattice theory. Our emphasis is on those parts of the subject that are useful in the treatment of projective geometry given in Part II, and on those aspects that are interesting from the point of view of Distance Geometry (to which Part III may serve as an introduction). But apart from those considerations, lattice theory exemplifies the essential unity of geometry and algebra, and consequently is appropriately considered in Part 1, whose *Leitmotiv* is unification by abstraction.

2.1   THE PRINCIPLE OF DUALITY

We recall (Section 1.4) that an abstract lattice $\mathscr{L}$ is an abstract set which is closed with respect to each of two associative, commutative, idempotent, binary operations, $\cdot$ and $+$, which are related by the condition that $a \cdot b = a$ if and only if $a + b = b$ $(a, b \in L)$. Nothing essential is altered in the above sentence if the two binary operations are interchanged, and it follows that corresponding to each logical consequence (theorem) of the defining properties of a lattice, there is another valid consequence of those properties whose statement may be obtained by interchanging $\cdot$ and $+$ in the statement of the previous consequence. A formal derivation or proof of the corresponding statement may be obtained automatically by effecting this interchange in each step of the proof of the first statement. The process is called *dualizing*; each of two such associated statements is called the *dual* of the other, and the assertion of their covalidity is the *principle of duality*. We shall have frequent recourse to that principle.

In making use of the principle of duality, the reader is asked to observe the following precautions. If notions *derived* from the primitive operations (that is, defined in terms of them) are present in a theorem or in its derivation, care must be used when dualizing to replace such notions by their duals, obtained by dualizing the definitions of the original notions. For example, the relation $a \prec b$ is in this context a derived (defined) notion; that is, $a \prec b \underset{\mathrm{D}}{\equiv} a \cdot b = a$ (where $\underset{\mathrm{D}}{\equiv}$ is read, "is, by definition") and its dual is a relation defined by the relation $a + b = a$. Since this latter relation is equivalent to $a \cdot b = b$, and by the above definition can be written $b \prec a$, we see that the dual of $a \prec b$ is $b \prec a$, which we may also write $a \succ b$. Thus each of the derived notions $\prec$, $\succ$ is the dual of the other.

It should be noted that the principle of duality might be rendered *invalid* by placing an additional restriction on the class of lattices. For example, in a lattice with a *first* element a (that is, a is an element such that $a \cdot x = a$ for every element x of the lattice), but no *last* element b (that is, an element b such that $x + b = b$ for every element x of the lattice) the principle of duality is clearly invalid, since the following valid consequence of its defining properties

The lattice has a first element.

does not have a valid dual.

Whenever an additional property is demanded of a lattice (for example, the distributive property $(a + b) \cdot c = a \cdot c + b \cdot c$) it is important to decide whether the duality principle still holds. It evidently will hold if and only if the property that is dual to the one introduced can be established in the new system. When we prove later, for example, that in a distributive lattice, $a \cdot b + c = (a + c) \cdot (b + c)$ for each three elements a, b, c of the lattice—a statement which is the dual of the assumed distributivity property (distributivity of multiplication over addition)—we can then conclude that the principle of duality is valid in the class of distributive lattices.

In all that follows, the lattice product $a \cdot b$ of two elements of a lattice is frequently written ab.

EXERCISES

1. How is the principle of duality proved valid in a lattice $\mathscr{L} = \{L; \prec)$ considered as a partially ordered set?

2. Dualize the relation $a \cdot b + b \cdot c = b = (a + b) \cdot (b + c)$. When is a relation self-dual?

3. Dualize the statement: if a, b, $c \in L$ and $a \prec c$, then $(a + b)c = a + bc$.

4. If a, b, $c \in L$ such that $b = (a + b) \cdot (b + c)$, does it follow from the principle of duality that $b = a \cdot b + b \cdot c$?

5. For what lattices $\mathscr{L}$, if any, is at least one of the relations
   $a \cdot b + b \cdot c = b = (a + b) \cdot (b + c)$,
   $a \cdot c + c \cdot b = c = (a + c) \cdot (c + b)$,
   $b \cdot a + a \cdot c = a = (b + a) \cdot (a + c)$,
   satisfied for every three elements a, b, c of L?
   Could two of the relations be satisfied by three pairwise distinct elements a, b, c of L?

6. What other mathematical systems do you know that admit a duality principle?

2.2   FINITE LATTICES.

SUM-RELATIVE AND
PRODUCT-RELATIVE COMPLEMENTS

A pictorial device is useful to define simple lattices containing just a few elements. For this purpose we adopt the convention of picturing elements of a lattice by small circles, and denoting them by small letters. If $a \neq b$ and the relation $a \prec x \prec b$ holds if and only if $a = x$ or $x = b$ (element b is then called the *immediate successor* of a, and a is the *immediate predecessor* of b), then

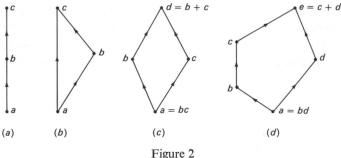

Figure 2

the two circles are placed at the ends of an arrow pointing toward b, and circle a is put at a lower level than circle b.

The following examples will make the procedure clear (Figure 2(a)–(d)).

Figure 2(a) depicts the lattice formed by an ordered set (chain) of three elements: $a \prec b \prec c, a \neq b \neq c \neq a$; Figure 2(b) violates one of our conventions (which one?) and hence does not represent any lattice; Figure 2(c) is a lattice of four pairwise distinct elements with $ab = ac = ad = bc = a$, and $b + c = d$.

The lattice in Figure 2(d) is of special interest. Clearly $(b + d)c = ec = c$, $bc + dc = b + a = b$, and consequently $(b + d)c \neq bc + dc$; that is, the lattice of Figure 2(d) is *not* distributive. In fact, since $b \prec c$ and $c = (b + d)c \neq b + dc = b + a = b$, the lattice is not even modular. We shall see later that every *nonmodular lattice contains the lattice of Figure* 2(d) *as a sublattice.*

DEFINITION   A subset of a lattice $\mathscr{L}$ that contains, with each two of its elements, their sum and product in $\mathscr{L}$ is called a *sublattice* of $\mathscr{L}$.

Remark.   By definition, a lattice $\mathscr{L}$ contains the greatest lower bound (product) and the least upper bound (sum) of each two of its elements, and mathematical induction easily establishes that corresponding to each finite subset $a_1, a_2, \ldots, a_n$ of elements of $\mathscr{L}$ there are elements $a_1 a_2 \cdots a_n$ and $a_1 + a_2 + \cdots + a_n$ of $\mathscr{L}$ such that (1) $a_1 a_2 \cdots a_n \prec a_i$ ($i = 1, 2, \ldots, n$), and follows every element with that property, and (2) $a_1 + a_2 + \cdots + a_n \succ a_i$ ($i = 1, 2, \ldots, n$), and precedes every element of $\mathscr{L}$ with that property. Thus $a_1 a_2 \cdots a_n$ and $a_1 + a_2 + \cdots + a_n$ are the (necessarily unique) greatest lower and least upper bounds, respectively, of the subset $\{a_1, a_2, \ldots, a_n\}$. An application of the above remark leads to the conclusion that *every finite lattice contains both a first and a last element.*

For lattices that have both a first element 0 and a last element 1, the notion of complementary elements was made relative, in two ways, by G. Bergmann in 1929. If $a \prec b$, one relative complement of a with respect to b, the *sum-relative complement*, is any element x such that $ax = 0$ and $a + x = b$. (If the first condition $ax = 0$ is omitted, then a sum-relative complement always exists; namely, the element b.) Another relative complement of a with respect to b, the *product-relative complement*, is any element y such that $b + y = 1$ and $by = a$. (A remark similar to the one made above explains the reason for the condition $b + y = 1$.)

Applied to $a \prec 1$, the sum-relative complement is any element x such that $ax = 0$ and $a + x = 1$; that is, *a sum-relative complement is simply a complement*. Similarly, applied to $0 \prec a$, a product-relative complement is seen to be an ordinary complement.

In the lattice $\mathscr{L}$ of all subsets of a line, for example, (with lattice sum and product, the ordinary set sum and product, respectively), let $a \prec b$. The sum-relative complement is *unique* and is the ordinary set complement of a *with respect to* b; for if $x \in \mathscr{L}$ such that $ax = 0$ and $a + x = b$, where 0 is the null set $\varnothing$, then, since this lattice is distributive,

$$b \cdot C(a) = (a + x)C(a) = x \cdot C(a)$$
$$= ax + x \cdot C(a) = (a + C(a))x = x,$$

where $C(a)$ denotes the set-complement of a.

To obtain the product-relative complement, suppose $y \in \mathscr{L}$ such that $y + b = 1$ and $by = a$, where 1 denotes the whole set. Then

$$a + C(b) = by + C(b) = (b + C(b))(y + C(b))$$
$$= y + C(b) = (y + C(b))(y + b)$$
$$= y + y \cdot C(b) + by = y,$$

and so the product-relative complement is unique and equals the set sum of a and the set complement of b.

Let us note that the two relative complements are complementary elements; for $bC(a)[a + C(b)] = 0$ and

$$bC(a) + a + C(b) = [b + C(b)] \cdot [C(a) + C(b) + a] = C(a) + C(b) + a = 1.$$

As shown in the following theorem, the mutual complementarity of sum-relative and product-relative complements is valid in every lattice in which they exist.

THEOREM 2.1 *If $a, b \in \mathscr{L}$ with $a \prec b$, each sum-relative complement x is a complement of each product-relative complement y.*

Proof. Clearly $xy = x(a + x)y = xby = ax = 0$, and $x + y = y + yb + x = y + a + x = y + b = 1$.

If a lattice with both first and last elements, contains for each two elements a, b with $a \prec b$ both sum-relative and product-relative complements, it follows from a previous remark that the lattice is *complemented* (that is, it contains a complement of each of its elements). But a complemented lattice need not contain sum-relative or product-relative complements for each pair of elements a, b with $a \prec b$. Consider, for example, the complemented lattice of Figure 2(d), with first element a and last element e. We have $b \prec c$, but no sum-relative complementary element exists, since the only elements satisfying the condition $bx = a$ are elements a and d, and neither satisfies the condition $b + x = c$. Also, b has no product-relative complement with respect to c.

EXERCISES

1. If a, b, c, d are elements of a lattice $\mathscr{L}$ show that
   (a) $a \prec c$ and $b \prec d$ imply $ab \prec cd$ and $a + b \prec c + d$,
   (b) $a \prec b$ implies $ac \prec bc$ and $a + c \prec b + c$,
   (c) $a \prec b \prec c$ and $a = c$ imply $a = b = c$,
   (d) $ab = a + b$ if and only if $a = b$,
   (e) $ac + bc \prec (a + b)c \prec c \prec ab + c \prec (a + c)(b + c)$.

2. How does a sublattice of $\mathscr{L}$ differ from a subset of $\mathscr{L}$ that is a lattice with respect to the two lattice operations of $\mathscr{L}$? Give an example.

3. Give an example to show that neither sum-relative nor product-relative complements are necessarily unique.

2.3   CONDITIONS FOR MODULAR LATTICES

Let a, b, c be elements of a lattice $\mathscr{L}$, with $a \prec c$. Then $ab \prec bc$, $ab + bc = bc$, $a + c = c$, $(a + c)b = bc$, and so $(a + c)b = ab + bc$. Also, $a \prec c \prec b + c$ implies $(b + c)a = a = a + ab = ba + ca$. But although $a \prec c$ yields $(a + b)c \succ a + bc$, the lattice of Figure 2(d) of the preceding section shows that the partial ordering symbol of the last relation cannot, in general, be replaced by equality. Hence the condition of modularity,

$$a \prec c \text{ implies } (a + b)c = ac + bc = a + bc,$$

is the only one of the three distributivities that represents any restriction on the lattice. We denote this modularity condition by $(\mathscr{M})$.

Remark 1.   The principle of duality is valid in a modular lattice, since the modularity condition $(\mathscr{M})$ is self-dual.

Remark 2.   Condition $(\mathscr{M})$ is equivalent to $(\mathscr{M}^*)$: $a, b, c \in \mathscr{L}, (ac + b)c = ac + bc$. For since $ac \prec c$, $(\mathscr{M})$ implies $(\mathscr{M}^*)$, and applying $(\mathscr{M}^*)$ for $a \prec c$ gives $(\mathscr{M})$.

Remark 3. A subgroup $\mathfrak{A}$ of an arbitrary group $\mathfrak{G}$ is an *invariant* subgroup of $\mathfrak{G}$ provided for each element α of $\mathfrak{A}$ and each element γ of $\mathfrak{G}$, $\gamma \circ \alpha \circ \gamma^{-1} \in \mathfrak{A}$, where $\circ$ denotes the group composition and γ^{-1} is the inverse of γ. *The set I of all invariant subgroups of $\mathfrak{G}$ forms a modular (nondistributive) lattice with respect to the partial ordering relation of set inclusion*; that is, we write $\mathfrak{A} \prec \mathfrak{B}$ provided the invariant subgroup $\mathfrak{A}$ of $\mathfrak{G}$ is a subset of the invariant subgroup $\mathfrak{B}$ of $\mathfrak{G}$.

It follows that the greatest lower bound $\mathfrak{A}\mathfrak{B}$ of two elements $\mathfrak{A}$, $\mathfrak{B}$ of I is their set product, and the least upper bound $\mathfrak{A} + \mathfrak{B}$ is the smallest invariant subgroup of $\mathfrak{G}$ that contains both $\mathfrak{A}$ and $\mathfrak{B}$; that is, $\mathfrak{A} + \mathfrak{B}$ denotes the set product of all invariant subgroups of $\mathfrak{G}$ that contain both $\mathfrak{A}$ and $\mathfrak{B}$. It is easily seen to consist of all those elements of the group $\mathfrak{G}$ that are products of an element of $\mathfrak{A}$ with an element of $\mathfrak{B}$. For clearly all these products must belong to $\mathfrak{A} + \mathfrak{B}$ and, on the other hand, they form a group that contains the groups $\mathfrak{A}$ and $\mathfrak{B}$. To prove the group closure property for the set of these products—all of the other group properties are obvious—let $\alpha_1 \circ \beta_1$ and $\alpha_2 \circ \beta_2$ be two such products, $\alpha_1, \alpha_2 \in \mathfrak{A}$, $\beta_1, \beta_2 \in \mathfrak{B}$.   Then

$$(\alpha_1 \circ \beta_1) \circ (\alpha_2 \circ \beta_2) = (\alpha_1 \circ \alpha_2) \circ [(\alpha_2^{-1} \circ \beta_1 \circ \alpha_2) \circ \beta_2],$$

and since $\alpha_1 \circ \alpha_2 \in \mathfrak{A}$, and $(\alpha_2^{-1} \circ \beta_1 \circ \alpha_2) \circ \beta_2 \in \mathfrak{B}$ (since $\mathfrak{B}$ is an invariant subgroup, $\alpha_2^{-1} \circ \beta_1 \circ \alpha_2 \in \mathfrak{B}$), then the product $(\alpha_1 \circ \beta_1) \circ (\alpha_2 \circ \beta_2)$ is seen to be the product of an element of $\mathfrak{A}$ with an element of $\mathfrak{B}$.

Now let $\mathfrak{A}$, $\mathfrak{B}$, $\mathfrak{C}$ be three elements of this lattice, with $\mathfrak{A} \prec \mathfrak{C}$. To show $(\mathfrak{A} + \mathfrak{B})\mathfrak{C} = \mathfrak{A} + \mathfrak{B}\mathfrak{C}$ it suffices to establish $(\mathfrak{A} + \mathfrak{B})\mathfrak{C} \prec \mathfrak{A} + \mathfrak{B}\mathfrak{C}$, since by part (e) of Exercise 1 of the preceding section, $\mathfrak{A} \prec \mathfrak{C}$ implies $\mathfrak{A} + \mathfrak{B}\mathfrak{C} \prec (\mathfrak{A} + \mathfrak{B})\mathfrak{C}$ in every lattice. If $\alpha \in \mathfrak{A}$, $\beta \in \mathfrak{B}$, $\gamma \in \mathfrak{C}$, the subgroup $(\mathfrak{A} + \mathfrak{B})\mathfrak{C}$ consists of those elements γ of $\mathfrak{C}$ that are products $\alpha \circ \beta$; that is, for all elements of

$(\mathfrak{A} + \mathfrak{B})\mathfrak{C}$, $\gamma = \alpha \circ \beta$ and so $\beta = \alpha^{-1} \circ \gamma$. Since $\mathfrak{A} \prec \mathfrak{C}$, $\alpha^{-1} \in \mathfrak{C}$ and so $\beta \in \mathfrak{C}$. Hence every element $\alpha \circ \beta$ belonging to $(\mathfrak{A} + \mathfrak{B})\mathfrak{C}$ is the product of an element α of $\mathfrak{A}$ with an element β belonging to both $\mathfrak{B}$ and $\mathfrak{C}$. Hence if $\mathfrak{A} \prec \mathfrak{C}$, then $(\mathfrak{A} + \mathfrak{B})\mathfrak{C} \prec \mathfrak{A} + \mathfrak{B}\mathfrak{C}$, and so the lattice is modular.

THEOREM 2.2   *If any lattice $\mathscr{L}$ has one of the following properties, it has all of them:*

(i) $a \prec c$ *implies* $(a + b)c = a + bc$,

(ii) $a \prec b$, $ac = bc$, $a + c = b + c$ *implies* $a = b$,

(iii) $\mathscr{L}$ *does not contain as a sublattice the lattice of Figure 2(d)*,

(iv) *if in any sublattice $\mathscr{L}'$ of $\mathscr{L}$ $ab \underset{\mathscr{L}'}{\longrightarrow} a$ and $ab \underset{\mathscr{L}'}{\longrightarrow} b$, then $a \underset{\mathscr{L}'}{\longrightarrow} a + b$ and $b \underset{\mathscr{L}'}{\longrightarrow} a + b$, where $x \underset{\mathscr{L}'}{\longrightarrow} y$ symbolizes the relation " y is the immediate successor of x in the sublattice $\mathscr{L}'$."*

Proof.   Suppose $\mathscr{L}$ has property (i) and let a, b, c be elements of $\mathscr{L}$ such that $a \prec b$, $ac = bc$, and $a + c = b + c$. Then $a = a + ac = a + bc = (a + c)b = (b + c)b = b$, and so (i) implies (ii). Property (ii) expresses a *weak cancellation law* for modular lattices.

To show that (ii) implies (iii) we establish the contrapositive proposition; namely, that *not* (iii) implies *not* (ii). This is immediate, for if $\mathscr{L}$ *does* contain the sublattice of Figure 2(d) then it contains elements a, b, c such that $a \prec b$, $ac = bc$, $a + c = b + c$, but $a \neq b$; that is, property (ii) does *not* hold.

Property (iii) implies property (i), because if $\mathscr{L}$ contains elements a, b, c with $a \prec c$ but $(a + b)c \neq a + bc$, then, since $a + bc \prec (a + b)c$, the five elements $a + bc$, $(a + b)c$, b, $a + b = a + bc + b$, $bc = b(a + b)c$ form a sublattice of the kind assumed absent. For $a + bc \prec (a + b)c$ implies $b(a + bc) \prec bc$, and since $a + bc \succ bc$, then $b(a + bc) \succ bc$. Hence $(a + bc)b = bc = b(a + b)c$, and in a similar way, $(a + b)c + b = a + b = a + bc + b$.

Hence the properties (i), (ii), (iii) are pairwise equivalent. Now property (iv) implies property (i), since if (i) does *not* hold, neither does (iii), and the presence of a sublattice $\mathscr{L}'$ of the kind represented in Figure 2(d) exhibits elements (for example, a and c of Figure 3) such that $ac = d \underset{\mathscr{L}'}{\longrightarrow} a$, $ac \underset{\mathscr{L}'}{\longrightarrow} c$ but a is *not* the immediate predecessor in $\mathscr{L}'$ of $a + c = e$, and so property (iv) does not hold.

It remains to show that (iii) implies (iv). Let a, b be distinct elements of a sublattice $\mathscr{L}'$ of a lattice $\mathscr{L}$ which has property (iii), and suppose $ab \underset{\mathscr{L}'}{\longrightarrow} a$, $ab \underset{\mathscr{L}'}{\longrightarrow} b$. If we assume the existence of an element c of $\mathscr{L}'$ such that $a \prec c \prec a + b$, $a \neq c \neq a + b$, the elements ab, a, b, $a + b$, c form a sublattice

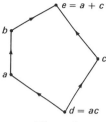

Figure 3

of $\mathscr{L}$ of five elements of the kind which, by property (iii), the lattice $\mathscr{L}$ does *not* contain.

First, the elements are pairwise distinct. For if $a = a + b$ then $ab \prec b \prec a$, which contradicts $ab \xrightarrow{\mathscr{L}} a$; similarly, $b = a + b$ is seen to be impossible. Other coincidences are even more obviously impossible.

Second, $bc = ab$ and $b + c = a + b$; for from $a \prec c \prec a + b$ we have $ab \prec bc \prec b$, and since $bc \in \mathscr{L}'$ and $ab \xrightarrow{\mathscr{L}} b$, then consequently $ab = bc$ or $bc = b$. But $bc = b$ gives $b + c = bc + c = c$, and $a \prec c$ gives $a + b \prec b + c = c$; and consequently $a + b = c$, contrary to what we have assumed concerning the element c. Hence $bc = ab$, and similarly, $b + c = a + b$. Thus the five elements ab, a, b, $a + b$, c form a sublattice of $\mathscr{L}$ of the type represented in Figure 2(d). Hence $a \xrightarrow{\mathscr{L}} a + b$. In a similar manner, it is shown that $b \xrightarrow{\mathscr{L}} a + b$.

COROLLARY   *If $\mathscr{L}$ is modular, $xy \to x$, $xy \to y$ imply $x \to x + y$, $y \to x + y$.*

It is interesting to observe that the form of *weak distributivity of multiplication over addition*, given by the modularity condition, property (i), is logically equivalent in any lattice to the *weak cancellation law*, which is expressed by property (ii).

Remark 4.   We may now prove the remark made in Section 1.6 that the lattice of all linear subspaces of a projective three-dimensional space is modular; for, by a fundamental relation that the reader may easily verify,

$$\dim(x + y) + \dim xy = \dim x + \dim y,$$

where x, y are any elements of the lattice and dim x denotes the dimension of the element x (for example, dim $x = -1, 0, 1, 2, 3$ according as x denotes the null set, a point, line, plane, or three-space, respectively).

Hence if a, b, c are elements of this lattice such that $a \prec b$, $ac = bc$, $a + c = b + c$, then

$$\dim(a + c) + \dim ac = \dim a + \dim c,$$
$$\dim(b + c) + \dim bc = \dim b + \dim c.$$

Hence $\dim a = \dim b$, which, since $a \prec b$, yields $a = b$.

EXERCISES

1. Show that in a complemented modular lattice, sum- and product-relative complements exist for each two elements a, b with $a \prec b$.

2. Noting that x, $y \in \mathscr{L}'$, ($\mathscr{L}'$ a sublattice of $\mathscr{L}$), $x \underset{\mathscr{L}'}{\nrightarrow} y$ does not imply $x \rightarrow y$, give an example of a *nonmodular* lattice $\mathscr{L}$ such that if x, $y \in \mathscr{L}$ with $xy \rightarrow x$ and $xy \rightarrow y$, then $x \rightarrow x + y$ and $y \rightarrow x + y$, where $x \rightarrow y$ denotes that y is the immediate successor of x in $\mathscr{L}$.

3. The lattice of invariant subgroups of a group $\mathfrak{G}$ has been shown to be modular. For what groups $\mathfrak{G}$ is that lattice distributive. ?

4. Let x, y denote elements of a modular lattice $\mathscr{L}$ such that $xy \rightarrow x$ and $xy \rightarrow y$. Prove that $x \rightarrow x + y$ by showing that $x \neq x + y$, and that if z is an element of $\mathscr{L}$ with $x \prec z \prec x + y$, then $x = z$ or $z = x + y$. In a similar manner $y \rightarrow x + y$ is established.

2.4   THE PRINCIPAL CHAIN THEOREM

A finite sequence x_0, x_1, x_2, ..., x_{k-1}, x_k of elements of a lattice is a *principal chain* joining x_0 and x_k, provided $x_0 \rightarrow x_1 \rightarrow x_2 \rightarrow \cdots \rightarrow x_{k-1} \rightarrow x_k$, where, we recall, the notation $a \rightarrow b$ signifies that $a \prec b$, $a \neq b$, and $a \prec c \prec b$ if and only if $a = c$ or $c = b$. The number k is the *length* of the chain. A principal chain is clearly a *maximal* chain between its end elements.

It is observed that the elements a and e of the lattice of Figure 2(d) are joined by two principal chains $a \rightarrow b \rightarrow c \rightarrow e$ and $a \rightarrow d \rightarrow e$, one of which has length 3 and the other has length 2. It is one of the most remarkable and useful theorems of lattice theory that such a phenomenon (that is, the existence of principal chains of *different lengths* joining the same two elements) cannot occur in modular lattices. This theorem, the analogue in lattice theory of the Jordan-Hölder theorem of group theory, is proved in this section. The following remark will be found useful in the proof of the theorem.

Remark. If $a, x, y \in \mathcal{L}$ and $a \to x, a \to y, x \neq y$, then $a = xy$. For $a \prec xy \prec x$ gives $a = xy$ or $xy = x$, and $a \prec xy \prec y$ implies $a = xy$ or $xy = y$, which (since $x \neq y$) yields $a = xy$.

THEOREM 2.3   *In a modular lattice all principal chains joining a given pair of elements have the same length.*

Proof.   If there exists a principal chain of length 1 joining elements a, b then $a \to b$, and it is clear that every principal chain joining a, b has length 1. This anchors an inductive hypothesis that if two elements are joined by a principal chain of length k, then every principal chain joining those two elements has length k.

Now let

$$a = x_0 \to x_1 \to x_2 \to \cdots \to x_k \to x_{k+1} = b$$

be a principal chain of length $k + 1$ joining a and b, and let

$$a = y_0 \to y_1 \to y_2 \to \cdots \to y_{n-1} \to y_n = b$$

be a principal chain of length n that joins a, b. The proof is completed when it is shown that $n = k + 1$. This is accomplished by establishing two assertions.

Assertion A.   *There exists at least one index* i $(1 \leq i \leq k + 1)$ *such that* $y_1 + x_{i-1} = x_i$.

Suppose the contrary, then $y_1 + x_k \neq x_{k+1}$, and since $x_k \prec y_1 + x_k \prec x_{k+1}$, then $x_k = y_1 + x_k$ and consequently $y_1 \prec x_k$.

From this result, $x_{k-1} \prec y_1 + x_{k-1} \prec x_k$, which, together with $y_1 + x_{k-1} \neq x_k$, gives $x_{k-1} = y_1 + x_{k-1}$, and hence $y_1 \prec x_{k-1}$. Repetition of this procedure yields the contradiction $y_1 \prec x_0$, and proves the assertion.

Let j denote the smallest of the indices $1, 2, \ldots, k + 1$ such that $y_1 + x_{j-1} = x_j$.

Assertion B.   *The sequence of elements*

$$y_1 = y_1 + x_0, y_1 + x_1, y_1 + x_2, \ldots, y_1 + x_{j-1} = x_j, x_{j+1}, \ldots, x_{k+1}$$

is a principal chain joining y_1 *and* $x_{k+1} = b$.

If $j = 1$, then $y_1 = x_1$ and we have the principal chain

$$y_1 \to x_2 \to x_3 \to \cdots \to x_{k+1} = b.$$

If $j \neq 1$, then $y_1 + x_0 \neq x_1$, and since $a \to y_1 = y_1 + x_0$, $a \to x_1$, use of the remark preceding the theorem gives $a = (y_1 + x_0)x_1$; that is, $x_1(y_1 + x_0) \to x_1$ and $x_1(y_1 + x_0) \to y_1 + x_0$. Since $\mathcal{L}$ is modular, it has property (iv) of Theorem 2.2, and so we may conclude that $x_1 \to y_1 + x_0 + x_1 = y_1 + x_1$ and $y_1 + x_0 \to y_1 + x_1$. Thus Assertion B is established in case $j = 2$ (that is, if $y_1 + x_1 = x_2$).

In the contrary case, $y_1 + x_1 \neq x_2$, $x_1 \to y_1 + x_1$, $x_1 \to x_2$ imply $x_1 = (y_1 + x_1)x_2$, and again property (iv) of $\mathcal{L}$ gives $y_1 + x_1 \to y_1 + x_1 + x_2 = y_1 + x_2$ and $x_2 \to y_1 + x_2$, which proves Assertion B in case $j = 3$.

Proceeding in this manner we obtain

$$y_1 = y_1 + x_0 \to y_1 + x_1 \to y_1 + x_2 \to \cdots \to y_1 + x_{j-1} = x_j,$$

and Assertion B is established.

Now the length of the principal chain

$$y_1 = y_1 + x_0 \to y_1 + x_1 \to \cdots \to y_1 + x_{j-1} = x_j \to x_{j+1} \to \cdots \to x_k \to x_{k+1} = b$$

joining the elements y_1 and b is k and hence, by the inductive hypothesis, every principal chain joining y_1 and b has length k. It follows that the principal chain $y_1 \to y_2 \to \cdots \to y_n = b$ has length k and so $n - 1 = k$ or $n = k + 1$, completing the proof of the theorem.

Since the nonmodular lattice of Figure 2(d) contains two elements a and e which are joined by two principal chains of different lengths, and since, by Theorem 2.2, that lattice is a sublattice of every nonmodular lattice, we have the following corollary.

COROLLARY   *A lattice is modular if and only if in any sublattice $\mathcal{L}'$ of $\mathcal{L}$ all principal chains in $\mathcal{L}'$ joining a given pair of elements of $\mathcal{L}'$ have the same length.*

Of course, the lengths of such chains depend on the elements that they join.

EXERCISES

1. Give an example of a nonmodular lattice $\mathcal{L}$ such that for every pair of distinct elements a, b of $\mathcal{L}$, all principal chains joining a, b have the same length.

2. Show that for each positive integer n exceeding 1, a modular lattice exists containing two elements that are the terminal elements of two principal chains, each of length n, that have only those two elements in common.

2.5 DISTRIBUTIVE LATTICES.
ELEMENTARY PROPERTIES

We call a lattice distributive provided for each three of its elements a, b, c,

$(\mathscr{D})$ $$(a + b)c = ac + bc.$$

Remark 1.   Clearly every distributive lattice is modular, since if $a \prec c$ the condition of distributivity $(\mathscr{D})$ reduces to that of modularity $(\mathscr{M})$. Since, as we have seen, the lattice of linear subspaces of a projective three-space is modular (Remark 4, Section 2.3) but not distributive (Section 1.6), the class of distributive lattices is a proper subclass of the class of modular lattices.

Remark 2.   *The duality principle is valid in distributive lattices.* It suffices to establish the dual of $(\mathscr{D})$:

$(\mathscr{D}^*)$ $$ab + c = (a + c)(b + c).$$

Use of $(\mathscr{D})$ gives

$$(a + c)(b + c) = a(b + c) + c(b + c) = ab + ac + c = ab + c.$$

Moreover, if $ab + c = (a + c)(b + c)$ is assumed for each three elements of a lattice $\mathscr{L}$, then

$$ac + bc = (a + bc)(c + bc) = (a + b)(a + c)(b + c)c = (a + b)c,$$

and so $(\mathscr{D})$ results.

Hence either of the two kinds of distributivity (multiplication over addition, and addition over multiplication) implies the other, and so may be used to define the class of distributive lattices.

Remark 3.   The lattice of five elements shown in Figure 4 is not distributive, since $(a + b)c = ec = c$, while $ac + bc = d + d = d$ and $c \neq d$.   Clearly, the nonmodular lattice of Figure 2(d) is also, a fortiori, nondistributive.

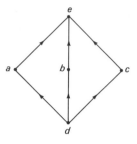

Figure 4

In Theorem 2.2 we proved that the weak distributivity property ($\mathcal{M}$) possessed by modular lattices is equivalent to the weak cancellation property (ii), and also equivalent to the absence of the lattice of five elements of Figure 2(d) as a sublattice. The next theorem shows that strengthening the modularity property to unrestricted distributivity validates a similarly un-restricted cancellation law.

THEOREM 2.4 *If any lattice $\mathcal{L}$ has any one of the following properties, it has all of them:*

(i) $a, b, c \in \mathcal{L}$, $ac = bc$, $a + c = b + c$ *implies* $a = b$,
(ii) $a, b, c \in \mathcal{L}$ *implies* $(a + b)c = ac + bc$,
(iii) $a, b, c \in \mathcal{L}$ *implies* $ab + c = (a + c)(b + c)$,
(iv) $\mathcal{L}$ *does not contain* as a sublattice either of the lattices shown *in Figures* 2(d) and 4.

Proof. By Remark 2, properties (ii) and (iii) are equivalent. To show property (i) implies property (ii), suppose that $ac = bc$ and $a + c = b + c$, imply $a = b$. Then $\mathcal{L}$ is modular (Theorem 2.2), and putting $p = (a + c)b + ac$, $q = (b + c)a + bc$, we have $pc = ac + bc = qc$, and (by duality) $p + c = q + c$. Consequently $p = q$, and hence $pq = p + q$. But $pq = ab + bc + ca$ and, dually, $p + q = (a + b)(b + c)(c + a)$. Then we have $ab + bc + ca = (a + b)(b + c)(c + a)$. Now $ac + bc = ac + bc + bac = (ac + bc + ab)c$, and $(a + b)c = (a + b)(a + c)(b + c)c$; consequently $(a + b)c = (ab + bc + ca)c = ac + bc$.

Property (ii) implies property (i), since $a = a(a + c) = a(b + c) = ab + ac = ab + bc = (a + c)b = (b + c)b = b$. Hence properties (i), (ii), (iii) are pairwise equivalent.

Property (iii) implies property (iv), for if $\mathscr{L}$ contains a sublattice of either of the kinds shown in Figure 2(d) or Figure 4, then clearly property (i) is not valid, and consequently neither is property (iii).

Finally, property (iv) implies property (i). For if property (i) does *not* hold, elements a_0, b_0, c_0 of $\mathscr{L}$ exist such that $a_0 c_0 = b_0 c_0, a_0 + c_0 = b_0 + c_0$, but $a_0 \neq b_0$. It follows that a_0, c_0 are not comparable, and neither are b_0, c_0. (If, for example, $a_0 \prec c_0$, then the above two equalities yield $a_0 = b_0 c_0, c_0 = b_0 + c_0$ and so $a_0 = b_0(b_0 + c_0) = b_0$. If $c_0 \prec a_0$, then $c_0 = b_0 c_0$ and so $c_0 \prec b_0$. Then $a_0 + c_0 = b_0 + c_0$ yields $a_0 = b_0$.)

The proof is completed by showing that if $\mathscr{L}$ does *not* contain the lattice of Figure 2(d) then it *does* contain the lattice of Figure 4 as a sublattice.

Now if Figure 2(d) is absent, then $\mathscr{L}$ is modular and it follows that a_0 and b_0 are not comparable; for if either $a_0 \prec b_0$ or $b_0 \prec a_0$, the modular cancellation law gives $a_0 = b_0$, contrary to $a_0 \neq b_0$ assumed above.

We now assert that the elements

$$a_0 b_0, \qquad a_0, \qquad b_0, \qquad (a_0 + b_0)(c_0 + a_0 b_0), \qquad a_0 + b_0$$

form a sublattice of $\mathscr{L}$ of the kind given in Figure 4. The reader will have no difficulty in showing (Exercise 3) that

$$a_0 b_0 \prec a_0 \prec a_0 + b_0, a_0 b_0 \prec (a_0 + b_0)(c_0 + a_0 b_0) \prec a_0 b_0,$$
$$a_0 b_0 \prec b_0 \prec a_0 + b_0,$$
$$a_0 b_0 = a_0(a_0 + b_0)(c_0 + a_0 b_0) = b_0(a_0 + b_0)(c_0 + a_0 b_0),$$
$$a_0 + b_0 = a_0 + (a_0 + b_0)(c_0 + a_0 b_0) = b_0 + (a_0 + b_0)(c_0 + a_0 b_0),$$

and the proof of the theorem is complete.

Remark 4.   Neither of the equivalent conditions $(\mathscr{D})$, $(\mathscr{D}^*)$ for distributivity of a lattice is self-dual (each is the dual of the other). A self-dual condition that is necessary and sufficient for distributivity (and which is, therefore, equivalent to $(\mathscr{D})$ and to $(\mathscr{D}^*)$) is $ab + bc + ca = (a + b)(b + c)(c + a)$. For the necessity is immediate, and it is verified at once that if this condition holds, then sublattices of Figures 2(d) and 4 are absent.

We establish now a few of the more readily obtained consequences of assuming that $\mathscr{L}$ is distributive. Other, less obvious, properties of distributive lattices will be developed in later sections.

Property 1.   *Relative complementary (and hence complementary) elements are unique whenever they exist.*

For from $x_1 a = 0 = x_2 a$ and $x_1 + a = b = x_2 + a$ follows (property (ii), Theorem 2.4) $x_1 = x_2$; similarly, $y_1 = y_2$ is a consequence of $y_1 b = a = y_2 b$, $y_1 + b = 1 = y_2 + b$.

In Section 2.2 we saw that in the case of the lattice of all subsets of a line, sum-relative and product-relative complements were expressible in terms of ordinary complements.

Property 2. In any distributive lattice, sum-relative and product-relative complements are expressible in terms of ordinary complements, whenever complements exist.

If $a \prec b$ then $a'b$ is the sum-relative complement of a with respect to b, and $a + b'$ is the product-relative complement; for $a(a'b) = 0$ and $a + a'b = (a + a')(a + b) = a + b = b$, where "primes" indicate complements, and

$$(a + b')b = ab + 0 = a,$$
$$b + (a + b') = a + (b + b') = a + 1 = 1.$$

Property 3. If $a \prec b$ and a', b' exist, then $b' \prec a'$.

From $a \prec b$, $ab = a$ and $a(a' + b') = ab' = (ab)b' = 0$. Also $a + (a' + b') = (a + a') + b' = 1 + b' = 1$. Hence $a' + b'$ is the complement of a; that is, $a' + b' = a'$ and since $b' \prec a' + b'$, then $b' \prec a'$.

Property 4. The de Morgan formulas for lattices,

$$(ab)' = a' + b',$$
$$(a + b)' = a'b',$$

are valid, whenever the complements exist.

For since $ab(a' + b') = aa'b + abb' = 0$ and $ab + a' + b' = (a + a')(b + b') + b' = (a + a' + b')(b + b') = 1$, then $a' + b'$ is the complement of ab; that is $a' + b' = (ab)'$.

Similarly, $(a + b)a'b' = 0$ and $a + b + a'b' = (a + b + a')(a + b + b') = 1$, and so $a'b' = (a + b)'$.

Property 5. In a distributive lattice, the complements of elements a, b exist if and only if the complements of their sum and their product exist; moreover,

(∗)
$$a' = (a + b)' + b(ab)',$$
$$b' = (a + b)' + a(ab)'.$$

For

$$a + (a + b)' + b(ab)' = (a + b' + a + b) \cdot [a + (ab)']$$
$$= (a + b)' + (a + b)(a + a' + b')$$
$$= (a + b)' + (a + b) = 1,$$

and

$$[(a + b)' + b(ab)'] = a(a + b)' + ab(ab)'$$
$$= aa'b' = 0.$$

Interchanging a and b in the first of formulas (∗) gives the second.

EXERCISES

1. Prove that the lattice of nonnegative real numbers, with $a \prec b$ provided $a \leq b$, is distributive.

2. Prove the distributivity of the lattice of natural numbers, with $a \prec b$ provided a is a divisor of b.

3. Establish the relations concluding the proof of Theorem 2.4; that is, prove that the elements $a_0 b_0$, a_0, b_0, $(a_0 + b_0)(c_0 + a_0 b_0)$, $a_0 + b_0$ of that theorem form a sublattice of $\mathscr{L}$ of the kind given in Figure 4.

2.6 NORMED LATTICES.
A UNION OF ALGEBRA AND GEOMETRY

In some lattices one may associate with each element x a nonnegative real number, denoted by $|x|$ and called the *norm* of x, such that (1) if $x \prec y$ and $x \neq y$, then $|x| < |y|$, and (2) for each two elements x, y of the lattice, $|x + y| + |xy| = |x| + |y|$. Though distinct elements may have equal norms, it follows from (1) that $|x| = |y|$ and $x \prec y$ imply $x = y$. This norm criterion for equality of elements proves to be very useful.

An explicit characterization of those lattices that admit a norm is not known, but the following theorem shows that only modular lattices have that property.

THEOREM 2.5   *Each normed lattice $\mathscr{L}$ is modular.*

Proof. If $\mathscr{L}$ is not modular, it contains the lattice of Figure 2(d) as a sublattice. Now this sublattice cannot be normed, for supposing otherwise we would have $|b + d| + |bd| = |b| + |d|$ and hence $|e| + |a| = |b| + |d|$; also $|c + d| + |cd| = |c| + |d|$, and so $|e| + |a| = |c| + |d|$. This yields $|b| = |c|$, which, since $b \prec c$, implies $b = c$.

Some examples of normed lattices are the following.

1.   The lattice $\mathscr{L}$ of natural numbers—where $a \oplus b =$ the least common multiple of a, b, and $ab =$ the greatest common divisor of a, b—is normed by putting $|a| = \log a$. For $|a| \geq 0$ for every $a \in \mathscr{L}$, and $a \prec b$ and $a \neq b$ imply $|a| < |b|$, since $a \prec b$ if and only if a is a divisor of b. We show now that $|a + b| + |ab| = |a| + |b|$.

Writing $a = \prod_p p^\alpha$, $b = \prod_p p^\beta$, where the products are taken over all primes p, and α, β are nonnegative integers with $\alpha = 0$ unless p divides a and $\beta = 0$ unless p divides b, it is clear that, putting $[a, b] = $ g.c.d(a, b) and $\{a, b\} = $ l.c.m.(a, b),

$$[a, b] = \prod_p p^{\min(\alpha, \beta)},$$

$$\{a, b\} = \prod_p p^{\max(\alpha, \beta)}.$$

And since

$$\{a, b\} \cdot [a, b] = a \times b,$$

then

$$\log\{a, b\} + \log[a, b] = \log a + \log b.$$

2.   The modular lattice of linear subspaces of a projective three-space is normed by putting $|a| = 1 + \dim a$. For example, let a, b denote two skew lines of the space. Then $a \oplus b =$ whole space, $ab = \varnothing$, the null space, $|a \oplus b| + |ab| = 4 + 0$ and $|a| + |b| = 2 + 2$. If a is a line that meets a plane b, not containing it, in an ideal point, then $|a + b| + |ab| = 4 + 1$, $|a| + |b| = 2 + 3$.

3.   The lattice of nonnegative real numbers, with $a \prec b$ defined as $a \leq b$, is normed by setting $|a| = a$, for then $a \oplus b = \max(a, b)$, $ab = \min(a, b)$ and the arithmetic sum of the norms of $a \oplus b$ and ab equals the arithmetic sum of the two numbers a, b.

One of the important properties of a normed lattice is that a distance may be defined for each two of its elements—the distance being intimately connected with the lattice operations—with respect to which the elements form a metric space (Section 1.7). Thus the set of elements of $\mathscr{L}$ bears the *algebraic structure* of a lattice and the *geometric structure* of a metric space, with the two structures closely related. This suggests a comparative study of geometric and lattice properties of the set. But first of all, let us see how the metric space arises.

DEFINITION   The *distance* $\delta(a, b)$ of two elements a, b of the normed lattice $\mathscr{L}$ is defined by

$$\delta(a, b) = |a + b| - |ab|.$$

THEOREM 2.6   *With respect to the distance* $\delta(a, b) = |a + b| - |ab|$, *the elements of a normed lattice $\mathscr{L}$ form a metric space.*

Proof.   Since $ab \prec a + b$, and $ab = a + b$ if and only if $a = b$, then $\delta(a, b) \geq 0$, and $\delta(a, b) = 0$ if and only if $a = b$. Clearly $\delta(a, b) = \delta(b, a)$, and it remains to establish the triangle inequality.

From $(a + b)c \prec (a + c)(b + c)$ follows

$$\begin{aligned}
|a + b| + |c| &= |a + b + c| + |(a + b)c| \\
&\leq |(a + c) + (b + c)| + |(a + c)(b + c)| \\
&= |a + c| + |b + c|,
\end{aligned}$$

and, since $ac + bc \prec ab + c$, then

$$\begin{aligned}
|ab| + |c| &= |ab + c| + |abc| \\
&\geq |ac + bc| + |(ac)(bc)| \\
&= |ac| + |bc|.
\end{aligned}$$

It follows that

$$|a + b| - |ab| \leq |a + c| - |ac| + |b + c| - |bc|.$$

That is, $\delta(a, b) \leq \delta(a, c) + \delta(c, b)$.

The metric space formed by the elements of a normed lattice $\mathscr{L}$, with distance $\delta(a, b) = |a + b| - |ab|$, is denoted by $\mathbf{M}(\mathscr{L})$. It is called the *associated metric space* of $\mathcal{L}$. We shall refer to the elements of the common ground-set of $\mathcal{L}$ and $\mathbf{M}(\mathcal{L})$ as elements of $\mathcal{L}$ or as elements of $\mathbf{M}(\mathcal{L})$, indiscriminately.

While each normed lattice $\mathcal{L}$ has associated with it a metric space $\mathbf{M}(\mathcal{L})$, that space (due to its genesis) has certain *special* properties that we shall expose later. It is, therefore, not the case that each metric space is the associated metric space of a normed lattice. We shall ascertain in another section those metric properties of a metric space that ensure that the space is the associated metric space of a normed lattice.

An interesting union of algebra and geometry is presented by a parallel study of the algebraic properties of $\mathcal{L}$ and the geometric properties of $\mathbf{M}(\mathcal{L})$. Certain lattice properties of $\mathcal{L}$, expressible wholly in terms of lattice sum and product, are metrically significant (expressible in terms of the metric of $\mathbf{M}(\mathcal{L})$), and vice versa. What metric property of $M(\mathcal{L})$, for instance, is logically equivalent to the algebraic property of $\mathcal{L}$ being distributive? Since the geometric structure of $\mathbf{M}(\mathcal{L})$ is quite simple (compared, for example, to that of the euclidean plane), a study of the space $\mathbf{M}(\mathcal{L})$ provides an easy introduction to abstract geometry. Another, quite different, introduction is furnished by *Boolean geometry*, which will be discussed in some detail in the next chapter.

As a first step in the direction of such a study, we examine in the next section the notion of betweenness as a geometric concept in $\mathbf{M}(\mathcal{L})$ and a lattice relation in $\mathcal{L}$.

EXERCISES

1. Show that the lattice of real numbers, with $a \prec b$ provided $a \leq b$, is normed by defining $|a| = e^a$.

2. If $a, x, y \in \mathcal{L}$, a normed lattice, prove that $\delta(a + x, a + y) + \delta(ax, ay) \leq \delta(x, y)$, and so each of the mappings $f(x) = a + x$, $g(x) = ax$, of $\mathcal{L}$ into itself is a *contraction*.

2.7 BETWEENNESS IN NORMED LATTICES

One of the most important concepts in geometry is a relation that one point q may have with respect to two points p, r, which is usually described by saying that the point q is between the points p and r. The reader might be aware that one of the defects of Euclid's *Elements* is that properties of the betweenness relation for points of a line were used without such a relation being either explicitly or axiomatically defined. The introduction into geometry of an axiomatization of betweenness is due to the German mathematician Moritz Pasch (1843-1931) whose postulates (or modifications of them) are now to be found in many postulational treatments of geometry.

In metric spaces the betweenness relation has been explicitly defined by K. Menger. For a parallel study of the notion in $\mathscr{L}$ and $\mathbf{M}(\mathscr{L})$ it is convenient to modify slightly that definition as follows.

DEFINITION   An element b of $\mathbf{M}(\mathscr{L})$ is *between* two elements a, c if and only if $\delta(a, b) + \delta(b, c) = \delta(a, c)$. The relation is denoted by writing $\mathscr{B}(a, b, c)$.

Remark 1.   In metric spaces, *metric* betweenness requires $a \neq b \neq c$, *in addition* to the equality stated in the above definition. In a lattice study of betweenness, however, it is more convenient to use the wider sense of the notion, which we call *betweenness* rather than *metric* betweenness.

The question arises: can the metric relation $\mathscr{B}(a, b, c)$ of three elements a, b, c of $\mathbf{M}(\mathscr{L})$ be expressed solely in terms of the lattice operations of $\mathscr{L}$? In other words, what is the lattice equivalent of betweenness $\mathscr{B}(a, b, c)$?

DEFINITION   An element b of a lattice $\mathscr{L}$ is *order between* two elements a, c of $\mathscr{L}$, symbolized by $\mathcal{O}(a, b, c)$, if and only if $a \prec b \prec c$ or $c \prec b \prec a$.

Remark 2.   If a, b, c are elements of a normed lattice, then $\mathcal{O}(a, b, c)$ implies $\mathscr{B}(a, b, c)$, but not conversely. For if $a \prec b \prec c$, then $a + b = b = bc$, $ab = ac = a$, $a + c = c = b + c$,

$$\delta(a, b) = |b| - |a|, \qquad \delta(b, c) = |c| - |b|, \qquad \delta(a, c) = |c| - |a|,$$

and so $\delta(a, c) = \delta(a, b) + \delta(b, c)$. The same conclusion follows from $c \prec b \prec a$.

On the other hand, consider the normed lattice $\mathscr{L}$ formed by the set of all ordered pairs (x_1, x_2) of nonnegative real numbers, with $(x_1, x_2) \prec (y_1, y_2)$ if and only if $x_1 \leq y_1$ and $x_2 \leq y_2$, and $|(x_1, x_2)| = x_1 + x_2$. The element $(2, 2)$ of $\mathscr{L}$ is between the elements $(1, 4)$ and $(3, 1)$, but *not* order between them. (Let the reader verify this.)

Thus, order betweenness is sufficient for betweenness, but not necessary.

Remark 3.   If *a, b, c* are elements of a normed lattice, then $\mathscr{B}(a, b, c)$ implies the lattice relation $ac \prec b \prec a + c$, but not conversely.

The first part of this remark follows immediately from the corollary of the next theorem (Exercise 1). To see that the necessary condition $ac \prec b \prec a + c$ for $\mathscr{B}(a, b, c)$ is not sufficient, let *c* denote a plane and *a, b* lines not on *c* which are concurrent in a point of *c*, where *a, b, c* are elements of the lattice of linear subspaces of a projective three-space. Then clearly $ac \prec b \prec a + c$, but $\delta(a, b) = 2$, $\delta(b, c) = 3 = \delta(a, c)$, and so $\mathscr{B}(a, b, c)$ does not subsist.

Remarks 2 and 3 lend additional interest to the following theorem.

THEOREM 2.7   *An element b of the associated metric space* $\mathbf{M}(\mathscr{L})$ *of a normed lattice* $\mathscr{L}$ *is between two elements a, c of* $\mathbf{M}(\mathscr{L})$ *if and only if*

(†)
$$a(b + c) \prec b \prec a + bc.$$

Proof.   We note first that the lattice relation (†) is equivalent to the relations

(∗)
$$a + b = a + bc, \qquad ab = a(b + c).$$

For $b \prec a + bc$ is equivalent to $a + bc + b = a + bc$, which is equivalent to $a + b = a + bc$, and the second relation of (∗) follows by duality.

Assuming now the lattice relation (†), then $ac \prec b \prec a + c$, and so $abc = ac$, $a + b + c = a + c$. Hence $|a + b| = |a + bc| = |a| + |bc| - |abc| = |a| + |b| + |c| - |b + c| - |ac|$, $|ab| = |a(b + c)| = |a| + |b + c| - |a + b + c| = |a| + |b| + |c| - |bc| - |a + c|$, and consequently

$$\delta(a, b) = |a + c| - |ac| - [|b + c| - |bc|] = \delta(a, c) - \delta(b, c).$$

That is, $\mathscr{B}(a, b, c)$ subsists.

If, conversely, we suppose $\mathscr{B}(a, b, c)$ valid, then

$$|a + b| - |ab| = |a + c| - |ac| - |b + c| + |bc|$$

$$\leqq |a + b + c| - |abc| - |b + c| + |bc|$$

$$= |a| + |bc| - |abc| - |a| - |b + c| + |a + b + c|$$

$$= |a + bc| - |a(b + c)|.$$

But since $|a + b| \geqq |a + bc|$ and $|ab| \leqq |a(b + c)|$, then

$$|a + b| - |ab| \geqq |a + bc| - |a(b + c)|,$$

and consequently $|a + b| - |ab| = |a + bc| - |a(b + c)|$. This equality implies $|a + b| = |a + bc|$ and $|ab| = |a(b + c)|$, and, since $a + bc \prec a + b$ and $ab \prec a(b + c)$, then $a + bc = a + b$ and $ab = a(b + c)$. But the last two equalities are, as noted in the first part of the proof, equivalent to $a(b + c) \prec b \prec a + bc$, and the proof is complete.

COROLLARY 1   *An element b of the associated metric space $\mathbf{M}(\mathscr{L})$ of a normed lattice $\mathscr{L}$ is between two elements a, c if and only if $\mathscr{G}(a, b, c)$ subsists, where $\mathscr{G}(a, b, c)$ denotes the lattice relation*

$$ab + bc = b = (a + b)(b + c).$$

Proof.   According to the preceding theorem, $\mathscr{B}(a, b, c)$ holds if and only if $a(b + c) \prec b \prec a + bc$ (that is, if and only if $(a + bc)b = b = b + a(b + c)$). Since $\mathscr{L}$ is modular, $(a + bc)b = ab + bc$ and $b + a(b + c) = (a + b)(b + c)$, and so $\mathscr{B}(a, b, c)$ is equivalent to $ab + bc = b = (a + b)(b + c)$.

COROLLARY 2   *If $a, b, c \in \mathbf{M}(\mathscr{L})$, $\mathscr{B}(a, b, c)$ is equivalent to $b = ac + xa + xc$ for some element x of $\mathbf{M}(\mathscr{L})$.*

For any element b of that form satisfies condition (†) of Theorem 2.7, and if an element b does satisfy (†), then $b = (a + bc)b = ab + bc$, and from $a(b + c) \prec b$ we have $a(b + c) + bc \prec b$. Since $ab + ac \prec a(b + c)$, then $ab + ac + bc \prec a(b + c) + bc \prec b$, and it follows that $b = ac + ba + bc$. Hence $b = ac + xa + xc$ for $x = b$.

Remark 4.   Dually, $\mathscr{B}(a, b, c)$ holds if and only if $b = (a + c)(x + a)(x + c)$ for an element x of $\mathscr{L}$.

Since $ab = (a + b)(ab + a)(ab + b)$ and $a + b = (a + b)(a + b + a)(a + b + b)$, then ab and $a + b$ are each metrically between a, b for every $a, b \in \mathbf{M}(\mathscr{L})$.

Remark 5.   If $a, b \in \mathbf{M}(\mathscr{L})$, then each of the betweenness relations $\mathscr{B}(a, ab, b)$, $\mathscr{B}(ab, b, a + b)$, $\mathscr{B}(b, a + b, a)$, $\mathscr{B}(a + b, a, ab)$ holds.

The second and fourth of these relations follow immediately from $\mathscr{G}(a, b, c)$ and the first and third have been established in the preceding remark.

Remark 6.   If $a \prec c$, then condition (†) of Theorem 2.7 becomes $a \prec b \prec c$, and hence $\mathscr{B}(a, b, c)$ holds if and only if $\mathcal{O}(a, b, c)$ does. Thus, an element is between two *comparable* elements if and only if it is order between them.

The self-dual relation $\mathscr{G}(a, b, c)$—which in *normed* lattices is the lattice equivalent of the betweenness relation $\mathscr{B}(a, b, c)$ in the associated metric space of the lattice—is meaningful in an *arbitrary* lattice $\mathscr{L}$. Let us consider, now, two additional lattice relations $\mathscr{G}^*(a, b, c)$, $\mathscr{G}^{**}(a, b, c)$ defined in any lattice $\mathscr{L}$, as follows:

$$\mathscr{G}^*(a, b, c): ab + bc = b = b + ac,$$
$$\mathscr{G}^{**}(a, b, c): b(a + c) = b = (a + b)(b + c).$$

In a distributive lattice, $\mathscr{G}(a, b, c) = \mathscr{G}^*(a, b, c) = \mathscr{G}^{**}(a, b, c)$.

THEOREM 2.8   *In any lattice* $\mathscr{L}$, $\mathscr{G}(a, b, c)$ *implies both* $\mathscr{G}^*(a, b, c)$ *and* $\mathscr{G}^{**}(a, b, c)$, *but* $\mathscr{G}^*(a, b, c)$ *implies* $\mathscr{G}(a, b, c)$, $\mathscr{G}^{**}(a, b, c)$ *implies* $\mathscr{G}(a, b, c)$, $\mathscr{G}^*(a, b, c)$ *implies* $\mathscr{G}^{**}(a, b, c)$, *or* $\mathscr{G}^{**}(a, b, c)$ *implies* $\mathscr{G}^*(a, b, c)$, *if and only if* $\mathscr{L}$ *is modular.*

Proof.   Assuming $\mathscr{G}(a, b, c)$, then $b + ac \prec b + a$ and $b + ac \prec b + c$ give $b \prec b + ac \prec (b + a)(b + c) = b$, and so $b = b + ac$ and $\mathscr{G}^*(a, b, c)$ holds. The duality principle yields $\mathscr{G}^{**}(a, b, c)$.

Suppose now that $\mathscr{L}$ is modular and assume that $\mathscr{G}^*(a, b, c)$ subsists. Since $b = ab + bc \prec b(a + c)$, then $b = b(a + c)$. By modularity,

$$(a + b)(b + c) = b + (a + b)c = b + (a + ab + bc)c = b + (a + bc)c$$
$$= b + ac + bc = b + ac = b,$$

and so $\mathscr{G}^{**}(a, b, c)$ and $\mathscr{G}(a, b, c)$ both are valid. By duality, $\mathscr{G}^{**}(a, b, c)$ implies $\mathscr{G}^*(a, b, c)$ and $\mathscr{G}(a, b, c)$.

If $\mathscr{L}$ is not modular, then it contains a sublattice of five elements isomorphic to the lattice of Figure 2(d). Then $\mathscr{G}^*(c, b, d)$ subsists, but *not* $\mathscr{G}(c, b, d)$ nor $\mathscr{G}^{**}(c, b, d)$; and $\mathscr{G}^{**}(b, c, d)$ is valid, but *neither* $\mathscr{G}^*(b, c, d)$ *nor* $\mathscr{G}(b, c, d)$ holds.

COROLLARY   *In a normed lattice,* $\mathscr{B}(a, b, c)$ *implies that* b *distributes multiplicatively over* $a + c$ *and additively over* ac.

Thus, in a modular lattice (and only in such a lattice) the relations $\mathscr{G}(a, b, c)$, $\mathscr{G}^*(a, b, c)$, $\mathscr{G}^{**}(a, b, c)$ are mutually equivalent and hence in normed lattices betweenness $\mathcal{B}(a, b, c)$ is represented by any one of those lattice relations. Each may be used to *define* a notion of betweenness in a *general* lattice, and a comparative study made of the properties of the different betweenness notions so obtained. This has recently been done by R. J. Bumcrot, who, in his University of Missouri dissertation, examined thirteen lattice relations, including $\mathcal{G}(a, b, c)$, $\mathcal{G}^*(a, b, c)$, and $\mathcal{G}^{**}(a, b, c)$, that can qualify as betweenness relations in a general lattice. What is meant by the phrase "can qualify as betweenness relations" is discussed in the next section.

EXERCISES

1. Show that $\mathcal{B}(a, b, c)$ implies $ac \prec b \prec a + c$, where $a, b, c \in \mathbf{M}(\mathcal{L})$.

2. Show that in a normed lattice $\mathcal{L}$, the relation $a(b + c) \prec c \prec a + bc$ is equivalent to $b(a + c) \prec c \prec b + ac$. Are these two relations equivalent in every lattice?

2.8 BETWEENNESS IN GENERAL LATTICES.
NORM CONDITIONS FOR DISTRIBUTIVITY

In the most general sense in which the notion is used, a betweenness relation is any ternary (three-place) relation $\mathcal{R}(a, b, c)$ defined for ordered triples of elements of an abstract set S which (1) is *symmetric with respect to the outer points* (that is, $\mathcal{R}(a, b, c)$ holds if and only if $\mathcal{R}(c, b, a)$ does) and (2) *has special inner point* (that is, $\mathcal{R}(a, b, c)$ and $\mathcal{R}(a, c, b)$ imply $b = c$). As more structure is given to the abstract set S, more properties are demanded of a betweenness relation. Thus, if S is partially ordered, it is natural to require that (3) $a \prec b \prec c$ should imply $\mathcal{R}(a, b, c)$; if S is a lattice, that (4) $\mathcal{R}(a, b, c)$ implies $ac \prec b \prec a + c$ and (5) $\mathcal{R}(a, ac, c)$, $\mathcal{R}(a, a + c, c)$ hold for all elements a, c; and if S is a metric space, then (6) the invariance of $\mathcal{R}(a, b, c)$ under congruences is required (that is, $\mathcal{R}(a, b, c)$ and $a, b, c \approx a', b', c'$ imply $\mathcal{R}(a', b', c')$).

Examining $\mathcal{G}(a, b, c)$, $\mathcal{G}^*(a, b, c)$, and $\mathcal{G}^{**}(a, b, c)$ with respect to the properties (1) to (5) listed above, it is observed that each of these relations is symmetric with respect to the outer points, and it is equally obvious that $\mathcal{G}(a, b, b)$, $\mathcal{G}^*(a, b, b)$, and $\mathcal{G}^{**}(a, b, b)$ subsist. Suppose $\mathcal{G}^*(a, b, c)$ and $\mathcal{G}^*(a, c, b)$ hold. Then from $ab + bc = b = b + ac$ and $ac + cb = c = c + ab$ we obtain

$$b = b + ac = b + ac + bc = b + c = ab + bc + c = ab + c = c,$$

and consequently $\mathscr{G}^*(a, b, c)$ (and dually $\mathscr{G}^{**}(a, b, c)$) has the special inner point property. Since $\mathscr{G}(a, b, c)$ implies $\mathscr{G}^*(a, b, c)$ in every lattice (Theorem 2.8), it follows that $\mathscr{G}(a, b, c)$ has the special inner point property also.

It may be immediately verified that $a \prec b \prec c$ implies $\mathscr{G}(a, b, c)$, $\mathscr{G}^*(a, b, c)$, and $\mathscr{G}^{**}(a, b, c)$, and it is clear that each of these relations implies $ac \prec b \prec a + c$. Finally, direct substitution shows that each of $\mathscr{G}(a, b, c)$, $\mathscr{G}^*(a, b, c)$, and $\mathscr{G}^{**}(a, b, c)$ subsists when $b = ac$ and when $b = a + c$, for all a, c.

Hence each of those relations "qualifies" as a betweenness relation in a general lattice.

In normed lattices the equivalent relations $\mathscr{G}$, $\mathscr{G}^*$, $\mathscr{G}^{**}$ are congruence invariants also, since $\mathscr{G}(a, b, c)$ is equivalent to $\mathscr{B}(a, b, c)$, which is obviously a congruence invariant.

We shall see later that if, for four pairwise distinct points a, b, c, d of a metric space, the betweenness relations $\mathscr{B}(a, b, c)$, $\mathscr{B}(a, d, b)$ hold, then so do the relations $\mathscr{B}(d, b, c)$, $\mathscr{B}(a, d, c)$, and conversely. It follows that in a normed lattice *each* of the lattice relations $\mathscr{G}(a, b, c)$, $\mathscr{G}^*(a, b, c)$, $\mathscr{G}^{**}(a, b, c)$ has this property; for example, $\mathscr{G}(a, b, c)$ and $\mathscr{G}(a, d, b)$ imply (and are implied by) $\mathscr{G}(d, b, c)$ and $\mathscr{G}(a, d, c)$. In *every* lattice, $\mathscr{G}(a, b, c)$ and $\mathscr{G}(a, d, b)$ imply $\mathscr{G}(d, b, c)$, but not necessarily $\mathscr{G}(a, d, c)$ (Exercise 1), but even this weaker transitivity is possessed by $\mathscr{G}^*$ and $\mathscr{G}^{**}$ only in modular lattices. For if a lattice is not modular, then, as we have seen, it contains the lattice of Figure 2(d) as a sublattice, and it is easily verified that for this sublattice $\mathscr{G}^*(b, c, e)$ and $\mathscr{G}^*(b, e, d)$ are valid, but the relation $\mathscr{G}^*(b, c, d)$ is not valid. A counterexample for $\mathscr{G}^{**}$ follows by duality.

We shall prove later (Theorem 2.19) that a lattice is distributive if and only if $\mathscr{G}(a, b, c)$ is equivalent to the relation $ac \prec b \prec a + c$. Since $\mathscr{G}(a, b, c)$ and $\mathscr{B}(a, b, c)$ are equivalent in normed lattices, it follows that a normed lattice is distributive if and only if $\mathscr{B}(a, b, c)$ is equivalent to $ac \prec b \prec a + c$; that is, if and only if $ac \prec b \prec a + c$ implies $\mathscr{B}(a, b, c)$, since the implication in the other direction is valid in any normed lattice.

Now $ac \prec b \prec a + c$ implies $\mathscr{B}(a, b, c)$ if and only if for each $x \in \mathscr{L}$, $\delta(a, ac + x(a + c)) + \delta(ac + x(a + c), c) = \delta(a, c)$, since $ac \prec ac + x(a + c) \prec a + c$, and every element b such that $ac \prec b \prec a + c$ may be so written. This is equivalent to

$$|a + c| - |ac|$$
$$= |a + x(a + c)| - |a(ac + x(a + c))| + |c + x(a + c)| - |c(ac + x(a + c))|$$
$$= |(a + c)(a + x)| - |ac + ax| + |(a + c)(c + x)| - |ac + cx|$$
$$= |a + c| + |a + x| - |a + c + x| - |ac| - |ax| + |acx|$$
$$\quad + |a + c| + |c + x| - |a + c + x| - |ac| - |cx| + |acx|;$$

and hence to

(∗) $2|a + c + x| - 2|acx| = |a + c| + |a + x| + |c + x| - |ac| - |ax| - |cx|.$

The norm condition (∗) is equivalent to each of the two norm conditions

(†) $|a + c + x| - |acx| = |a| + |c| + |x| - |ac| - |ax| - |cx|,$

  $|a + c + x| - |acx| = |a + c| + |a + x| + |c + x| - |a| - |c| - |x|.$

We have thus established the following theorem.

THEOREM 2.9 *A normed lattice is distributive if and only if for each three of its elements a, b, c any one of the following three norm conditions is satisfied:*

$$2|a + b + c| - 2|abc| = |a + b| + |b + c| + |a + c| - |ab| - |bc| - |ac|,$$
$$|a + b + c| - |abc| = |a| + |b| + |c| - |ab| - |bc| - |ac|,$$
$$|a + b + c| - |abc| = |a + b| + |b + c| + |a + c| - |a| - |b| - |c|.$$

EXERCISES

1. Show that in every lattice $\mathscr{L}$, the relations $\mathscr{G}(a, b, c)$ and $\mathscr{G}(a, d, b)$ imply $\mathscr{G}(d, b, c)$, but not necessarily $\mathscr{G}(a, d, c)$.

2. If $\mathscr{L}$ is a distributive lattice, prove that $\mathscr{G}(a, b, c)$ is equivalent to $ac \prec b \prec a + c$.

2.9 MAPPINGS OF LATTICES.
ELEMENTARY PROPERTIES

If to each element x of a lattice $\mathscr{L}$ there corresponds an element $f(x)$ of a lattice $\mathscr{L}^*$, the correspondence f defines a *mapping of $\mathscr{L}$ into $\mathscr{L}^*$*. The mapping f is *onto* $\mathscr{L}^*$ provided $x^* \in L^*$ implies the existence of at least one element x of $\mathscr{L}$ such that $f(x) = x^*$; and if the element x if unique, then f is *biuniform* or *one-to-one*. In this case a mapping f^{-1} of $\mathscr{L}^*$ onto $\mathscr{L}$ is defined by putting $f^{-1}(x^*) = x$ provided $x^* = f(x)$. The mapping f^{-1} is called the *inverse* of f.

A lattice $\mathscr{L}$ is *order-imbeddable* in a lattice $\mathscr{L}^*$ provided there exists a one-to-one, order-preserving mapping f of $\mathscr{L}$ into $\mathscr{L}^*$ (that is, if $a, b \in \mathscr{L}$ and $a \prec b$, then $f(a), f(b) \in \mathscr{L}^*$ and $f(a) \prec f(b)$). Note that an order-preserving mapping f need not be either meet-preserving or join-preserving (that is,

$f(a) \prec f(b)$ does not necessarily imply that either $f(ab) = f(a)f(b)$ or $f(a + b) = f(a) + f(b)$). For though $f(ab), f(a), f(b)$ are elements of $\mathscr{L}^*$ and $f(ab) \prec f(a), f(ab) \prec f(b)$ imply $f(ab) \prec f(a)f(b)$, the element $f(a)f(b)$ of $\mathscr{L}^*$ need not be the image of any element of $\mathscr{L}$ by f, and even if it is the image of a unique element of $\mathscr{L}$ (even in the event that f is a one-to-one, order-preserving mapping of $\mathscr{L}$ onto $\mathscr{L}^*$), that element need not be ab.

THEOREM 2.10   *Let f denote a one-to-one mapping of a lattice $\mathscr{L}$ onto a lattice $\mathscr{L}^*$, which is order-preserving from $\mathscr{L}$ to $\mathscr{L}^*$, and has its inverse f^{-1} order-preserving from $\mathscr{L}^*$ to $\mathscr{L}$. Then f preserves both sums and products.*

Proof.   If $a, b \in \mathscr{L}$, then $ab \prec a$, $ab \prec b$ imply $f(ab) \prec f(a), f(ab) \prec f(b)$, and so $f(ab) \prec f(a)f(b)$. Now $z \prec f(a)$ and $z \prec f(b)$ imply $f^{-1}(z) \prec a$ and $f^{-1}(z) \prec b$ (since f^{-1} is order-preserving), and so $f^{-1}(z) \prec ab$. Then $z \prec f(ab)$ and it follows from the definition of lattice product in terms of the precedence relation that $f(ab) = f(a)f(b)$.

Similarly, it is shown that $f(a + b) = f(a) + f(b)$.

   A mapping f of $\mathscr{L}$ onto $\mathscr{L}^*$ that preserves both sums and products is an *isomorphism*, and the two lattices are said to be *isomorphic*.

THEOREM 2.11   *A one-to-one mapping of a lattice $\mathscr{L}$ onto a lattice $\mathscr{L}^*$ that preserves products (sums) preserves sums (products) also (and hence is an isomorphism).*

Proof.   We show that if f preserves products, then both f and f^{-1} are order-preserving, and consequently f preserves sums (Theorem 2.10). For if $a, b \in \mathscr{L}$ and $a \prec b$, then $ab = a$ and $f(a) = f(ab) = f(a)f(b)$, so $f(a) \prec f(b)$; and if $a^*, b^* \in \mathscr{L}^*$ and $a^* \prec b^*$, then $a^* = a^*b^*$, where $a^* = f(a), b^* = f(b)$. Since $f(ab) = f(a)f(b) = a^*b^*$, then $ab = f^{-1}(a^*b^*) = f^{-1}(a^*) = a$, and so $a \prec b$.

It follows from Theorem 2.10 that f preserves both sums and products.

   A similar argument is used in case it is assumed that f preserves sums.

THEOREM 2.12   *Let f be a one-to-one mapping of a normed lattice $\mathscr{L}$ onto a normed lattice $\mathscr{L}^*$. If f preserves both products and norms, then f is a congruence of $\mathbf{M}(\mathscr{L})$ with $\mathbf{M}(\mathscr{L}^*)$.*

Proof.   If $a, b \in \mathbf{M}(\mathscr{L})$, then

$$\delta(a, b) = |a + b| - |ab| = |f(a + b)| - |f(ab)|$$
$$= |f(a) + f(b)| - |f(a)f(b)| \qquad \text{(by Theorem 2.11)}$$
$$= \delta(f(a), f(b)),$$

and so f is a congruence of $\mathbf{M}(\mathscr{L})$ with $\mathbf{M}(\mathscr{L}^*)$.

THEOREM 2.13   *Let f be a one-to-one mapping of a normed lattice $\mathscr{L}$ onto a normed lattice $\mathscr{L}^*$. If f preserves products and is a congruence of $\mathbf{M}(\mathscr{L})$ with $\mathbf{M}(\mathscr{L}^*)$, then f preserves norms, modulo a constant.*

Proof.   If $a, b \in \mathscr{L}$, then

$$|a + b| - |a| = \delta(a + b, a) = \delta(f(a + b), f(a)) = |f(a + b) + f(a)| - |f(a)|$$
$$= |f(a + b)| - |f(a)|,$$
$$|a + b| - |b| = \delta(a + b, b) = \delta(f(a + b), f(b)) = |f(a + b) + f(b)| - |f(b)|$$
$$= |f(a + b)| - |f(b)|,$$

for since $a \prec a + b$ and $b \prec a + b$, then $f(a) \prec f(a + b)$, $f(b) \prec f(a + b)$, since f is order-preserving, by the argument of Theorem 2.11.

Hence $|f(b)| - |b| = |f(a)| - |a| = \text{constant}$.

THEOREM 2.14   *Let f be a one-to-one mapping of a normed lattice $\mathscr{L}$ onto a normed lattice $\mathscr{L}^*$. If f is norm-preserving, modulo a constant, and is a congruence of $\mathbf{M}(\mathscr{L})$ with $\mathbf{M}(\mathscr{L}^*)$, then f is an isomorphism of $\mathscr{L}$ with $\mathscr{L}^*$.*

Proof.   According to Theorem 2.10 it suffices to show that f and f^{-1} are order-preserving. Let $a, b \in \mathscr{L}$ and $a \prec b$. Then $\delta(a, b) = |a + b| - |ab| = |b| - |a| = |f(b)| - |f(a)|$. Now

$$\delta(f(a), f(b)) = |f(a) + f(b)| - |f(a)f(b)|$$
$$= |f(a)| + |f(b)| - 2|f(a)f(b)|.$$

Since f is a congruence of $\mathbf{M}(\mathscr{L})$ with $\mathbf{M}(\mathscr{L}^*)$, $\delta(f(a), f(b)) = \delta(a, b)$, and consequently $|f(a)f(b)| = |f(a)|$. This equality, together with $f(a)f(b) \prec f(a)$, gives $f(a) = f(a)f(b)$, and consequently $f(a) \prec f(b)$. Hence f preserves order, and it is shown similarly that $f(a) \prec f(b)$ implies $a \prec b$, completing the proof of the theorem.

EXERCISES

1. Give an example of a one-to-one mapping f of a lattice $\mathscr{L}$ into a lattice $\mathscr{L}^*$ that is order-preserving, but whose inverse is not order-preserving.

2. Give an example of a one-to-one mapping of a lattice $\mathscr{L}$ into a lattice $\mathscr{L}^*$ which is order-preserving, preserves meets, but does not preserve joins.

2.10 CONVEX EXTENSION OF NORMED LATTICES

A normed lattice $\mathscr{L}$ is called *convex* provided its associated metric space $\mathbf{M}(\mathscr{L})$ is *metrically convex* (that is, if $a, c \in \mathbf{M}(\mathscr{L})$, $a \neq c$, at least one element b of $\mathbf{M}(\mathscr{L})$ has the property: $a \neq b \neq c$, $\delta(a, b) + \delta(b, c) = \delta(a, c)$). Note that b is *metrically* between a and c.

The question arises: if $\mathscr{L}$ is any normed lattice, does there exist a *convex* normed lattice $\mathscr{L}^*$ which contains the normed lattice $\mathscr{L}$ as a sublattice (sums, products, and norms the same in $\mathscr{L}$ and $\mathscr{L}^*$)? Any such lattice $\mathscr{L}^*$ is called a *convex* extension of $\mathscr{L}$. Following Aronszajn, we show in this section that the question may be answered in the affirmative.

THEOREM 2.15 *Every normed lattice $\mathscr{L}_0$ has a convex extension.*

Proof. In the set $\mathscr{L}_0 \times \mathscr{L}_0$ of all ordered pairs $[a, b]$ of elements of $\mathscr{L}_0$, put $[a, b] = [c, d]$ if and only if $a = c$ and $b = d$, and define sum, product, and norm as follows:

(1) $$[a, b] \cdot [c, d] = [ac, bd],$$

(2) $$[a, b] + [c, d] = [a + c, b + d],$$

(3) $$|[a, b]| = \tfrac{1}{2}(|a| + |b|).$$

It is immediately clear that $\mathscr{L}_0 \times \mathscr{L}_0$ is closed with respect to each of these operations, that each operation is associative, commutative, and idempotent, and that $[a, b] \cdot [c, d] = [a, b]$ if and only if $[a, b] + [c, d] = [c, d]$. Hence $\mathscr{L}_0 \times \mathscr{L}_0$ is a lattice (Definition of Section 1.4). It is equally clear that $|[a, b]|$ defined in (3) has the two properties (a), (b) required of a norm (Section 2.7), and hence $\mathscr{L}_0 \times \mathscr{L}_0$ is a normed lattice. Identifying, now, the element x of $\mathscr{L}_0$ with the element $[x, x]$ of $L_0 \times L_0$, it is seen that $\mathscr{L}_0$ is a sublattice of $\mathscr{L}_0 \times \mathscr{L}_0$; for example, $x = [x, x]$, $y = [y, y]$, $x + y = [x, x] + [y, y] = [x + y, x + y]$, and $|[x, x]| = \tfrac{1}{2}(|x| + |x|) = |x|$. The lattice $\mathscr{L}_0 \times \mathscr{L}_0$ is an extension of $\mathscr{L}_0$ but it is not the desired lattice because, in general, it is not convex. Let $\mathscr{L}_1$ denote the normed lattice $\mathscr{L}_0 \times \mathscr{L}_0$.

The procedure used to obtain the normed lattice $\mathscr{L}_1$ from the normed lattice $\mathscr{L}_0$ is applied to $\mathscr{L}_1$ to obtain the normed lattice $\mathscr{L}_2 = \mathscr{L}_1 \times \mathscr{L}_1$, and by complete mathematical induction we obtain the infinite sequence $\mathscr{L}_0 \subset \mathscr{L}_1 \subset \cdots \subset \mathscr{L}_n \subset \cdots$ with $\mathscr{L}_i$ a sublattice of $\mathscr{L}_{i+1} (i = 0, 1, 2, \ldots)$, with sums, products, and norms the same in $\mathscr{L}_i$ and $\mathscr{L}_{i+1}$.

Consider now the set sum

$$\mathscr{L}^* = \bigcup_{i=0}^{\infty} \mathscr{L}_i.$$

If $a, b \in \mathscr{L}^*$ there is a smallest index n such that $a, b \in \mathscr{L}_n$. Defining sum, product, and norm for a, b in $\mathscr{L}^*$ as their sum, product, and norm, respectively, in $\mathscr{L}_n$, it is clear that $\mathscr{L}^*$ is a normed lattice which contains $\mathscr{L}_0$ as a sublattice, with sums, products, and norms in $\mathscr{L}_0$ the same as in $\mathscr{L}^*$.

The proof is concluded by showing that $\mathscr{L}^*$ is convex. Let a, c be two distinct elements of $\mathscr{L}^*$, and n an index such that $a, c \in \mathscr{L}_n$. Then $[a, c] \in \mathscr{L}_{n+1}$ and hence $[a, c] \in \mathscr{L}^*, a \neq [a, c] \neq c$. Now

$$\begin{aligned}
\delta(a, [a, c]) &= |a + [a, c]| - |a \cdot [a, c]| \\
&= |[a, a] + [a, c]| - |[a, a] \cdot [a, c]| \\
&= |[a, a + c]| - |[a, ac]| \\
&= \tfrac{1}{2}(|a + c| - |ac|) = \tfrac{1}{2}\delta(a, c).
\end{aligned}$$

Similarly, it is seen that $\delta([a, c], c) = \frac{1}{2} \delta(a, c)$, and so $\delta(a, [a, c]) + \delta([a, c], c) = \delta(a, c)$, and $\mathscr{L}^*$ is convex.

Remark 1.   Since the norm of an element of $\mathscr{L}_0$ equals its norm as an element of $\mathscr{L}^*$, then the distance of any two elements of $\mathscr{L}_0$ equals their distance as elements of $\mathscr{L}^*$. Hence $M(\mathscr{L}_0)$ is *imbedded congruently* in $M(\mathscr{L}^*)$.

Remark 2.   If lattice $\mathscr{L}_0$ is distributive, then also $\mathscr{L}_1, \mathscr{L}_2, \ldots$ are distributive, and so is $\mathscr{L}^*$.

2.11   COMPLETION OF NORMED LATTICES

An infinite sequence $p_1, p_2, \ldots, p_n, \ldots$ of elements of a metric space is called a *Cauchy sequence* provided for each $\varepsilon > 0$ there exists a natural number N such that the distance $\delta(p_i, p_j) < \varepsilon$ whenever $i > N$ and $j > N$; that is $\lim_{i,j \to \infty} \delta(p_i, p_j) = 0$. A metric space is called *complete* whenever each

Cauchy sequence of its elements has a limit; that is, if $\{p_n\}$, $n = 1, 2, \ldots$, is a Cauchy sequence of elements, then an element p of the space exists such that $\lim_{n \to \infty} pp_n = 0$.

Remark. If an infinite sequence $\{p_n\}$ of elements of a metric space has a limit, then the sequence is a Cauchy sequence. For if $\lim_{n \to \infty} p_n = p$, $\delta(p_i, p_j) \leq \delta(p, p_i) + \delta(p, p_j)$ for every pair of indices i, j and consequently

$$\lim_{i, j \to \infty} \delta(p_i, p_j) = 0.$$

Calling a normed lattice *metrically complete* whenever its associated metric space is complete, the question arises: if $\mathscr{L}$ is any normed lattice, does there exist a metrically complete normed lattice $\bar{\mathscr{L}}$ that contains $\mathscr{L}$ as a sublattice (sums, products, and norms in $\mathscr{L}$ the same as in $\bar{\mathscr{L}}$)? We obtain in this section an affirmative answer to this question, and prove the existence of a unique "smallest" such lattice $\bar{\mathscr{L}}$, which we call *the completion* of $\mathscr{L}$.

The completion procedure by which we obtain $\bar{\mathscr{L}}$ from $\mathscr{L}$ is so fundamental in mathematics (being essentially the process used to define the real numbers as Cauchy sequences of rationals, as well as the method employed by Hausdorff to imbed an arbitrary metric space in a complete metric space) that we are presenting it in all necessary detail. Each student of mathematics should have the experience, at least once, of examining every step of such a useful device. In the present case, the fact that we are dealing with lattices endows the procedure with additional complication and interest.

THEOREM 2.16   *Every normed lattice $\mathscr{L}$ has a unique completion $\bar{\mathscr{L}}$ in which it is dense.*

Proof.   In the set C of *all* Cauchy sequences $\{a_n\}$ of elements $a_1, a_2, \ldots, a_n, \ldots$ of $\mathscr{L}$, define

(1) $\{a_n\} = \{b_n\}$ provided $\lim_{n \to \infty} \delta(a_n, b_n) = 0,$

(2) $\{a_n\} \cdot \{b_n\} = \{a_n b_n\},$

(3) $\{a_n\} + \{b_n\} = \{a_n + b_n\},$

(4) $|\{a_n\}| = \lim |a_n|.$

Assertion A.   *The sequences $\{a_n b_n\}$, $\{a_n + b_n\}$ are Cauchy sequences, and $\lim_{n \to \infty} |a_n|$ exists.*

Consider the sequence $\{a_n b_n\}$, where $\{a_n\}$ and $\{b_n\}$ are Cauchy sequences. Clearly

$$0 \leq \delta(a_i b_i, a_j b_j) = |a_i b_i + a_j b_j| - |a_i b_i a_j b_j|$$
$$= |a_i b_i| + |a_j b_j| - 2|a_i a_j b_i b_j|$$
$$= |a_i| + |b_i| - |a_i + b_i| + |a_j| + |b_j| - |a_j + b_j|$$
$$- 2|a_i a_j| - 2|b_i b_j| + 2|a_i a_j + b_i b_j|,$$

and hence

$$0 \leq \delta(a_i b_i, a_j b_j)$$
$$= \delta(a_i, a_j) + \delta(b_i, b_j) + 2|a_i a_j + b_i b_j| - |a_i + b_i| - |a_j + b_j|.$$

Since $a_i a_j + b_i b_j \prec a_i + b_i$ and $a_i a_j + b_i b_j \prec a_j + b_j$,

$$2|a_i a_j + b_i b_j| \leq |a_i + b_i| + |a_j + b_j|,$$

and hence

$$0 \leq \delta(a_i b_i, a_j b_j) \leq \delta(a_i, a_j) + \delta(b_i, b_j).$$

Consequently

$$0 \leq \lim_{i,j \to \infty} \delta(a_i b_i, a_j b_j) \leq \lim_{i,j \to \infty} \delta(a_i, a_j) + \lim_{i,j \to \infty} \delta(b_i, b_j) = 0,$$

and so $\{a_n b_n\}$ is a Cauchy sequence.

Similarly, it is shown that $\{a_n + b_n\}$ is a Cauchy sequence.

To prove that $\lim_{n \to \infty} |a_n|$ exists for each Cauchy sequence $\{a_n\}$, we have $|a_i| \leq |a_i + a_j|$, $|a_i a_j| \leq |a_j|$, whence

$$\delta(a_i, a_j) = |a_i + a_j| - |a_i a_j| \geq |a_i| - |a_j|,$$

and similarly, $\delta(a_i, a_j) \geq |a_j| - |a_i|$. Hence

$$\text{abs. val. } (|a_i| - |a_j|) \leq \delta(a_i, a_j),$$

and since $\{a_n\}$ is a Cauchy sequence, $\delta(a_i, a_j) \to 0$ as $i, j \to \infty$.

It follows that $\lim_{i,j \to \infty}$ abs. val. $(|a_i| - |a_j|) = 0$, and so the sequence $|a_1|, |a_2|, \ldots, |a_n|, \ldots$ is a Cauchy sequence of real numbers. Hence $\lim_{n \to \infty} |a_n|$ exists, and is nonnegative.

Assertion B. *With respect to equality, sum, product, and norm defined by* (1), (2), (3), (4), *the set C is a normed lattice* $\overline{\mathscr{L}}$.

Assertion A shows that C is closed with respect to each of the operations of sum and product, and each operation is obviously associative, commutative, and idempotent.

We show that $\{a_n\} \cdot \{b_n\} = \{a_n\}$ if and only if $\{a_n\} + \{b_n\} = \{b_n\}$; that is, $\{a_n b_n\} = \{a_n\}$ if and only if $\{a_n + b_n\} = \{b_n\}$.

Now $\delta(a_n b_n, a_n) = |a_n| - |a_n b_n|$, since $a_n b_n \prec a_n$, and

$$\delta(a_n + b_n, b_n) = |a_n + b_n| - |b_n| = |a_n| + |b_n| - |a_n b_n| - |b_n|$$
$$= |a_n| - |a_n b_n| = \delta(a_n b_n, a_n).$$

Hence $\lim_{n \to \infty} \delta(a_n + b_n, b_n) = 0$ if and only if $\lim_{n \to \infty} \delta(a_n b_n, a_n) = 0$, which, by use of (1), gives the desired result.

A partial ordering is defined in C by

$$\{a_n\} \prec \{b_n\} \equiv \{a_n\} \cdot \{b_n\} = \{a_n\}.$$

To show that (4) defines a norm, let $\{a_n\} \prec \{b_n\}$ and $\{a_n\} \neq \{b_n\}$. Then $\{a_n b_n\} = \{a_n\}$, but $\lim_{n \to \infty} \delta(a_n, b_n) > 0$. Hence

$$\lim_{n \to \infty} [|a_n| + |b_n| - 2|a_n b_n|] > 0,$$

and since, as shown above, the limit of each summand exists,

(*)
$$\lim_{n \to \infty} |a_n| + \lim_{n \to \infty} |b_n| - 2 \lim_{n \to \infty} |a_n b_n| > 0.$$

Now from $\{a_n b_n\} = \{a_n\}$ follows $\lim_{n \to \infty} \delta(a_n b_n, a_n) = 0$; that is,

$$\lim_{n \to \infty} [|a_n| - |a_n b_n|] = 0, \quad \text{or} \quad \lim_{n \to \infty} |a_n b_n| = \lim_{n \to \infty} |a_n|.$$

Substitution in (*) yields $\lim_{n \to \infty} |b_n| > \lim_{n \to \infty} |a_n|$, and so $|\{a_n\}| < |\{b_n\}|$. Then condition (a) for a norm is satisfied.

To show that condition (b) is fulfilled we have

$$|\{a_n\} + \{b_n\}| + |\{a_n\} \cdot \{b_n\}| = |\{a_n + b_n\}| + |\{a_n b_n\}|$$
$$= \lim_{n \to \infty} |a_n + b_n| + \lim_{n \to \infty} |a_n b_n|$$
$$= \lim_{n \to \infty} [|a_n + b_n| + |a_n b_n|]$$
$$= \lim_{n \to \infty} [|a_n| + |b_n|]$$
$$= |\{a_n\}| + |\{b_n\}|.$$

Hence Assertion B is established.

Identifying the element x of $\mathscr{L}$ with the Cauchy sequence $x_1, x_2, \ldots,$ $x_n, \ldots$, where $x_i = x$, $i = 1, 2, \ldots$, it is seen that the normed lattice $\bar{\mathscr{L}}$ contains the lattice $\mathscr{L}$ as a sublattice with sums, products, and norms of elements of $\mathscr{L}$ the same in $\mathscr{L}$ as in $\bar{\mathscr{L}}$.

Assertion C.   The lattice $\mathscr{L}$ is dense in the lattice $\bar{\mathscr{L}}$.

We must show that if $\{a_n\} \in \bar{\mathscr{L}}$ and $\{a_n\}$ does not equal an element of $\mathscr{L}$, then $\{a_n\}$ is an accumulation element of $\mathscr{L}$ (that is, $\{a_n\}$ is an accumulation element of the $M(\mathscr{L})$, which is a subset of $M(\bar{\mathscr{L}})$. For $\varepsilon > 0$, we seek an element $\{b_n\}$ of $\mathscr{L}$, $b_n = b$, $n = 1, 2, \ldots$, such that $0 < \delta(\{a_n\}, \{b_n\}) < \varepsilon$.

Since $\{a_n\} \in \bar{\mathscr{L}}$, it is a Cauchy sequence and consequently there corresponds to the given $\varepsilon > 0$ a natural number N such that whenever $i > N$ and $j > N$, then $\delta(a_i, a_j) < \varepsilon$. Let i_0 be a fixed index, $i_0 > N$, and consider the element $\{b_n\}$, $b_n = a_{i_0}$ $(n = 1, 2, \ldots)$ of $\bar{\mathscr{L}}$ which is identified with the element a_{i_0} of $\mathscr{L}$. Clearly

$$\delta(\{a_n\}, \{b_n\}) = |\{a_n + b_n\}| - |\{a_n b_n\}|$$
$$= \lim_{n \to \infty} [|a_n + b_n| - |a_n b_n|] = \lim_{n \to \infty} \delta(a_n, b_n) < \varepsilon$$

The proof of Assertion C is completed when it is shown that

$$\delta(\{a_n\}, \{b_n\}) = \lim_{n \to \infty} \delta(a_n, b_n) \neq 0.$$

Now $\lim_{n \to \infty} \delta(a_n, b_n) = 0$ would imply that the element a_{i_0} is the limit of the sequence $a_1, a_2, \ldots, a_n \ldots$. This is indeed possible, but if our choice of the index i_0 had been so unfortunate, then, remembering that *any index exceeding N may be chosen*, there surely exists one, say i_1, such that $a_{i_1} \neq a_{i_0}$—for in the contrary case $a_n = a_{i_0}$ for every $n > N$, and then, according to our definition of equality, $\{a_n\}$ equals an element of $\mathscr{L}$, contrary to its selection. The element $\{b_n\}$, $b_n = a_{i_1}(n = 1, 2, \ldots)$ is, then, an element of $\mathscr{L}$ that satisfies our desires, and the Assertion C is established.

Assertion D.   The normed lattice $\bar{\mathscr{L}}$ is complete.

Let $\alpha_1, \alpha_2, \ldots, \alpha_m, \ldots$ be a Cauchy sequence of elements of $\bar{\mathscr{L}}$, $\alpha_k = \{a_n^{(k)}\}$ $(k = 1, 2, \ldots)$. Since $\mathscr{L}$ is dense in $\bar{\mathscr{L}}$, to each positive k and element α_k of the sequence corresponds an element $\beta_k = \{b_n^{(k)}\}$ of $\mathscr{L}$, $b_n^{(k)} = b^{(k)}(n = 1, 2, \ldots)$, such that $\delta(\alpha_k, \beta_k) < 1/k$; that is,

$$\lim_{k \to \infty} \delta(\alpha_k, \beta_k) = 0.$$

Now the sequence $\beta_1, \beta_2, \ldots, \beta_n, \ldots$ is a Cauchy sequence; for since $M(\mathcal{L})$ is a metric space, application of the triangle inequality gives

$$\delta(\beta_i, \beta_j) \leq \delta(\beta_i, \alpha_i) + \delta(\alpha_i, \alpha_j) + \delta(\alpha_j, \beta_j)$$

$$\leq \delta(\alpha_i, \alpha_j) + \frac{1}{i} + \frac{1}{j},$$

and since $\alpha_1, \alpha_2, \ldots, \alpha_n, \ldots$ is a Cauchy sequence, $\delta(\alpha_i, \alpha_j) \to 0$ as $i, j \to \infty$. Since the elements of the Cauchy sequence $\beta_1, \beta_2, \ldots, \beta_n, \ldots$ are elements of $\mathcal{L}$, then the sequence itself is an element of $\overline{\mathcal{L}}$. Letting $\beta = \{\beta_n\}$ ($n = 1, 2, \ldots$), the proof is completed by showing that $\lim_{k \to \infty} \alpha_k = \beta$.

Clearly

$$\delta(\beta, \beta_k) = \delta(\{\beta_n\}, \{b_n^{(k)}\}) = |\{\beta_n\} + \{b_n^{(k)}\}| - |\{\beta_n\}\{b_n^{(k)}\}|$$

$$= \lim_{n \to \infty} [|\beta_n + b_n^{(k)}| - |\beta_n b_n^{(k)}|] = \lim_{n \to \infty} \delta(\beta_n, b_n^{(k)}),$$

and consequently

$$\lim_{k \to \infty} \delta(\beta, \beta_k) = \lim_{k \to \infty} \lim_{n \to \infty} \delta(\beta_n, b_n^{(k)}) = 0,$$

for since $\beta_1, \beta_2, \ldots, \beta_n, \ldots$ is a Cauchy sequence, $\lim_{k,n \to \infty} \delta(\beta_n, \beta_k) = 0$.

From the metricity of $M(\mathcal{L})$.

$$\delta(\beta, \alpha_k) \leq \delta(\beta, \beta_k) + \delta(\beta_k, \alpha_k).$$

Consequently

$$\lim_{k \to \infty} \delta(\beta, \alpha_k) \leq \lim_{k \to \infty} \delta(\beta, \beta_k) + \lim_{k \to \infty} \delta(\beta_k, \alpha_k) = 0,$$

and so $\lim_{k \to \infty} \alpha_k = \beta$.

We remark, finally, that since $\overline{\mathcal{L}}$ consists of $\mathcal{L}$ and all of the accumulation elements of $\mathcal{L}$ (since each such element is the limit of a Cauchy sequence of elements of $\mathcal{L}$ (Theorem 1.3)) then $\overline{\mathcal{L}}$ is unique.

EXERCISES

1. If $\{a_n\} = \{b_n\}$, show that $\{a_n\} = \{a_n b_n\}$.
   Is the converse valid?

2. If $\{a_n\}, \{b_n\}, \{c_n\} \in \overline{\mathcal{L}}$ and $\{a_n\} = \{b_n\}$, does $\{a_n c_n\} = \{b_n c_n\}$?

2.12 COMPLETE CONVEX EXTENSION OF NORMED LATTICES

The convex extension $\mathscr{L}^*$ of a normed lattice $\mathscr{L}$, obtained in Section 2.10 may not be complete, and the completion $\bar{\mathscr{L}}$ of $\mathscr{L}$ constructed in the preceding section may not be convex. The question arises: if $\mathscr{L}$ is any normed lattice, does there exist a metrically complete and convex normed lattice $\bar{\mathscr{L}}^*$ that contains $\mathscr{L}$ as a sub-lattice, with sums, products, norms of elements in $\mathscr{L}$ the same as in $\bar{\mathscr{L}}^*$? This section is devoted to giving an affirmative answer to this question.

THEOREM 2.17 *Each normed lattice $\mathscr{L}$ has a complete convex extension.*

Proof. Let $\bar{\mathscr{L}}^*$ denote the completion of $\mathscr{L}^*$, the convex extension of $\mathscr{L}$. We show that $\bar{\mathscr{L}}^*$ is convex.

If $a, c \in \bar{\mathscr{L}}^*$, $a \neq c$, then since $\mathscr{L}^*$ is dense in $\bar{\mathscr{L}}^*$, for each natural number n, elements a_n, c_n of $\mathscr{L}^*$ exist such that $\delta(a, a_n) < 1/n$, $\delta(c, c_n) < 1/n$. Hence $\lim_{n \to \infty} a_n = a$, $\lim_{n \to \infty} c_n = c$.

Now $b_n = [a_n, c_n] \in \mathscr{L}^*$, $a_n \neq b_n \neq c_n$, and $\delta(a_n, b_n) = \delta(b_n, c_n) = \frac{1}{2}\delta(a_n, b_n)$, $n = 1, 2, \ldots$ (see the last part of the proof of Theorem 2.15).

Assertion. *The sequence* $\{b_n\} = \{[a_n, c_n]\}(n = 1, 2, \ldots)$ *is a Cauchy sequence.*
We have

$$\begin{aligned}
\delta([a_i, c_i], [a_j, c_j]) &= |[a_i, c_i] + [a_j, c_j]| - |[a_i, c_i] \cdot [a_j, c_j]| \\
&= |[a_i + a_j, c_i + c_j]| - |[a_i a_j, c_i c_j]| \\
&= \tfrac{1}{2}[|a_i + a_j| + |c_i + c_j| - |a_i a_j| - |c_i c_j|] \\
&= \tfrac{1}{2}[\delta(a_i, a_j) + \delta(c_i, c_j)].
\end{aligned}$$

Now since the sequences $\{a_n\}$, $\{b_n\}$ $(n = 1, 2, \ldots)$ have limits, they are Cauchy sequences, and consequently

$$\lim_{i, j \to \infty} \delta(a_i, a_j) = \lim_{i, j \to \infty} \delta(c_i, c_j) = 0.$$

Hence

$$\lim_{i, j \to \infty} \delta([a_i, c_i], [a_j, c_j]) = 0,$$

and so the sequence $\{b_n\} = \{[a_n, c_n]\}(n = 1, 2, \ldots)$ is a Cauchy sequence of elements of the metrically complete metric space $\mathbf{M}(\mathscr{L}^*)$. It follows that $\mathbf{M}(\bar{\mathscr{L}}^*)$ contains one element b such that $b = \lim_{n \to \infty} b_n$.

Finally, from $\delta(a_n, b_n) = \frac{1}{2}\delta(a_n, c_n) = \delta(b_n, c_n)$ $(n = 1, 2, \ldots)$ follows

$$\lim_{n\to\infty} \delta(a_n, b_n) = \frac{1}{2}\lim_{n\to\infty} \delta(a_n, c_n) = \lim_{n\to\infty} \delta(b_n, c_n).$$

That is (Theorem 1.2),

$$\delta(a, b) = \frac{1}{2}\delta(a, c) = \delta(b, c),$$

and b is metrically between the elements a, c. Thus $\mathscr{L}^*$ is convex, and the theorem is proved.

2.13   METRIC CHARACTERIZATION OF NORMED LATTICES. STUDY OF THE ASSOCIATED METRIC SPACE

We have seen that associated with each normed lattice $\mathscr{L}$ there is a metric space $\mathbf{M}(\mathscr{L})$ obtained by attaching to each two elements x, y of $\mathscr{L}$ the number $\delta(x, y) = |x + y| - |xy|$ as distance. It is natural to conjecture that the metric space $\mathbf{M}(\mathscr{L})$ is not a "general" metric space, but has certain special properties due to its origin. It is the purpose of this section to ascertain those properties. Expressed in another way, we wish to characterize metrically those metric spaces that are the associated metric spaces of normed lattices, and thus to obtain a geometric characterization of normed lattices. We restrict ourselves to the consideration of normed lattices with a first element.

Consider now the associated metric space $\mathbf{M}(\mathscr{L})$ of a normed lattice $\mathscr{L}$ with first element o. There is no loss of generality in assuming that $|o| = 0$, for if $|o| = C$, a new norm $\mu(x)$ may be defined, $\mu(x) = |x| - C$, for which $\mu(o) = 0$, *and the new associated metric space is the same as the original one.*

We observe, first, that $\mathscr{B}(o, a, b)$ is valid in $\mathbf{M}(\mathscr{L})$ if and only if $a \prec b$; for $\mathscr{B}(o, a, b)$ holds if and only if $oa + ab = a = (o + a)(a + b)$, and since o is the first element of $\mathscr{L}$, the latter relation is equivalent to $a \prec b$.

Assertion I.   *The following five-point betweenness transitivities are valid in* $\mathbf{M}(\mathscr{L})$:

(i) $\mathscr{B}(a, c, b)$ and $\mathscr{B}(o, x, a)$ and $\mathscr{B}(o, x, b)$ imply $\mathscr{B}(o, x, c)$,

(ii) $\mathscr{B}(a, c, b)$ and $\mathscr{B}(o, a, x)$ and $\mathscr{B}(o, b, x)$ imply $\mathscr{B}(o, c, x)$.

Proof.   By the preceding observation, $\mathscr{B}(o, x, a)$ and $\mathscr{B}(o, x, b)$ imply $x \prec a$ and $x \prec b$. Hence $x \prec a + c$, $x \prec b + c$, and so $x \prec (a + c)(b + c)$. Since $\mathscr{B}(a, c, b)$ holds, then $c = (a + c)(b + c)$; that is, $x \prec c$. Using the observation again, we have $\mathscr{B}(o, x, c)$ and (i) is established.

To obtain (ii), $\mathscr{B}(o, a, x)$ and $\mathscr{B}(o, b, x)$ yield $a \prec x$ and $b \prec x$, and so $ac \prec x$, $bc \prec x$. Hence $ac + bc \prec x$, and from $\mathscr{B}(a, c, b)$ we have $c = ac + bc$. It follows that $c \prec x$ and consequently $\mathscr{B}(o, c, x)$ holds.

Assertion II. *If $\mathscr{B}(a, b) = [x \in \mathscr{L} \mid \mathscr{B}(a, x, b)]$, $a, b \in \mathscr{L}(a \neq b)$, there exist $x^*, y^* \in B(a, b)$ such that for every $x \in B(a, b)$, the relations $\mathscr{B}(o, x^*, x)$ and $\mathscr{B}(o, x, y^*)$ hold.*

Proof. We show that $x^* = ab$ and $y^* = a + b$.

By remark 5, Section 2.8, $ab \in B(a, b)$ and $a + b \in B(a, b)$. If $x \in B(a, b)$, then $ax + xb = x = (a + x)(x + b)$. The second equality gives $ab \prec x$, and hence $\mathscr{B}(o, ab, x)$; and from the first equality, $x \prec a + b$, and consequently $\mathscr{B}(o, x, a + b)$.

Thus, the associated metric space $\mathbf{M}(\mathscr{L})$ of any normed lattice with a first element contains a point o with respect to which the metric properties listed in Assertions I and II are valid.

Conversely, we assume now that a metric space $\mathbf{M}$, with metric δ, contains a point o with respect to which the space has the properties stated in Assertions I and II. We will show that there exists a normed lattice $\mathscr{L}$, with a first element, such that $\mathbf{M} = \mathbf{M}(\mathscr{L})$.

Consider the element set of $\mathbf{M}$ and define, for $a, b \in \mathbf{M}$, $a \prec b \equiv \mathscr{B}(o, a, b)$. We show that with respect to this relation, $\mathbf{M}$ is a lattice $\mathscr{L}$ with a first element.

The relation $\prec$ is a partial ordering of the elements of $\mathbf{M}$.

(a) For each element, $a \prec a$; for since $\delta(o, a) + \delta(a, a) = \delta(o, a)$, then $B(o, a, a)$ holds.

(b) If $a, b \in \mathbf{M}$ such that $a \prec b$ and $b \prec a$, then $a = b$, and conversely; for from $\delta(o, a) + \delta(a, b) = \delta(o, b)$ and $\delta(o, b) + \delta(b, a) = \delta(o, a)$ follows $\delta(a, b) = 0$, and hence $a = b$. The converse follows from (a).

(c) If $a, b\, c \in \mathbf{M}$, and $a \prec b$, $b \prec c$, then $a \prec c$. From the definition of the relation $\prec$, its *transitivity* is equivalent to the following four-point *transitivity* of the betweenness relation: $\mathscr{B}(o, a, b)$ and $\mathscr{B}(o, b, c)$ implies $B(o, a, c)$—a happy agreement in terminology! Let us establish this transitivity.

As a consequence of the two betweenness relations,

$$\delta(o, a) + \delta(a, b) = \delta(o, b),$$
$$\delta(o, b) + \delta(b, c) = \delta(o, c).$$

Since the space is metric, $\delta(o, a) + \delta(a, c) = \delta(o, c) + \mu$, $\mu \geq 0$. Addition of the first two equalities gives $\delta(o, a) + \delta(a, b) + \delta(b, c) = \delta(o, c)$. Hence

$\delta(o, a) + \delta(a, c) = \delta(o, a) + \delta(a, b) + \delta(b, c) + \mu$, or $\delta(a, c) \geq \delta(a, b) + \delta(b, c)$. But since **M** is metric, $\delta(a, c) \leq \delta(a, b) + \delta(b, c)$, and so $\delta(a, c) = \delta(a, b) + \delta(b, c)$; that is, $\mu = 0$, and consequently $\delta(o, a) + \delta(a, c) = \delta(o, c)$. Hence $\mathscr{B}(o, a, c)$ holds and $a \prec c$.

Remark. It will be useful later to observe now that (1) our argument has derived $\mathscr{B}(a, b, c)$ also, and (2) the four-point betweenness transitivity that we have just established is valid for *metric* betweenness (that is, when $\mathscr{B}(x, y, z)$ means $\delta(x, y) + \delta(y, z) = \delta(x, z)$ *and* $x \neq y \neq z$) in which sense the notion is used in distance geometry.

We establish next that **M** is a lattice $\mathscr{L}$ with respect to the partial order relation $\prec$, by showing that if $a, b \in$ **M**, the element $x^* = $ g.l.b.(a, b) and the element $y^* = $ l.u.b.(a, b), where x^*, y^* are the elements whose existence we are now assuming (Assertion II).

First, $x^* \prec a$ and $x^* \prec b$; for since $\mathscr{B}(o, x^*, x)$ holds for each element x satisfying $\mathscr{B}(a, x, b)$, and a, b satisfy that relation, then $\mathscr{B}(o, x^*, a)$ and $\mathscr{B}(o, x^*, b)$ are valid, and the desired conclusion follows. Second, if $z \prec a$ and $z \prec b$, then $z \prec x^*$; for then $\mathscr{B}(o, z, a)$, $\mathscr{B}(o, z, b)$ hold, and, together with $\mathscr{B}(a, x^*, b)$, they imply by Assertion I, $\mathscr{B}(o, z, x^*)$, and so $z \prec x^*$. Hence $x^* = $ g.l.b.$(a, b) = ab$.

Similarly, $a \prec y^*$ and $b \prec y^*$; for since $\mathscr{B}(o, x, y^*)$ holds for each element x satisfying $\mathscr{B}(a, x, b)$, and a, b satisfy that relation, then $\mathscr{B}(o, a, y^*)$ and $\mathscr{B}(o, b, y^*)$ are valid, and the desired conclusion follows. Further, from $a \prec z$ and $b \prec z$ follow $\mathscr{B}(o, a, z)$ and $\mathscr{B}(o, b, z)$ which, together with $\mathscr{B}(a, y^*, b)$, give $\mathscr{B}(o, y^*, z)$ ((ii), Assertion I) and consequently $y^* \prec z$.

Hence **M** is a lattice $\mathscr{L}$ with respect to the partial order relation $\prec$; moreover, since $\mathscr{B}(o, o, x)$ holds for every x of $\mathscr{L}$, $o \prec x$, $x \in \mathscr{L}$, and so o is the first element of $\mathscr{L}$.

The lattice $\mathscr{L}$ is normed by putting

$$|x| = \delta(o, x), \qquad x \in \mathscr{L}.$$

For if $a, b \in \mathscr{L}$, $a \prec b$, and $a \neq b$, then since $\mathscr{B}(o, a, b)$ holds, $\delta(o, a) + \delta(a, b) = \delta(o, b)$, and $\delta(a, b) \neq 0$ implies $\delta(o, a) < \delta(o, b)$; that is, $|a| < |b|$.

To establish the second property of the norm, $ab \prec a$, $ab \prec b$ imply $\mathscr{B}(o, ab, a)$ and $\mathscr{B}(o, ab, b)$; that is,

$$\delta(o, ab) + \delta(ab, a) = \delta(o, a),$$
$$\delta(o, ab) + \delta(ab, b) = \delta(o, b).$$

Adding, and taking into account the relation $\mathscr{B}(a, ab, b)$,

$(*)$ $\qquad\qquad |ab| = \delta(o, ab) = \tfrac{1}{2}[|a| + |b| - \delta(a, b)].$

In similar fashion the relations $\mathscr{B}(o, a, a + b)$, $\mathscr{B}(o, b, a + b)$, and $\mathscr{B}(a, a + b, b)$ yield

$(**)$ $\qquad\qquad |a + b| = \delta(o, a + b) = \tfrac{1}{2}[|a| + |b| + \delta(a, b)],$

and hence

$$|a + b| + |ab| = |a| + |b|.$$

Let $\delta^*(a, b)$ denote the distance of elements a, b in the associated metric space $\mathbf{M}(\mathscr{L})$. Then

$$\delta^*(a, b) = |a + b| - |ab|,$$

which, by $(*)$, $(**)$ becomes $\delta^*(a, b) = \delta(a, b)$. Hence $\mathbf{M}(\mathscr{L})$ is the original metric space $\mathbf{M}$.

We have thus characterized, by means of the metric properties appearing in Assertions I and II, those metric spaces which are associated metric spaces of normed lattices with a first element. Thus normed lattices (with first elements) are characterized by the *geometric* properties of their associated metric spaces. We state the result we have established in the following theorem.

THEOREM 2.18   *A metric space* $\mathbf{M}$ *is the associated metric space* $\mathbf{M}(\mathscr{L})$ *of a normed lattice* $\mathscr{L}$ *with a first element if and only if* $\mathbf{M}$ *contains a point o such that, if* $a, b, c \in \mathbf{M}$ *then*

(I) *for every point x of* $\mathbf{M}$,

(i) $\mathscr{B}(o, x, a)$ *and* $\mathscr{B}(o, x, b)$ *and* $\mathscr{B}(a, c, b)$ *implies* $\mathscr{B}(o, x, c)$,

(ii) $\mathscr{B}(o, a, x)$ *and* $\mathscr{B}(o, b, x)$ *and* $\mathscr{B}(a, c, b)$ *implies* $\mathscr{B}(o, c, x)$, *and*

(II) *between-points* x^*, y^* *of* a, b *exist such that, if x is any between-point of* a, b, *then the betweenness relations* $\mathscr{B}(o, x^*, x)$ *and* $\mathscr{B}(o, x, y^*)$ *subsist.*

DEFINITION   If a, b, c, p, q are five points of a metric space $\mathbf{M}$ such that $\mathscr{B}(p, a, q)$, $\mathscr{B}(p, c, q)$, and $\mathscr{B}(a, b, c)$ subsist, but $\mathscr{B}(p, b, q)$ does not, the five points are said to form a θ-*figure*, and the space $\mathbf{M}$ is said to admit a θ-*figure*. If five points a, b, c, p, q of a metric space $\mathbf{M}$ are such that $\mathscr{B}(p, q, a)$, $\mathscr{B}(p, q, c)$, and $\mathscr{B}(a, b, c)$ subsist, but $\mathscr{B}(p, q, b)$ does not, the five points form a *tailed* Δ-*figure*, with origin p, and the space $\mathbf{M}$ is said to admit a *tailed* Δ-*figure*.

Remark. Property I of Theorem 2.18 can be stated: the metric space **M** does not admit either a θ-figure or a tailed Δ-figure.

EXERCISES

1. Illustrate the developments of this section for the lattice formed by the set of all ordered pairs (x_1, x_2) of nonnegative real numbers, with $(x_1, x_2) \prec (y_1, y_2)$ provided $x_1 \leqq y_1$ and $x_2 \leqq y_2$, with $|(x_1, x_2)| = x_1 + x_2$.

2. Give an example of a metric space that has Property I of Theorem 2.18 but not Property II, and one that has Property II but not Property I.

3. Give examples of metric spaces that admit (i) both a θ-figure and a tailed Δ-figure; (ii) a θ-figure, but not a tailed Δ-figure; (iii) a tailed Δ-figure, but not a θ-figure.

2.14 GEOMETRICAL CHARACTERIZATION OF DISTRIBUTIVE LATTICES

We have seen that the lattice $\mathscr{L}$ of linear subspaces of a projective three-space can be normed by setting $|x| = 1 + \dim x$, $x \in \mathscr{L}$ (Section 2.7). If x, y denote two skew lines of that lattice, which are intersected by two skew lines, a, b, it is easily seen that $\mathscr{B}(x, a, y)$ and $\mathscr{B}(x, b, y)$ hold. Let c denote a line that intersects $x, a,$ and b but not y. Then $\mathscr{B}(a, c, b)$ holds, *but $\mathscr{B}(x, c, y)$ does not*; that is, in the metric space $\mathbf{M}(\mathscr{L})$ the five-point betweenness transitivity,

$$(*) \qquad \mathscr{B}(x, a, y) \text{ and } \mathscr{B}(x, b, y) \text{ and } \mathscr{B}(a, c, b) \text{ implies } \mathscr{B}(x, c, y)$$

is not valid, and so $\mathbf{M}(\mathscr{L})$ contains a θ-figure.

Now the lattice $\mathscr{L}$ is not distributive, and (if one is endowed with great insight!) it might be conjectured that the lack of distributivity is due to the presence of θ-figures in $\mathbf{M}(\mathscr{L})$.

In any lattice, if $\mathscr{G}(a, b, c)$ subsists, then $ac \prec b \prec a + c$, and in a distributive lattice the converse is valid. For if $ac \prec b \prec a + c$, then $ac + b = b$, and so $b = (a + b)(b + c)$, and $b(a + c) = b$ gives $b = ab + bc$. Thus if $\mathscr{L}$ is a distributive lattice, the relations $\mathscr{G}(a, b, c)$ and $ac \prec b \prec a + c$ are equivalent.

Conversely, if $\mathscr{G}(a, b, c)$ and $ac \prec b \prec a + c$ are equivalent in a lattice $\mathscr{L}$, then $\mathscr{L}$ is distributive; for it cannot contain a sublattice isomorphic to the lattice of Figure 2(d), since in that lattice $bd \prec c \prec b + d$, but $\mathscr{G}(b, c, d)$ does *not* hold; and it cannot contain a lattice isomorphic to the lattice of Figure 4, since in that lattice we have $ac \prec b \prec a + c$ but *not* $\mathcal{G}(a, b, c)$. We have thus established the following theorem.

THEOREM 2.19 *A lattice $\mathcal{L}$ is distributive if and only if for each three elements a, b, c of $\mathcal{L}$, the relations $ac \prec b \prec a + c$ and $ab + bc = b = (a + b)(b + c)$ are equivalent.*

Since, in a normed lattice, the relation $ab + bc = b = (a + b)(b + c)$ is equivalent to the betweenness relation $\mathcal{B}(a, b, c)$, we have proved the following.

THEOREM 2.20 *A normed lattice $\mathcal{L}$ is distributive if and only if an element b is between two elements a, c if and only if $ac \prec b \prec a + c$.*

Assume now that the relations $\mathcal{G}(a, b, c)$ and $ac \prec b \prec a + b$ are equivalent in a lattice $\mathcal{L}$. Then $\mathcal{L}$ has the five-point transitivity

(∗) $\mathcal{G}(a, x, b)$ and $\mathcal{G}(a, y, b)$ and $\mathcal{G}(x, c, y)$ implies $\mathcal{G}(a, c, b)$.

The three $\mathcal{G}$ relations yield

$$ab \prec x \prec a + b,$$
$$ab \prec y \prec a + b,$$
$$xy \prec c \prec x + y,$$

from which it follows at once that

$$ab \prec xy \prec a + b,$$
$$ab \prec x + y \prec a + b,$$

and hence $ab \prec c \prec a + b$. By our assumption, the last relation implies $\mathcal{G}(a, c, b)$.

Conversely, if a lattice $\mathcal{L}$ has the five-point transitivity (∗), then $\mathcal{G}(a, b, c)$ is equivalent to $ac \prec b \prec a + b$. Since $\mathcal{G}(a, b, c)$ implies $ac \prec b \prec a + c$ in any lattice, we show that $ac \prec b \prec a + c$ implies $\mathcal{G}(a, b, c)$ whenever $\mathcal{L}$ has the five-point transitivity (∗). The relations $\mathcal{G}(a, ac, c)$ and $\mathcal{G}(a, a + c, c)$ are immediate, and relation

$$\mathcal{G}(ac, b, a + c): \quad acb + b(a + c) = b = (ac + b)(b + a + c)$$

is valid, since we are assuming $ac \prec b \prec a + c$. Hence the assumed five-point transitivity gives $\mathcal{G}(a, b, c)$.

We have established the following theorem.

THEOREM 2.21   *The relations $\mathcal{G}(a, b, c)$ and $ac \prec b \prec a + c$ are equivalent in a lattice $\mathcal{L}$ if and only if $\mathcal{L}$ has the five-point transitivity:*

$$\mathcal{G}(x, a, y) \text{ and } \mathcal{G}(x, c, y) \text{ and } \mathcal{G}(a, b, c) \text{ implies } \mathcal{G}(x, b, y).$$

This theorem, combined with Theorem 2.19, gives the desired following result.

THEOREM 2.22   *A lattice $\mathcal{L}$ is distributive if and only if the relation $\mathcal{G}(a, b, c)$ has the transitivity property:*

$$\mathcal{G}(x, a, y) \text{ and } \mathcal{G}(x, c, y) \text{ and } \mathcal{G}(a, b, c) \text{ implies } \mathcal{G}(x, b, y).$$

Since in a normed lattice $\mathcal{L}$ the relation $\mathcal{G}(x, y, z)$ is equivalent to the betweenness relation $\mathcal{B}(x, y, z)$ in $\mathbf{M}(\mathcal{L})$, the preceding theorem gives the following interesting geometrical necessary and sufficient condition that a normed lattice be distributive.

THEOREM 2.23   *A normed lattice $\mathcal{L}$ is distributive if and only if its associated metric space $\mathbf{M}(\mathcal{L})$ has the five-point metric betweenness transitivity:*

$$\mathcal{B}(x, a, y) \text{ and } \mathcal{B}(x, c, y) \text{ and } \mathcal{B}(a, b, c) \text{ implies } \mathcal{B}(x, b, y),$$

and hence does not admit a θ-figure.

Thus, for normed lattices, the algebraic property of distributivity is equivalent to an important geometric property of the associated metric space of the lattice; namely, the absence in that space of θ-figures.

Combining Theorem 2.18 and Theorem 2.23 yields the following metric characterization of those metric spaces that are associated metric spaces of normed *distributive* lattices.

THEOREM 2.24   *A metric space M is the associated metric space of a normed, distributive lattice with a first element if and only if (1) M does not admit a θ-figure, (2) M contains a point o such that if $a, b \in M$, between-points x^*, y^* of a, b exist such that $B(o, x^*, x)$ and $B(o, x, y^*)$ subsist, where x is any between-point of a, b, and (3) M does not admit a tailed Δ-figure with origin o.*

EXERCISES

1. Show that a metric space M is the associated metric space of a normed Boolean algebra if and only if, in addition to possessing properties (1), (2), (3) of Theorem 2.24, it has the following two properties:
   (4)   There exists at least one point 1 of M, distinct from o, such that $B(o, x, 1)$ subsists for every point x of M.
   (5)   If $x \in M$, there exists at least one point x' of M such that $B(x, o, x')$ and $B(x, 1, x')$ subsist.

2. Show that a metric space with properties (1), (2), (3), (4), (5) does not admit any tailed Δ-figures.

2.15   σ-COMPLETE LATTICES.
SEQUENTIAL TOPOLOGY

In a lattice $\mathscr{L}$, each two elements have a "product" and a "sum," and an induction shows that the same is true for any finite number of elements of the lattice. If, however, $\{x_n\}$ $(n = 1, 2, \ldots)$ is an infinite sequence of elements of $\mathscr{L}$, there may not exist any element of $\mathscr{L}$ that "precedes" each element of the sequence and "follows" every element that has that property; nor need there be any element that "follows" every element of the sequence and "precedes" every element that does. In case such elements do exist, we denote them by $\prod x_n$ and $\sum x_n$, respectively, and if they exist for every infinite sequence of elements of $\mathscr{L}$, the lattice is said to be σ-*complete*. If such elements exist for *every* subset of $\mathscr{L}$, denumerable or not, the lattice is called *complete*.

We assume that $\mathscr{L}$ is a σ-complete lattice, and observe that the elements $\prod x_n$, $\sum x_n$ are unique and depend only on the element-set of the sequence $\{x_n\}$ and not on the particular serialization of the elements that is effected by the sequence.

Remark 1.   If $\{x_n\}$, $\{y_n\}$ $(n = 1, 2, \ldots)$ are such that for every y_k there exists an x_{j_k} with $x_{j_k} \prec y_k$, then $\prod x_n \prec \prod y_n$, since $\prod x_n \prec x_{j_k} \prec y_k$ for every $k = 1, 2, \ldots$.

If $\{x_{i_k}\}$ is a subsequence of $\{x_n\}$, then $\prod x_n \prec x_{i_k}$ for every index i_k and hence $\prod x_n \prec \prod x_{i_k}$. Similarly, $\sum x_{i_k} \prec \sum x_n$.

Remark 2.   If $c \in \mathscr{L}$, then $c \cdot \prod x_n = \prod c x_n$; for clearly $\prod c x_n \prec c x_n \prec x_n$ $(n = 1, 2, \ldots)$, so $\prod c x_n \prec \prod x_n$, and $\prod c x_n \prec c x_n \prec c$, so $\prod c x_n \prec c$, and consequently $\prod c x_n \prec c \cdot \prod x_n$. But $\prod x_n \prec x_n$ implies $c \cdot \prod x_n \prec c x_n$ $(n = 1, 2, \ldots)$, and so $c \cdot \prod x_n \prec \prod c x_n$, from which $c \cdot \prod x_n = \prod c x_n$ follows.

In a similar manner

$$\sum(c + x_n) = c + \sum x_n, \quad c + \prod x_n \prec \prod(c + x_n), \quad \sum cx_n \prec c \cdot \sum x_n$$

are obtained. Notice that the last two relations give "semidistributivity" of addition over multiplication and of multiplication over addition, respectively. In case *both distributivities* are valid (that is, $c + \prod x_n = \prod(c + x_n)$ and $\sum cx_n = c \cdot \sum x_n$), the lattice is called σ-distributive. Contrary to the finite case, either one of the distributivities may hold without the other.

A means of associating an element x of a σ-complete lattice with an infinite sequence $\{x_n\}$ of elements as a "limit of the sequence" ($\lim_{n \to \infty} x_n = x$) is due to the Russian mathematician Kantorovich, who applied to lattices a procedure that the French mathematician E. Borel (1871-1956) had used earlier for sets.

Corresponding to each infinite sequence $\{x_n\}$ $(n = 0, 1, 2, \ldots)$ are two sequences,

$$\pi x_n = \prod_{k=0}^{\infty} x_{n+k}, \qquad \sigma x_n = \sum_{k=0}^{\infty} x_{n+k},$$

where $n = 0, 1, 2, \ldots$. The limit inferior of $\{x_n\}$ (lim inf x_n) and limit superior (lim sup x_n) are defined by the expressions

$$\lim \inf x_n = \sum_{n=0}^{\infty} \pi x_n = \sum_{n=0}^{\infty} \prod_{k=0}^{\infty} x_{n+k},$$

$$\lim \sup x_n = \prod_{n=0}^{\infty} \sigma x_n = \prod_{n=0}^{\infty} \sum_{k=0}^{\infty} x_{n+k},$$

and *in the event* lim inf $x_n = a = $ lim sup x_n, then a is called the *limit* of sequence $\{x_n\}$, and we write lim $x_n = a$.

Remark 3. The terms "limit inferior" and "limit superior" are justified by observing that for any sequence $\{x_n\}$ $(n = 0, 1, 2, \ldots)$, lim inf $x_n \prec$ lim sup x_n; for since $\pi x_n \prec x_n \prec \sigma x_n$, then for $m \geq n$, $\pi x_m \prec \sigma x_m \prec \sigma x_n$, while if $m \leq n$, then $\pi x_m \prec \pi x_n \prec \sigma x_n$. Hence for each two indices m, n, we have $\pi x_m \prec \sigma x_n$ and consequently $\sum_{n=0}^{\infty} \pi x_n \prec \prod_{n=0}^{\infty} \sigma x_n$; that is, lim inf $x_n \prec$ lim sup x_n. To show that lim x_n exists, one need only show that lim sup $x_n \prec$ lim inf x_n.

We list now some properties of the limit of a sequence.

1. If lim x_n exists, it is unique.

2. If $x_n = a (n = 0, 1, 2, \ldots)$, then lim $x_n = a$.

3. If $x_0 \prec x_1 \prec \cdots \prec x_n \cdots$, then

$$\lim \inf x_n = x_0 + x_1 + \cdots = (x_0 + x_1 + \cdots)(x_1 + x_2 + \cdots)(x_2 + x_3 + \cdots) \cdots$$

$$= \lim \sup x_n,$$

and hence $\lim x_n = \sum_{k=0}^{\infty} x_k$.

Similarly, if $\cdots \prec x_n \prec \cdots \prec x_2 \prec x_1 \prec x_0$, then $\lim x_n = \prod x_k$.

Hence in a σ-lattice, *monotone sequences have limits*.

4. The sequences $\{\pi x_n\}$, $\{\sigma x_n\}$ ($n = 0, 1, 2, \ldots$) associated with an arbitrary sequence $\{x_n\}$ are monotone and hence $\lim \pi x_n = \sum_{n=0}^{\infty} \pi x_n = \lim \inf x_n$, and $\lim \sigma x_n = \prod \sigma x_n = \lim \sup x_n$. Hence $\lim x_n$ exists if and only if $\lim \pi x_n = \lim \sigma x_n$.

5. If $\lim x_n = a$ and $\{y_n\}$ is a subsequence of $\{x_n\}$, then $\lim y_n = a$. For every index n, the sequence $y_n, y_{n+1}, \ldots$ is a subsequence of the sequence $x_n, x_{n+1}, \ldots$, and consequently $\pi x_n \prec \pi y_n$ and $\sigma y_n \prec \sigma x_n$ ($n = 0, 1, 2, \ldots$). Hence

$$\lim x_n = \lim \inf x_n = \sum_{n=0}^{\infty} \pi x_n \prec \sum_{n=0}^{\infty} \pi y_n = \lim \inf y_n,$$

$$\lim x_n = \lim \sup x_n = \prod_{n=0}^{\infty} \sigma x_n \succ \prod_{n=0}^{\infty} \sigma y_n = \lim \sup y_n,$$

and consequently

$$\lim \sup y_n \prec \lim x_n \prec \lim \inf y_n.$$

Using property 3, it follows that

$$\lim \inf y_n = \lim x_n = \lim \sup y_n.$$

Remark 4. It follows from properties 1, 2, 5 that a σ-complete lattice $\mathscr{L}$ is a Fréchet L-space with respect to the Kantorovich definition of limit of an infinite sequence (see Section 1.7).

6. If $\lim x_n = a$ and $\mathscr{L}$ is a σ-complete and σ-distributive lattice, then for each element b of $\mathscr{L}$, $\lim b x_n = ab$ and $\lim (b + x_n) = a + b$. We leave the proofs to the reader.

The topology of a σ-complete lattice $\mathscr{L}$ based upon the Kantorovich definition of the limit of an infinite sequence of elements of $\mathscr{L}$ is called the Kantorovich sequential topology. We shall study this sequential topology in the next chapter, where $\mathscr{L}$ is a σ-complete Boolean algebra.

EXERCISES

1. Prove the six properties of the limit of an infinite sequence of elements of a σ-complete lattice listed above.

2. Give an example of a σ-complete lattice in which multiplication distributes over addition, but addition does not distribute over multiplication.

3. Show that if $\mathscr{L}$ is not σ-distributive, property 6 of $\lim x_n$ is not necessarily valid.

REFERENCES

Section 2.7   The equivalence of $B(a, b, c)$ and $G(a, b, c)$ is due to V. Glivenko, *Theorie des Structures*, Paris, 1938. The relations $G^*(a, b, c)$ and $G^{**}(a, b, c)$ were introduced in L. M. Blumenthal and D. Ellis, "Notes on lattices," *Duke Math. J.*, **16**: 585–590 (1949).

Section 2.9   These theorems were first proved in the paper by L. M. Blumenthal and D. Ellis referred to above.

Sections 2.10 to 2.12   V. Glivenko, *Théorie des Structures*, Paris, 1938.

CHAPTER 3

Distance Geometry and Topology Over a Boolean Algebra

3.1 INTRODUCTORY REMARKS

In this chapter we study two abstract geometries that are less than fifteen years old. In 1952, L. M. Blumenthal began the investigation on the space obtained by associating with each two elements a, b of a Boolean algebra $\mathscr{B}$ (Section 1.6) the element $d(a, b) = ab' + a'b$ *of the algebra* as "distance," where the "prime" of an element denotes its (unique) complement. We shall see that distance so defined has all of the *formal* properties of a metric, and so the system $\mathbf{B} = \{\mathscr{B}; d\}$ is referred to as a *Boolean metric space*. It is an important member of the class of Boolean metric spaces to be defined later.

Congruence of two figures of a Boolean metric space is defined in the usual manner; in particular, two figures F_1, F_2 of $\mathbf{B}$ are congruent if there exists a one-to-one, distance-preserving correspondence γ between their elements, and we write $F_1 \underset{\gamma}{\approx} F_2$. In case $\mathbf{B} \underset{\gamma}{\approx} \mathbf{B}$, then γ is called a *motion*.

All figures of **B** fall into equivalence classes—two figures being in the same class provided they are congruent—and a Boolean distance geometry is the study of all properties that a figure of a Boolean metric space has in common with every figure in its equivalence class. Hence, in Boolean distance geometry, the fundamental transformations are those that preserve distance—namely, the congruences.

It should be noted that the operator $d(a, b) = ab' + a'b$ that associates with the elements a, b their distance $d(a, b)$ is well known in another connection. It is the standard means of introducing in a Boolean algebra $\mathscr{B}$ an *algebraic* addition $\oplus$ (distinct from the Boolean addition) that makes the double composition system $\{\mathscr{B}; \oplus, \cdot\}$ (where the dot denotes the Boolean multiplication) an *algebraic (Boolean) ring*, with a unit. From the algebraic viewpoint, the Boolean distance geometry we are going to develop in this chapter is the study of the invariants of a Boolean ring with unit element under the group of transformations that preserve only ring addition.

We begin by establishing theorems of order (linearity) and obtaining properties of segments. Later, we shall deal with concepts involving continuity, based on the introduction in **B** of the Kantorovich sequential topology (Section 2.15), and study some of the topological properties of the resulting space.

EXERCISES

1. In a Boolean algebra $\mathscr{B}$ show that
   $$d(a, b) = (a + b)(a' + b') = (a + b)(ab)',$$
   $$d'(a, b) = ab + a'b',$$
   $$d(a, b) = d(a', b'),$$
   $$d(o, a) = a,$$
   where o denotes the first element of $\mathscr{B}$.

2. If $a, b \in \mathbf{B}$ and $c = d(a, b)$, show that $d(a, c) = b$ and $d(b, c) = a$.

3. If $a, c \in \mathbf{B}$, prove that there is exactly one element b of **B** such that $d(a, b) = c$. Hence each element of **B** is uniquely determined by its distance from any one element of **B**, and so **B** does not contain an isosceles triple of pairwise distinct points.

3.2 A BOOLEAN METRIC.

GROUP OF MOTIONS AND CONGRUENCE ORDER OF BOOLEAN SPACE B

The reader will recall that a Boolean algebra $\mathscr{B}$ is a distributive lattice that is complemented; that is, $\mathscr{B}$ contains with each element a an element a' such that $a + a' = 1$, and $a \cdot a' = 0$, where 0, 1 denote the first and last elements of $\mathscr{B}$, respectively. According to Property 1, Section 2.5, the element a' (the complement of element a) is unique. Properties 3, 4, 5, of Section 2.5 are, of course, valid in $\mathscr{B}$.

We call $d(x, y) = xy' + x'y$, $(x, y \in \mathbf{B})$, the *distance function* or *metric* in $\mathbf{B}$, the latter term being justified by the following theorem.

THEOREM 3.1 *The distance function $d(x, y)$ in $\mathbf{B}$ has the following properties:*

(i) $d(x, y) = 0$, *if and only $x = y$,*

(ii) $d(x, y) = d(y, x)$,

(iii) $d(x, z) \prec d(x, y) + d(y, z)$, $(x, y, z) \in \mathbf{B}$.

Proof. Clearly, $d(x, x) = xx' + x'x = 0$; and, if $d(x, y) = 0$, then $xy' = 0$, and (taking complements) $x' + y = 1$, since $(y')' = y$. Now also $x'y = 0$, and we have

$$x + x' = 1 = x' + y; \quad xx' = 0 = x'y.$$

Since $\mathscr{B}$ is a distributive lattice, it follows that $x = y$ (Theorem 2.4), and so (i) is established.

Property (ii) is an immediate consequence of the commutativity of lattice addition and lattice multiplication.

Property (iii), the "triangle inequality", results from

$$[d(x, y) + d(y, z)] \cdot d(x, z) = (xy' + x'y + yz' + y'z)(xz' + x'z)$$
$$= xy'z' + xyz' + x'yz + x'y'z$$
$$= xz'(y + y') + x'z(y + y')$$
$$= xz' + x'z = d(x, z).$$

That is, $d(x, z) \prec d(x, y) + d(y, z)$.

Thus, the Boolean metric $d(x, y)$ has all of the properties of an ordinary metric (Section 1.7) except, of course, that it is *not* a real number, and the partial order relation replaces the relation $\leqq$.

A congruent (distance-preserving) mapping f of **B** *onto* itself is a *motion*. If $f(a) = f(b)$, then $d(a, b) = d(f(a), f(b)) = d(f(a), f(a)) = 0$, and it follows from (i) of the preceding theorem that $a = b$. Hence a congruent mapping is one-to-one. The following theorems reveal the very simple nature of motions in **B**.

THEOREM 3.2   *Any congruent mapping of* **B** *into itself is a motion.*

Proof.   Let f denote such a mapping, and put $f(f(x)) = y$, $x \in$ **B**. Then, since f is a congruence, $d(x, f(x)) = d(f(x), y)$, and so $f(x)$ is equidistant from x and y. It follows (Exercise 3, Section 3.1) that $x = y = f(f(x))$, and the congruence f interchanges x and $f(x)$. (Such mappings are called involutory.) Hence if $x \in$ **B**, then $x = f(f(x))$, and consequently x is the image by f of an element—namely, $f(x)$—of **B**; that is, the mapping f is *onto*, and hence is a motion.

THEOREM 3.3   *If f is a motion of* **B**, *then for each x $(x \in B)$, $f(x) = d(a, x)$, where $a = f(0)$.*

Proof.   Putting $g(x) = d(a, x)$, we show first that $g(x)$ is a motion. Now $g(x)$ is clearly a mapping of **B** into itself, and since

$$
\begin{aligned}
d(g(x), g(y)) &= d(ax' + a'x, ay' + a'y) \\
&= (a' + x)(a + x')(ay' + a'y) + (ax' + a'x)(a' + y)(a + y') \\
&= a'x'y + axy' + a'xy' + ax'y \\
&= (a + a')(xy' + x'y) = d(x, y),
\end{aligned}
$$

then $g(x)$ is a congruence and hence, by the preceding theorem, $g(x)$ is a motion. Moreover, since

$$g(0) = d(a, 0) = a = f(0),$$

each of the two motions f and g transforms 0 into a, and hence they must be identical. For in the contrary case, an element b of **B** exists such that $f(b) \neq g(b)$. But then

$$d(a, f(b)) = d(0, b) = d(g(0), g(b)) = d(a, g(b)).$$

That is, the element a is equidistant from the elements $f(b)$, $g(b)$, which implies $f(b) = g(b)$.

THEOREM 3.4 *There is exactly one motion that takes an arbitrary element a into a prescribed element b.*

Proof. Applying the argument used in the proof of the preceding theorem, we see that the mapping $g(x) = d(d(a, b), x)$ is a motion, and since $g(a) = d(d(a, b), a) = b$, that motion sends element a into element b. It is the only motion that has that property. (Why?)

COROLLARY *The group of motions of* **B** *is simply transitive.*

THEOREM 3.5 *If E, F are subsets of* **B**, *any congruence g between their elements can be extended to a motion (that is, any two congruent subsets of* **B** *are freely superposable).*

Proof. If $a \in E$ and $b = g(a)$, $(b \in F)$, consider the unique motion $f(x) = d(d(a, b), x)$ of **B** that maps a into b.

If $y \in E$, then $f(y) = g(y)$, since the elements $f(y)$ and $g(y)$ are equidistant from b (each has distance $d(a, y)$ from b). Hence the motion f is the (unique) extension of the congruence g.

Let $\mathscr{B}$ denote a given Boolean algebra, and let S denote any abstract set. The system $\{S; \mathscr{B}; d\}$ denotes the $\mathscr{B}$-metric space obtained by attaching to each ordered pair of elements p, q of S, an element $d(p, q)$ of $\mathscr{B}$, such that (i) $d(p, q) = 0$ if and only if $p = q$, (ii) $d(p, q) = d(q, p)$, and (iii) if $p, q, r \in S$, $d(p, r) \prec d(p, q) + d(q, r)$. The space **B** that is studied in this chapter is a special system $\{S; \mathscr{B}: d\}$ in two respects: (1) $S = \mathscr{B}$, and (2) the element $d(p, q)$ of $\mathscr{B}$ that is associated with the elements p, q of $\mathscr{B}$ as distance is the element $pq' + p'q$. When S, $\mathscr{B}$, and d are general, we have the general class of Boolean metric spaces.

THEOREM 3.6 *The space* **B** *has congruence order 3 with respect to the class of Boolean metric spaces; that is, if Σ is any Boolean metric space such that each 3 points of Σ are congruent with 3 points of* **B**, *then Σ is congruent with a subset of* **B**.

Proof. Let α, a denote arbitrary but fixed elements of Σ, **B**, respectively, If $\zeta \in \Sigma$, then α, $\zeta \approx a^*$, x^*, elements of **B**, and a unique motion of **B** exists that takes a^* into a (Theorem 3.4). That motion carries x^* into an element x, with α, $\zeta \approx a$, x, and we have a mapping of Σ into **B**.

To show that this mapping is a congruence, let ζ, η, be any two elements of Σ, with corresponding elements x, y respectively, in **B**. By hypothesis,

$$\alpha, \zeta, \eta \approx \bar{a}, \bar{x}, \bar{y}, \text{ elements of } \mathbf{B},$$

and the unique motion of **B** that takes $\bar{a}$ into a, evidently carries $\bar{x}$ into x and $\bar{y}$ into y. Hence α, ζ, $\eta \approx a$, x, y, and consequently $d(\zeta, \eta) = d(x, y)$.

Remark.   The space **B** does not have congruence order 2 with respect to the class of Boolean metric spaces, for if $S = \{\alpha_1, \alpha_2, \alpha_3\}$ and $d(\alpha_i, \alpha_j) = 1$ $(i, j = 1, 2, 3; i \neq j)$, $d(\alpha_i, \alpha_i) = 0$ $(i = 1, 2, 3)$, where 0, 1 are the first and last elements, respectively, of $\mathscr{B}$, then clearly the Boolean metric space $\Sigma = \{S; \mathscr{B}, d\}$ has each two of its elements congruent with two elements of **B**, but Σ is not congruent with a subset of **B**, since **B** does not contain an equilateral triple of pairwise distinct points (Exercise 3, Section 3.1).

EXERCISES

1. Give other binary operators $g(x, y)$, defined in a given Boolean algebra $\mathscr{B}$, that can qualify as Boolean metrics—that is, have properties (i), (ii), (iii) of Theorem 3.1.

2. In addition to having properties (i), (ii), (iii) of Theorem 3.1, prove that the operator $d(x, y)$ is a group composition on $\mathscr{B}$.

3. Show that if the Boolean bilinear operator

$$f(x, y) = Axy + Bxy' + Cx'y + Dx'y', \qquad (A, B, C, D \in \mathscr{B})$$

is equal to 0 if and only if $x = y$, then $f(x, y) = d(x, y)$.

4. Show that every motion f is complement-preserving (that is, $f(x') = f'(x)$, for every element x of **B**) but is neither sum-preserving nor product-preserving nor order-preserving.

5. Prove that each motion f of **B** is involutory; that is, if $y = f(x)$, then $x = f(y)$.

3.3 $\mathscr{L}$-LINEAR AND $\mathscr{D}$-LINEAR m-TUPLES

Of basic importance in many axiomatic studies of a geometry is the notion of betweenness and the property of linearity to which it gives rise. Since a Boolean metric space is a distance space, we utilize here the procedure adopted in those distance spaces for which distance is a nonnegative real number, and define an element b of **B** to be *between* elements a, c of **B** provided $a \neq b \neq c$ and $d(a, b) + d(b, c) = d(a, c)$. It follows that if b is between a, c, then $a \neq b \neq c \neq a$. The reader's attention is directed to the condition $a \neq b \neq c$, which was not required in our earlier study (see Section 2.7; in particular, Remark 1). Writing $\mathscr{L}(a, b, c)$ to denote that b is between a, c in the sense just defined, $(a, b, c \in \mathscr{B})$, the question arises: what ternary relation in the lattice $\mathscr{B}$ is equivalent to the ternary relation $\mathscr{L}$ that we have defined in the space **B**? We have already answered such a question for the case of a *normed lattice* $\mathscr{L}$ and the ordinary metric space associated with it (Section 2.7). But we are confronted here with a very different situation, and it is, therefore, rather surprising that the ternary lattice relation that answered the previous question should also provide the answer to the present question. This is established in the following theorem.

THEOREM 3.7   *In a Boolean metric space, $\mathscr{L}(a, b, c)$ is equivalent to*

$$(*) \qquad ab + bc = b = (a + b)(b + c), \qquad a \neq b \neq c.$$

Proof.   Assuming (*), then $a = c$ implies $ab = b = a + b$, and the contradiction $a = b$ follows. Hence (*) implies that a, b, c are pairwise distinct.

From $ac \prec (a + b)(b + c) = b$ follows $ac = (ab)(bc)$ and, dually, $a + c \succ ab + bc = b$ yields $a + c = (a + b) + (b + c)$. Hence

$$d(a, c) = ac' + a'c = (a + c)(ac)'$$
$$= [(a + b) + (b + c)] \cdot [(ab)(bc)]'$$
$$= (a + b)(ab)' + (b + c)(bc)' + ac' + a'c.$$

That is, $d(a, c) \succ d(a, b) + d(b, c)$. But from the triangle inequality, $d(a, c) \prec d(a, b) + d(b, c)$, and consequently, $d(a, c) = d(a, b) + d(b, c)$. Thus (*) implies $\mathscr{L}(a, b, c)$.

Assuming $\mathscr{L}(a, b, c)$, then $a \neq b \neq c$, and

$$(\dagger) \qquad (a + b)(ab)' + (b + c)(bc)' = (a + c)(ac)'.$$

Adding $(a + b)ab$, $(b + c)bc$ to the left side of (†), and their respective equals ab, bc to the right side gives (applying distributivity)

$$a + b + b + c = (a + c)(ac)' + ab + bc.$$

Since $b \prec a + b + b + c$ and $(a + c)(ac)' + ab + bc \prec a + c$, then $b \prec a + c$ or $b = b(a + c) = ab + bc$.

Complementing (†) and applying the dual procedure yields

$$b = (a + b)(b + c),$$

and completes the proof of the theorem.

COROLLARY   In a Boolean metric space, $\mathscr{L}(a, b, c)$, is equivalent to $b = ab + bc = (a + b)(b + c) = b(a + c) = b + ac, a \neq b \neq c$.

DEFINITION   An m-tuple of pairwise distinct elements of a Boolean metric space is $\mathscr{L}$-linear provided a labeling $p_1, p_2, \ldots, p_m$ of the elements exists such that for $1 \leq i_1 < i_2 < i_3 \leq m$, the relation $\mathscr{L}(p_{i_1}, p_{i_2}, p_{i_3})$ subsists. In this case we write $\mathscr{L}(p_1, p_2, \ldots, p_m)$.

DEFINITION   An m-tuple of **B** is $\mathscr{D}$-linear provided a labeling $p_1, p_2, \ldots, p_m$ exists such that

$$d(p_1, p_m) = \sum_{i=1}^{m-1} d(p_i, p_{i+1}).$$

In this case we write $\mathscr{D}(p_1, p_2, \ldots, p_m)$.

In every semimetric space (see 5 of Section 1.7), $\mathscr{L}$-linearity of an m-tuple implies its $\mathscr{D}$-linearity, but not conversely; in ordinary metric spaces, the two kinds of linearity are mutually equivalent (see Exercises). In a Boolean metric space, however, the notions are mutually equivalent only for triples, though clearly $\mathscr{L}$-linearity implies $\mathscr{D}$-linearity for all m-tuples.

Remark.   If a, b are noncomparable elements of **B** (that is, neither $a \prec b$ nor $b \prec a$), then **B** contains elements ab and $a + b$, and the four elements $a, b, ab, a + b$ are pairwise distinct. Since $d(a, a + b) = a'b$, $d(a + b, ab) = ab' + a'b$, $d(ab, b) = a'b$, then

$$d(a, a + b) + d(a + b, ab) + d(ab, b) = ab' + a'b = d(a, b),$$

and $\mathscr{D}(a, a + b, ab, b)$ subsists.

On the other hand, the quadruple $a, b, ab, a + b$ is not $\mathscr{L}$-linear, though each of the four triples contained in the quadruple is $\mathscr{L}$-linear.

DEFINITION   Four elements of **B** are *pseudo-$\mathscr{L}$-linear* provided the four elements are not $\mathscr{L}$-linear, but each of the four triples they contain is $\mathscr{L}$-linear.

The quadruple $a, b, ab, a + b$ is an example of a *pseudo-$\mathscr{L}$-linear* quadruple. It has two metric features that seem quite special, but which are (as we shall soon show) possessed by every such quadruple. Those properties are (1) each distance equals its "opposite" distance, and (2) the quadruple is $\mathscr{D}$-linear. But, the special *lattice* structure of the quadruple is not characteristic of pseudo-$\mathscr{L}$-linear quadruples.

THEOREM 3.8   *In a Boolean metric space* **B**, *the metric betweenness relation $\mathscr{L}$ has the following transitive property*:

$$\mathscr{L}(a, b, c) \text{ and } \mathscr{L}(a, d, b) \text{ is equivalent to } \mathscr{L}(d, b, c) \text{ and } \mathscr{L}(a, d, c).$$

Proof.   Assuming $\mathscr{L}(a, b, c)$, $\mathscr{L}(a, d, b)$, and applying Theorem 3.7,

$$(*) \qquad \begin{aligned} ab + bc = b = (a + b)(b + c), \qquad & a \neq b \neq c \neq a, \\ ad + db = d = (a + d)(d + b), \qquad & a \neq d \neq b \neq a. \end{aligned}$$

Now $db + bc > abd + abc = ab + ac$, since $ac < b$ and $ab < d$. Hence

$$db + bc > ab + bc + ac = b + ac = b,$$

which, together with $db + bc < b$, gives $db + bc = b$.

Dualizing yields $b = (d + b)(b + c)$, and so

$$db + bc = b = (d + b)(b + c), \qquad d \neq b \neq c.$$

That is, $\mathscr{L}(d, b, c)$ subsists. (The reader will notice that the above argument is valid in a *general* lattice.)

Since **B** is a distributive lattice, the first one of the relations $(*)$ gives $b(a + c) = b = b + ac$, and so $ac < b < a + c$, and from the second relation, $ab < d < a + b$. From $a < a + c$ and $b < a + c$ follows $a + b < a + c$, and consequently $d < a + c$. Also $ac < a$ and $ac < b$ yield $ac < ab$, and so $ac < d$. Hence $ac < d < a + c$; that is, $d(a + c) = d = d + ac$, or $ad + dc = d = (d + a)(d + c)$.

In order to conclude $\mathscr{L}(a, d, c)$, we need $a \neq d \neq c$. The first inequality is given in the second of the relations (*); and if $d = c$, then $\mathscr{L}(a, b, c)$ and $\mathscr{L}(a, c, b)$ hold, which is impossible since $a \neq b \neq c \neq a$. Hence $\mathscr{L}(a, b, c)$, $\mathscr{L}(a, d, b)$ imply $\mathscr{L}(d, b, c)$, $\mathscr{L}(a, d, c)$.

The implication in the other direction follows immediately from the symmetry of the $\mathscr{L}$ relation with respect to the "outer" points, and from what has just been established.

THEOREM 3.9   *An m-tuple of B is $\mathscr{L}$-linear if and only if a labeling p_1, $p_2, \ldots, p_m$ exists such that $\mathscr{L}(p_1, p_2, p_k)$, $\mathscr{L}(p_2, p_j, p_k)$ subsist $(3 \leq j < k \leq m)$.*

Proof.   The necessity is trivial. To prove the sufficiency, consider indices i_1, i_2, i_3, with $1 \leq i_1 < i_2 < i_3 \leq m$. Now $\mathscr{L}(p_{i_1}, p_{i_2}, p_{i_3})$ subsists by hypothesis, in case $i_1 = 2$ or $i_2 = 2$. If $i_1 = 1$ and $i_2 \neq 2$, then $\mathscr{L}(p_1, p_2, p_{i_3})$ and $\mathscr{L}(p_2, p_{i_2}, p_{i_3})$ imply $\mathscr{L}(p_1, p_{i_2}, p_{i_3})$ (Theorem 3.8). For $i_1 \neq 1, 2$, and $i_2 \neq 2$, $\mathscr{L}(p_1, p_2, p_{i_2})$ and $\mathscr{L}(p_2, p_{i_1}, p_{i_2})$ imply $\mathscr{L}(p_1, p_{i_1}, p_{i_2})$, and $\mathscr{L}(p_1, p_{i_2}, p_{i_3})$ results from $\mathscr{L}(p_1, p_2, p_{i_3})$ and $\mathscr{L}(p_2, p_{i_2}, p_{i_3})$.

Hence, for all indices i_1, i_2, i_3 $(1 \leq i_1 < i_2 < i_3 \leq m)$, $\mathscr{L}(p_{i_1}, p_{i_2}, p_{i_3})$ holds, and consequently so does $\mathscr{L}(p_1, p_2, \ldots, p_m)$.

EXERCISES

1. Show that in every semimetric space, $\mathscr{L}$-linearity of an m-tuple implies $\mathscr{D}$-linearity of the m-tuple, but not conversely.

2. Prove that $\mathscr{L}$-linearity and $\mathscr{D}$-linearity are equivalent for m-tuples of an ordinary metric space.

3. Prove that in an ordinary metric space $\mathscr{L}(a, b, c)$ and $\mathscr{L}(a, d, b)$ is equivalent to $\mathscr{L}(d, b, c)$ and $\mathscr{L}(a, d, c)$.

3.4   SPECIAL QUADRUPLES.
PTOLEMAIC PROPERTY OF BOOLEAN SPACE B

Pseudo-$\mathscr{L}$-linear quadruples were defined in the preceding section, and pseudo-$\mathscr{D}$-linear quadruples are defined in an analogous manner. We obtain both a lattice and a metric characterization of the first kind of quadruples, and show that the second kind does not exist in **B**.

If $p, q, r, s \in \mathbf{B}$ with $\mathcal{L}(p, q, r)$, $\mathcal{L}(q, r, s)$, $\mathcal{L}(r, s, p)$ and $\mathcal{L}(s, p, q)$ holding, we write $\mathcal{C}(p, q, r, s)$.

THEOREM 3.10 *A quadruple of* $\mathbf{B}$ *is pseudo-*$\mathcal{L}$-*linear if and only if a labeling* p, q, r, s *exists such that* $\mathcal{C}(p, q, r, s)$ *holds.*

Proof. The condition is sufficient, for if $\mathcal{C}(p, q, r, s)$ subsists, each triple contained in the quadruple p, q, r, s is $\mathcal{L}$-linear, but the quadruple is not $\mathcal{L}$-linear because $\mathcal{L}(p, q, r)$ and $\mathcal{L}(q, r, s)$ are satisfied by only two orderings of the elements of the quadruple (namely, p, q, r, s and s, r, q, p) and in neither order is the quadruple $\mathcal{L}$-linear.

To establish the necessity, let $Q = \{p, q, r, s\}$ denote a pseudo-$\mathcal{L}$-linear quadruple of $\mathbf{B}$ with the labeling assumed so that $\mathcal{L}(p, q, r)$ holds. We seek to prove $\mathcal{L}(q, r, s)$, $\mathcal{L}(r, s, p)$, and $\mathcal{L}(s, p, q)$.

Now if $\mathcal{L}(p, s, q)$, then $\mathcal{L}(p, s, r)$ and $\mathcal{L}(s, q, r)$ follow (Theorem 3.8) and so $\mathcal{L}(p, s, q, r)$ holds, contrary to the assumed pseudo-$\mathcal{L}$-linearity of Q. To dispose of the possibility $\mathcal{L}(p, q, s)$, let us consider the triple q, r, s. If $\mathcal{L}(q, r, s)$, then $\mathcal{L}(p, q, s)$ gives $\mathcal{L}(p, q, r, s)$; and $\mathcal{L}(q, s, r)$ with $\mathcal{L}(p, q, r)$ yields $\mathcal{L}(p, q, s, r)$. Finally, if $\mathcal{L}(r, q, s)$ holds, consider p, r, s. Now $\mathcal{L}(p, r, s)$ and $\mathcal{L}(r, q, s)$ give $\mathcal{L}(p, r, q)$, contradicting $\mathcal{L}(p, q, r)$; $\mathcal{L}(p, s, r)$ and $\mathcal{L}(p, q, s)$ imply $\mathcal{L}(q, r, s)$, in contradiction to the assumption that $\mathcal{L}(r, q, s)$ holds; and $\mathcal{L}(r, p, s)$ with $\mathcal{L}(p, q, r)$ implies $\mathcal{L}(r, q, p, s)$, a contradiction. It follows that neither $\mathcal{L}(p, s, q)$ nor $\mathcal{L}(p, q, s)$ subsists, and consequently $\mathcal{L}(s, p, q)$ does.

We have so far established $\mathcal{L}(p, q, r)$ and $\mathcal{L}(s, p, q)$. Turning to the triple q, r, s, $\mathcal{L}(r, s, q)$ and $\mathcal{L}(p, q, r)$ imply $\mathcal{L}(p, q, s, r)$; and $\mathcal{L}(r, q, s)$ and $\mathcal{L}(s, p, q)$ give $\mathcal{L}(r, q, p, s)$. These two contradictions imply $\mathcal{L}(q, r, s)$. Examining the remaining triple p, r, s it is seen that $\mathcal{L}(p, r, s)$ and $\mathcal{L}(p, q, r)$ give $\mathcal{L}(p, q, r, s)$, and $\mathcal{L}(r, p, s)$ with $\mathcal{L}(q, r, s)$ yields $\mathcal{L}(q, r, p, s)$. Hence $\mathcal{L}(r, s, p)$ must hold, and the proof is complete.

COROLLARY 1 *If a quadruple of* $\mathbf{B}$ *has all triples* $\mathcal{L}$-*linear, then a labeling* p, q, r, s *of the elements of the quadruple exists such that either* $\mathcal{L}(p, q, r, s)$ *or* $\mathcal{C}(p, q, r, s)$ *holds.*

COROLLARY 2 *If a quadruple of* $\mathbf{B}$ *has all triples* $\mathcal{L}$-*linear, and a labeling* p, q, r, s *of the elements of the quadruple exists so that* $\mathcal{L}(p, q, r)$ *and* $\mathcal{L}(p, q, s)$ *hold, then either* $\mathcal{L}(p, q, r, s)$ *or* $\mathcal{L}(p, q, s, r)$ *subsists.*

Proof. If the quadruple is pseudo-$\mathscr{L}$-linear and $\mathscr{L}(p, q, r)$ holds, then (as the argument of the preceding theorem shows) $\mathscr{L}(s, p, q)$ subsists and not $\mathscr{L}(p, q, s)$. It follows that no labeling will result in $\mathscr{C}(p, q, r, s)$ holding.

THEOREM 3.11 *A necessary and sufficient condition that four pairwise distinct elements of* **B** *form a psuedo-$\mathscr{L}$-linear quadruple is that a labeling a, b, c, d of the elements exists such that $a + c = b + d$ and $ac = bd$.*

Proof. By the previous theorem, four elements of **B** form a pseudo-$\mathscr{L}$-linear quadruple provided a labeling of the elements exists such that $\mathscr{L}(a, b, c)$, $\mathscr{L}(b, c, d)$, $\mathscr{L}(c, d, a)$, and $\mathscr{L}(d, a, b)$ subsist. Those betweenness relations hold, according to the Corollary of Theorem 3.7, if and only if

$$(*) \quad \begin{aligned} b &= ab + bc = (a + b)(b + c) = b(a + c) = b + ac, \\ c &= bc + cd = (b + c)(c + d) = c(b + d) = c + bd, \\ d &= cd + da = (c + d)(d + a) = d(c + a) = d + ac, \\ a &= da + ab = (d + a)(a + b) = a(d + b) = a + bd. \end{aligned}$$

Using the first and third of relations $(*)$, we obtain $b + d \prec a + c$, and the second and fourth relations yield $a + c \prec b + d$. It follows that $a + c = b + d$. Similarly, the first and third relations yield $ac \prec bd$, and from the second and fourth we get $bd \prec ac$; that is, $ac = bd$.

Conversely, if $a + c = b + d$ and $ac = bd$, the relations $(*)$ subsist. For example, $c + bd = c + ac = c$, and $c(b + d) = c(a + c) = c$. Hence

$$c = c(b + d) = bc + cd = c + bd = (b + c)(c + d),$$

and the second of relations $(*)$ is established.

In a similar manner, the other three relations are established, and the elements form a pseudo-$\mathscr{L}$-linear quadruple with $\mathscr{C}(a, b, c, d)$ subsisting.

COROLLARY *Four pairwise distinct elements of* **B** *are pseudo-$\mathscr{L}$-linear provided a labeling p, q, r, s of the elements exists such that $d(p, q) = d(r, s)$, $d(q, r) = d(p, s)$, and $d(p, r) = d(q, s) = d(p, q) + d(q, r)$.*

Proof. According to the preceding theorem, four pairwise distinct elements of **B** are pseudo-$\mathscr{L}$-linear if and only if a labeling p, q, r, s exists such that $p + r = q + s$ and $pr = qs$.

Assuming such a labeling exists (and hence the quadruple is pseudo-$\mathscr{L}$-linear with $\mathscr{C}(p, q, r, s)$ holding), then

$$d(p, r) = (p + r)(pr)' = (q + s)(qs)' = d(q, s),$$

and

$$d(p, q) + d(q, r) = d(p, r), d(q, r) + d(r, s) = d(q, s) = d(p, r).$$

Hence

(1) $$d(p, q) + d(q, r) = d(r, s) + d(q, r).$$

Now

$$d(p, q) \cdot d(q, r) = (pq' + p'q) \cdot (qr' + q'r)$$
$$= prq' + p'r'q = qsq' + q's'q = 0,$$

and similarly, $d(r, s) \cdot d(q, r) = 0$. Hence

(2) $$d(p, q) \cdot d(q, r) = d(r, s) \cdot d(q, r),$$

and relations (1), (2) yield (Theorem 2.4)

$$d(p, q) = d(r, s).$$

The equality $d(q, r) = d(p, s)$ is obtained in a similar manner.

Conversely, the relations $d(p, q) = d(r, s)$, $d(q, r) = d(p, s)$, $d(p, r) = d(p, q) + d(q, r) = d(q, s)$ are easily seen to imply $\mathscr{C}(p, q, r, s)$ and the pseudo-$\mathscr{L}$-linear nature of the quadruple follows.

THEOREM 3.12   *The Boolean space* **B** *does not contain a pseudo-$\mathscr{D}$-linear quadruple.*

Proof.   If a quadruple has all of its triples $\mathscr{D}$-linear, then all of its triples are $\mathscr{L}$-linear and a labeling of its elements may be chosen so that either $\mathscr{L}(p, q, r, s)$ or $\mathscr{C}(p, q, r, s)$ holds. If $\mathscr{L}(p, q, r, s)$ holds, then obviously $\mathscr{D}(p, q, r, s)$ does also, and the quadruple is $\mathscr{D}$-linear. If $\mathscr{C}(p, q, r, s)$ subsists, then by the preceding Corollary,

$$d(p, q) + d(q, s) + d(s, r) = d(p, q) + d(p, r) + d(p, q) = d(p, r),$$

and $\mathscr{D}(p, q, s, r)$ holds.

THEOREM 3.13   *The Boolean space* **B** *is ptolemaic; that is, if* $p, q, r, s \in$ **B,** *then*

$$d(p, q) \cdot d(r, s) + d(p, r) \cdot d(q, s) > d(p, s) \cdot d(q, r).$$

Proof.   Consider the motion $f(x) = d(p, x)$, which sends p into 0 and q, r, s into a, b, c, respectively. Then

$$d(p, q) = d(0, a) = a,$$
$$d(p, r) = d(0, b) = b,$$
$$d(p, s) = d(0, c) = c,$$

and we have

$$
\begin{aligned}
d(0, a) \cdot d(b, c) + d(0, b) \cdot d(a, c) &= a(bc' + b'c) + b(ac' + a'c) \\
&= abc' + ab'c + a'bc \\
&> c(ab' + a'b) \\
&= d(0, c) \cdot d(a, b).
\end{aligned}
$$

Remark.   For each quadruple of **B**, the three products of "opposite" distances satisfy the triangle inequality.

EXERCISES

1. If $p, q, r, s \in$ **B**, find a necessary and sufficient condition in order that

$$d(p, q) \cdot d(r, s) + d(p, r) \cdot d(q, s) = d(p, s) \cdot d(q, r).$$

2. Four elements of **B** form a *convex tripod* provided a labeling p, q, r, s of the elements exists such that $\mathscr{L}(q, p, r)$, $\mathscr{L}(q, p, s)$ and $\mathscr{L}(r, p, s)$ subsist. The element p is the *vertex* of the tripod. Show that the elements q, r, s are not $\mathscr{L}$-linear. Show that p, q, r, s form a convex tripod with vertex p if and only if $d(p, q) \cdot d(p, r) = d(p, r) \cdot d(p, s) = d(p, s) \cdot d(p, q) = 0$.

3. Give an example of an ordinary metric space that is not ptolemaic.

4. Give an example of a semimetric ptolemaic space that is not metric.

3.5 FUNDAMENTAL LINEARITY THEOREM FOR BOOLEAN SPACE **B**

In the preceding section, we have characterized pseudo-$\mathscr{L}$-linear quadruples and shown that pseudo-$\mathscr{D}$-linear quadruples do not exist. In this section, we continue our study of linearity in a space **B**, by extending the latter result to m-tuples, and proving that an m-tuple of **B** is pseudo-$\mathscr{L}$-linear only if $m = 4$. That is, every subset of **B** that contains *more than* four pairwise distinct elements is $\mathscr{L}$-linear whenever each three of the elements has that property.

We establish first some lemmas concerning quintuples.

LEMMA 3.1   *If all triples of a quintuple are $\mathscr{L}$-linear, at least one quadruple is $\mathscr{L}$-linear.*

Proof.   Assuming the contrary, we may label the quintuple p, q, r, s, t, with $\mathscr{C}(p, q, r, s)$ holding. Considering the $\mathscr{L}$-linear triple p, q, t, clearly $\mathscr{L}(p, t, q)$ and $\mathscr{L}(p, q, r)$ imply $\mathscr{L}(p, t, q, r)$, and $\mathscr{L}(p, q, t)$ with $\mathscr{L}(p, q, r)$ implies $\mathscr{L}(p, q, t, r)$ or $\mathscr{L}(p, q, r, t)$ by Corollary 2, Theorem 3.10. It follows tha $\mathscr{L}(t, p, q)$ must subsist.

A similar consideration of the $\mathscr{L}$-linear triple p, s, t shows that $\mathscr{L}(p, t, s)$ must hold. But $\mathscr{L}(p, t, s)$ and $\mathscr{L}(s, p, q)$ which is implied by $\mathscr{C}(p, q, r, s)$, yield $\mathscr{L}(s, t, p, q)$, and the lemma is proved.

LEMMA 3.2   *If all triples of a quintuple are $\mathscr{L}$-linear, and the labeling is chosen so that $\mathscr{L}(p, q, r, s)$ subsists, then $\mathscr{C}(t, p, q, r)$ does not hold.*

Proof.   In the contrary case, $\mathscr{L}(r, t, p)$ subsists, which, together with $\mathscr{L}(p, r, s)$, gives $\mathscr{L}(p, t, r, s)$. But $\mathscr{L}(q, s, t)$ with $\mathscr{L}(q, r, s)$ gives $\mathscr{L}(r, s, t)$; $\mathscr{L}(q, t, s)$ with $\mathscr{L}(t, p, q)$ gives $\mathscr{L}(q, p, s)$; and $\mathscr{L}(s, q, t)$ with $\mathscr{L}(t, p, q)$ gives $\mathscr{L}(s, p, t)$. In each case, the conclusion contradicts an earlier linearity, and consequently, the triple q, s, t cannot be $\mathscr{L}$-linear, in contradiction to the hypothesis.

LEMMA 3.3   *If all triples of a quintuple p, q, r, s, t are $\mathscr{L}$-linear ·with $\mathscr{L}(p, q, r, s)$ holding, then the quintuple is $\mathscr{L}$-linear if any one of the relations $\mathscr{L}(p, t, q)$, $\mathscr{L}(q, t, r)$, $\mathscr{L}(r, t, s)$ holds.*

The proof is clear.

THEOREM 3.14   *If all triples of a quintuple are $\mathscr{L}$-linear, then the quintuple is $\mathscr{L}$-linear.*

Proof.   According to Lemma 3.1, the quintuple contains at least one $\mathscr{L}$-linear quadruple, say $\mathscr{L}(p, q, r, s)$. If t denotes the remaining point of the quintuple, then the quintuple is $\mathscr{L}$-linear in case $\mathscr{L}(p, t, q)$ holds (Lemma 3.3).

If $\mathscr{L}(p, q, t)$ subsists, then by Corollary 2, Theorem 3.10, either $\mathscr{L}(p, q, r, t)$ or $\mathscr{L}(p, q, t, r)$ holds. In the latter case, $\mathscr{L}(q, t, r)$ is valid and hence (by the preceding lemma), the quintuple is $\mathscr{L}$-linear. In the former case, consider r, s, t. If $\mathscr{L}(r, t, s)$, then $\mathscr{L}(p, q, r, t, s)$, and $\mathscr{L}(r, s, t)$ with $\mathscr{L}(p, q, r, t)$ yields $\mathscr{L}(p, q, r, s, t)$. Finally, if $\mathscr{L}(s, r, t)$ then $\mathscr{L}(s, r, q)$ yields $\mathscr{L}(s, r, t, q)$ or $\mathscr{L}(s, r, q, t)$, and either alternative contradicts $\mathscr{L}(q, r, t)$. We have disposed, now, of the case $\mathscr{L}(p, q, t)$.

It remains to consider the possibility $\mathscr{L}(t, p, q)$.

By Lemma 3.2 $\mathscr{C}(t, p, q, r)$ does not hold, and consequently $\mathscr{L}(t, p, q, r)$ subsists. The relations $\mathscr{L}(t, p, q)$, $\mathscr{L}(p, q, s)$ imply $\mathscr{L}(t, p, q, s)$ or $\mathscr{C}(t, p, q, s)$. The latter relation implies $\mathscr{L}(q, s, t)$ which, together with $\mathscr{L}(q, r, s)$, yields $\mathscr{L}(q, r, t)$, in contradiction to $\mathscr{L}(t, p, q, r)$; and $\mathscr{L}(t, p, q, r, s)$ results from the former relation, together with $\mathscr{L}(p, r, s)$.

COROLLARY   *If all triples of a quintuple are $\mathscr{L}$-linear, and the labeling is selected so that $\mathscr{L}(p, q, r, s)$ holds, then, denoting the fifth point by t, (a) $\mathscr{L}(p, t, q)$ implies $\mathscr{L}(p, t, q, r, s)$, (b) $\mathscr{L}(t, p, q)$ implies $\mathscr{L}(t, p, q, r, s)$, and (c) $\mathscr{L}(p, q, t)$ yields $\mathscr{L}(p, q, t, r, s)$ or $\mathscr{L}(p, q, r, t, s)$ or $\mathscr{L}(p, q, r, s, t)$.*

The proof makes use of Lemma 3.3 and an examination of the proof of the foregoing theorem.

LEMMA 3.4   *If the m-tuple $p_1, p_2, \ldots, p_m$ has all of its triples $\mathscr{L}$-linear $(m \geqq 6)$, and $\mathscr{L}(p_1, p_2, \ldots, p_{k-1}, p_k)$, $\mathscr{L}(p_{k-1}, p_k, p_{k+1}, \ldots, p_m)$ subsist for an index k $(3 \leqq k \leqq m)$, then $\mathscr{L}(p_1, p_2, \ldots, p_m)$ holds.*

Proof.   Let i_1, i_2, i_3 denote indices $(1 \leqq i_1 < i_2 < i_3 \leqq m)$. For $1 \leqq i_1 < k - 1$, and $k < i_2 < i_3 \leqq m$, $\mathscr{L}(p_{i_1}, p_{k-1}, p_k)$ and $\mathscr{L}(p_{k-1}, p_k, p_{i_2}, p_{i_3})$ follow from the hypothesis, and part (b) of the preceding corollary yields

$$\mathscr{L}(p_{i_1}, p_{k-1}, p_{i_2}, p_{i_3}).$$

Hence $\mathscr{L}(p_{i_1}, p_{i_2}, p_{i_3})$ holds for $1 \leqq i_1 < k - 1$ and $k < i_2 < i_3 \leqq m$, and if the indices do not satisfy these inequalities, the desired betweenness relation is implied at once by the hypothesis of the lemma.

The previous lemmas and theorems have now put us in position to prove the following principal theorem of this section.

THEOREM 3.15 *If all triples of an m-tuple are $\mathscr{L}$-linear and $m > 4$, then the m-tuple is $\mathscr{L}$-linear.*

Proof.   The validity of the theorem is assured by Theorem 3.14 in case $m = 5$. This anchors an inductive argument.

Consider an $(m + 1)$-tuple, $m \geq 5$, with all triples $\mathscr{L}$-linear, and assume the labeling so that $\mathscr{L}(p_1, p_2, \ldots, p_m)$ holds. Labeling p_{m+1} the remaining point of the $(m + 1)$-tuple, then either (1) $\mathscr{L}(p_1, p_{m+1}, p_2)$ or (2) $\mathscr{L}(p_{m+1}, p_1, p_2)$ or (3) $\mathscr{L}(p_1, p_2, p_{m+1})$ subsists.

Now (1) with $\mathscr{L}(p_1, p_2, p_3)$ yields $\mathscr{L}(p_1, p_{m+1}, p_2, p_3)$, which, with $\mathscr{L}(p_2, p_3, \ldots, p_m)$, gives $\mathscr{L}(p_1, p_{m+1}, p_2, p_3, \ldots, p_m)$ by Lemma 3.4, and that lemma, applied to $\mathscr{L}(p_{m+1}, p_1, p_2)$ and $\mathscr{L}(p_1, p_2, \ldots, p_m)$, yields the desired result in case (2) holds.

Finally, if (3) $\mathscr{L}(p_1, p_2, p_{m+1})$ subsists, then by (c) of the corollary of Theorem 3.14, we have (i) $\mathscr{L}(p_1, p_2, p_{m+1}, p_3, p_4)$, or (ii) $\mathscr{L}(p_1, p_2, p_3, p_{m+1}, p_4)$, or (iii) $\mathscr{L}(p_1, p_2, p_3, p_4, p_{m+1})$.

Clearly, (i) leads to the desired conclusion. If (ii) holds, $\mathscr{L}(p_3, p_{m+1}, p_4)$ and $\mathscr{L}(p_3, p_4, p_5)$ yield $\mathscr{L}(p_3, p_{m+1}, p_4, p_5)$ which, together with

$$\mathscr{L}(p_1, p_2, p_3, p_{m+1}),$$

gives $\mathscr{L}(p_1, p_2, p_3, p_{m+1}, p_4, p_5)$. The last relation combined with $\mathscr{L}(p_4, p_5, \ldots, p_m)$ yields $\mathscr{L}(p_1, p_2, p_3, p_{m+1}, p_4, p_5, \ldots, p_m)$, the desired result.

Considering, finally, the possibility (iii), the relations $\mathscr{L}(p_3, p_4, p_{m+1})$, $\mathscr{L}(p_3, p_4, p_5)$ imply $\mathscr{L}(p_3, p_4, p_{m+1}, p_5)$ or $\mathscr{L}(p_3, p_4, p_5, p_{m+1})$ (Corollary 2, Theorem 3.10); that is, $\mathscr{L}(p_1, p_2, p_3, p_4, p_{m+1}, p_5)$ or $\mathscr{L}(p_1, p_2, p_3, p_4, p_5, p_{m+1})$ holds. If $m = 5$, the proof is complete. If $m > 5$, $\mathscr{L}(p_3, p_4, p_{m+1})$ and $\mathscr{L}(p_3, p_4, p_5, p_6)$ imply the desired conclusion if $m = 6$ or

$$\mathscr{L}(p_3, p_4, p_5, p_6, p_{m+1})$$

does not hold.

Continuing in this manner, we eventually obtain $\mathscr{L}(p_1, p_2, \ldots, p_m, p_{m+1})$, in case the $\mathscr{L}$-linearity of the $(m + 1)$-tuple is not established at an earlier stage.

Remark 1.   Since pseudo-$\mathscr{L}$-linear quadruples exist in a Boolean metric space **B**, the restriction $m > 4$ of the theorem is essential to its validity.

Remark 2.   On the other hand, any m-tuple of **B** is $\mathscr{D}$-linear if all of its triples are $\mathscr{D}$-linear. For if $m > 4$, the m-tuple is $\mathscr{L}$-linear and hence $\mathscr{D}$-linear; if $m = 4$ and the m-tuple is pseudo-$\mathscr{L}$-linear, it is still $\mathscr{D}$-linear (Theorem 3.12). Of course, if $m \leq 3$, the $\mathscr{D}$-linearity of the m-tuple is immediate.

3.6   SEGMENTS IN BOOLEAN SPACE **B**

In seeking to define "segment" in a Boolean metric space **B**, several possibilities arise for consideration. Thus, (1) a (directed) segment might be identified with the lattice interval $[x, y]$ where $[x, y] = [z \in \mathbf{B} \mid x \prec z \prec y]$, or (2) a segment might be defined to be the set $(x, y) = [z \in \mathbf{B} \mid xy \prec z \prec x + y]$. The latter set is identical with the set $[z \in \mathbf{B} \mid x = z$ or $z = y$ or $\mathscr{L}(x, z, y)]$. Both notions (1) and (2) have disadvantages which are serious enough to disqualify them. If, for example, notion (1) is adopted, then two distinct elements x, y are joined by a segment only if they are comparable (that is, if either $x \prec y$ or $y \prec x$). (We wish each two distinct elements of **B** to be end points of a segment.) Notion (2) is, indeed, free from this defect, but the segments that it defines are, in general, too "fat"—they contain "more" elements than we want them to have. The notion of segment that we adopt in what follows seems entirely suitable for the geometric theory that we seek to develop.

We recall that a *chain* is any ordered subset Γ of **B** (that is, if $x, y \in \Gamma$, then $x \prec y$ or $y \prec x$). If α, $\beta \in \Gamma$ such that $\alpha \prec x \prec \beta$ for every $x \in \Gamma$, then α, β are called the first and last elements, respectively, of Γ, and we write $\Gamma = \Gamma_\alpha^\beta$. A chain Γ_α^β is maximal provided it is not a proper subset of any chain with first element α and last element β. If the chain Γ_α^β is *maximal* then it contains every element x of **B** that is comparable with *each* of its elements and such that $\alpha \prec x \prec \beta$.

DEFINITION.   A subset of **B** is a *segment* provided it is a congruent image of a maximal chain with first and last elements.

We write $S_a^b = f(\Gamma_\alpha^\beta)$, where f is a congruence, $a = f(\alpha)$, $b = f(\beta)$, and call S_a^b a segment with first element a and last element b. We shall show later that the orientation of S_a^b so defined is an intrinsic property of the segment.

The reader might observe that the above manner of approaching the definition of a *Boolean segment* is very similar to the usual procedure of defining a *metric segment* as a congruent image of a line segment (which is, indeed, a maximal chain between its endpoints in the lattice of real numbers). (A metric segment is defined in Section 5.3.)

THEOREM 3.16   *A subset S of* **B** *that contains elements a, b (a ≠ b) is a segment (with first element a and last element b) if and only if there exists a maximal chain* Γ_α^β *and a motion F such that* $F(\Gamma_\alpha^\beta) = S$, *with* $F(\alpha) = a$, $F(\beta) = b$ *(and consequently* $F(x) = d(d(a, \alpha), x)$).

Proof.   Since each motion is a congruence, the condition is sufficient. On the other hand, if S is a segment S_a^b, then $S_a^b \underset{f}{\approx} \Gamma_\alpha^\beta$, by definition, and by Theorem 3.5 the congruence f can be extended to a motion F.

Remark.   Since each motion F of **B** is involutory (Exercise 3, Section 3.2), $F(\Gamma_\alpha^\beta) = S_a^b$ is equivalent to $\Gamma_\alpha^\beta = F(S_a^b)$.

If $S_a^b = F(\Gamma_\alpha^\beta)$, consider the motion $G(x) = d(d(a, b), x)$ that sends a into b (and hence sends b into a). Then $G(S_a^b) = GF(\Gamma_\alpha^\beta) = H(\Gamma_\alpha^\beta)$, where H is a motion, and hence $G(S_a^b)$ is a segment. Since $H(\alpha) = GF(\alpha) = G(a) = b$ and $H(\beta) = a$, then $G(S_a^b)$ is a segment with *first* element b and *last* element a. We write $G(S_a^b) = {}^*S_b^a$ and call ${}^*S_b^a$ the *dual* of S_a^b. Clearly, S_a^b is the dual of ${}^*S_b^a$, for from $G(S_a^b) = {}^*S_b^a$ follows $G({}^*S_b^a) = S_a^b$, and G is the only motion that sends b into a.

THEOREM 3.17   *The dual segments* S_a^b, ${}^*S_b^a$ *have only their end elements a, b in common.*

Proof.   We have $S_a^b = F(\Gamma_\alpha^\beta)$, where $F(x) = d[d(a, \alpha), x]$, and ${}^*S_b^a = G(S_a^b)$, where $G(x) = d[d(a, b), x]$.
   If $c \in S_a^b \cap {}^*S_b^a$ there exist elements $\bar{c}$, $\bar{\gamma}$ of S_a^b, Γ_α^β, respectively, such that

$$d[d(a, b), \bar{c}] = c = d[d(a, \alpha), \bar{\gamma}],$$

and since $\bar{c} \in S_a^b$, there is an element γ of Γ_α^β such that

$$\bar{c} = d[d(a, \alpha), \gamma].$$

   Then

$$d[d(a, b), \bar{c}] = d[d(a, b), d(d(a, \alpha), \gamma)] = d[d(a, b), d(a, d(\alpha, \gamma))]$$
$$= d[d(a, d(a, b)), d(\alpha, \gamma)] = d[b, d(\alpha, \gamma)] = d[d(\alpha, b), \gamma],$$

and so $c = d[d(b, \alpha), \gamma] = d[d(a, \alpha), \bar{\gamma}]$.

For a given $\gamma(\gamma \in \Gamma_\alpha^\beta)$, the space **B** contains at most one element $\bar{\gamma}$ that satisfies the last equation (Exercise 3, Section 3.1). Since

$$d[d(b, d(\gamma, a)), d(a, \alpha)] = d[d(\gamma, d(a, b)), d(a, \alpha)]$$
$$= d[\gamma, d(d(a, b), d(a, \alpha))]$$
$$= d[\gamma, d(b, \alpha)],$$

then $\bar{\gamma} = d[b, d(\gamma, a)] = d[\gamma, d(a, b)] = d[\gamma, d(\alpha, \beta)]$ is the unique solution of the equation.

Assertion.   If $\gamma \in \Gamma_\alpha^\beta$, then $\bar{\gamma} = d[\gamma, d(\alpha, \beta)] \in \Gamma_\alpha^\beta$ if and only if $\gamma = \alpha$ or $\gamma = \beta$.

Clearly, if $\gamma = \alpha$, then $\bar{\gamma} = \beta$, and if $\gamma = \beta$, then $\bar{\gamma} = \alpha$, so the condition is sufficient.

To show the necessity, suppose $\bar{\gamma} = d[\gamma, d(\alpha, \beta)] \in \Gamma_\alpha^\beta$. Then (1) $\gamma \prec \bar{\gamma}$ or (2) $\bar{\gamma} \prec \gamma$.

In case (1), then

$$\gamma \prec d[\gamma, d(\alpha, \beta)] = \gamma d'(\alpha, \beta) + \gamma' d(\alpha, \beta),$$

and taking the product with γ yields $\gamma \prec \gamma d'(\alpha, \beta)$, from which $\gamma = \gamma d'(\alpha, \beta)$. Hence $\gamma \prec d'(\alpha, \beta) = \alpha\beta + \alpha'\beta' = \alpha + \beta'$, since $\alpha \prec \beta$. Taking the product with β, we obtain $\gamma \prec \alpha$, and consequently $\gamma = \alpha$.

In case (2), we have $\gamma \succ d[\gamma, d(\alpha, \beta)] = \gamma d'(\alpha, \beta) + \gamma' d(\alpha, \beta)$, and taking the product with γ' yields $0 \succ \gamma' d(\alpha, \beta)$. Hence $\gamma' d(\alpha, \beta) = 0$, $\gamma + \gamma' d(\alpha, \beta) = \gamma$, $(\gamma + \gamma')(\gamma + d(\alpha, \beta)) = \gamma$, $\gamma + d(\alpha, \beta) = \gamma$, and so $d(\alpha, \beta) \prec \gamma$. The last relation gives $\alpha'\beta \prec \gamma$, from which $\alpha + \alpha'\beta \prec \alpha + \gamma = \gamma$, or $\alpha + \beta \prec \gamma$; that is $\beta \prec \gamma$, and since $\gamma \prec \beta$, it follows that $\gamma = \beta$, completing the proof of the Assertion.

Hence the only element c such that

$$d[\bar{\gamma}, d(a, \alpha)] = c = d[\gamma, d(b, \alpha)], \quad \gamma, \bar{\gamma} \in \Gamma_\alpha^\beta$$

are obtained for $\gamma = \alpha$, $\bar{\gamma} = \beta$ and $\gamma = \beta$, $\bar{\gamma} = \alpha$. In the first case, $c = d[\alpha, d(b, \alpha)] = b$ (also $c = d[\beta, d(a, \alpha)] = d[\beta, d(b, \beta)] = b$, since the motion $d[x, d(a, b)]$ that sends α into β is the same as the motion $d[x, d(a, b)]$ that sends a into b), and in the second case $c = d[\alpha, d(a, \alpha)] = a$ (also, $c = d[\beta, d(b, \alpha)] = d[b, d(\alpha, \beta)] = d[b, d(a, b)] = a$).

Consequently, the segments S_a^b, $*S_b^a$ have only their end elements in common, and the proof of the theorem is complete.

THEOREM 3.18   *If $a, b \in$ **B** $(a \neq b)$, there exists at least one segment S_a^b.*

Proof. The motion $F(x) = d(a, x)$ sends a into 0 and b into $d(a, b)$. Invoking the axiom of choice, there exists at least one maximal chain $\Gamma_0^{d(a, b)}$, and $F(\Gamma_0^{d(a, b)})$ is a segment S_a^b.

The following theorem establishes some properties that Boolean segments have in common with line segments.

THEOREM 3.19   *If S_a^b denotes a Boolean segment with end elements a, b, then (i) for every positive integer m, each m-tuple of S_a^b is $\mathscr{L}$-linear, (ii) the end elements are unique (iii) if c, $d \in S_a^b$ ($c \neq d$), the set of all elements of S_a^b (metrically) between c, d is a segment with end elements c, d, and (iv) if $x \in S_a^b$ ($a \neq x \neq b$), then $\mathscr{L}(a, x, b)$ subsists.*

Proof.   Since every m-tuple of a chain is $\mathscr{L}$-linear and $\mathscr{L}$-linearity is a congruence invariant, property (i) is valid. If $x \in S_a^b$ ($a \neq x \neq b$), then a, x, $b \approx \alpha$, γ, β of $\Gamma_\alpha^\beta = f(S_a^b)$, and since $\alpha \prec \gamma \prec \beta$ ($\alpha \neq \gamma \neq \beta$), then $\mathscr{L}(\alpha, \gamma, \beta)$, holds, and consequently so does $\mathscr{L}(a, x, b)$. Hence (iv) is established.

Suppose c, d are end elements of S_a^b, each distinct from a, b. Then S_a^b is congruent with a maximal chain whose end elements correspond in that congruence to c, d. It follows from (iv) that $\mathscr{L}(a, c, b)$, $\mathscr{L}(a, d, b)$, $\mathscr{L}(c, a, d)$, and $\mathscr{L}(c, b, d)$ subsist; that is, $\mathscr{C}(a, c, b, d)$ holds, contrary to the $\mathscr{L}$-linearity of the quadruple a, b, c, d, which is implied in (i) (Theorem 3.10).

If $a = c$ but $d \neq b$, then $\mathscr{L}(a, d, b)$ and $\mathscr{L}(a, b, d)$ hold, which is impossible. The proof of (ii) is complete.

To establish (iii), note that the motion f that maps S_a^b onto Γ_α^β, with $\alpha = f(a)$, $\beta = f(b)$, sends c, d into elements γ, δ, respectively, and all those elements of S_a^b that are metrically between c, d are mapped onto a sub-chain of Γ_α^β with end-elements γ, δ. It is clear that this sub-chain contains all the elements of Γ_α^β that are order-between γ, δ. This implies that the subchain is maximal, and so property (iii) is established.

Remarks.   It follows from parts (i) and (iv) of the preceding theorem that if p, $q \in S_a^b$ (a, $b \neq p \neq q \neq a$, b), then $\mathscr{L}(a, p, q)$ or $\mathscr{L}(q, p, b)$ subsists. If two segments S_a^b, S_c^d are related by a congruence g, then $g(a) = c$ or $g(a) = d$. This is obvious if $S_a^b = \{a, b\}$. If $S_a^b \neq \{a, b\}$ and $c \neq g(a) \neq d$, then $\mathscr{L}(c, g(a), d)$ subsists, and the elements $g(c)$, a, $g(d)$ of S_a^b are in the relation $\mathscr{L}(g(c), a, g(d))$. This violates part (iv) of the preceding theorem in case $g(c) = b$ or $g(d) = b$, and in the contrary case, the first of these remarks implies $\mathscr{L}(a, g(c), g(d))$ or $\mathscr{L}(g(d), g(c), b)$. The first alternative contradicts

$\mathcal{L}(g(c), a, g(d))$, and combining the second with $\mathcal{L}(g(c), a, g(d))$ yields $\mathcal{L}(g(d), a, b)$, in contradiction to $\mathcal{L}(a, g(d), b)$.

Though a segment is *superposable* (congruent by a motion) with a maximal chain, no two elements of a segment need be comparable, since a motion is not, in general, order-preserving. The questions arise: when is a segment a chain? When does a motion transform a chain into a chain?

To answer the first question, let S_a^b be a segment with $a \prec b$. If $x \in S_a^b$, then $d(a, x) + d(x, b) = d(a, b)$; that is, $ax' + a'x + bx' + b'x = ab' + a'b$. Now $a \prec b$ implies $ab' = 0$, $ax' \prec bx'$, $b' \prec a'$, $b'x \prec a'x$, and so we have $bx' + a'x = a'b$.

Hence $bx' \prec a'b$, $abx' = 0$, and since $a \prec b$, then $ax' = 0$. It follows that $ax' + x = x$ and so $a + x = x$; that is, $a \prec x$. Starting with $a'x \prec a'b$, it is proved in a similar manner that $x \prec b$. Consequently if $x \in S_a^b$ and $a \prec b$, then $a \prec x \prec b$.

Now if x, y are distinct elements of S_a^b, each different from a, b, we may (by a previous remark) assume the labeling so that $\mathcal{L}(a, x, y)$ holds. Then the relations $a \prec y$ and $d(a, x) + d(x, y) = d(a, y)$ imply (as above) that $a \prec x \prec y$.

We have, therefore, proved the following result.

THEOREM 3.20   *A segment S_a^b is a chain if and only if its end elements a, b are comparable.*

An answer to the second of the two questions posed above is provided by the following theorem.

THEOREM 3.21   *A motion $F(x) = d[d(a, \alpha), x]$ that sends α into a, carries a chain C_α^β (not necessarily maximal), with first element α, into a chain C_a^b, with first element a, if and only if for all elements x of C_α^β the product ax' is constant.*

Proof.   Clearly $F(C_\alpha^\beta)$ is a chain C_a^b, with first element $F(\alpha) = a$ and last element $F(\beta) = b$, if and only if $F(x)$ is order-preserving on C_α^β; that is, if and only if $x, y \in C_\alpha^\beta$ ($x \neq y$), then $x \prec y$ implies $F(x) \prec F(y)$.

An easy computation shows that $F(x) \prec F(y)$ if and only if

$$(*) \qquad\qquad a\alpha + a'\alpha'x + ay' = a\alpha + a'\alpha'x + ax'.$$

Multiplication of (*) by x', taking into account $\alpha \prec x \prec y$ and relations easily derived from it, yields $ay' = ax'$, which establishes the necessity. But if the last equality holds, then so does (*), and the proof of the theorem is complete.

COROLLARY   *The motion $F(x) = d(\alpha, x)$ carries a chain C_α^β into a chain $C_0^{d(\alpha, \beta)}$.*

THEOREM 3.22   *If a motion $F = d[d(\alpha, a), x]$ carries a maximal chain Γ_α^β into a chain C_a^b, then C_a^b is a maximal chain between a and b.*

Proof.   If C_a^b is not maximal between a and b, then an element y of the space exists such that the set union $C_a^b \cup \{y\}$ is a chain with end elements a, b, and $y \notin C_a^b$. Since each motion is involutory, $F(C_a^b \cup \{y\}) = \Gamma_\alpha^\beta \cup \{F(y)\}$.

Now $a \prec y \prec b$ implies $b' \prec y' \prec a'$, and consequently $\alpha b' \prec \alpha y' \prec \alpha a'$. Since $F(C_a^b) = \Gamma_\alpha^\beta$, a chain, then by the preceding theorem, $\alpha x'$ is constant for all elements x of C_a^b; that is, $\alpha b' = \alpha y' = \alpha a'$, and it follows that $\alpha x'$ is constant for all elements of the chain $C_a^b \cup \{y\}$. Consequently, $\Gamma_\alpha^\beta \cup \{F(y)\}$ is a chain.

Clearly $\alpha \prec F(y) \prec \beta$, for since $\mathscr{L}(a, y, b)$ holds, then so does $\mathscr{L}(\alpha, F(y), \beta)$, and $\alpha \prec \beta$ implies the desired relation (see the argument preceding the proof of Theorem 3.20). Hence $\Gamma_\alpha^\beta \cup \{F(y)\}$ is a chain *with end elements* α, β, and since Γ_α^β is a maximal such chain, then $F(y) \in \Gamma_\alpha^\beta$. It follows that $y \in C_a^b$, in contradiction to a previous statement.

THEOREM 3.23   *A subset S of $\mathbf{B}$ containing distinct elements a, b is a segment S_a^b (with the first element a and last element b) if and only if the motion $F(x) = d(a, x)$ carries S into a maximal chain $\Gamma_0^{d(a, b)}$.*

Proof.   If $F(S) = \Gamma_0^{d(a, b)}$, $F(a) = 0$, $F(b) = d(a, b)$, then $S = F(\Gamma_0^{d(a, b)})$, with $F(0) = a$, $F(d(a, b)) = b$, and $S = S_a^b$ is a segment with first element a and last element b by Theorem 3.16.

Conversely, if $S = S_a^b$, then a motion $G(x)$ and a maximal chain Γ_α^β exist such that $G(\Gamma_\alpha^\beta) = S_a^b$, with $G(\alpha) = a$, $G(\beta) = b$. According to the previous theorem and the corollary preceding it, the motion $H(x) = d(\alpha, x)$ carries the maximal chain Γ_α^β into the maximal chain $\Gamma_0^{d(\alpha, \beta)}$. Considering the product $F = GH$ we have

$$F(\Gamma_0^{d(a, b)}) = GH(\Gamma_0^{d(\alpha, \beta)}) = G(\Gamma_\alpha^\beta) = S_a^b,$$

with $F(0) = a$. Consequently, the motion F is defined by $F(x) = d(a, x)$, and the theorem is proved.

We can now show that the *orientation* of a segment S_a^b is an intrinsic property if and only if $S_a^b \neq \{a, b\}$. Suppose $S_a^b = F(\Gamma_0^{d(\alpha, \beta)})$, with $F(0) = a$, and also $S_a^b = G(\Gamma_0^{d(a, b)})$, with $G(0) = b$. Then $G = HF$, where H is the motion that sends a into b; that is $G(\Gamma_0^{d(a,\,b)}) = {}^*S_b^a$ which, as we have seen, has nothing in common with S_a^b except its end elements. Hence the orientation of a segment can be changed if and only if the segment consists of its two end elements.

Let S_a^b denote a segment and $\Gamma_0^{d(a,\,b)} = F(S_a^b)$, where F is the motion $F(x) = d(a, x)$. An m-tuple $E = \{p_1, p_2, \ldots, p_m\}$ of S_a^b is said to be in *normal order* provided

$$F(p_1) \prec F(p_2) \prec \cdots \prec F(p_m),$$

and the *length* $\lambda(E)$ of a normally ordered m-tuple E is given by

$$\begin{aligned}
\lambda(E) &= d(p_1, p_2) + d(p_2, p_3) + \cdots + d(p_{m-1}, p_m) \\
&= d[F(p_1), F(p_2)] + d[F(p_2), F(p_3)] + \cdots + d[F(p_{m-1}), F(p_m)] \\
&= d[F(p_1), F(p_m)] \\
&= d(p_1, p_m).
\end{aligned}$$

DEFINITION   The length $\lambda(S_a^b)$ of a segment S_a^b is the least upper bound (lattice sum) of the lengths of all normally ordered (finite) subsets of S_a^b.

THEOREM 3.24   *If S_a^b denotes a segment, $\lambda(S_a^b) = d(a, b)$.*

Proof.   If $E = \{p_1, p_2, \ldots, p_m)$ is any normally ordered subset of S_a^b, then

$$\lambda(E) = d(p_1, p_m) \prec d(a, p_1) + d(p_1, p_m) + d(p_m, b) = d(a, b).$$

Since the set $\{a, b\}$ is itself a normally ordered subset of S_a^b, then l.u.b. $\underset{E \subset S_a^b}{\lambda(E)} = d(a, b)$. and consequently $\lambda(S_a^b) = d(a, b)$.

Thus, Boolean segments share with metric segments the important property of having their lengths equal to the distance of their end points (see Section 9.5). But although this property characterizes metric segments among metric arcs, we shall soon see that it does *not* characterize Boolean segments among Boolean arcs; that is, a Boolean arc whose length equals the distance of its end elements is not necessarily a segment.

THEOREM 3.25   *If S_a^b, S_b^c are segments and $\mathscr{L}(a, b, c)$ holds, then $S_a^b \cup S_b^c$ is a segment S_a^c.*

Proof. We show first that $S_a^b \cap S_b^c = \{b\}$. If $x \in S_a^b \cap S_b^c$, $x \neq b$, then $x \neq a$, since $a \in S_b^c$ implies $\mathcal{L}(b, a, c)$, contradicting $\mathcal{L}(a, b, c)$. Similarly, it is seen that $x \neq c$. Then $x \in S_a^b$ yields $\mathcal{L}(a, x, b)$ which, with $\mathcal{L}(a, b, c)$, gives $\mathcal{L}(x, b, c)$ in contradition to $\mathcal{L}(b, x, c)$ which holds, since $x \in S_b^c$ and $b \neq x \neq c$.

By Theorem 3.23, the motion $F(x) = d(a, x)$ carries S_a^b into a maximal chain $\Gamma_0^{d(a, b)}$, and it carries S_b^c into $F(S_b^c)$ which has only the element $d(a, b)$ in common with $\Gamma_0^{d(a, b)}$. Now if $x, y \in S_b^c (b, c \neq x \neq y \neq b, c)$, then $x^* = F(x)$, $y^* = F(y)$ are distinct elements of $F(S_b^c)$, each distinct from $F(b)$, $F(c)$. Then $\mathcal{L}(a, b, x)$ and $\mathcal{L}(a, x, c)$ (consequences of $\mathcal{L}(a, b, c)$ and $\mathcal{L}(b, x, c)$), imply $\mathcal{L}(0, F(b), F(x))$ and $\mathcal{L}(0, F(x), F(c))$, and consequently $F(b) \prec x^* \prec F(c)$. Similarly, $F(b) \prec y^* \prec F(c)$.

Clearly, either $\mathcal{L}(b, x, y)$ or $\mathcal{L}(b, y, x)$ subsists. If the first alternative holds, then so does $\mathcal{L}(F(b), x^*, y^*)$; that is, $F(b)x^* + x^* y^* = x^* = (F(b) + x^*) \times (x^* + y^*)$. But since $F(b) \prec x^*$ and $F(b) \prec y^*$, then $F(b)x^* \prec x^* y^*$, and it follows that $x^* y^* = x^*$; that is, $x^* \prec y^*$.

In the same way, the relation $y^* \prec x^*$ follows from the second alternative, $\mathcal{L}(b, y, x)$.

We have proved that $F(S_b^c)$ is a chain $C_{F(b)}^{F(c)}$, with first element $F(b) = d(a, b)$ and last element $F(c) = d(a, c)$.

Now since S_b^c is a segment, a motion G and a maximal chain Γ_β^γ exists with $G(S_b^c) = \Gamma_\beta^\gamma$, $G(b) = \beta$, $G(c) = \gamma$. Then the motion FG carries the maximal chain Γ_β^γ into the chain $C_{F(b)}^{F(c)}$ which is, by Theorem 3.22, a maximal chain $\Gamma_{d(a, b)}^{d(a, c)}$ between its end elements. It follows that the set union $\Gamma_0^{d(a, b)} \cup \Gamma_{d(a, b)}^{d(a, c)}$ is a maximal chain $\Gamma_0^{d(a, c)}$, and consequently $S_a^b \cup S_b^c$ is a segment.

COROLLARY   *If $a, b, c \in \mathbf{B}$ and $\mathcal{L}(a, b, c)$ holds, then a segment S_a^c exists that contains b.*

THEOREM 3.26   *If each three elements of an m-tuple are linear ($\mathcal{L}$-linear or $\mathcal{D}$-linear) and $m > 4$, then the m-tuple is contained in a segment.*

Proof. By Theorem 3.15 the m-tuple is $\mathcal{L}$-linear and so a labeling $p_1, p_2, \ldots,$ p_m of the elements of the m-tuple exists such that $\mathcal{L}(p_1, p_2, \ldots, p_m)$ holds. Applying the preceding theorem $m - 2$ times establishes that $S_{p_1}^{p_2} \cup S_{p_2}^{p_3} \cup \cdots$ $\cup S_{p_{m-1}}^{p_m}$ is a segment $S_{p_1}^{p_m}$ that contains the given m-tuple.

Since a pseudo-$\mathcal{L}$-linear quadruple is not contained in any segment (part (i) of Theorem 3.19), the restriction $m > 4$ is indispensable.

EXERCISES

1. The concepts of $\mathscr{L}$-linearity and $\mathscr{D}$-linearity have been defined for finite subsets of a space **B**. How may these notions be defined for arbitrary subsets of **B**?

2. Give an example of a $\mathscr{D}$-linear quadruple that is not contained in a segment.

3. If the Boolean algebra forming the ground set of a Boolean space **B** is that of the set of all subsets of the euclidean plane, interpret all of the notions defined in the preceding sections of this chapter. In particular, how do pseudo-$\mathscr{L}$-linear quadruples appear in this context?

3.7   CHARACTERIZATION OF THE BOOLEAN METRIC

The mapping $d(x, y) = xy' + x'y$ of ordered pairs of elements of **B** *onto* **B** has, as we have seen, all of the formal properties of a metric. (It differs from an ordinary metric only in the respect that $d(x, y)$ is an element of a Boolean algebra $\mathscr{B}$ instead of being a nonnegative real number.) The question arises: how many essentially different Boolean metrics can be defined in a Boolean algebra?

In exercise 2 of Section 3.2 the reader was asked to show that the Boolean metric $d(x, y)$ is a group composition over the Boolean algebra $\mathscr{B}$. We prove in the following theorem that it is the *only* Boolean metric with that property.

THEOREM 3.27 *Let $\mathscr{B}$ denote a Boolean algebra and g any mapping on $\mathscr{B} \times \mathscr{B}$ to $\mathscr{B}$ with respect to which $\mathscr{B}$ is a group. If g has the formal properties of a metric—that is, $g(x, y) = 0$ if and only if $x = y$, $g(x, y) = g(y, x)$ and $g(x, z) \prec g(x, y) + g(y, z)$ $(x, y, z \in \mathscr{B})$—then $g(x, y) = xy' + x'y = d(x, y)$ for all elements x, y of $\mathscr{B}$.*

Proof.   If e denotes the element of $\mathscr{B}$ that is the unit element of the group, then $g(0, e) = 0$, and consequently $e = 0$; that is, 0 is the unit element of the group, and so $g(0, x) = x$ for every element x of $\mathscr{B}$. Applying the "triangle inequality" to the elements $0, x, x'$ gives $g(0, x) \prec g(0, x') + g(x, x')$, whence $x \prec x' + g(x, x')$. From the preceding relation, $x \prec xg(x, x')$ and since $xg(x, x') \prec x$, then $x = xg(x, x')$ and it follows that $x \prec g(x, x')$. In a similar manner, it is seen that $x' \prec g(x, x')$, and hence $1 \prec g(x, x')$; that is, $g(x, x') = 1 = g(x', x)$.

Using now the associativity of the group composition g we have

$$g(x, 1) = g(x, g(x, x')) = g(g(x, x), x') = g(0, x') = x',$$

and from that result follows

$$g(x, y') = g(x, g(y, 1)) = g(g(x, y), 1) = g'(x, y).$$

Hence $g(x', y') = g'(x', y) = g'(y, x') = g(y, x) = g(x, y)$. (Note that from the symmetric property of a metric, $\mathscr{B}$ is an abelian group with respect to g.)

From the triangle inequality applied to $0, x, y$ and to $0, x', y'$, we get $g(x, y) \prec x + y$ and $g(x', y') \prec x' + y'$. But $g(x, y) = g(x', y')$ and consequently $g(x, y) \prec (x + y)(x' + y') = d(x, y)$.

Then

$$
\begin{aligned}
d(g(x, y), d(x, y)) &= d(x, y)g'(x, y) + d'(x, y)g(x, y) \\
&= d(x, y)g'(x, y) + 0 \\
&= d(x, y)g(x, y') \\
&\prec d(x, y)d(x, y') \\
&= d(x, y)d'(x, y) \\
&= 0,
\end{aligned}
$$

and hence $g(x, y) = d(x, y)$.

It is of interest to point out certain subgroups of $\mathscr{B}$ (with respect to the group composition d). First, the set $\{0, x\}$ ($x \in \mathscr{B}$) is a subgroup of order 2, since $d(0, x) = x$ and $d(x, x) = 0$.

If $a, b \in \mathscr{B}, 0 \neq a \neq b \neq 0$, the four elements $0, a, b, d(a, b)$ form a subgroup of order 4 (let the reader show that $a \neq d(a, b) \neq b$), and if $c \in \mathscr{B}$, distinct from each element of this subgroup, then

$$\{0, a, b, c, d(a, b), d(a, c), d(b, c), d(c, d(a, b))\}$$

is a subgroup of order 8. These eight elements are the vertices of a " rectangular parallelopiped " in $\mathscr{B}$ since, for example,

$$c = d(a, d(a, c)) = d(b, d(b, c)) = d(d(a, b), d(c, d(a, b))),$$

with similar relations holding between the 24 remaining distances (that is, the 28 distances determined by the eight elements fall into 7 sets of 4 mutually equal distances).

More generally, if G is any subgroup of $\mathscr{B}$ and $x \in \mathscr{B}$, the set union $G \cup G^*$, where $G^* = [d(x, g) | g \in G]$, is a subgroup of $\mathscr{B}$. The set G^* is the translate of the group G through the distance x. Clearly, each subgroup of $\mathscr{B}$ has order 2^n (n finite or infinite) and has the geometrical structure described above for the case $n = 3$.

EXERCISES

1. Show that $d(x, y)$ is the only weakly associative binary operation definable in the Boolean algebra $\mathscr{B}$.

2. If A, B, C, D are fixed elements of a Boolean algebra $\mathscr{B}$, the function $f(x, y) = Axy + Bxy' + Cx'y + Dx'y'$ is a Boolean function of x, $y(x, y \in \mathscr{B})$. Show that $d(x, y)$ is the only Boolean function of x, y that equals 0 if and only if $x = y$.

3.8   CONVEX BOOLEAN METRIC SPACE.
AN IMBEDDING THEOREM

We call the Boolean metric space $\mathbf{B} = \{\mathscr{B}; d\}$ *convex* provided $a, c \in \mathscr{B}, a \neq c$, implies the existence of an element b of $\mathbf{B}$ such that $\mathscr{L}(a, b, c)$ holds (that is, $a \neq b \neq c$ and $d(a, b) + d(b, c) = d(a, c)$). An element x of $\mathbf{B}$ is an atom provided $0 \prec y \prec x$ implies $0 = y$ or $y = x$, where 0 denotes the first element of $\mathscr{B}$.

THEOREM 3.28   *A Boolean metric space $\mathbf{B} = \{\mathscr{B}; d\}$ is convex if and only if the Boolean algebra $\mathscr{B}$ is atom-free (that is, if $x \in \mathscr{B}, 0 \neq x$, there exists an element y of $\mathscr{B}$ such that $0 \neq y \neq x$ and $0 \prec y \prec x$).*

Proof.   Suppose $\mathbf{B}$ is convex and let c denote an element of $\mathbf{B}$ distinct from 0. Then there exists an element b of $\mathbf{B}$ such that $0 \neq b \neq c$ and $d(0, b) + d(b, c) = d(0, c)$. From the last equality, $b + b'c = c$, and so $0 \prec b \prec c$ which, together with $0 \neq b \neq c$, shows that c is not an atom.

On the other hand, let a, c be distinct elements of a Boolean metric space $\mathbf{B} = \{\mathscr{B}; d\}$ with $\mathscr{B}$ atom-free. The motion $f(x) = d(x, a)$ carries a into 0 and c into $f(c) = d(c, a) \neq 0$, Now $\mathscr{B}$ contains an element b such that $0 \prec b \prec f(c)$. $0 \neq b \neq f(c)$. and hence $\mathscr{L}(0, b, f(c))$ holds. Since betweenness is a congruence invariant, then $\mathscr{L}(f(0), f(b), f(f(c)))$ holds; that is, $\mathscr{L}(a, f(b), c)$ subsists, with $a \neq f(b) \neq c$, and so $\mathbf{B}$ is convex.

Remark.   Since for every element x of $\mathbf{B}$, $0 \prec x \prec 1$, then $\mathscr{L}(0, x, 1)$ holds for each x distinct from 0, 1. Hence neither $\mathscr{L}(0, 1, x)$ nor $\mathscr{L}(x, 0, 1)$ subsists for any element x of $\mathbf{B}$. The relation $\mathscr{L}(a, 0, b)$ is valid if and only if $ab = 0$, $a \neq 0 \neq b$, and $\mathscr{L}(a, 1, b)$ holds if and only if $a + b = 1$, $a \neq 1 \neq b$. Finally, it is clear that $\mathscr{L}(a, x, a')$ subsists for every element x of $\mathbf{B}$ distinct from a, a'.

We have proved that every normed lattice has a convex extension (Theorem 2.15). The argument used to establish that theorem may be applied (*mutatis mutandis*) to establish the following theorem.

THEOREM 3.29 *Each Boolean metric space* $\mathbf{B} = \{\mathscr{B}; d\}$ *may be extended to a convex Boolean space whose underlying Boolean algebra contains* $\mathscr{B}$ *as a subalgebra.*

Proof. In view of the observation that immediately precedes this theorem, a sketch of the proof will suffice.

First, it is clear that the set of all ordered pairs $[x, y]$ of elements x, y of $\mathscr{B}$ forms a Boolean metric space $\mathbf{B}_1$, with product, sum, complementation, and distance defined as follows:

$$[x, y] \cdot [u, v] = [xu, yv],$$
$$[x, y] + [u, v] = [x + u, y + v],$$
$$[x, y]' = [x', y'],$$
$$d([x, y], [u, v]) = [x, y]' \cdot [u, v] + [x, y] \cdot [u, v]'$$
$$= [d(x, u), d(y, v)].$$

Identifying the element x *of* $\mathbf{B}$ *with the element* $[x, x]$ *of* $\mathbf{B}_1$ *imbeds* $\mathbf{B}$ *isomorphically (and hence congruently) in* $\mathbf{B}_1$.

Repetitions of this procedure yield an infinite sequence

$$\mathbf{B}, \mathbf{B}_1, \mathbf{B}_2, \ldots, \mathbf{B}_n, \ldots$$

of Boolean metric spaces, each of which is an extension of each space preceding it in the sequence. Let $\mathbf{B}^*$ denote the set-sum of all spaces in the sequence, with addition, multiplication, complementation, and distance defined for elements of $\mathbf{B}^*$ in the same way in which those operations were defined in the $\mathbf{B}_k$ of smallest index k that contains them. Then $\mathbf{B}^*$ is a Boolean metric space, $\mathbf{B}^* = \{\mathscr{B}^*; d^*\}$, with $\mathscr{B}$ a subalgebra of $\mathscr{B}^*$, the ground set of $\mathbf{B}^*$, and $\mathbf{B}^*$ is an extension of $\mathbf{B}$. It remains to show that $\mathbf{B}^*$ is convex.

If $\alpha, \beta \in \mathbf{B}^*$, $\alpha \neq \beta$, consider the element $[\alpha, \beta]$ of $\mathbf{B}^*$. Clearly, $\alpha = [\alpha, \alpha] \neq [\alpha, \beta] \neq [\beta, \beta] = \beta$, and

$$d^*(\alpha, [\alpha, \beta]) = d([\alpha, \alpha], [\alpha, \beta]) = [d(\alpha, \alpha), d(\alpha, \beta)] = [0, d(\alpha, \beta)],$$
$$d^*([\alpha, \beta], \beta) = d([\alpha, \beta], [\beta, \beta]) = [d(\alpha, \beta), d(\beta, \beta)] = [d(\alpha, \beta), 0].$$

Consequently,

$$d^*(\alpha, [\alpha, \beta]) + d^*([\alpha, \beta], \beta) = [d(\alpha, \beta), d(\alpha, \beta)] = d(\alpha, \beta) = d^*(\alpha, \beta).$$

Hence the space **B*** is convex, and the theorem is proved.

The following purely algebraic theorem is an immediate corollary.

COROLLARY   *Every Boolean algebra is a subalgebra of an atom-free Boolean algebra.*

Proof.   Since **B*** is convex, its underlying Boolean algebra is atom-free (Theorem 3.28).

EXERCISES

1. In terms of the sequential topology defined in Section 2.15, show that if $\mathscr{B}$ is σ-complete, each of the Boolean operations is continuous; that is, if $\{x_n\}$, $\{y_n\}$ are infinite sequences of elements of $\mathscr{B}$, and $x, y \in \mathscr{B}$ such that $\lim_{n\to\infty} x_n = x$, $\lim_{n\to\infty} y_n = y$, then $\lim_{n\to\infty} x_n y_n = xy$, $\lim_{n\to\infty} x_n + y_n = x + y$, $\lim_{n\to\infty} x'_n = x'$.

2. Show that the distance function $d(x.y)$ in **B** is continuous (with respect to the sequential topology) and that the sequential topology is a metric topology (that is, $\lim_{n\to\infty} x_n = x$ if and only if $\lim_{n\to\infty} d(x_n, x) = 0$.).

3. Prove that each maximal chain of a convex Boolean metric space **B** is convex (that is, if c, d are distinct elements of a maximal chain Γ, then Γ contains an element p such that $\mathscr{L}(c, p, d)$ subsists).

4. Let $\mathbf{B} = \{\mathscr{B}; d\}$ be a Boolean metric space over the *complete* Boolean algebra $\mathscr{B}$ (that is, $\mathscr{B}$ contains a greatest lower bound $\prod_{x \in X} x$ and a least upper bound $\sum_{x \in X} x$ for every subset X of $\mathscr{B}$). Show that each maximal chain of **B** is complete.

   Hint: Show that, if $X \subset \Gamma(a, b)$, a maximal chain, then $\prod X = \prod_{x \in X} x$ is comparable with each element of $\Gamma(a, b)$.

3.9   SOME TOPOLOGICAL PROPERTIES OF MAXIMAL CHAINS

In the remaining sections of this chapter, it is assumed that $\mathbf{B} = \{\mathscr{B}; d\}$ is a convex Boolean metric space over the complete Boolean algebra $\mathscr{B}$ (Section 2.15), with the metric topology induced by the distance function d (Exercise 2,

Section 3.8). The space **B** is a Fréchet **L**-space (Section 2.15), and the usual topological notions are readily defined. For example, an element p is an accumulation element of a subset E of **B** provided E contains an infinite sequence $\{p_n\}$ of elements with $p_n \neq p$ $(n = 1, 2, \ldots)$, and $\lim_{n \to \infty} p_n = p$; a subset E of **B** is *closed* if it contains all of its accumulation elements, and *compact* provided every infinite subset of E has an accumulation element (not necessarily contained in E).

THEOREM 3.30   *Each complete and convex chain of* **B** *is maximal.*

Proof.   Let $C(a, b)$ denote a complete and convex chain of **B**. We show that $C(a, b)$ is not a proper subset of any chain $C^*(a, b)$ with first element a and last element b.

Assuming the contrary, then $C(a, b) \subset C^*(a, b)$, and $C^*(a, b)$ contains an element x such that (i) $x \notin C(a, b)$, (ii) $a \prec x \prec b$, and (iii) x is comparable with every element of $C(a, b)$.

Putting

$$C(a, x) = [p \in C(a, b) \,|\, p \prec x],$$
$$C(x, b) = [p \in C(a, b) \,|\, x \prec p],$$

it follows that $C(a, b) = C(a, x) \cup C(x, b)$.

Since $C(a, b)$ is complete $\sum C(a, x)$ and $\prod C(x, b)$ are elements of $C(a, b)$, and clearly

$$\sum C(a, x) \prec x \prec \prod C(x, b).$$

The contradiction we are seeking is obtained by showing that $\sum C(a, x) = \prod C(x, b)$.

If the above equality does not hold, then the convexity of $C(a, b)$ yields $y \in C(a, b)$ such that $\sum C(a, x) \prec y \prec \prod C(x, b)$ and $\sum C(a, x) \neq y \neq \prod C(x, b)$. But this implies $y \notin C(a, x)$ and $y \notin C(x, b)$; that is, $y \notin C(a, b)$, and the theorem is proved.

COROLLARY   *A chain of a complete, convex Boolean metric space* **B** *is maximal if and only if it is complete and convex.*

Proof.   The preceding theorem is combined with the results stated in Exercises 3 and 4 of Section 3.8 to yield the corollary.

THEOREM 3.31   *A maximal chain of a complete and convex Boolean metric space* **B** *is closed and compact.*

Proof. If x is an accumulation element of a maximal chain $\Gamma(a, b)$ and $x = \lim_{n \to \infty} x_n$, $x_n \in \Gamma(a, b)$ $(n = 1, 2, \ldots)$, then x_n is comparable with every element of $\Gamma(a, b)$. Hence for each element c of $C(a, b)$ either $cx_n = c$ or $cx_n = x_n$, and it easily follows by continuity of "product" that $cx = c$ or $cx = x$, and so x is comparable with every element of $\Gamma(a, b)$. Since, moreover, it is clear that $a \prec x \prec b$, the relation $x \in \Gamma(a, b)$ follows from the maximal property of $\Gamma(a, b)$, and we conclude that $\Gamma(a, b)$ is closed.

To prove that $\Gamma(a, b)$ is compact, we let X denote any infinite subset of $\Gamma(a, b)$, and show the existence of a strictly monotone infinite sequence of elements of X. Since all such sequences have limits (Section 2.15), the existence of an accumulation element of X will be established. (Of course, the Axiom of Choice is employed to select such an infinite sequence.)

Select any element x_1 of X such that $a_1 = a \prec x_1 \prec b = b_1$, $a_1 \neq x_1 \neq b_1$. At least one of the subchains $\Gamma(a_1, x_1)$, $\Gamma(x_1, b_1)$ contains infinitely many elements of X. Denote one such by $\Gamma(a_2, b_2)$ and select $x_2 \in X$ such that $x_2 \in \Gamma(a_2, b_2)$, $a_2 \neq x_2 \neq b_2$. If elements $x_1, x_2, \ldots, x_n$ of X have been obtained in the above manner, then at least one of the subchains $\Gamma(a_n, x_n)$, $\Gamma(x_n, b_n)$ contains infinitely many elements of X. Denote one such by $\Gamma(a_{n+1}, b_{n+1})$ and select $x_{n+1} \in X$ such that $x_{n+1} \in \Gamma(a_{n+1}, b_{n+1})$, $a_{n+1} \neq x_{n+1} \neq b_{n+1}$. Mathematical induction then yields an infinite sequence $x_1, x_2, \ldots, x_n, \ldots$ of pairwise different elements of X.

In case a natural number N exists such that for every index $n \prec N$, $a_n \prec x_{n+1} \prec x_n$, then clearly the infinite sequence $\{x_n\}$ is strictly monotone (decreasing) after a finite number of terms, and our objective has been attained.

If, on the other hand, for each natural number N an index n exists, $n > N$, such that $x_n \prec x_{n+1} \prec b_n$, then clearly the infinite sequence $\{x_n\}$ contains a strictly monotone (increasing) subsequence, and the desired result is obtained.

THEOREM 3.32   *If* $C(a, b)$ *is a connected subchain of a maximal chain* $\Gamma(a, b)$, *then* $C(a, b) = \Gamma(a, b)$.

Proof. If $x \in \Gamma(a, b)$ and $x \notin C(a, b)$, define

$$P(x) = [y \in \Gamma(a, b) \,|\, y \prec x],$$
$$Q(x) = [y \in \Gamma(a, b) \,|\, x \prec y].$$

Now each of the sets $P(x) \cap C(a, b)$, $Q(x) \cap C(a, b)$ is nonempty and closed in $C(a, b)$ (Exercise 2). Furthermore, the two sets are disjoint and their set sum is $C(a, b)$. This violates the connectedness of $C(a, b)$ (Theorem 1.11) and establishes the theorem.

COROLLARY.   *A connected chain is maximal.*

Proof.   If $C(a, b)$ is a connected chain and $\Gamma(a, b)$ is a maximal chain containing it, then $C(a, b) = \Gamma(a, b)$ by the theorem.

EXERCISES

1. Prove that two maximal chains are homeomorphic if there exists a biuniform, order-preserving mapping of one onto the other.

2. Show that each of the sets $P(x) \cap C(a, b)$, $Q(x) \cap C(a, b)$ is closed in $C(a, b)$

3.10   BOOLEAN METRIC SPACES OVER SEPARABLE BOOLEAN ALGEBRAS

We have shown that each connected chain of a complete convex Boolean space **B** is maximal. The converse will be proved valid in a class of spaces **B** defined below.

DEFINITION   A chain C of a complete Boolean algebra $\mathscr{B}$ is *separable* provided

(i)
$$\prod C = \prod_{c \in C} c = \prod_{i=1}^{\infty} x_i, \ x_i \in C,$$

(ii)
$$\sum C = \sum_{c \in C} c = \sum_{i=1}^{\infty} y_i, \ y_i \in C \qquad (i = 1, 2, \ldots).$$

The space $\mathbf{B} = \{\mathscr{B}; d\}$ is separable provided every chain of $\mathscr{B}$ is separable.

LEMMA 3.5   *If C is a closed, separable chain of* **B**, *then* $\prod C \in C$ *and* $\sum C \in C$.

Proof.   Since C is separable, $\prod C = \prod_{i=1}^{\infty} x_i$, $x_i \in C$, and hence a monotone decreasing sequence of elements of C exists with limit $\prod C$. Since C is closed, then $\prod C \in C$. Similarly, it is seen that $\sum C \in C$.

THEOREM 3.33   *If the Boolean space* **B** *is separable, every maximal chain* $\Gamma(a, b)$ *of* **B** *is connected.*

Proof.   Suppose $\Gamma(a, b)$ is not connected, and put $\Gamma(a, b) = A \cup B$, where $A \neq \varnothing$, $B \neq \varnothing$, $A \cap B = \varnothing$, and each of the sets A, B is closed in $\Gamma(a, b)$.

Since, by Theorem 3.31, $\Gamma(a, b)$ is closed, then A and B are closed subsets of $\Gamma(a, b)$. (Let the reader verify this.) Furthermore, each of the sets A, B is a separable chain, and so, by the preceding lemma, $\prod B = \beta$, $\beta \in B$.

If $\beta \neq a$, put $A^* = [x \in \Gamma(a, b) \mid x \prec \beta, x \neq \beta]$. Then the nonnull set A^* is a closed, separable chain, and consequently $\sum A^* = \alpha$, $\alpha \in A^*$. Since $A^* \subset A$, then $\alpha \in A$.

Now if $\alpha \neq \beta$, then the convexity of $\Gamma(a, b)$ implies the existence of an element γ of $\Gamma(a, b)$ such that $\alpha \prec \gamma \prec \beta$ and $\alpha \neq \gamma \neq \beta$. But this is impossible, for since $\gamma \prec \beta$ and $\gamma \neq \beta$, then $\gamma \in A^*$ and consequently $\gamma \prec \alpha$. Hence $\alpha = \beta$ and the sets A, B are not disjoint.

The assumption $\beta \neq a$ is, then, untenable, and so $\beta = a$ and $a \in B$.

Now if $\prod A = \zeta$, then since $\zeta \in A$ and $a \in B$ it follows that $\zeta \neq a$. Putting

$$B^* = [x \in \Gamma(a, b) \mid x \prec \zeta, x \neq \zeta],$$

the nonnull set B^* is a closed, separable chain and so $\sum B^* = \eta$, $\eta \in B^*$. Clearly $\eta \prec \zeta$, and if $\eta \neq \zeta$, the convexity of $\Gamma(a, b)$ leads to a contradiction, as before. But if $\eta = \zeta$, then, since $B^* \subset B$, the product $A \cap B \neq \varnothing$, contrary to what was assumed regarding the sets A, B, and the proof is complete.

THEOREM 3.34   *If* **B** *is complete, convex, and separable, then every closed and convex chain* $C(a, b)$ *of* **B** *is maximal.*

Proof.   Since $C(a, b)$ is closed, then it is complete (Lemma 3.5), and a complete, convex chain is maximal (Theorem 3.30).

THEOREM 3.35   *In a complete, convex, separable Boolean metric space* **B** $= \{\mathscr{B}; d\}$, *a chain is maximal if and only if it is closed and convex.*

Proof.   The theorem combines the result established in Exercise 3, Section 3.8, with Theorems 3.31 and 3.34.

3.11   ARCS IN A BOOLEAN METRIC SPACE **B**

A subset $A(\alpha, \beta)$ is an *arc* of **B** provided it is a homeomorph f of a maximal chain $\Gamma(a, b)$ of **B**, with $\alpha = f(a)$ and $\beta = f(b)$. Since $\Gamma(a, b)$ is closed, compact, connected (and contains more than one element), then $\Gamma(a, b)$ is a continuum, and consequently $A(\alpha, \beta) = f[\Gamma(a, b)]$ is also a continuum (Exercise 1). Furthermore, $A(\alpha, \beta)$ is *irreducible between its end points* α, β (that is, no proper subset of $A(\alpha, \beta)$ that contains α, β is a continuum) (Exercise 2). It is clear that every maximal chain is an arc. It is assumed that the space is complete, convex, and separable.

THEOREM 3.36   *If* $\Gamma(a_1, b_1)$, $\Gamma(a_2, b_2)$ *are maximal chains and* $f[\Gamma(a_1, b_1)]$ $= \Gamma(a_2, b_2)$, *with* f *a homeomorphism and* $f(a_1) = a_2, f(b_1) = b_2$, *then* f *is order-preserving.*

Proof.   If f is not order-preserving, then $c_1, d_1 \in \Gamma(a_1, b_1)$ exist $(c_1 \neq d_1)$ such that $c_1 \prec d_1$ but $f(d_1) \prec f(c_1)$. Clearly $c_1 \neq a_1$ and $d_1 \neq b_1$.

Since the set $X = [x \in \Gamma(a_1, b_1) | f(d_1) \prec f(x)]$ contains c_1, it is not null, and since X is a closed, separable, chain, then $p = \prod X \in X$. Clearly $p \neq a_1$, and since $p \in X$, then $f(d_1) \prec f(p)$, with $f(d_1) \neq f(p)$ (for if $f(d_1) = f(p)$, then $p = d_1$, which is impossible since $c_1 \in X$ and $c_1 \prec d_1$).

If $P(p) = [x \in \Gamma(a_1, b_1) | x \prec p, x \neq p]$, then $\sum P(p) = p$ follows from the convexity of $\Gamma(a_1, b_1)$, and since $P(p)$ is separable, $p = \sum_{i=1}^{\infty} p_i$, $p_i \in P(p)$ $(i = 1, 2, \ldots)$. The sequence $\{p_i\}$ contains a monotone increasing subsequence $\{p_{n_i}\}$, with $\lim_{i \to \infty} p_{n_i} = p$. Since f is continuous in $\Gamma(a_1, b_1)$, it follows that $\lim_{i \to \infty} f(p_{n_i}) = f(p)$. Now $f(p) \succ f(d_1)$, and hence an element p_{n_k} of $\{p_{n_i}\}$ exists such that $f(p_{n_k}) \succ f(d_1)$, and $f(p_{n_k}) \neq f(d_1)$. But this implies that $p_{n_k} \in X$, which contradicts $p_{n_k} \in P(p)$, for the last relation implies $p_{n_k} \prec p$, while p (as the greatest lower bound of all elements of the set X) precedes every element of X.

COROLLARY   *If* $f[\Gamma(a, b)] = \Gamma(a, b)$, *with* f *a homeomorphism and* $f(a) = a$, $f(b) = b$, *then* f *is order-preserving.*

DEFINITION   A finite subset $P(\alpha_0, \alpha_1, \ldots, \alpha_n)$ of an arc $A(\alpha, \beta) = f[\Gamma(a, b)]$ is *normally ordered* with respect to the homeomorphism f and a maximal chain $\Gamma(a, b)$ provided $p_0 \prec p_1 \prec \cdots \prec p_n$, where $p_i \in \Gamma(a, b)$ and $f(p_i) = \alpha_i$ $(i = 0, 1, 2, \ldots, n)$.

THEOREM 3.37   *If* $P(\alpha_0, \alpha_1, \ldots, \alpha_n)$ *is a normally ordered subset of an arc* $A(\alpha, \beta)$ *with respect to a homeomorphism* f *and a maximal chain* $\Gamma(a, b)$, *then* $P(\alpha_0, \alpha_1, \ldots, \alpha_n)$ *is also normally ordered with respect to any homeomorphism* g *and maximal chain* $\Gamma(c, d)$, *where* $A(\alpha, \beta) = g[\Gamma(c, d)]$, $g(c) = \alpha$, $g(d) = \beta$.

The proof is obtained as an easy application of Theorem 3.36.

It follows that the property of being a normally ordered subset of an arc is an intrinsic property of the subset—it is quite independent of any particular homeomorphism relating the arc to any particular maximal chain. This circumstance gives rise to the following definition of arc length as an intrinsic property of an arc.

DEFINITION   If $A(\alpha, \beta)$ is an arc, its *length* $\lambda(A)$ is defined by

$$\lambda(A) = \sum \lambda(P),$$

where the summation is extended over all finite, normally ordered subsets $P = \{\alpha_0, \alpha_1, \ldots, \alpha_n\}$ of A, and

$$\lambda(P) = d(\alpha_0, \alpha_1) + d(\alpha_1, \alpha_2) + \cdots + d(\alpha_{n-1}, \alpha_n).$$

Clearly, the length of an arc is an *intrinsic* property of the arc. It is independent of both the homeomorphism f and the maximal chain $\Gamma(a, b)$ that give rise to the arc.

DEFINITIONS   Any set of unordered pairs of elements of a subset X of **B** is called a *duplex* $\mathfrak{D}$ of X. A finite sequence $x_1, x_2, \ldots, x_n$ of elements of X *belongs to* a duplex $\mathfrak{D}$ of X provided for each $i = 1, 2, \ldots, n$, the pair $(x_i, x_{i+1}) \in \mathfrak{D}$. The sequence $x_1, x_2, \ldots, x_n (x_i \in X)$ *connects* p *with* q $(p, q \in X)$ provided $x_1 = p$ and $x_n = q$; and X is *connected with respect to a duplex* $\mathfrak{D}$ of X provided each two elements p, q of X are connected by a finite sequence of elements belonging to the duplex $\mathfrak{D}$. The *linear content* $\gamma(X)$ of a subset X of **B** is defined by

$$\gamma(X) = \sum d(a, b),$$

the summation extending over all pairs of elements a, b of X. The *duplex content* of a duplex $\mathfrak{D}$ is $\delta(\mathfrak{D}) = \sum_{(a, b) \in \mathfrak{D}} d(a, b)$.

THEOREM 3.38   *The linear content of a subset* X *of* **B** *equals the duplex content of any duplex* $\mathfrak{D}$ *of* X *with respect to which* X *is connected.*

Proof. It is clear that $\delta(\mathfrak{D}) \prec \gamma(X)$. If $p, q \in X$ and $x_1, x_2, \ldots, x_n$ is a sequence of elements that belongs to $\mathfrak{D}$ that connects $p = x_1$ and $q = x_n$, repeated use of the triangle inequality yields

$$d(p, q) \prec d(p, x_2) + d(x_2, x_3) + \cdots + d(x_{n-1}, q) \prec \delta(\mathfrak{D}).$$

Hence $\gamma(X) = \sum_{p, q \in X} d(p, q) \prec \delta(\mathfrak{D})$, and so $\delta(\mathfrak{D}) = \gamma(X)$.

Remark. If $X = \{x_1, x_2, \ldots, x_n\}$ is any finite subset of elements of **B** $(n > 1)$ a *polygon* $P_\pi(X)$ *over* X is the ordered set $(x_{i_1}, x_{i_2}, \ldots, x_{i_n})$, where $\pi = i_1, i_2, \ldots, i_n$ is a permutation of the indices $1, 2, \ldots, n$. Clearly, the number of polygons over X equals the number of permutations of the elements of X, namely $n!$, and X is connected with respect to any duplex that consists of all consecutive pairs of elements in any one polygon $P_\pi(X)$ over X. It follows from the preceding theorem that $\gamma(X) = \delta[P_\pi(X)]$.

THEOREM 3.39   *If* $A(\alpha. \beta)$ *denotes any arc of* **B**, $\lambda(A) = \gamma(A)$.

Proof. Noting that a normally ordered subset of n elements of $A(\alpha, \beta)$ $(n > 1)$, is a polygon P over the set X of its vertices, then by the previous theorem $\lambda(P) = \delta(P) = \gamma(X)$. Then $\lambda(A) = \sum \lambda(P) = \sum_{X \subset A} \gamma(X)$, the summation extended over all *finite* subsets X of $A(\alpha, \beta)$ containing at least two elements. Clearly $\gamma(X) \prec \sum_{x, y \in A} d(x, y)$ and consequently

$$\sum_{X \subset A} \gamma(X)' \prec \sum_{x, y \in A} d(x, y).$$

For $x, y \in A(\alpha, \beta)$,

$$d(x, y) \prec \gamma(X) \prec \sum_{X \subset A} \gamma(X),$$

and so $\sum_{x, y \in A} d(x, y) \prec \sum_{X \subset A} \gamma(X)$.

It follows that

$$\gamma(A) = \sum_{x, y \in A} d(x, y) = \sum_{X \subset A} \gamma(X) = \lambda(A).$$

THEOREM 3.40   *If* p *is any element of an arc* A *of* **B**, *then* $\lambda(A) = \sum_{x \in A} d(p, x)$.

Proof. The arc A is connected with respect to the duplex $\mathfrak{D}$ of all pairs (p, x), $x \in A$. Hence

$$\lambda(A) = \gamma(A) = \delta(\mathfrak{D}) = \sum_{x \in A} d(p, x).$$

COROLLARY $\lambda[A(\alpha, \beta)] = \sum_{x \in A} d(\alpha, x)$.

EXERCISES

1. Show that a homeomorph of a continuum of **B** is a continuum (compare Corollary, Theorem 1.14).

2. Prove that an arc is an irreducible continuum between its end points.

3. Show that arc length is a congruence invariant.

4. Show that arc length is additive; that is, if $\gamma \in A(\alpha, \beta)$, then

$$\lambda[A(\alpha, \beta)] = \lambda[A(\alpha, \gamma)] + \lambda[A(\gamma, \beta)].$$

3.12   FURTHER STUDY OF SEGMENTS

We have seen that maximal chains of a complete, convex, separable Boolean metric space **B** are convex, closed, compact, and connected. Since a segment is a congruent image of a maximal chain, it follows that each segment is convex, and since each congruence is a homeomorphism, each segment S_α^β is closed, compact, and connected. Also, $\lambda[S_\alpha^\beta] = d(\alpha, \beta)$; that is, the length of a segment is the distance of its endpoints. We shall see that this property does *not* characterize segments among arcs, but we shall show that the property of being convex does. Several lemmas point the way to that characterization.

LEMMA 3.6   *If F denotes any nonempty closed subset of* **B**, *and if F contains for each element* $x(x \neq 0)$ *an element* x^* *such that* $0 \prec x^* \prec x$, $0 \neq x^* \neq x$, *then* $0 \in F$.

Proof.   If $a \in F$, let $\Gamma(0, a)$ denote a maximal chain in $F^* = F \cup \{0\}$; that is, $\Gamma(0, a)$ is an ordered subset of F^* that contains each element of F^* that is comparable with every element of $\Gamma(0, a)$, and $0 \prec x \prec a$ for every $x \in \Gamma(0, a)$. Let Γ' denote the set obtained by deleting the element 0 from $\Gamma(0, a)$. Now if $0 \notin F$, then Γ' is closed in F, and since F is closed, so is Γ'.

Since Γ' is an ordered, closed subset of a separable Boolean space **B**, Γ' is complete, and so $\Pi\Gamma' = b$, $b \in \Gamma'$. But $\Gamma' \subset F$, and consequently $b \in F$. It follows that $b \neq 0$, and according to the hypothesis, F contains an element b^* such that $0 \prec b^* \prec b$, $0 \neq b^* \neq b$.

Clearly, b^* is comparable with every element of $\Gamma(0, a)$ and so $b^* \in \Gamma(0, a)$. But this is impossible, since $b^* \neq 0$ and $b^* \prec b = \Pi\Gamma'$, $b^* \neq b$. Hence $0 \notin F$ is untenable, so the lemma is proved.

COROLLARY *If $p \in \mathbf{B}$ and F is a nonempty, closed subset of $\mathbf{B}$ such that $q \in F$, $q \neq p$, implies the existence of $q^* \in F$ such that $\mathscr{L}(p, q^*, q)$ holds, then $p \in F$.*

Proof. Consider the motion $f(x) = d(p, x)$, $x \in \mathbf{B}$, which carries p into 0, and the set F into $f(F)$. Since the betweenness relation is a congruence invariant, it follows that the nonempty closed set $f(F)$ contains for each of its elements x an element x^* such that $\mathscr{L}(0, x^*, x)$ holds (that is $0 \prec x^* \prec x$, $0 \neq x^* \neq x$). Then, by the previous theorem, $0 \in f(F)$ and consequently $p \in f[f(F)] = F$.

LEMMA 3.7   *A closed and convex subset F of B is connected.*

Proof. Assuming the contrary, write $F = X \cup Y$, where X, Y are closed, nonnull, disjoint sets.

If $x_0 \in X$ there exists an element y_0 of Y such that no element of Y is between x_0 and y_0 (in the strict sense) (Corollary, Lemma 3.6). Hence the set

$$Z^*(x_0, y_0) = [p \in F \,|\, p = x_0 \quad \text{or} \quad \mathscr{L}(x_0, p, y_0)]$$

is a nonempty subset of X, and $Z^*(x_0, y_0)$ is, moreover, *closed*, since $Z^*(x_0, y_0) \cup \{y_0\}$ is closed and y_0 is clearly not an accumulation element of the latter set.

Now for each element p of $Z^*(x_0, y_0)$, there exists $p^* \in Z^*(x_0, y_0)$ such that $\mathscr{L}(y_0, p^*, p)$ holds. For since F is convex it contains an element p^* such that $\mathscr{L}(p, p^*, y_0)$ subsists, and that relation together with $\mathscr{L}(x_0, p, y_0)$ implies $\mathscr{L}(x_0, p^*, y_0)$ (hence $p^* \in Z^*(x_0, y_0)$) and $\mathscr{L}(y_0, p^*, p)$ in case $x_0 \neq p$. The existence of p^*, $p^* \in Z^*(x_0, y_0)$, such that $\mathscr{L}(y_0, p^*, p)$ holds, follows at once from the convexity of F in case $p = x_0$.

Applying the preceding corollary to the set $Z^*(x_0, y_0)$ and the element y_0 yields $y_0 \in Z^*(x_0, y_0)$ which implies $y_0 \in X \cap Y$, contrary to $X \cap Y = \varnothing$.

COROLLARY *An arc $A(\alpha, \beta)$ does not contain any proper subset that is closed, convex, and contains α, β.*

Proof. By the preceding lemma, such a subset is connected, and the result follows from Exercise 2, Section 3.11.

DEFINITION An element p of an arc $A(\alpha, \beta)$ is an *interior* element of the arc provided $\alpha \neq p \neq \beta$.

LEMMA 3.8 *Each interior element of a convex arc $A(\alpha, \beta)$ is metrically between α and β.*

Proof. Let $Z^*(\alpha, \beta) = [p \in A(\alpha, \beta) \,|\, p = \alpha, p = \beta, \text{ or } \mathscr{L}(a, p, \beta)]$. The continuity of the distance function implies that $Z^*(\alpha, \beta)$ is a closed set. We show now that $Z^*(\alpha, \beta)$ is a convex subset of $A(\alpha, \beta)$.

If $p, r \in Z^*(\alpha, \beta)$ $(p \neq r)$, the convexity of $A(\alpha, \beta)$ implies the existence of an element q of $A(\alpha, \beta)$ such that $\mathscr{L}(p, q, r)$ holds. If $\alpha \neq p$ and $r \neq \beta$, then $\mathscr{L}(\alpha, p, \beta)$ and $\mathscr{L}(a, r, \beta)$ subsist, and the last three betweenness relations imply $\mathscr{L}(\alpha, q, \beta)$. (Why?)

Hence $q \in Z^*(\alpha, \beta)$, which is, therefore, convex. Let the reader complete the proof of the convexity of $Z^*(\alpha, \beta)$ by examining the cases $p = \alpha$ and $r = \beta$.

Since $Z^*(\alpha, \beta)$ is a closed and convex subset of $A(\alpha, \beta)$, and α, $\beta \in Z^*(\alpha, \beta)$, it follows from the corollary of Lemma 3.7 that $Z^*(\alpha, \beta)$ coincides with $A(\alpha, \beta)$, and the lemma follows from the definition of the set $Z^*(\alpha, \beta)$.

LEMMA 3.9 *If x_0, y_0 are any two distinct interior points of a convex arc $A(\alpha, \beta)$, then either $\mathscr{L}(\alpha, x_0, y_0)$ or $\mathscr{L}(\alpha, y_0, x_0)$ subsists.*

Proof. Consider the set

$$A_{x_0}^*(\alpha, \beta) = [p \in A(\alpha, \beta) \,|\, p = \alpha \text{ or } p = \beta \text{ or } p = x_0 \text{ or } \mathscr{L}(\alpha, p, x_0) \text{ or }$$
$$\mathscr{L}(x_0, p, \beta)]$$

That set is closed; we shall show it is also convex.

If $p, r \in A_{x_0}^*(\alpha, \beta)$ and $p \neq r$, suppose $\mathscr{L}(\alpha, p, x_0)$ and $\mathscr{L}(\alpha, r, x_0)$ hold. Since $A(\alpha, \beta)$ is convex, it contains an element q with $\mathscr{L}(p, q, r)$ holding. Combining this with the two preceding betweennesses yields $\mathscr{L}(\alpha, q, x_0)$, and so $q \in A_{x_0}^*(\alpha, \beta)$.

If $\mathcal{L}(x_0, p, \beta)$ and $\mathcal{L}(x_0, r, \beta)$ hold, they combine with $\mathcal{L}(p, q, r)$ to imply $\mathcal{L}(x_0, q, \beta)$, and again $q \in A_{x_0}^*(\alpha, \beta)$.

If $\mathcal{L}(\alpha, p, x_0)$ and $\mathcal{L}(x_0, r, \beta)$ subsist, the first relation combines with $\mathcal{L}(\alpha, x_0, \beta)$ (Lemma 3.8) to give $\mathcal{L}(p, x_0, \beta)$, which together with $\mathcal{L}(x_0, r, \beta)$ yields $\mathcal{L}(p, x_0, r)$, and so the element x_0 of $A_{x_0}^*(\alpha, \beta)$ is between the element p, r.

The other possibilities (such as $p = \alpha$, $r = x_0$) offer even less difficulty.

We are now ready to prove the theorem characterizing segments as convex arcs.

THEOREM 3.41   *An arc $A(\alpha, \beta)$ is a segment if and only if it is convex.*

Proof.   The necessity has already been remarked. We turn to the sufficiency. Let $A(\alpha, \beta)$ denote a convex arc and consider the motion $f(x) = d(\alpha, x)$ that carries $A(\alpha, \beta)$ onto the set $f[A(\alpha, \beta)]$, where $f(\alpha) = 0$ and $f(\beta) = d(\alpha, \beta)$. We show that $f[A(\alpha, \beta)]$ is a maximal chain between 0 and $d(\alpha, \beta)$.

If $f(x)$, $f(y)$ are distinct points of $f[A(\alpha, \beta)]$, each different from 0 and $d(\alpha, \beta)$, then x, y are distinct interior points of $A(\alpha, \beta)$ ad so $\mathcal{L}(\alpha, x, y)$ or $\mathcal{L}(\alpha, y, x)$ holds. Consequently, either $\mathcal{L}(0, f(x), f(y))$ or $\mathcal{L}(0, f(y), f(x))$ is valid, which implies $f(x) \prec f(y)$ or $f(y) \prec f(x)$. Moreover, $f(x) \prec f(\beta) = d(\alpha, \beta)$, since $\mathcal{L}(\alpha, x, \beta)$ subsists and hence so does $\mathcal{L}(0, f(x), f(\beta))$. Thus $f[A(\alpha, \beta)]$ is a chain.

Finally, since $A(\alpha, \beta)$ is closed, compact, and convex, so is the chain $f[A(\alpha, \beta)]$, and it follows from Theorem 3.34 that $f[A(\alpha, \beta)]$ is a maximal chain between its endpoints. Hence $A(\alpha, \beta)$ is a segment with end points α, β.

3.13   SEGMENT-LIKE ARCS

Is was shown in a preceding section that the length of a segment of the space **B** is the distance of its end points, and it was remarked that this property does not characterize segments among arcs. An arc $A(\alpha, \beta)$ such that $\lambda[A(\alpha, \beta)] = d(\alpha, \beta)$ is called segment-like. Clearly, every segment is a segment-like arc, but as we now show, there exist segment-like arcs that are not segments. This is a peculiarity of Boolean metric spaces **B** that is not shared by ordinary metric spaces.

Let an open subset E of **B** be called *regular* provided it is the interior of its closure (that is, provided E is the set sum of all open subsets of cl E). It is easily shown (Exercise 1) that the set of all regular open subsets of the closed interval [0, 1] of the straight line forms a complete, separable, atom-free Boolean algebra with respect to the Boolean operations $+$ and $\cdot$ defined as follows.

If X and Y denote regular open subsets of [0, 1],

$$X \cdot Y = X \cap Y,$$
$$X + Y = \text{int}(\text{cl}(X \cup Y)).$$

Let $\Gamma(a, b)$ be the maximal chain of this Boolean algebra consisting of $a =$ the null set $\varnothing$, $b =$ the open interval (0, 1) and all open intervals (0, t), $0 < t \leq 1$. The arc $A(\alpha, \beta)$ we are seeking is obtained as the homeomorph of $\Gamma(a, b)$ by the mapping f defined by

$$f\ [(0, t)] = (0, t),\, 0 < t \leq \tfrac{1}{3}$$
$$= (t - \tfrac{1}{3}, t),\, \tfrac{1}{3} \leq t \leq \tfrac{2}{3}$$
$$= (1 - t, t),\, \tfrac{2}{3} \leq t \leq 1,$$
$$f(\varnothing) = \varnothing.$$

Let the reader verify (Exercise 2) that (i) f is a homeomorphism (and hence $A(\alpha, \beta)$ is an arc), (ii) $\lambda(A[\alpha, \beta]) = (0, 1)$, the distance of its end points, but (iii) the arc $A(\alpha, \beta)$ is *not* a segment, since it is not convex; for example, $A(\alpha, \beta)$ does not contain any element metrically between the end point α and the point $f[(0, \tfrac{2}{3})] = (\tfrac{1}{3}, \tfrac{2}{3})$.

DEFINITION   The *excess* $\mathscr{E}_p$ of an arc $A(\alpha, \beta)$ at an element p of the arc is defined by

$$\mathscr{E}_p = [d(\alpha, p) + d(p, \beta)] \cdot d'(\alpha, \beta).$$

The excess $\mathscr{E}[A(\alpha, \beta)]$ of arc $A(\alpha, \beta)$ is defined by

$$\mathscr{E}[A(\alpha, \beta)] = \sum_{p \in A} \mathscr{E}_p.$$

The concepts just defined are intended to give a kind of Boolean measure of the extent by which an arc fails to be segment-like. It is observed that if p is an interior point of $A(\alpha, \beta)$, then $\mathscr{E}_p = 0$ if and only if $\mathscr{L}(\alpha, p, \beta)$ holds. For if $\mathscr{E}_p = 0$, then $d(\alpha, p) + d(p, \beta) \prec d(\alpha, \beta)$, and $\mathscr{L}(\alpha, p, \beta)$ follows from the triangle inequality. On the other hand, if $\mathscr{L}(\alpha, p, \beta)$ holds, then $d(\alpha, p) + d(p, \beta) = d(\alpha, \beta)$, and so clearly $\mathscr{E}_p = 0$.

Remark.   Direct computation yields

$$\mathcal{E}_p = \alpha\beta p' + \alpha'\beta' p.$$

Notice that $\mathcal{E}_\alpha = \mathcal{E}_\beta = 0$, for every arc $A(\alpha, \beta)$.

THEOREM 3.42   *For each arc $A(\alpha, \beta)$ of **B**,*

$$\mathcal{E}[A(\alpha, \beta)] = \lambda[A(\alpha, \beta)] \cdot d'(\alpha, \beta).$$

Proof.   The following computation yields the desired result:

$$\mathcal{E}[A(\alpha, \beta)] = \sum_{p \in A} \mathcal{E}_p$$

$$= \sum_{p \in A} [d(\alpha, p) + d(p, \beta)] \cdot d'(\alpha, \beta)$$

$$= \sum_{p \in A} [d(\alpha, p) \cdot d'(\alpha, \beta) + d(p, \beta) \cdot d'(\alpha, \beta)]$$

$$= \lambda[A(\alpha, \beta)] \cdot d'(\alpha, \beta) + \lambda[A(\alpha, \beta)] \cdot d'(\alpha, \beta)$$

$$= \lambda[A(\alpha, \beta)] \cdot d'(\alpha, \beta).$$

COROLLARY   *An arc $A(\alpha, \beta)$ of **B** is segment-like if and only if*

$$\mathcal{E}[A(\alpha, \beta)] = 0.$$

THEOREM 3.43   *An arc $A(\alpha, \beta)$ of **B** is segment-like if and only if each interior element of $A(\alpha, \beta)$ is metrically between its end points α, and β.*

Proof.   Using the corollary of the preceding theorem, $A(\alpha, \beta)$ is segment-like if and only if $\mathcal{E}[A(\alpha. \beta)] = 0$, and since $\mathcal{E}[A(\alpha, \beta)] = \sum_{p \in A} \mathcal{E}_p$, then $A(\alpha, \beta)$ is segment-like if and only if $\mathcal{L}(\alpha, p, \beta)$ holds for every interior point of the arc.

EXERCISES

1. Show that the set of all regular open subsets of the interval $[0, 1]$ of the straight line forms a complete, separable, atom-free Boolean algebra with respect to the Boolean operations $+$ and $\cdot$ defined in this Section.

2. Verify the statements (i), (ii), and (iii) concerning the segment-like arc defined in this Section.

3. Show that the segment-like arc referred to in Exercise 2 is the set sum of three segments.

REFERENCES

Sections 3.1 to 3.7 L. M. Blumenthal, "Boolean geometry I," *Rend. Circ. Mat. Palermo*, Ser II. vol. I (1952), pp. 343–361.

Sections 12.8 to 12.13   L. M. Blumenthal and C. J. Penning, "Boolean geometry II," *Rend. Circ. Mat. Palermo*, Ser. II, vol. X (1961), pp. 175–192.

PART 2

PROJECTIVE AND RELATED STRUCTURES

CHAPTER 4

An Axiomatic Basis for Projective Geometry

4.1. INTRODUCTION

Large parts of Euclidean geometry deal with perpendicularity and congruence. But one may disregard these relations and concentrate on the results of projecting, joining, and intersecting figures. A study of this particular aspect of the Euclidean space was initiated around 1650 and, during the following two centuries, developed into a theory of spaces that generalized and extended the Euclidean space. This theory is called projective geometry—sometimes also the geometry of joining and intersecting.

Oddly enough, the traditional foundations of this branch of geometry were not formulated in terms of those operations. Not until 1928 was the development of projective and other geometries from algebraic postulates concerning two operations (referred to as joining and intersecting) initiated in a paper by Menger. Each such *algebra of geometry* is analogous to the theory of abstract fields, which likewise is based on assumptions about two operations (referred to as addition and multiplication). The algebra of projective geometry contains Boole's algebra of classes as the special case in which the two operations are related by two distributive laws. Several basic assumptions of the algebra of projective geometry are, moreover, identical with assumptions about the greatest common divisor and the least common multiple, which were formulated at the turn of the century by Dedekind in an algebra of moduls.

The postulates about an abstract field are satisfied, in particular, by the real and the complex numbers. Those of the algebra of geometry hold for *flats* (that is, points, lines, planes, and so on) of projective and certain other spaces. A simple example is the algebra of the flats in a projective plane, Π_2, which are defined as follows: In the three-dimensional Euclidean space E_3, one chooses a point O and considers only the O-lines and the O-planes— that is, the lines and planes through O. One defines the *points* of Π_2 as the O-lines, and the *lines* of Π_2 as the O-planes in E_3. In E_3, any two distinct O-lines, L_1 and L_2, span a O-plane. The line in Π_2 corresponding to this O-plane is considered as the *join* of the two points of Π_2 corresponding to L_1 and L_2. Each pair of distinct O-planes meets in an O-line. The corresponding point is considered as the *meet* of the two lines in Π_2 that correspond to the two O-planes. Thus in Π_2 the join of any two distinct points is a line; and the meet of any two distinct lines is a point. (In this last respect, the projective plane differs from the Euclidean plane, which contains pairs of lines that have no common point.) It will be noted that the perpendicularity and congruence properties of E_3 have not been used in these definitions of the flats of Π_2 and their joins and meets.

The algebra of geometry, however, requires more than joining distinct points and intersecting distinct lines. A join as well as a meet must be associated with *each* pair of elements. This is achieved by the following five stipulations:

(1) The join as well as the meet of any element X with itself is X.

(2) If a point and a line are incident, then their join is that line, and their meet is that point.

(3) Besides the points and lines, there exist a flat V (the *vacuum*) and a flat U (the *universe*) such that the join of V and any flat X is X, and the meet of V and X is V; and that the meet of U and X is X, and the join of U and X is U.

(4) The meet of distinct points is V; the join of distinct lines is U.

(5) A point and a line that are not incident have the join U and the meet V.

In Π_2, points and lines play identical roles. The elements V and U may be considered as corresponding to the point O and the entire space E_3, respectively.

More generally, one obtains an example of flats in an n-dimensional projective space Π_n by choosing a point O in the E_{n+1} and considering the O-flats in the E_{n+1}—that is, the lines and higher-dimensional linear subspaces of the E_{n+1} that contain O. The m-dimensional flats of the Π_n are defined as the $(m+1)$-dimensional O-flats of the E_{n+1}. The O-dimensional and $(n-1)$-dimensional flats of the Π_n are called *points* and *hyperplanes*, respectively. The joins and the meets of the flats in Π_n can be defined in terms of the joins and meets of the corresponding flats in E_{n+1}.

Whereas in these examples the flats (and, notably, flats of various dimensions) as well as their joins and meets have been explicitly defined, the algebra of projective geometry operates with *undefined* concepts: a single set of elements of an unspecified nature, which is the domain of two undefined binary operations. The elements of the set correspond to the flats of various dimensions in the examples; the operations correspond to the joining and intersecting of flats. The most important feature of the theory is the fact that simple algebraic assumptions about the operations yield a classification of the elements. One can single out certain elements that warrant the names of points and hyperplanes. More generally, with many an element, (and under an additional assumption, with *each* element) an integer can be associated that can justifiably be called the *dimension* of the element.

In earlier presentations of geometry, elements of different dimensions were introduced either as different primitive concepts (for example, by Hilbert, who started out with three classes of elements of the three-dimensional space: points, lines, and planes) or as certain sets of points (for example, by Veblen). The former method has the disadvantage that it necessitates the introduction of a new primitive concept for each higher dimension; the latter procedure assigns to points a distinguished role in the foundations of projective geometry that they do not play in the theory developed from these foundations, where hyperplanes play precisely the same role as points; moreover, it is based on complicated definitions, and introduces set theory that can be dispensed with.

In the early 1930's, the aforementioned algebra of projective geometry was incorporated in what G. Birkhoff called lattice theory—more specifically, in the theory of *modular lattices* (so called because they satisfy the postulates of Dedekind's algebra of moduls). About 1930, F. R. Klein introduced lattices under the name *Verbände*. In O. Ore's investigations they are called *structures.* The elements of these mutually equivalent theories are presented in Part 1 of this book.

This chapter contains a development of projective geometry on an even simpler and more general algebraic basis: all that is assumed about the set with the two binary operations is that it contains two elements which, under the two operations, behave like the vacuum and the universe, and that the operations satisfy a single postulate: a dual pair of "Projectivity Laws," formulated by Menger in 1932. Such a set is called a *projective structure* since the elements that are joins of points or meets of hyperplanes have the main properties of the flats in a projective space. By adjoining the assumption that there exists one single configuration consisting of points and hyperplanes in certain mutual relations, one obtains a complete foundation of projective geometry.

4.2. PROJECTIVE STRUCTURES

By a *projective structure* we mean a set $\mathfrak{S}$ with two binary operations, denoted by $\cup$ and $\cap$, which satisfy the following two assumptions:

POSTULATE I $\mathfrak{S}$ *contains two elements, V and U, such that for each element X of* $\mathfrak{S}$,

(I_1) $V \cup X = X$ and $V \cap X = V$;

(I_2) $U \cap X = X$ and $U \cup X = U$.

POSTULATE II   *For any three elements X, Y, Z of* $\mathfrak{S}$,

(II_1) $X \cup ((X \cup Y) \cap Z) = X \cup ((X \cup Z) \cap Y)$;

(II_2) $X \cap ((X \cap Y) \cup Z) = X \cap ((X \cap Z) \cup Y)$.

The fundamental relations between the two operations expressed in Postulate II will be referred to as the *Projectivity Laws*. V and U are called the *distinguished elements*. The first parts of I_1 and I_2 state that V and U are left-neutral with regard to the operations $\cap$ and $\cup$; the second, that they are left-annihilators of $\cap$ and $\cup$, respectively.

Postulates I_1 and I_2 as well as II_1 and II_2 are dual to one another; that is to say, interchanging $\cup$ and $\cap$ as well as V and U transforms one half into the other.

It will now be shown that a projective structure has all properties of a modular lattice except possibly one: the operations are commutative, absorptive, idempotent, and modular, and they satisfy some special cases of the associative law, but they are not necessarily associative.

THEOREM 4.1   *The operations are commutative: for each two elements X and Y of $\mathfrak{S}$, $X \cup Y = Y \cup X$ and $X \cap Y = Y \cap X$.*

Setting $X = U$ in Postulate II_2 and using I_2, one obtains $Y \cup Z = Z \cup Y$. Dually, setting $X = V$ in II_1 yields the commutativity of the operation $\cap$.

THEOREM 4.2   *The elements V and U are the only distinguished elements.*

If $V' \cup X = X$ for each X, then $V' \cup V = V$. From $V \cup V' = V'$ and $V' \cup V = V \cup V'$ it follows that $V' = V$. In a similar manner, it is seen that only one element U satisfies I_2.

THEOREM 4.3   *The operations are absorptive; that is, for each two elements X and Y of $\mathfrak{S}$ the following Absorption Laws are satisfied:*

$$X \cup (X \cap Y) = X \qquad \text{and} \qquad X \cap (X \cup Y) = X.$$

Setting $Z = V$ in Postulate II_1 and using commutativity and I_1 yields the first Absorption Law. The second law is dual to the first.

THEOREM 4.4   *The operations are idempotent; that is, for each element X of $\mathfrak{S}$*

$$X \cup X = X \qquad \text{and} \qquad X \cap X = X.$$

The assertion results from the application of the first Absorption Law to $Y = U$, and of the second to $Y = V$.

THEOREM 4.5   *The operations are alternative; that is, they are associative for each triple of elements of which two are identical:*

(a) $X \cup (X \cup Y) = (X \cup X) \cup Y$,
(b) $(Y \cup X) \cup X = Y \cup (X \cup X)$,
(c) $(X \cup Y) \cup X = X \cup (Y \cup X)$.

The dual formulas hold for $\cap$.

Setting $Z = U$ in Postulate II_1 yields $X \cup (X \cup Y) = X \cup Y$, which, because of idempotency, is equal to $(X \cup X) \cup Y$. By commutativity, (b) follows from (a), and (c) follows directly from commutativity.

THEOREM 4.6 *The two operations are associative for any three elements of which at least one is distinguished.*

The simple proof is left to the reader.

THEOREM 4.7 *The operations satisfy the Absorption Equivalence:*

$$X \cup Y = Y \quad \textit{if and only if} \quad X \cap Y = X.$$

This important equivalence is an immediate consequence of the Absorption Laws in Theorem 4.3.

EXERCISES

1. In a set containing more than one element, let $\cup$ and $\cap$ be two binary operations such that

    $$X \cup Y = X = X \cap Y \quad \text{for each } X \text{ and } Y.$$

    Prove that the operations satisfy the Projectivity Laws and are idempotent and associative but neither commutative nor absorptive.

4.3. THE PART RELATION AND MODULARITY

The Absorption Equivalence makes it possible to define in every projective structure *a part relation* as follows.

DEFINITION *If* X, $Y \in \mathfrak{S}$, *then* $X \subseteq Y$ *if and only if* $X \cup Y = Y$. We write $Y \supseteq X$ synonymously with $X \subseteq Y$.

THEOREM 4.8 *The part relation has the following properties:*

(1) $X \subseteq X$,
(2) $X \subseteq Y$ *and* $Y \subseteq X$ *if and only if* $X = Y$.

Property (1) expresses the idempotency of the operations, and (2) is an immediate consequence of the definition of the part relation. (The question of the transitivity of the part relation will be discussed later.)

THEOREM 4.9   *The part relation is connected with the operations by the following laws and implications:*

(1) $V \subseteq X \subseteq X \subseteq U$, *for each* X,

(2) $X \cap Y \subseteq X$ *and* $X \subseteq X \cup Y$, *for each* X *and* Y,

(3) *if* $X \subseteq Z_1$ *and* $X \subseteq Z_2$, *then* $X \subseteq Z_1 \cap Z_2$,
   if $Z_1 \subseteq X$ *and* $Z_2 \subseteq X$, *then* $Z_1 \cup Z_2 \subseteq X$,

(4) *if* $X \cap Y \subseteq Z \subseteq Y$, *then* $Z \cup X \subseteq Y \cup X$,
   if $Y \subseteq Z \subseteq X \cup Y$, *then* $Y \cap X \subseteq Z \cap X$.

Statements (1) and (2) express Postulate I, idempotency, and the absorption laws. The first of the implications (3) may be formulated as follows: if $Z_1 \cap X = X$ and $X \cup Z_2 = Z_2$, then $(Z_1 \cap Z_2) \cup X = Z_1 \cap Z_2$. Setting $Z_1 \cap Z_2 = T$, one must show that $T \cup X = T$. Using the definition of T, the second Projectivity Law (II$_2$) and the assumptions $Z_1 \cap X = X$ and $X \cup Z_2 = Z_2$, one obtains

$$Z_1 \cap (T \cup X) = Z_1 \cap ((Z_1 \cap Z_2) \cup X) = Z_1 \cap ((Z_1 \cap X) \cup Z_2)$$
$$= Z_1 \cap (X \cup Z_2) = Z_1 \cap Z_2 = T.$$

By the first Projectivity Law, $T \cup ((T \cup X) \cap Z_1) = T \cup ((T \cup Z_1) \cap X)$. Since $Z_1 \cap (T \cup X)$ has been shown to be T, the element on the left side is $T \cup T = T$. On the right side, by the Absorption Law, $T \cup Z_1 = Z_1$, whence $T = T \cup (Z_1 \cap X) = T \cup X$. The second implication (3) is dual to the first and can be proved dually.

If Z_1, Z_2, Z_3 are parts of X, then by two applications of (3) one concludes that $(Z_1 \cup Z_2) \cup Z_3 \subseteq X$.

The following result, which is one of the most important tools in the development of the theory, is obtained by induction.

LEMMA 4.1   *If* $Z_i \subseteq X$ *for* $i = 1, \ldots, n$, *then*

$$(\ldots ((Z_1 \cup Z_2) \cup Z_3) \cup \cdots \cup Z_{n-1}) \cup Z_n \subseteq X.$$

(Since the operations of a projective structure are not assumed to be associative, the parentheses on the left side cannot be dispensed with.) It goes without saying that the dual of the Lemma is also valid.

Before demonstrating (4) we prove the following theorem, which is due to Helen Skala, as is the proof of (3) in Theorem 4.9.

THEOREM 4.10   *The operations are modular; that is,*

$$\text{if } X \subseteq Z, \text{ then } (X \cup Y) \cap Z = X \cup (Y \cap Z).$$

If $X \subseteq Z$, then since $X \subseteq X \cup Y$, it follows from (3) of Theorem 4.9, that $X \subseteq (X \cup Y) \cap Z$. Therefore by Postulate II_1 and commutativity one obtains

$$(X \cup Y) \cap Z = X \cup ((X \cup Y) \cap Z) = X \cup ((X \cup Z) \cap Y) = X \cup (Y \cap Z).$$

To prove (4) of Theorem 4.9, assume $X \cap Y \subseteq Z$. It follows that $Z = Z \cup (X \cap Y)$. Assuming $Z \subseteq Y$, by modularity one obtains $Z \cup (X \cap Y) = (Z \cup X) \cap Y$. Hence $X \cup Z = X \cup ((Z \cup X) \cap Y) = X \cup (Y \cap (Z \cup X))$ Since $X \subseteq Z \cup X$, modularity yields $X \cup (Y \cap (X \cup Z)) = (X \cup Y) \cap (X \cup Z)$, and so $X \cup Z \subseteq X \cup Y$.

If $X \subseteq Z$ and $X \neq Z$, then X is said to be a *proper part* of Z—in symbols, $X \subset Z$. If $X \subset Z$, and $X \subseteq Y \subseteq Z$ implies $X = Y$ or $Y = Z$, we write $X \to Z$ (see Sections 2.2 and 2.4) and call X a *maximal proper part* of Z. Thus $X \to Z$ symbolizes that no element distinct from X and Z can be interpolated between X and Z.

THEOREM 4.11   *The operations satisfy the following Interpolation Equivalence:*

$$X \cap Y \to Y \text{ if and only if } X \to X \cup Y.$$

Because of the duality of the two implications that make up this equivalence, it is sufficient to derive $X \cap Y \to Y$ from $X \to X \cup Y$. Assume $X \cap Y \subseteq Z \subseteq Y$. Because of (4), Theorem 4.9, it follows that $Z \cup X \subseteq Y \cup X$. From $X \subseteq Z \cup X \subseteq Y \cup X$ (in view of $X \to X \cup Y$) one concludes that either (1) $Z \cup X = X$ or (2) $Z \cup X = X \cup Y$. In Case (1), $Z \subseteq X$ and $Z \subseteq Y$ (by (3) of Theorem 4.9) imply $Z \subseteq X \cap Y$. Because of the assumption $X \cap Y \subseteq Z$, one concludes that in Case (1) $Z = X \cap Y$. In Case (2),

$$Z = Z \cup (Y \cap X) = Z \cup ((Z \cup Y) \cap X) = Z \cup ((Z \cup X) \cap Y)$$
$$= Z \cup ((X \cup Y) \cap Y) = Z \cup Y = Y.$$

Thus, under the assumption that $X \to X \cup Y$, it has been proved that $X \cap Y \to Y$, which concludes the proof of the theorem.

A projective structure consisting of at most four elements is associative and hence is a modular lattice. This follows from Theorem 4.6, since each triple of elements of such a structure includes at least one distinguished element.

In general, the operations of a projective structure are not associative. In the following examples, only the nondistinguished elements will be listed. It is understood that each structure contains U and V which satisfy Postulate I, and that the operations (being projective) are commutative, idempotent, and absorptive so that, for example, $A \cup B = B$ implies $A \cap B = B \cap A = A$.

Example 1. The set $\mathfrak{S}_1$ contains A, B, C such that

$$A \cup B = B, \qquad B \cup C = C, \qquad A \cup C = U, \qquad A \cap C = V.$$

Clearly $(A \cup B) \cup C \neq A \cup (B \cup C)$. Moreover, $A \subseteq B$ and $B \subseteq C$, yet $A \nsubseteq C$; and $A \subseteq B$, yet $A \nsubseteq B \cup C$.

Example 2. The set $\mathfrak{S}_2$ is obtained from $\mathfrak{S}_1$ by setting $B = B_1$ and adjoining elements $B_2, \ldots, B_n$ such that

$$A \cup B_i = B_i, \qquad B_i \cup C = C, \qquad B_i \cup B_j = C, \qquad B_i \cap B_j = A$$

for $i \neq j$. One readily verifies that $B_1 \cap B_2 \nsubseteq B_1 \cup B_2$. Moreover, $A \subseteq B_1$ and $A \subseteq B_2$ and yet $A \nsubseteq B_1 \cup B_2$; and $B_1 \subseteq C$ and $B_2 \subseteq C$ and yet $B_1 \cap B_2 \nsubseteq C$.

Example 3. The set $\mathfrak{S}_3$ contains A, B, C, D such that

$$A \cup B = B, \qquad B \cup C = C, \qquad C \cup D = D, \qquad D \cup A = A,$$
$$A \cup C = B \cup D = U, \qquad A \cap C = B \cap D = V.$$

Example 4. The set $\mathfrak{S}_4$ contains $P_1, \ldots, P_n$, L, and H $(n \geqslant 2)$ such that

$$P_i \cup P_j = P_i \cup L = L, \; P_i \cup H = U, \; L \cup H = H, \; P_i \cap P_j = P_i \cap H = V$$

for $i, j = 1, \ldots, n; \; i \neq j$.

EXERCISES

1. Prove that in each projective structure $X \cup Y = Z$ implies $X \subseteq Z$. Dualize this statement.

2. Verify the Projectivity Laws in Examples 1 to 4. (In Example 1, only six equalities have to be checked.)

3. Prove that each nonassociative projective structure containing exactly five elements is isomorphic to $\mathfrak{S}_1$.

4. Prove that a set containing $A_1, \ldots, A_n$ $(n > 4)$ such that

$$A_i \cup A_{i+1} = A_{i+1} \ (i = 1, \ldots, n-1), \ A_n \cup A_1 = A_1,$$

$$A_i \cup A_j = U, \ A_i \cap A_j = V \qquad \text{for } |i - j| \neq 1,$$

is not a projective structure.

5. Demonstrate that in a nonassociative projective structure, the set of all parts of an element is not necessarily a projective structure. (*Hint:* In $\mathfrak{S}_2$, B_1 and B_2 are parts of C. What about their meet?)

6. Show that a projective structure does not necessarily satisfy the self-dual law:

$$((X \cup Y) \cap (Y \cup Z)) \cap (Z \cup X) = ((X \cap Y) \cup (Y \cap Z)) \cup (Z \cap X).$$

(*Hint:* What follows from this law in the case $Y = V$?)

4.4   POINTS AND HYPERPLANES

An element P is called a *point* if $P \neq V$ and P has no part other than P and V; that is to say, if $V \subseteq X \subseteq P$ implies $X = P$ or $X = V$. Thus P is a point if P has exactly one proper part (which of course must be V); in symbols, if $V \to P$. Dually, an element H is called a *hyperplane* if $H \neq U$ and $U \supseteq X \supseteq H$ implies $X = H$ or $H = U$; that is, if $H \to U$.

Euclid's *Elements* begins with the words "*A point is that which has no parts.*" But nowhere in his book did (or could) Euclid utilize this definition, since he never formulated adequate assumptions about parts. Only in the algebra of geometry was Euclid's definition provided with a postulational foundation and incorporated in a deductive theory.

The family of all subsets of a straight line, with $\cup$ and $\cap$ denoting union and intersection, is a projective structure (and even a Boolean algebra). The sets consisting of one element are the points of this structure; the sets containing all but one element are its hyperplanes. The family of all *open* subsets of the straight line is a projective structure containing hyperplanes but no points; the structure consisting of all *closed* subsets contains points but no hyperplanes. The nonassociative structure $\mathfrak{S}_3$ defined in the preceding section contains neither points nor hyperplanes. The structures $\mathfrak{S}_1$ and $\mathfrak{S}_2$ contain one point each.

If U is a point, then V is a hyperplane and the system consists of these two elements only. If each nondistinguished element is a point, then the points are also the hyperplanes. The simplest example is a structure containing only one nondistinguished element.

Although Postulates I and II do not imply the existence of points and hyperplanes, they do imply that any points and hyperplanes that may exist have the properties of points and hyperplanes in projective spaces.

Since $P \cap W \subseteq P$ for each element W, the definition of points immediately yields the following simple but very useful

LEMMA 4.2 (ON POINTS)   *If P is a point, then for any element W, either $P \cap W = P$ and $P \subseteq W$, or $P \cap W = V$ and $P \not\subseteq W$.*

If, in the Modularity Implication, Y is a point P then, according to the preceding lemma, $X \subseteq Z$ implies $(X \cup P) \cap Z = X$ or $(X \cup P) \cap Z = X \cup P$. Hence, if $X \subseteq Z \subseteq X \cup P$, then $Z = X$ or $Z = X \cup P$. Dually, if $X \supseteq Z \supseteq X \cap H$, then $Z = X$ or $Z = X \cap H$. Of course, these implications (which for $X = V$ and $X = U$ are the definitions of points and hyperplanes) are of interest only if $X \subset X \cup P$ and $X \supset X \cap H$, in which case they yield

THEOREM 4.12 (FUNDAMENTAL PROPERTY OF POINTS AND HYPERPLANES) *If $P \not\subseteq X$, then $X \to X \cup P$, and if $H \not\supseteq X$, then $W \cap H \to W$.*

In the sequel, it will be convenient to set

$$X \cup ((X \cup Y) \cap Z) = [X; Y, Z]_1 \text{ or, simply, } [X; Y, Z]$$
$$X \cap ((X \cap Y) \cup Z) = [X; Y, Z]_2 .$$

The Projectivity Laws then read

$$[X; Y, Z] = [X; Z, Y] \quad \text{and} \quad [X; Y, Z]_2 = [X; Z, Y]_2 .$$

The fundamental property of points has the following important consequence:

THEOREM 4.13   *If P is a point then*

(i) $(X \cup P) \cap Z = X \cap Z$ *if and only if* $P \subseteq X$ *or* $P \nsubseteq X \cup Z$;

(ii) $Z \subseteq (Z \cup Y) \cup P$ *for any two elements* Y *and* Z;

(iii) $Z \subseteq X$ *implies* $Z \subseteq X \cup P$.

If

(∗) $(X \cup P) \cap Z = X \cap Z,$

then $X = X \cup (X \cap Z) = [X; P, Z] = [X; Z, P]$. Hence $(X \cup Z) \cap P \subseteq X$ and, because of Lemma 4.2, $P \subseteq X$ or $P \nsubseteq X \cup Z$. Conversely, $P \subseteq X$ clearly implies (∗). It remains to derive (∗) from $P \nsubseteq X \cup Z$. The assumption implies $(X \cup Z) \cap P = V$ and

$$X = X \cup V = [X; Z, P] = [X; P, Z] = X \cup ((X \cup P) \cap Z),$$

thus $(X \cup P) \cap Z \subseteq X$, which, combined with $(X \cup P) \cap Z \subseteq Z$, yields $(X \cup P) \cap Z \subseteq X \cap Z$. In order to prove $(X \cup P) \cap Z \supseteq X \cap Z$, in view of $X \cap Z \subseteq Z$ it is sufficient to show that $X \cap Z \subseteq X \cup P$. Setting $(P \cup X) \cap (X \cap Z) = Y$ one must prove $T = X \cap Z$. Now

$$P \cup Y = [P; X, X \cap Z] = [P; X \cap Z, X] = P \cup (((X \cap Z) \cup P) \cap X).$$

In view of $X \cap Z \subseteq X$, by modularity the last element equals $P \cup ((X \cap Z) \cup (P \cap X))$, and this (regardless of whether $P \cap X$ is V or P) equals $P \cup (X \cap Z)$. Consequently

$$Y \subseteq X \cap Z \subseteq P \cup (X \cap Z) = P \cup Y.$$

According to Theorem 4.12, either $X \cap Z = Y$, as claimed, or $X \cap Z = P \cup Y$. In the latter case, $Y = (P \cup Y) \cap (P \cup X)$, which, by modularity and since $Y \cap (P \cup X) = Y$, equals $P \cup (Y \cap (P \cup X)) = P \cup Y = X \cap Z$; hence again $X \cap Z = Y$. This completes the proof of (i), from which one obtains (ii) by setting $X = Z \cup Y$. In view of $X = X \cup Z$, (iii) is a corollary of (ii).

EXERCISES

1. State and prove the dual of the theorems and lemmas in this section.

2. As a corollary of the fundamental property of points, prove that if an element X is a maximal proper part of exactly one element Z, then each point (if any) that is not a part of X is a part of Z.

3. Using Examples 1 and 2 in the preceding section, show that, under the assumptions of Exercise 2, points that are parts of X are not necessarily points of Z.

4. Let R be the set of all rational points in the straight line. The family of all sets that are both open and closed in R is a projective structure without either points of hyperplanes.

5. Verify in a Euclidean plane or space the fundamental property of points (Theorem 4.12), but show that the fundamental property of hyperplanes must be restricted to cases where $W \cap H \neq V$.

4.5   REGULAR ELEMENTS

The *join of the sequence* $(X_1, \ldots, X_n)$ of elements of a projective structure can be defined inductively.

$$X_1 \cup X_2 \cup \cdots \cup X_{n-1} \cup X_n = (X_1 \cup X_2 \cup \cdots \cup X_{n-1}) \cup X_n$$
$$= (\cdots (X_1 \cup X_2) \cup \cdots \cup X_{n-1}) \cup X_n.$$

Dually, one defines the *meet of a sequence*. Because of the idempotency of the operations, the definition yields the following

COROLLARY   *If* $X_1 = X_2$, *then* $X_1 \cup X_2 \cup X_3 \cup \cdots \cup X_n = X_1 \cup X_3 \cup \cdots \cup X_n.$

In nonassociative projective structures these definitions are of limited significance. If $(X_1 \cup X_2) \cup X_3 \neq X_1 \cup (X_2 \cup X_3)$, then, notwithstanding the commutativity of the operations, $X_1 \cup X_2 \cup X_3$ cannot be considered as the join of the set $\{X_1, X_2, X_3\}$. Helen Skala has proved, however, that for sequences of *points* the operation $\cup$ is associative. For example, for any three points P_1, P_2, P_3 it can be shown that

$$P_1 \cup P_2 \cup P_3 = P_1 \cup P_3 \cup P_2 = P_2 \cup P_1 \cup P_3 = P_2 \cup P_3 \cup P_1$$
$$= P_3 \cup P_1 \cup P_2 = P_3 \cup P_2 \cup P_1,$$

and so this element (if the three points are distinct) may be considered as the join of the set $\{P_1, P_2, P_3\}$. If $P_1 = P_2$ or $P_1 = P_2 = P_3$, then $P_1 \cup P_2 \cup P_3$ is the join of the set $\{P_1, P_3\}$ or $\{P_1\}$, respectively.

Dually, the operation $\cap$ is associative for sequences of *hyperplanes*. Because of duality, it is sufficient to study the joins of sequences of points. The following theorem is of basic importance for this study.

THEOREM 4.14   *For any sequence of points* $(P_1, \ldots, P_n)$,

$$P_i \subseteq P_1 \cup \cdots \cup P_n \qquad (i = 1, \ldots, n).$$

If one denotes the element on the right side by S and sets $S' = P_1 \cup \cdots \cup P_{n-1}$, then $S = S' \cup P_n$ and $P_n \subseteq S$. Since the assertion is trivial if $n = 1$, it may be proved under the assumption that it holds for $n - 1$. By this assumption, $P_i \subseteq S'$ for $i = 1, \ldots, n - 1$. But then

$$S' = S' \cup P_i = P_1 \cup \cdots \cup P_i \cup \cdots \cup P_{n-1} \cup P_i.$$

From the last assertion of Theorem 4.13, it follows that

$$P_i \subseteq (S' \cup P_i) \cup P_n = S' \cup P_n = S \qquad (i = 1, \ldots, n - 1).$$

LEMMA 4.3   *If* $A \subseteq P_1 \cup \cdots \cup P_n = S$ *and* $i_1, \ldots, i_n$ *is a permutation of* $1, \ldots, n$, *then* $A \cup P_{i_1} \cup \cdots \cup P_{i_n} \subseteq S$.

Since, by the preceding theorem, each P_{i_h} is a part of S, and $A \subseteq S$ by assumption, the assertion follows from Lemma 4.1.

THEOREM 4.15   *If* $i_1, \ldots, i_n$ *is any permutation of* $1, \ldots, n$, *then*

$$P_{i_1} \cup \cdots \cup P_{i_n} = P_1 \cup \cdots \cup P_n.$$

The element on the left side is a part of S (on the right side) by the preceding lemma (for $A = V$). Since $1, \ldots, n$ is a permutation of $i_1, \ldots, i_n$, S is also a part of the element on the left side.

If P_i and P_j in the sequence $(P_1, \ldots, P_n)$ are equal, consider the permutation of $1, \ldots, n$ which interchanges 1 with i, and 2 with j. By virtue of the corollary of the inductive definition, the join of the sequence remains unchanged if P_j is omitted. This procedure may be repeated until the sequence is replaced by one with distinct elements, say $(P_{i_1}, \ldots, P_{i_m})$, where $\mathfrak{P} = \{P_{i_1}, \ldots, P_{i_m}\}$ is the set of all elements in the sequence $(P_1, \ldots, P_n)$.

Let $\mathfrak{P}$ be any finite set of points. If $(P_1, \ldots, P_n)$ is any serialization of $\mathfrak{P}$, then the element $P_1 \cup \cdots \cup P_n$ (which by Theorem 4.15 is independent of the choice of the serialization) is called *the join of the set* $\mathfrak{P}$ and is denoted by $\bigcup \mathfrak{P}$. What has just been proved can then be formulated as follows.

THEOREM 4.16 *The join of a finite sequence of points is equal to the join of the set of all points in the sequence. Each element of a finite set of points is a part of the join of the set.*

An element that is the join of a finite set of points will be said to be *regular*. In particular, each point is a regular element. Moreover, since V is the join of the empty set of points, V is called regular. In the nonassociative structure $\mathfrak{S}_4$ in Section 4.3 all elements except H and U are regular. In $\mathfrak{S}_1$ and $\mathfrak{S}_2$ only V and A are regular. In $\mathfrak{S}_3$ no element except V is regular.

In the study of the set of all regular elements, the following assertion, which in a way sharpens Lemma 4.3, will be useful.

LEMMA 4.4 *If $A \subseteq P_1 \cup \cdots \cup P_n = S$, then*

$$A \cup P_1 \cup \cdots \cup P_n = S.$$

That $A \cup (P_1 \cup \cdots \cup P_n) = S$ immediately follows from the assumption $A \subseteq S$. But $A \cup (P_1 \cup \cdots \cup P_n)$ is not the element on the right side of the equality asserted in the lemma. The latter is, by the definition of the join of a sequence, $(\ldots (A \cup P_1) \cup \cdots \cup P_{n-1}) \cup P_n$. That *this* element is a part of S follows from Lemma 4.3. It remains to show that S is a part of this element. If $n = 1$, then A, which is a part of P_1, is either P_1 or V, and the assertion holds in either case. Assuming its validity for $n - 1$ and setting

$$S' = A \cup P_1 \cup \cdots \cup P_{n-1},$$

one has $P_h \subseteq S' = S' \cup P_h$ for $h = 1 \ldots, n - 1$ and hence, by Lemma 4.3,

$$P_h \subseteq (S' \cup P_h) \cup P_i = S' \cup P_i = S \qquad (h = 1, \ldots, n - 1).$$

Furthermore, since $P_n \subseteq S$, Lemma 4.1 yields the assertion for n.

THEOREM 4.17 *In the set of all regular elements of a projective structure, the part relation is transitive; that is, if X, Y, Z are regular, then $X \subseteq Y$ and $Y \subseteq Z$ imply $X \subseteq Z$.*

Let the regular elements be

$$X = P_1 \cup \cdots \cup P_i, \qquad Y = Q_1 \cup \cdots \cup Q_j, \qquad Z = R_1 \cup \cdots \cup R_k,$$

where the P_h, Q_h, R_h are points. According to Lemma 4.4, $X \subseteq Y$ implies

$$Y = X \cup Q_1 \cup \cdots \cup Q_j = P_1 \cup \cdots \cup P_i \cup Q_1 \cup \cdots \cup Q_j,$$

and $Y \subseteq Z$ implies

$$Z = Y \cup R_1 \cup \cdots \cup R_k =$$
$$P_1 \cup \cdots \cup P_i \cup Q_1 \cup \cdots \cup Q_j \cup R_1 \cup \cdots \cup R_k.$$

By Theorem 4.14 $P_h \subseteq Z$ for $h = 1, \ldots, i$. From Lemma 4.2 it follows that $X = P_1 \cup \cdots \cup P_i \subseteq Z$. Thus the part relation is transitive.

THEOREM 4.18 *If $\mathfrak{P}'$ is a subset of $\mathfrak{P}$, then $\bigcup \mathfrak{P}' \subseteq \bigcup \mathfrak{P}$.*

If $\mathfrak{P} = \{P_1, \ldots, P_k\}$, then by Theorem 4.11 each P_i is a part of $\bigcup \mathfrak{P}$ and consequently so is $\bigcup \mathfrak{P}'$.

THEOREM 4.19 *If X is regular, say, the join of m points, and $Y \subseteq X$, then there exist elements $Y_1, \ldots, Y_{k-1}$ ($k \leq m$) such that*

$$Y = Y_0 \to Y_1 \to \cdots \to Y_{k-1} \to Y_k = X.$$

Let X be the join of the set $\{P_1, \ldots, P_m\}$. In the chain

$$Y \subseteq Y \cup P_1 \subseteq Y \cup P_1 \cup P_2 \subseteq \cdots \subseteq Y \cup P_1 \cup P_2 \cup \cdots \cup P_{m-1} \subseteq X,$$

by Theorem 4.12, each element is either equal to, or a maximal proper part of, its successor. Hence the chain contains a subchain that satisfies the theorem.

EXERCISES

1. Under the assumption that the element A is regular, prove the assertion of Lemma 4.4 without using Lemma 4.3.

2. Prove that $\bigcup (\mathfrak{P} \cap \mathfrak{Q}) \subseteq \bigcup \mathfrak{P} \cap \bigcup \mathfrak{Q}$, and demonstrate by examples that the two elements are not in general equal.

3. Dualize the concept of regular element and all theorems in this section.

4.6 BASIC SETS

If in a Euclidean or projective space a point lies on the join of $n-1$ other points, then the n points are said to be *dependent*. Collinearity of three points and coplanarity of four points are the simplest examples of dependence. Of particular importance in ordinary geometry as well as in the theory of projective structures are *independent* points (none of which is on the join of the others). We say that such points constitute a *basic set*. In other words, a point set $\mathfrak{P}$ is *basic* if each point and the join of the other points have the meet V. The empty set will be considered as basic. (The point sets considered in this section are finite.)

It should be noted that in the sequel the symbols $\cap$ and $\cup$ are used in two meanings: for the meet and the join of two elements of a projective structure as well as for the intersection and the union of two sets of elements. (This duplicity first occurred in Exercise 2 in the preceding section.) Similarly, the symbol $\subseteq$ will denote the part relation in the realm of elements as well as the subset relation in the realm of sets of elements.

For $\mathfrak{P}$ to be basic it is sufficient that each point and the join of *certain* ones of the other points of $\mathfrak{P}$ have the meet V. On the other hand, a necessary condition for $\mathfrak{P}$ to be basic is that the joins of *any two disjoint subsets* of $\mathfrak{P}$ have the meet V.

THEOREM 4.20   *In order that a set $\mathfrak{P}_n$ of n points be basic it is sufficient that for at least one serialization of $\mathfrak{P}_n$ (say, $(P_1, \ldots, P_n)$), the following $n-1$ conditions hold:*

(C_i) $(P_1 \cup P_2 \cup \cdots \cup P_i) \cap P_{i+1} = V$ $(i = 1, 2, \ldots, n-1)$.

If $\mathfrak{P}$ is basic, then for any two sets $\mathfrak{Q}$ and $\mathfrak{R}$ with the union $\mathfrak{P}$

$$\bigcup \mathfrak{Q} \cap \bigcup \mathfrak{R} = \bigcup (\mathfrak{Q} \cap \mathfrak{R}).$$

In particular, if $\mathfrak{Q}$ and $\mathfrak{R}$ are disjoint, then $\bigcup \mathfrak{Q} \cap \bigcup \mathfrak{R} = V$.

Set $S_i' = P_1 \cup \cdots \cup P_{i-1} \cup P_{i+1} \cup \cdots \cup P_{n-1}$ and $S_i = S_i' \cup P_n$. Since (C_1) is sufficient for $\mathfrak{P}_2$ to be basic, the first assertion will be proved under the assumption that it is valid for $\mathfrak{P}_{n-1}$. Then, if $(P_1, \ldots, P_{n-1}, P_n)$ satisfy $(C_1), \ldots, (C_{n-2}), (C_{n-1})$, the first $n-2$ conditions imply that the set $\{P_1, \ldots, P_{n-1}\}$ is basic. Hence

$$P_i \cap S_i' = V \qquad (i = 1, \ldots, n-1).$$

Since (C_{n-1}) implies that $P_n \not\subseteq S_i' \cup P_i$, it follows from Theorem 4.13 that

$$P_i \cap S_i = P_i \cap (S_i' \cup P_n) = P_i \cap S_i' = V \qquad (i = 1, \ldots, n-1).$$

Thus, in view of (C_{n-1}), the set $\mathfrak{P}_n$ is basic.

To prove the second assertion, set $\mathfrak{Q} \cap \mathfrak{R} = \mathfrak{T} = \{T_1, \ldots, T_k\}$, and assume that $\mathfrak{R}$ contains $k + i$ points, $\mathfrak{R} = \{T_1, \ldots, T_k, R_1, \ldots, R_i\}$. Since for $i = 0$ the assertion is trivial, its validity may be assumed for $i - 1$. Set

$$\mathfrak{R}' = \{T_1, \ldots, T_k, R_1, \ldots, R_{i-1}\}, \qquad R' = \bigcup \mathfrak{R}', \ R = \bigcup \mathfrak{R}, \ Q = \bigcup \mathfrak{Q}.$$

Clearly R_i is not an element of the set $\mathfrak{R}' \cup \mathfrak{Q}$. Since $\mathfrak{P}$ is basic, $R_i \not\subseteq \bigcup(\mathfrak{R}' \cup \mathfrak{Q})$, which, by Theorem 4.18, equals $\bigcup \mathfrak{R}' \cup \bigcup \mathfrak{Q} = R' \cup Q$. Hence Theorem 4.13 and the inductive assumption yield

$$R \cap Q = (R' \cup R_i) \cap Q = R' \cap Q = \bigcup \mathfrak{T}.$$

This completes the proof of the second assertion, which, in turn, yields the third assertion for empty $\mathfrak{T}$.

THEOREM 4.21   *Each point set $\mathfrak{P}$ contains a basic subset $\mathfrak{P}'$ such that* $\bigcup \mathfrak{P}' = \bigcup \mathfrak{P}$.

If $\mathfrak{P} = \{P_1, \ldots, P_n\}$, choose $P_{i_1} = P_1$. If $P_{i_1}, \ldots, P_{i_k}$ have been chosen and if there is at least one point of $\mathfrak{P}$ that is not a part of $P_{i_1} \cup \cdots \cup P_{i_k}$, then choose such a point $P_{i_{k+1}}$ until a set $\mathfrak{P}'$ is reached such that no point of $\mathfrak{P}$ is left that is not in $\bigcup \mathfrak{P}'$. From $P_i \subseteq \bigcup \mathfrak{P}'$ for $i = 1, \ldots, n$, it follows that $\bigcup \mathfrak{P} \subseteq \bigcup \mathfrak{P}'$; and clearly $\bigcup \mathfrak{P}' \subseteq \bigcup \mathfrak{P}$. By its construction, the set $\mathfrak{P}'$ satisfies the conditions which, according to the preceding theorem, are sufficient for $\mathfrak{P}'$ to be basic.

THEOREM 4.22   *Each subset of a basic set is basic.*

The assertion will be proved by assuming that the set $\{P_1, \ldots, P_m\}$ is not basic and showing that each set $\mathfrak{P}' = \{P_1, \ldots, P_m, P_{m+1}, \ldots, P_{m+n}\}$ is not basic. Supposing that $P_1 \subseteq P_2 \cup \cdots \cup P_m$, one can assume that

$$P_1 \subseteq P_2 \cup \cdots \cup P_m \cup P_{m+1} \cup \cdots \cup P_{m+k} = S_{m+k}.$$

Then $S_{m+k} = S_{m+k} \cup P_1$. From Lemma 4.4 it follows that

$$P_1 \subseteq (S_{m+k} \cup P_1) \cup P_{m+k+1} = S_k \cup P_{m+k+1} = S_{m+k+1}.$$

Thus by induction one obtains $P_1 \subseteq S_{m+n}$, and so $\mathfrak{P}'$ is not basic.

THEOREM 4.23 *If $\mathfrak{P}$ is a basic set of m points, and $\mathfrak{Q}$ is a set of n points such that $\bigcup\mathfrak{Q} = \bigcup\mathfrak{P}$, then $\mathfrak{Q}$ contains a basic subset $\mathfrak{Q}'$ of m points such that $\bigcup\mathfrak{Q}' = \bigcup\mathfrak{P}$. Any two basic sets of points with the same join contain equally many elements.*

Set $Y = \bigcup\mathfrak{P} = \bigcup\mathfrak{Q}$, and let $P_1, \ldots, P_m$ denote the points of $\mathfrak{P}$, and $Q_1, \ldots, Q_n$ the points of $\mathfrak{Q}$. Inductively, one can make the assumptions (satisfied for $k = 0$):

(α_k) Y is the join of the basic set $\{P_1, \ldots, P_{n-k-1}, P_{n-k}, Q_{i_1}, \ldots, Q_{i_k}\}$.

If $\mathfrak{P}_k = \{P_1, \ldots, P_{n-k-1}, Q_{i_1}, \ldots, Q_{i_k}\}$ and $Y_k = \bigcup\mathfrak{P}_k$, then $Y_k \to Y_k \cup P_{n-k} = Y$, and $\mathfrak{Q}$ contains a point Q that is not a part of Y_k. Since $Y_k \subset Y_k \cup Q \subseteq Y_k \cup P = Y$, it follows from Theorem 4.12 that $Y_k \cup Q_k = Y$, and from Theorem 4.20 that $\mathfrak{P}_k \cup \{Q\}$ is basic. Setting $Q = Q_{i_{k+1}}$ one sees that $\mathfrak{P}_k \cup \{Q_{i_{k-1}}\}$ satisfies (α_{k+1}). In this way, one obtains (α_n), which is the first assertion of the theorem. Clearly, $\mathfrak{Q}$ must contain at least m points. If $\mathfrak{Q}$ is also basic, then by letting $\mathfrak{P}$ and $\mathfrak{Q}$ change roles, one sees that $m = n$.

THEOREM 4.24 *If $\mathfrak{P}$ and $\mathfrak{Q}$ are basic sets of m and n points, respectively, and $X = \bigcup\mathfrak{P}$ is a part of $Y = \bigcup\mathfrak{Q}$, then there exists a subset $\mathfrak{Q}'$ of $\mathfrak{Q}$ consisting of $n - m$ points such that $\mathfrak{P}$ and $\mathfrak{Q}'$ are disjoint, $\mathfrak{P} \cup \mathfrak{Q}'$ is basic, and $Y = \bigcup(\mathfrak{P} \cup \mathfrak{Q}')$.*

Set $X = X_0$. If a part X_{k-1} of Y has been defined and $X_{k-1} \neq Y$, then there exists a point Q_k in $\mathfrak{Q}$ such that $Q_k \not\subseteq X_{k-1}$. Set $X_k = X_{k-1} \cup Q_k$. The sets X_k so defined are basic by Theorem 4.20. There must be an integer k such that $X_k = Y$. Since Y is the join of a basic set of n points, it follows from Theorem 4.23 that this k is $n - m$.

The theory of basic sets was first derived from the Projectivity Laws by Otto Schreiber (Doctoral Thesis, Vienna 1932).

EXERCISE

1. Dualize the concept of basic sets of points and all theorems in this section.

4.7   THE SET OF ALL REGULAR ELEMENTS OF A PROJECTIVE STRUCTURE

The theory developed in this section is the work of Dr. H. Skala. It makes use of the following statement about the meet of maximal proper parts of an element.

LEMMA 4.5   *If $Z \to X$, $Z' \to X$, $Z \neq Z'$, and there exists a point P that is a part of X but not a part of Z, then $Z \cap Z' \to Z$. Moreover, if $P \nsubseteq Z'$, then $Z \cap Z' \to Z'$.*

Let W be any element such that $Z \cap Z' \subseteq W \subseteq Z$. It must be proved that $W = Z \cap Z'$ or $W = Z$. Since $P \nsubseteq Z \to X$, Theorem 4.12 implies $Z \cup P = X$. Hence, by Theorem 4.13,

$$W \cap X = W \cap (Z \cup P) = W \cap Z = W,$$

and consequently $W \subseteq X$. Furthermore, since $Z' \subseteq X$, it follows that $Z' \subseteq Z' \cup W \subseteq X$. Because of $Z' \to X$ one has $Z' \cup W = Z'$ or $Z' \cup W = X$. In the first case, $W \subseteq Z'$; and since $W \subseteq Z$, one has $W \subseteq Z \cap Z'$ which, in view of the assumption $Z \cap Z' \subseteq W$, implies $W = Z \cap Z'$. In the second case, $Z = X \cap Z = (Z' \cup W) \cap Z$; and, in view of $W \subseteq Z$, modularity yields $(W \cup Z') \cap Z = W \cup (Z' \cap Z)$. This last element equals W because of $Z \cap Z' \subseteq W$, whence $W = Z$. This completes the proof.

If an element that is the join of two distinct points is called a *line*, then the preceding lemma has the following consequence:

LEMMA 4.6   *Each proper part of a line is either a point or V.*

Let Y be a maximal proper part of the line $L = P_1 \cup P_2$, where $P_1 \neq P_2$. At least one of the two points is not a part of Y—say $P_2 \nsubseteq Y$. Clearly $P_2 \nsubseteq P_1$. If $Y = P_1$, then Y is a point. If $Y \neq P_1$, then, by the preceding lemma, (1) $P_1 \cap Y \to P_1$, whence $P_1 \cap Y = V$, and (2) $P_1 \cap Y \to Y$; that is, $V \to Y$, and Y is again a point. Since each maximal proper part of L is a point, the only nonmaximal proper part of L is V.

LEMMA 4.7   *If Q is a point, L a line, and Z any regular element such that $Q \subseteq L$ and $Q \nsubseteq Z$, then $Q \subseteq Z \cup L$.*

From $Q \nsubseteq Z \cup L$ it follows that $(Z \cup L) \cap Q = V$, whence by the First Projectivity Law

(*) $\qquad Z = Z \cup V = Z \cup ((Z \cup L) \cap Q) = Z \cup ((Z \cup Q) \cap L).$

From the assumptions it follows that $Q \subseteq (Z \cup Q) \cap L \subseteq L$. Consequently, according to the preceding lemma, $(Z \cup Q) \cap L$ is either Q or L. In the first case, (*) yields $Z = Z \cup Q$ against the assumption $Q \nsubseteq Z$. In the second case, $Z = Z \cup L$ and thus $L \subseteq Z$. Furthermore, since $Q \subseteq L$ and Z is regular, it follows from Theorem 4.17 that $Q \subseteq Z$, which contradicts the assumption. In either case, $Q \nsubseteq L \cup Z$ leads to a contradiction.

THEOREM 4.25 *If $\mathfrak{P}$ is a basic set of m points, and $Y \to \bigcup \mathfrak{P}$, then Y is the join of m − 1 points.*

By Lemma 4.6, the assertion holds for $m = 2$. It will be proved for $m + 1$ under the assumption that it holds for m.

Let $\mathfrak{P} = \{P_1, \ldots, P_m, P_{m+1}\}$ be basic and $Y \to \bigcup \mathfrak{P} = X$. Some point of $\mathfrak{P}$, say P_{m+1}, is not a part of Y. Nor is P_{m+1} a part of $X' = P_1 \cup \cdots \cup P_m \to X$. Hence, according to Lemma 4.5, $Y \cap X' \to X'$. Setting $Y \cap X' = Z$, by the inductive assumption one has $Z = R_1 \cup \cdots \cup R_{m-1}$, where $\{R_1, \ldots, R_{m-1}\}$ is a basic set. According to Theorem 4.24, there exist two points, Q_1 and Q_2, such that $X = R_1 \cup \cdots \cup R_{m-1} \cup Q_1 \cup Q_2$. Setting $Q_1 \cup Q_2 = L$ from Lemma 4.7 one obtains $Q_1 \subseteq Z \cup L$ and $Q_2 \subseteq Z \cup L$. Since $Z \subseteq Z \cup L$ it follows that $X \subseteq Z \cup L$. Clearly, also $Z \cup L \subseteq X$. Thus $X = Z \cup L$. Therefore, in view of $Z \subseteq Y$, modularity yields $Y = X \cap Y = (Z \cup L) \cap Y = Z \cup (L \cap Y)$. Being a part of L, the element $L \cap Y$ is either V or L or a point. The first two cases are impossible, since they would imply $Y = Z$ and $Y = Z \cup L = X$, respectively. Hence there is a point P such that $Y = Z \cup P$, and Y is the join of m points.

THEOREM 4.26 *Each part of a regular element is regular.*

In view of Theorem 4.19, the assertion follows from the preceding theorem.

THEOREM 4.27 *In the set of all regular elements, the operations $\cup$ and $\cap$ are associative.*

Since $\bigcup\mathfrak{P} \cup \bigcup\mathfrak{Q} \subseteq \bigcup(\mathfrak{P} \cup \mathfrak{Q})$ for any two point sets $\mathfrak{P}$ and $\mathfrak{Q}$, it follows from Theorem 4.26 that the join of any two regular elements is regular; and by the same theorem the meet of any two regular elements is regular. In order to prove the associativity of $\cup$, it suffices to show that $X \cup (Y \cup Z) \subseteq (X \cup Y) \cup Z$ for any three regular elements X, Y, Z. Now $X \subseteq X \cup Y$ and $X \cup Y \subseteq (X \cup Y) \cup Z$, and since the three elements are regular, it follows from Theorem 4.17 that $X \subseteq (X \cup Y) \cup Z$. Similarly, $Y \subseteq (X \cup Y) \cup Z$, and obviously $Z \subset (X \cup Y) \cup Z$. Hence $X \cup (Y \cup Z) \subseteq (X \cup Y) \cup Z$. The associativity of $\cap$ can be proved similarly.

COROLLARY *If Y and Z are regular and $Y' \subseteq Y$ and $Z' \subseteq Z$, then $Y' \cup Z' \subseteq Y \cup Z$ and $Y' \cap Z' \subseteq Y \cap Z$.*

Using associativity and commutativity of $\cup$ and the assumptions $Y \cup Y' = Y$ and $Z \cup Z' = Z$, one concludes that indeed

$$(Y \cup Z) \cup (Y' \cup Z') = (Y \cup Y') \cup (Z \cup Z') = Y \cup Z.$$

THEOREM 4.28 *For any two finite point sets $\mathfrak{P}$ and $\mathfrak{Q}$*

$$\bigcup\mathfrak{P} \cup \bigcup\mathfrak{Q} = \bigcup(\mathfrak{P} \cup \mathfrak{Q}).$$

If $\mathfrak{P} \cup \mathfrak{Q}$ is basic, then

$$\bigcup\mathfrak{P} \cap \bigcup\mathfrak{Q} = \bigcup(\mathfrak{P} \cap \mathfrak{Q}).$$

The first assertion is a corollary of the associativity of $\cup$. In order to prove the second, let $\mathfrak{P} \cap \mathfrak{Q}$ be the set $\{R_1, \ldots, R_k\}$ and suppose that $\mathfrak{Q}$ contains $k + i$ points. Since for $i = 0$ the assertion is trivial we assume its validity for $i - 1$. Assume

$$\mathfrak{Q} = \{R_1, \ldots, R_k, Q_1, \ldots, Q_{i-1}, Q_i\}$$

and set

$$\mathfrak{Q}' = \{R_1, \ldots, R_k, Q_1, \ldots, Q_{i-1}\}, \ \ Y = \bigcup\mathfrak{Q}, \ \ Y' = \bigcup\mathfrak{Q}', \ \ X = \bigcup\mathfrak{P}.$$

Since Q_i is not an element of $\mathfrak{Q}' \cup \mathfrak{P}$, and $\mathfrak{Q} \cup \mathfrak{P}$ is basic,

$$Q_i \nsubseteq \bigcup(\mathfrak{Q}' \cup \mathfrak{P}) = \bigcup\mathfrak{Q}' \cup \bigcup\mathfrak{P} = Y' \cup X.$$

Hence from Theorem 4.13 and the inductive assumption it follows that

$$\bigcup \mathfrak{P} \cap \bigcup \mathfrak{Q} = Y \cap X = (Y' \cup Q_i) \cap X = Y' \cap X = \bigcup \mathfrak{Q}' \cap \bigcup \mathfrak{P}$$
$$= \bigcup(\mathfrak{Q}' \cap \mathfrak{P}) = \bigcup(\mathfrak{Q} \cap \mathfrak{P}).$$

If $\mathfrak{P}$ and $\mathfrak{Q}$ are disjoint, then $\bigcup(\mathfrak{P} \cap \mathfrak{Q}) = V$. Using Theorem 4.22 one thus obtains (as an immediate consequence of the last theorem) the following necessary condition for basic sets, which sharpens the property by which basic sets are defined.

COROLLARY *If $\mathfrak{R}$ is a basic set, then for any two disjoint subsets $\mathfrak{P}$ and $\mathfrak{Q}$ of $\mathfrak{R}$,*

$$\bigcup \mathfrak{P} \cap \bigcup \mathfrak{Q} = V.$$

If $Y \subseteq Z$, then two elements W and W' are said to be (Y, Z)-*complementary* if $W \cap W' = Y$ and $W \cup W' = Z$. Either of the elements W and W' (which are parts of Z) is also called a (Y, Z)-*complement* of the other. Elements that are (V, U)-complementary are said to be *complementary*.

THEOREM 4.29 *If Z is regular and $Y \subseteq W \subseteq Z$, then there exists a (Y, Z)-complement of W. If X is any (not necessarily regular) element such that Z, W, and Y are parts of X and $Y \subseteq W \subseteq Z$, then each (Y, Z)-complement of W is a part of X.*

Before going into the proof of the theorem one should observe that the assumed regularity of Z implies that W and Y are regular. But if X is non-regular the assumption that $Z \subseteq X$, even if Z is regular, does not imply that W and Y are parts of X. In the nonassociative structure $\mathfrak{S}_4$ in Section 4.3, the P_i are points, L is regular and a part of H; yet the P_i are not parts of H. Hence, in the theorem, along with $Z \subseteq X$ it is *assumed* that $W \subseteq X$ and $Y \subseteq X$, and both assumptions will be used in its proof, which follows.

Let $\mathfrak{Y}$ be a basic set whose join is the regular element Y. By Theorem 4.25 there exist (1) a set $\mathfrak{W}$ disjoint from $\mathfrak{Y}$ such that $\mathfrak{Y} \cup \mathfrak{W}$ is basic and $W = \bigcup(\mathfrak{Y} \cup \mathfrak{W})$; (2) a set $\mathfrak{Z}$ disjoint from $\mathfrak{Y} \cup \mathfrak{W}$ and such that $\mathfrak{Y} \cup \mathfrak{W} \cup \mathfrak{Z}$ is basic and $Z = \bigcup(\mathfrak{Y} \cup \mathfrak{W} \cup \mathfrak{Z})$. Set $\bigcup(\mathfrak{Y} \cup \mathfrak{Z}) = W'$. Then

$$W \cup W' = \bigcup(\mathfrak{Y} \cup \mathfrak{W}) \cup \bigcup(\mathfrak{Y} \cup \mathfrak{Z}) = \bigcup(\mathfrak{Y} \cup \mathfrak{W} \cup \mathfrak{Z}) = Z.$$

By Theorem 4.28,

$$W \cap W' = \bigcup(\mathfrak{Y} \cup \mathfrak{W}) \cap \bigcup(\mathfrak{Y} \cup \mathfrak{Z}) = \bigcup \mathfrak{Y} = Y.$$

Hence W' is a (Y, Z)-complement of W. In order to prove the second assertion of the theorem, let X be an element such that $Z \subseteq X$, $W \subseteq X$, $Y \subseteq X$ and let W^* by any (Y, Z)-complement of W. Because of $Z \subseteq X$ and $W \subseteq X$, by modularity,

$$Z = Z \cap X = (W \cup W^*) \cap X = W \cup (W^* \cap X).$$

In view of $W^* \subseteq Z$ and $Y \subseteq X$, the Second Projectivity Law yields

$$W^* = W^* \cap Z = W^* \cap ((W^* \cap X) \cup W) = W^* \cap ((W^* \cap W) \cup X)$$
$$= W^* \cap (Y \cup X) = W^* \cap X.$$

Thus $W^* \subseteq X$. This completes the proof of the theorem.

Although the join of any two parts of an element X is a part of X, their meet need not be, as demonstrated by the nonassociative structure $\mathfrak{S}_2$ in Section 4.3.

LEMMA 4.8 *If Y and Z are regular parts of X, then $Y \cap Z \subseteq X$.*

Setting $Y \cap Z = T$ one must prove that $T \subseteq X$. By the preceding theorem, there exists a (V, Y)-complement T' of T. First it will be shown that $T' \subseteq X$. By the corollary of Theorem 4.27, it follows from $T' \subseteq Y$ that $T' \cap Z \subseteq Y \cap Z = T$, and hence

$$T' \cap Z = (T' \cap Z) \cap T = (T' \cap T) \cap Z = V \cap Z = V.$$

Furthermore, one has $T' \cup Z = T' \cup (T \cup Z) = (T' \cup T) \cup Z = Y \cup Z$. Since $Z \subseteq X$, modularity yields

$$Y \cup Z = (Y \cup Z) \cap X = (Z \cup T') \cap X = Z \cup (T' \cap X).$$

Using $T' \subseteq Y \subseteq Y \cup Z$, the preceding result, and the Second Projectivity Law one obtains

$$T' = T' \cap (Y \cup Z) = T' \cap ((T' \cap X) \cup Z)$$
$$= T' \cap ((T' \cap Z) \cup X) = T' \cap (V \cup X) = T' \cap X.$$

Thus $T' \subseteq X$, as claimed. Since V and Y also are parts of X and $V \subseteq T' \subseteq Y$, the preceding theorem implies that each (V, Y)-complement of T'—in particular, T—is a part of X.

Let X be any element of a projective structure $\mathfrak{S}$, and denote by $\mathfrak{L}_X$ the set of all regular parts of X. If Y and Z belong to $\mathfrak{L}_X$, then so do both $Y \cup Z$ and $Y \cap Z$ (the latter by Theorem 4.28), and these elements are parts of X by Lemmas 4.1 and 4.8. Hence $\mathfrak{L}_X$ is a modular lattice containing V. Suppose $\mathfrak{L}_X$ contains a *universal element of its own* in the sense of an element U' such that $U' \cap Z = Z$ for each element Z of $\mathfrak{L}_X$. Then $U' \cap X = X$ and thus $X \subseteq U'$. But U', being an element of $\mathfrak{L}_X$, is also a part of X, whence $U' = X$. Hence $\mathfrak{L}_X$ contains a universal element, namely X, if and only if X is regular. A lattice $\mathfrak{L}$ is said to be *relatively complemented* if, for any three elements Y, W, Z such that $Y \subseteq W \subseteq Z$, $\mathfrak{L}$ contains a (Y, Z)-complement W' of W. According to Theorem 4.29, $\mathfrak{L}_X$ has this property. One thus arrives at the following fundamental result.

THEOREM 4.30   *For any element X of a projective structure $\mathfrak{S}$, the set of all regular parts of X is a relatively complemented modular lattice containing V. The set contains a universal element of its own, namely X, if and only if X is regular. In particular, the set of all regular elements of $\mathfrak{S}$ is a relatively complemented modular lattice, which is identical with $\mathfrak{S}$ if and only if U is regular.*

Example 1.   Let O be a point in the Hilbert space. The set of all linear subspaces that contain O is a projective structure; in fact, it is a modular lattice. Its regular elements are the finite-dimensional linear subspaces of the Hilbert space containing O. This sublattice includes neither the Hilbert space, U, nor even a universe of its own.

Example 2.   In the projective structure $\mathfrak{S}_4$ in Section 4.3, the lattice of all regular parts includes all elements except H and U. This lattice includes V and a universe of its own, L.

EXERCISES

1. Let $\mathfrak{S}$ be a projective structure with distinct distinguished elements U and V. Show that by adjoining to $\mathfrak{S}$ a *superuniverse* U^* such that $U^* \neq U$ and $U^* \cap X = X$ for each X (in particular, for $X = U$, whereas $U \cap X = X$ holds only for all $X \neq U^*$), one obtains a projective structure whose universe is not regular. What if an *infravacuum* (that is, the dual of a superuniverse) is adjoined to $\mathfrak{S}$?

2. Dualize all theorems and lemmas in this section.

3. Prove that for any three elements X, Y, Z of a modular lattice (as for any three regular elements of a projective structure) the following so-called *exchange equivalences* hold:

$$(X \cup Y) \cap Z = X \cap Z \text{ if and only if } (X \cup Z) \cap Y = X \cap Y,$$
$$(X \cap Y) \cup Z = X \cup Z \text{ if and only if } (X \cap Z) \cup Y = X \cup Y.$$

As a corollary, derive Theorem 4.13. Show that the exchange equivalence does not hold in Example 3 of Section 4.3.

4. Show that $\bigcup(\mathfrak{P} \cap \mathfrak{Q})$ is a part, but may be a proper part, of $\bigcup\mathfrak{P} \cap \bigcup\mathfrak{Q}$.

4.8 DIMENSION

By Theorem 4.21, each regular element X of a projective structure is the join of a basic point set. By Theorem 4.23, the number of points in all basic sets with the join X is the same. This number, diminished by 1, may be called the *dimension* of X—briefly, dim X. (The reason for subtracting 1 is that lines and planes, which are joins of pairs and triples of points, are traditionally called one- and two-dimensional, respectively.) According to this definition, each point is 0-dimensional, and dim $V = -1$.

The main properties of dimension are consequences of

LEMMA 4.9 *If $W = X \cap Y$ and $W = \bigcup\mathfrak{W}$, $X = \bigcup\mathfrak{X}$, $Y = \bigcup\mathfrak{Y}$, then there exist three disjoint point sets $\mathfrak{W}' \subseteq \mathfrak{W}$, $\mathfrak{X}' \subseteq \mathfrak{X}$, $\mathfrak{Y}' \subseteq \mathfrak{Y}$ such that*

(1) $W = \bigcup\mathfrak{W}'$, $X = \bigcup(\mathfrak{W}' \cup \mathfrak{X}')$, $Y = \bigcup(\mathfrak{W}' \cup \mathfrak{Y}')$,
(2) $\mathfrak{W}' \cup \mathfrak{X}' \cup \mathfrak{Y}'$ *is a basic set.*

By Theorems 4.21 and 4.24, there exist (1) a basic subset $\mathfrak{W}'$ of $\mathfrak{W}$ such that $\bigcup\mathfrak{W}' = \mathfrak{W}$, and (2) subsets $\mathfrak{X}'$ and $\mathfrak{Y}'$ of $\mathfrak{X}$ and $\mathfrak{Y}$, respectively, disjoint from $\mathfrak{W}'$ and such that $\mathfrak{W}' \cup \mathfrak{X}'$ and $\mathfrak{W}' \cup \mathfrak{Y}'$ are basic and have the joins X and Y, respectively. Let $(P_1, \ldots, P_n)$ be a serialization of $\mathfrak{Y}'$, and set $T_k = W \cup P_1 \cup \cdots \cup P_k$ for $k = 1, \ldots, n$. Clearly, $W \subseteq X \cap T_k$. Since $T_k \subseteq Y$, it follows from Theorem 4.27 that $X \cap T_k \subseteq X \cap Y = W$ for $k = 1, \ldots, n$ and hence $X \cap T_k \subseteq X \cap T_{k+1} = X \cap (T_k \cup P_{k+1})$. In view of $P_{k+1} \nsubseteq T_k$, Theorem 4.13 implies $P_{k+1} \nsubseteq X \cup T_k$. Consequently, by Theorem 4.20, the set $\mathfrak{W}' \cup \mathfrak{X}' \cup \mathfrak{Y}'$ is basic. This concludes the proof of the lemma.

THEOREM 4.31   *If X and Y are regular elements of a projective structure, then*

(1) dim $X <$ dim Y if $X \subset Y$, and dim $X =$ dim $Y - 1$ if X is a maximal proper part of Y,

(2) dim $X +$ dim $Y = \dim(X \cup Y) + \dim(X \cap Y)$ (*Dimension Formula*).

The first assertion follows immediately from the definition of dimension, the second from the preceding lemma. For since $\mathfrak{X}'$ and $\mathfrak{Y}'$ are disjoint from $\mathfrak{W}'$, which is the intersection of $\mathfrak{W}' \cup \mathfrak{X}'$ and $\mathfrak{W}' \cup \mathfrak{Y}'$, it follows that if $|\mathfrak{Z}|$ denotes the number of points in the point set $\mathfrak{Z}$,

$$|\mathfrak{W}' \cup \mathfrak{X}'| + |\mathfrak{W}' \cup \mathfrak{Y}'| = |\mathfrak{W}' \cup \mathfrak{X}' \cup \mathfrak{Y}'| + |\mathfrak{W}'|.$$

Since dim $\bigcup \mathfrak{Z} = |\mathfrak{Z}| - 1$, equation (2) follows immediately.

COROLLARY   *If* dim $X =$ dim Y *and either* $X \subseteq Y$ *or* $Y \subseteq X$, *then* $X = Y$.

EXERCISES

1. Prove that $\dim(X \cup Y) \leqq$ dim $X +$ dim $Y + 1$ for any two regular elements.

2. If dim $X =$ dim $Y = n$ and $\dim(X \cup Y) = 2n + 1$, then X and Y are said to be a pair of *skew* elements. Note that X and Y may well be two n-dimensional parts of a $(2n + 1)$-dimensional element without being skew.

3. Prove that if Z is regular and W' is any (Y, Z)-complement of W, then

$$\dim W' = \dim Y + \dim Z - \dim W.$$

Hence any two (Y, Z)-complements of W have the same dimension.

4. Prove that for any three regular elements X, Y, Z

$$\dim(X \cup Y \cup Z) = \dim X + \dim Y + \dim Z$$
$$- \dim(X \cap Y) - \dim(Y \cap Z) - \dim(Z \cap Y) + \dim(X \cap Y \cap Z).$$

(*Hint:* Begin with a basic set whose join is $X \cap Y \cap Z$.)

5. Derive the first assertion of Theorem 4.13 for regular elements using the concept of dimension.

4.9   DUALIZATION

The results of the preceding Sections can be dualized. In comparing the results with their dual counterparts it is convenient to call the regular elements ∪-*regular*. Dually, an element X is said to be ∩-regular if there exists a (finite) set $\mathfrak{H}$ of hyperplanes such that $X = \bigcap \mathfrak{H}$. Each element having a ∩-regular part is ∩-regular. For each element X of a projective structure, the set of all ∩-regular elements having X as a part is a relatively complemented modular lattice containing U. This lattice contains a vacuum of its own (that is, an element such that $V' \cup X = X$ for each element X of the lattice) if and only if X is ∩-regular. In this case $V' = X$.

The set $\mathfrak{H}$ of hyperlanes is basic if no element of $\mathfrak{H}$ has the meet of the other elements as a part. Each subset of a basic set is basic. Each set of hyperplanes contains a basic subset with the same meet. Any two basic sets of hyperplanes with the same meet, X, contain equally many elements. This number diminished by 1 may be called the ∩-*dimension* of X—briefly, $\dim_\cap X$. Hyperplanes have the ∩-dimension 0, and $\dim_\cap U = -1$. Clearly the ∩-dimension satisfies the dimension formula (2) of Theorem 4.31.

Connecting the two dimensions with one another it is convenient to write ∪-dim X for dim X.

THEOREM 4.32   *The universe U is ∪-regular if and only if the vacuum V is ∩-regular. If, in a projective structure, U is ∪-regular, then for each element X*

$$\dim_\cup X + \dim_\cap X = \dim_\cup U + \dim_\cup V = \dim_\cap U + \dim_\cap V = \dim_\cup U - 1.$$

If $\dim U = n - 1$ and X is the join of the basic set $\mathfrak{P} = \{P_1, \ldots, P_k\}$, then by Theorem 4.24 there exists a set $\mathfrak{Q} = \{Q_1, \ldots, Q_{n-k}\}$ such that $\mathfrak{P} \cup \mathfrak{Q}$ is basic and $U = \bigcup (\mathfrak{P} \cup \mathfrak{Q})$. If $H_i = X \cup Q_1 \cup \cdots \cup Q_{i-1} \cup Q_{i+1} \cup \cdots \cup Q_{n-k}$, then by iterated application of Theorem 4.20 the set $\{H_1, \ldots, H_{n-k}\}$ is seen to be basic and its meet is X. The second assertion of the theorem thus follows from the definitions of the dimensions. For $X = V$ one has $k = 0$, and V is the meet of a basic set of n hyperplanes.

If X and X' are complementary, then

$$\dim_\cup X + \dim_\cup X' = \dim_\cup U - 1.$$

One thus obtains the following

COROLLARY   *If X and X' are complementary, then $\dim_\cap X = \dim_\cup X'$.*

EXERCISES

1. Prove that if U is the join of the basic set $\{P_1, \ldots, P_n\}$ and that if $H_i = P_1 \cup \cdots \cup P_{i-1} \cup P_{i+1} \cup \cdots \cup P_n$ $(i = 1, \ldots, n)$, then

$$H_1 \cap \cdots \cap H_k = P_{k+1} \cup \cdots \cup P_n \ (k = 1, \ldots, n-1).$$

Dualize this statement.

4.10 ASSOCIATIVITY IN PROJECTIVE STRUCTURES

In a projective structure some special cases of associativity imply the associativity of the operations. More generally, let S be a set such that with any two elements x and y an element xy is associated. The binary operation is said to be *quasiassociative* if

$$(xy)(yz) = x(y(yz)) = ((xy)y)z \text{ for each } x, y, z \text{ in } S.$$

LEMMA 4.10 *An idempotent, quasiassociative operation is associative.*

From $y = yy = y(yy)$ it follows that

$$xy = x(y(yy)) = (xy)(yy) = (xy)y$$

for any two elements x and y. Similarly, $yz = y(yz)$. Hence

$$(xy)z = ((xy)y)z = (xy)(yz) = x(y(yz)) = x(yz).$$

In terms of the binary operation, a binary relation ρ can be introduced by setting

$$x\rho y \text{ if and only if } xy = x.$$

The relation ρ is $\subseteq$ for the binary operation $\cap$, and $\supseteq$ for $\cup$. The binary operation is said to be *ρ-associative* if $x\rho y$ implies $(xy)z = x(yz)$ for each z.

LEMMA 4.11 *An idempotent, commutative, ρ-associative operation is associative.*

The idempotency of the operation implies that $x\rho x$ for each x. Hence commutativity and ρ-associativity yield

$$(xy)x = x(xy) = (xx)y = xy \text{ and } (xy)z = ((xy)y)z = (xy)(yz) = x(yz).$$

THEOREM 4.33   *In a projective structure,*

(1) *the following three statements are mutually equivalent:*
    α) *the operation* $\cup$ *is associative,*
    β) $X \subseteq Y$ *implies* $X \subseteq Y \cup Z$ *for each* Z,
    γ) *the relation* $\subseteq$ *is transitive;*
(2) *if one of the operations* $\cup$, $\cap$ *is associative, then so is the other;*
(3) *it is sufficient for the associativity of the operations that one of them be quasiassociative or* $\subseteq$*-associative or* $\supseteq$*-associative.*

$(\alpha \to \beta)$. $X \subseteq Y$ by (α) implies $X \cup (Y \cup Z) = (X \cup Y) \cup Z = Y \cup Z$, whence $X \subseteq Y$ implies $X \subseteq Y \cup Z$.

$(\beta \to \gamma)$. $X \subseteq Y$ by (β) implies $X \subseteq Y \cup Z$. If $Y \subseteq Z$, then $X \subseteq Z$.

$(\gamma \to \alpha)$. $Y \subseteq Y \cup Z$ and $Y \cup Z \subseteq X \cup (Y \cup Z)$ by (γ) imply (δ) $Y \subseteq X \cup (Y \cup Z)$. Similarly, (δ') $Z \subseteq X \cup (Y \cup Z)$. By Lemma 4.10, (δ) combined with $X \subseteq X \cup (Y \cup Z)$ yields (δ'') $X \cup Y \subseteq X \cup (Y \cup Z)$. By Lemma 4.10, (δ'') and (δ') yield $(X \cup Y) \cup Z \subseteq X \cup (Y \cup Z)$. In the same way one obtains the opposite inclusion and hence (α).

This concludes the proof of (1). Since the relations $\subseteq$ and $\supseteq$ corresponding to $\cap$ and $\cup$, respectively, are inverse, if one is transitive, then so is the other. Hence (2) follows from (1) and its dual. (3) follows from (2) and the remarks preceding the theorem.

EXERCISES

1. Under the assumptions of Lemma 4.11, formulate properties of the binary operations which are equivalent to the transitivity of the relation ρ.

2. Examine the implications: if $x\rho y$ then $x\rho yz$ for each z; if $x\rho y$, then $wx \, \rho \, y$ for each w. (Notice the completely different character of these two implications.)

3. Find the properties of the part relation in a projective structure, other than those mentioned in the theorem, which are equivalent to associativity of the operations.

4. The *converse of modularity* is the implication: If $(X \cup Y) \cap Z = X \cup (Y \cap Z)$, then $X \subseteq Z$. Prove that in a projective structure the converse of modularity holds for any three elements X, Y, Z if and only if the part relation is transitive or, what amounts to the same, the operations $\cup$ and $\cap$ are associative.

4.11 FINITE-DIMENSIONAL PROJECTIVE STRUCTURES AND THEIR PRODUCTS

A projective structure is called *finite-dimensional* if it satisfies the following self-dual assumption.

POSTULATE III *The universe, U, is $\cup$-regular (that is, the join of a finite set of points) or the vacuum, V, is $\cap$-regular (that is, the meet of a finite set of hyperplanes.)*

According to Theorem 4.32, either alternative implies the other. By the dimension of a finite-dimensional structure one means $\dim_\cup U$ or $\dim_\cap V$. From Theorem 4.26 it also follows that in a finite-dimensional structure each element is both $\cup$- and $\cap$-regular and has finite $\cup$- and $\cap$-dimensions. Each -1-dimensional structure is isomorphic to $\mathfrak{B}$ consisting of $V (= U)$ only. Of course, when considering several -1-dimensional structures simultaneously one may denote them by $\mathfrak{B}_1$, $\mathfrak{B}_2$, ..., and their elements by V_1, V_2, ... Each 0-dimensional structure consists of two elements: V and $U (\neq V)$. Here U is a point, whence the elements of a 0-dimensional structure are often denoted by V and P or V' and P' or the like.

In view of the corollary to Theorem 4.32, Postulate III might be replaced by the following equivalent assumption.

POSTULATE III' *There exist two complementary elements such that either both are $\cup$-regular or both are $\cap$-regular.*

Examples of finite-dimensional projective structures include, for each finite set S, the Boolean algebra of all subsets of S with union and intersection as join and meet, with S and the empty set playing the roles of U and V, respectively. If S has exactly n elements, then $\dim U = n - 1$ and the structure contains exactly 2^n elements including n points corresponding to the elements of S. If $Y \subseteq W \subseteq Z$, then W has exactly one (Y,Z)-complement. For each $k < m \leqslant n - 1$, any two m-dimensional elements have equally many parts that are k-dimensional.

The Cartesian product—briefly, *product*—of two projective structures $\mathfrak{S}'$ and $\mathfrak{S}''$ is the set $\mathfrak{S}' \times \mathfrak{S}''$ of all pairs (X', X'') such that X' and X'' belong to $\mathfrak{S}'$ and $\mathfrak{S}''$, respectively, with join and meet defined as follows:

$$(X', X'') \cup (Y', Y'') = (X' \cup Y', X'' \cup Y''),$$
$$(X', X'') \cap (Y', Y'') = (X' \cap Y', X'' \cap Y'')$$

The Projectivity Laws are satisfied, and (V', V'') and (U', U'') are distinguished elements. Hence $\mathfrak{S}' \times \mathfrak{S}''$ is a projective structure; and clearly,

$$(X', X'') \subseteq (Y', Y'') \text{ if and only if } X' \subseteq Y' \text{ and } X'' \subseteq Y''$$

A substructure $\mathfrak{S}^*$ of $\mathfrak{S}$ is called *complete* if $\mathfrak{S}^*$ consists of an element of $\mathfrak{S}$ and all its parts. One readily proves

LEMMA 4.12   *Each complete substructure $\mathfrak{S}^*$ of $\mathfrak{S}' \times \mathfrak{S}''$ is the product of complete substructures of $\mathfrak{S}'$ and $\mathfrak{S}''$.* (One factor of $\mathfrak{S}^*$ may be $\mathfrak{B}$. If $\mathfrak{S}^*$ consists of (V', V'') alone, then both factors are $\mathfrak{B}$.)

The elements (P', V'') and (V', P'') for all points P' of $\mathfrak{S}'$, and P'' of $\mathfrak{S}''$, are the points of $\mathfrak{S}' \times \mathfrak{S}''$, while (H', U'') and (U', H'') are the hyperplanes. Suppose dim $\mathfrak{S}' = k'$ and dim $\mathfrak{S}'' = k''$, and assume

$$U' = P'_1 \cup \cdots \cup P'_{k'+1} \qquad \text{and} \qquad U'' = P''_1 \cup \cdots \cup P''_{k''+1}.$$

Then the universe of $\mathfrak{S}' \times \mathfrak{S}''$ is

$$(U', U'') = (P'_1, V'') \cup \cdots \cup (P'_{k'+1}, V'') \cup (V', P''_1) \cup \cdots \cup (V', P''_{k''+1}),$$

where the set of the $k' + k'' + 2$ points in the join is readily seen to be basic. Hence

$$\dim(\mathfrak{S}' \times \mathfrak{S}'') = k' + k'' + 1 = \dim \mathfrak{S}'' + \dim \mathfrak{S}'' + 1.$$

By induction, one obtains

THEOREM 4.34   (PRODUCT THEOREM FOR DIMENSION)

$$\dim(\mathfrak{S}_1 \times \cdots \times \mathfrak{S}_m) = \dim \mathfrak{S}_1 + \cdots + \dim \mathfrak{S}_m + m - 1.$$

If all factors except $\mathfrak{S}_i$ are -1-dimensional, then the product has the dimension of $\mathfrak{S}_i$. In fact, $\mathfrak{B} \times \cdots \times \mathfrak{B} \times \mathfrak{S}_i \times \mathfrak{B} \times \cdots \times \mathfrak{B}$ clearly is isomorphic to $\mathfrak{S}_i$.

If $\mathfrak{S}'$ and $\mathfrak{S}''$ are finite-dimensional structures $\neq \mathfrak{B}$, then they include points P' and P'', respectively, and $\mathfrak{S}' \times \mathfrak{S}''$ contains the element (P', P''), which is a *line* (that is, one-dimensional) and has, besides (V', V''), only two parts: the points (P', V'') and (V', P''). One thus obtains

THEOREM 4.35 *Each product of finite-dimensional structures $\neq \mathfrak{B}$ includes a line having, besides the vacuum, only two points as parts.*

COROLLARY *In the poduct of two 0-dimensional structures, the universe is a line with exactly two points.*

The one-dimensional (Boolean) structure consisting of V, U, and two points Q_1, Q_2 is not by definition a product of structures. But it obviously is isomorphic to the product of the 0-dimensional structures consisting of V', P', and V'', P'' by virtue of the correspondence

$$V, Q_1, Q_2, U \leftrightarrow (V', V''), (P', V''), (V', P''), (P', P'')$$

A structure that is isomorphic to the product of at least two structures $\neq \mathfrak{B}$ is said to be *composite*; the structures $\neq \mathfrak{B}$ that are not composite are called *simple*. Whereas the one-dimensional Boolean structure was shown to be composite, each one-dimensional nonBoolean structure $\mathfrak{M}$ (including three or more points Q_1, Q_2, Q_3, ...) is simple. For if $\mathfrak{M}$ were composite, then by Theorem 4.35 $\mathfrak{M}$ would include a one-dimensional element with exactly two points, whereas the (only) line in $\mathfrak{M}$ has at least three points as parts. Each n-dimensional Boolean structure is the product of $n + 1$ structures of dimension 0.

Simple structures are analogous to prime numbers in arithmetic, $\mathfrak{B}$ and 1 corresponding to one another. ($\mathfrak{B}$ and 1 are considered as neither composite nor simple or prime.) The following is an analogue of the theorem that each integer n (> 1) is either prime or the product of prime numbers (some of which may be equal) and that if

$$n = p_1 \cdot p_2 \cdots \cdot p_n = p_1' \cdot p_2' \cdots \cdot p_n',$$

then $n' = n$ and the p_j' are a permutation of the p_i.

THEOREM 4.36 *Each finite-dimensional projective structure is either simple or isomorphic to the product of simple structures (some of which may be isomorphic to one another). If $\mathfrak{S}$ is isomorphic to $\mathfrak{S}_1 \times \cdots \times \mathfrak{S}_m$ as well* *as to* $\mathfrak{S}_1' \times \cdots \times \mathfrak{S}_{m'}'$, *where all* $\mathfrak{S}_i$ *and* $\mathfrak{S}_j'$ *are simple, then* $m' = m$ *and* $\mathfrak{S}_k$ *is isomorphic to* $\mathfrak{S}_{i_k}'$ $(k = 1, \ldots, m)$ *for some permutation* $i_1, \ldots, i_m$ *of* $1, \ldots, m$.

If $\mathfrak{S}$ is not simple let m be the greatest number such that $\mathfrak{S}$ is isomorphic to the product of m simple structures. According to Theorem 4.34 $m \leqslant 1 +$ dim $\mathfrak{S}$. If $\mathfrak{S}$ is isomorphic to $\mathfrak{S}_1 \times \cdots \times \mathfrak{S}_m$, then each $\mathfrak{S}_i$ is simple. For if, say, $\mathfrak{S}_m$ were isomorphic to the product of two factors, then $\mathfrak{S}$ would be isomorphic to the product of at least $m + 1$ factors.

The second assertion is trivial for $m = 1$ and thus may be assumed to be valid for $m + 1$. Being isomorphic to $\mathfrak{S}$, the products $\mathfrak{P}$ of the $\mathfrak{S}_i'$, and $\mathfrak{P}'$ of the $\mathfrak{S}_j'$ are isomorphic to one another. Let ι be an isomorphic mapping of $\mathfrak{P}$ on $\mathfrak{P}'$. Since $\mathfrak{S}_m$ is isomorphic to the complete substructure $\mathfrak{B} \times \cdots \times \mathfrak{B} \times \mathfrak{S}_m$ of $\mathfrak{P}$, Lemma 4.12 implies that the image of $\mathfrak{S}_m$ by ι is a product of substructures of $\mathfrak{S}_1', \ldots, \mathfrak{S}_{m'}'$. Since $\mathfrak{S}_m$ is simple, $m' - 1$ of these factors must be $\mathfrak{B}$, and only one factor, say the substructure of $\mathfrak{S}_{i_m}'$, is isomorphic to $\mathfrak{S}_m$. The inverse of ι maps $\mathfrak{S}_{i_m}'$ on the product of substructures of $\mathfrak{S}_1, \ldots, \mathfrak{S}_{m-1}, \mathfrak{S}_m$; and since $\mathfrak{S}_i'$ is simple, all factors except one of them are $\mathfrak{B}$. This one factor clearly must be $\mathfrak{S}_m$. Hence $\mathfrak{S}_m$ and $\mathfrak{S}_{i_m}'$ are isomorphic. By the inductive assumption, $m' - 1 = m - 1$ and $\mathfrak{S}_{i_1}', \ldots, \mathfrak{S}_{i_m-1}', \mathfrak{S}_{i_m+1}', \ldots, \mathfrak{S}_{i_m-1}'$ is a permutation of $\mathfrak{S}_1, \ldots, \mathfrak{S}_{m-1}$. This concludes the proof of the theorem.

Each two-dimensional composite structure $\mathfrak{S}$ is either Boolean or the product of a O-dimensional and a simple one-dimensional structure. Hence $\mathfrak{S}$ includes, besides V and U, (1) a point P and a line L such that $P \nsubseteq L$; (2) at least three points $Q_1, Q_2, Q_3, \ldots$, such that $P \neq Q_i \subseteq L$; (3) at least three lines $M_i = P \cup Q_i$, whence $L \neq M_i \supseteq P$.

According to the Product Theorem for Dimension, each composite three-dimensional structure is either Boolean or of one of three types depending on whether the factors have the dimensions $0, 0, 1$ or $0, 2$ or $1, 1$, all factors of positive dimension being simple. An example of the third type is a structure whose nondistinguished elements are (1) two lines L and M, each containing at least three points $P_1, P_2, P_3, \ldots$ and $Q_1, Q_2, Q_3, \ldots$ as parts; (2) the lines $P_i \cup Q_j$ for all i and j; (3) the planes $L \cup Q_j$ and $M \cup P_i$ for all i and j.

EXERCISES

1. Let $\mathfrak{S}$ be a composite two-dimensional structure including a point P and a line L containing as parts at least three points but not P. Show that $\mathfrak{S}$ is

isomorphic to the product of a 0-dimensional structure and the one-dimensional structure consisting of the parts of L.

2. List all types of composite four-dimensional structures.

3. Prove that a projective structure in which each line has exactly two points as parts is Boolean.

4. Is a noncomplete substructure of $\mathfrak{S}_1 \times \cdots \times \mathfrak{S}_m$ necessarily the product of substructures of $\mathfrak{S}_1, \ldots, \mathfrak{S}_m$?

4.12   PROJECTIVE SPACES

For each element X of a projective structure $\mathfrak{S}$ and each nonnegative integer k, let $[X]_k$ be the set of all k-dimensional parts of X. One calls $\mathfrak{S}$ *homogeneous* if for each pair of elements, Y and Y', having equal dimensions and for each k there is a one-to-one correspondence between the sets $[Y]_k$ and $[Y']_k$, expressed by $[Y]_k \sim [Y']_k$. *Each nonBoolean composite structure is nonhomogeneous.* For $\mathfrak{S}$, being composite, includes a line L such that $[L]_0$ consists of exactly two points and, being nonBoolean, includes a line M such that $[M]_0$ contains more than two points.

The term *space* will be reserved for structures that are homogeneous. But homogeneity need not be postulated. It can be guaranteed by a much weaker assumption. All that has to be assumed about a structure of dimension > 1 is the existence of a finite set $\mathfrak{C}$ consisting of points and hyperplanes in certain relations. These elements will be denoted by the letters A and B (rather than P and H) in order to emphasize their unique role; and the set $\mathfrak{C}$ will be called a *fundamental configuration* of the structure. All that is postulated is the existence of a single such configuration, but there may of course be many. No assumption will be made about the (geometrically uninteresting) one-dimensional structures, whose hyperplanes are points. To Postulates I, II, III for a finite-dimensional projective structure we add the following fourth and last assumption.

POSTULATE IV   *If the projective structure is n-dimensional $(n \geqslant 2)$, then there exists a set $\mathfrak{C}$ consisting of two basic n-tuples, $\{A_1, \ldots, A_n\}$ of points and $\{B_1, \ldots, B_n\}$ of hyperplanes, such that the join, B_0, of the A_i (which is a hyperplane) and the meet, A_0, of the B_j (which is a point) satisfy the following conditions:*

(α) $A_0 \subseteq B_0$;
(β) *all n-tuples $\{A_0, A_{i_1}, \ldots, A_{i_{n-1}}\}$ and $\{B_0, B_{i_1}, \ldots, B_{i_{n-1}}\}$ are basic.*

The n-dimensional structures satisfying Postulate IV for some $n \geq 2$ and the one-dimensional structures including more than two points are called *projective* spaces. Thus a projective plane is a two-dimensional projective structure containing two points A_1, A_2 and two lines B_1, B_2 such that (α) the meet, A_0, of the two lines is a part of the join, B_0, of the two points; (β) A_0 is distinct from A_1 and A_2; B_0, from B_1 and B_2. Clearly, B_0 is a line having the three points A_0, A_1, A_2 as parts.

Hyperplanes of a three-dimensional structure are called *planes*. A triple of points (of planes) is nonbasic if and only if the elements are *collinear*— that is, if there exists a line having the three points are parts (being a part of the three planes). The elements of a basic triple of points or of planes are therefore said to be *noncollinear*. The fundamental configuration of a three-dimensional space consists of three noncollinear points A_1, A_2, A_3 and three noncollinear planes B_1, B_2, B_3 such that (α) the point A_0, which is the meet of the three planes, is a part of the plane B_0, which is the join of the three points; (β) all six triples $\{A_0, A_i, A_j\}$ and $\{B_0, B_i, B_j\}$ are noncollinear.

If the join of a finite set of points is a hyperplane and/or the meet of a finite set of hyperplanes is a point, then the structure is finite-dimensional and the two sets, if they are basic, contain the same number of elements, which is equal to the dimension of the structure. Postulates III and IV may, therefore, be replaced by a single assumption, namely,

POSTULATE III*   *There exists a finite set $\mathfrak{C}$ which is the union of a set $\mathfrak{A}$ of points and a set $\mathfrak{B}$ of hyperplanes such that*

(1) *$\bigcup \mathfrak{A}$ is a hyperplane, B_0, and $\bigcap \mathfrak{B}$ is a point, A_0;*
(2) *$A_0 \subseteq B_0$;*
(3) *each set consisting of A_0 and all points but one of $\mathfrak{A}$, or of B_0 and all hyperplanes but one of $\mathfrak{B}$, is basic.*

In proving that projective spaces are homogeneous it is convenient to relativize the idea of fundamental configurations and to introduce such sets in each element of dimension > 1.

A configuration $\mathfrak{C}_k$ in a k-dimensional element X $(k > 1)$ is the union of two k-tuples of parts of X:

$$\mathfrak{P} = \{P_1, \ldots, P_k\} \quad \text{and} \quad \mathfrak{H} = \{H_1, \ldots, H_k\}$$

such that dim $P_i = 0$ and dim $H_i = k - 1$ $(i = 0, \ldots, k)$ and that

($1'_k$) $\mathfrak{P}$ is basic, whence $H_0 = P_1 \cup \cdots \cup P_k$ is $(k - 1)$-dimensional;

($1''_k$) $\mathfrak{H}$ is basic, whence $P_0 = H_1 \cap \cdots \cap H_k$ is a point;

(2_k) $P_0 \subseteq H_0$;

($3'_k$) $\mathfrak{P}_{i_k} = \{P_0, P_{i_1}, \ldots, P_{i_{k-1}}\}$ is basic;

($3''_k$) $\mathfrak{H}_{i_k} = \{H_0, H_{i_1}, \ldots, H_{i_{k-1}}\}$ is basic;

for each permutation $i_1, \ldots, i_{k-1}, i_k$ *of* $1, \ldots, k - 1, k.$
  By $\mathfrak{C}_1$ is meant a triple of points.

Postulate IV assumes the existence of a configuration $\mathfrak{C}_n$ in an n-dimensional space for $n > 1$; and each one-dimensional space contains a $\mathfrak{C}_1$.

THEOREM 4.37 *The element* H_0 *of a configuration* $\mathfrak{C}_k$ *contains a configuration* $\mathfrak{C}_{k-1}$.

At least one point of $\mathfrak{P}$ is not a part of H_k, say $P_k \not\subseteq H_k$. One can then define
  (a) $(k - 2)$-dimensional elements $H_i^* = H_i \cap H_0$ $(i = 1, \ldots, k - 1)$. It follows that $\mathfrak{H}^* = \{H_1^*, \ldots, H_{k-1}^*\}$ is basic and $\bigcap \mathfrak{H}^* = P_0$. Thus ($1''_{k-1}$) holds.
  (b) $P_0^* = P_0$ and $H_0^* = H_k \cap H_0$. Hence $P_0^* \subseteq H_0^*$, that is, (2_{k-1}) and also ($3''_{k-1}$) hold.
  (c) $P_j^* = (P_j \cup P_k) \cap H_0$ $(j = 1, \ldots, k - 1)$. Then ($1'_{k-1}$) and ($3'_{k-1}$) hold. For from $P_j^* \subseteq P_{i_1}^* \cup \cdots \cup P_{i_{k-2}}^*$ it would follow that

$$P_j \subseteq P_j^* \cup P_k \subseteq P_{i_1} \cup \cdots \cup P_{i_{k-2}} \cup P_k,$$

contrary to assumptions ($1'_k$) and ($3'_k$).
  Thus $\mathfrak{P}^*$ and $\mathfrak{H}^*$ constitute a $\mathfrak{C}_{k-1}$ in H_0.
  A finite-dimensional element X of a projective structure is called *spacelike* if X contains a configuration $\mathfrak{C}_k$ where $k = \dim X$. It thus has been proved that

  The element H_0 *of a configuration* $\mathfrak{C}_k$ *is spacelike.*

In a projective structure, let X be a k-dimensional element containing a configuration $\mathfrak{C}_k$ consisting of k-tuples $\mathfrak{P}$ and $\mathfrak{H}$ with $\bigcup \mathfrak{P} = H_0$. Since $\mathfrak{H}$ is basic, each element

$$L_i = H_1 \cap \cdots \cap H_{i-1} \cap H_{i+1} \cap \cdots \cap H_k \qquad (i = 1, \ldots, k)$$

is one-dimensional. Moreover, $L_i \nsubseteq L_1 \cup \cdots \cup L_{i-1}$. For in view of $L_i \subseteq H_j$ for $j \neq i$, from $L_i \subseteq L_1 \cup \cdots \cup L_{i-1}$, it would follow that $L_i \subseteq H_i$ contrary to the definition of L_i and assumption $(1''_k)$. One readily proves the following

Remark.   If Q_i is any point such that $P_0 \neq Q_i \subseteq L_i$, then the set $\{Q_1, \ldots, Q_k\}$ is basic.

Since dim $X >$ dim H_0 there exists a point R such that $R \subseteq X$ and $R \nsubseteq H_0$. Setting

$$H' = P_1 \cup \cdots \cup P_{n-1} \cup R \qquad \text{and} \qquad H'' = P_2 \cup \cdots \cup P_n \cup R,$$

one has dim $H' =$ dim $H'' = k - 1$ and $H' \neq H'' \neq H_0 \neq H'$. Hence $P_0 \nsubseteq H'$ and $P_0 \nsubseteq H''$. Now let H be any $(k - 1)$-dimensional part $\neq H_0$ of X. Clearly, $H \neq H'$ and/or $H \neq H''$, say $H \neq H'$. For each $i = 1, \ldots, k$ the element $Q_i = L_i \cap H'$ is a point $\neq P_0$. By the preceding remark, the set $\mathfrak{Q} = \{Q_1, \ldots, Q_k\}$ is basic and $\bigcup \mathfrak{Q} = H'$. Since $H \neq H'$, for at least one point Q_1, which will be called, briefly, Q, one has $Q \nsubseteq H$. Moreover, $Q \nsubseteq H_0$. In fact, $Q_i \nsubseteq H_0$ for each i, since $Q_i \subseteq H_0$ implies $L_i = P \cup Q_i \subseteq H_0$ contrary to assumption $(3'_k)$. There thus exists a point Q such that $Q \nsubseteq H$ and $Q \nsubseteq H_0$. Associating with each j-dimensional part K of H_0 the (j-dimensional) element $(K \cup Q) \cap H$ one defines a one-to-one mapping of $[H_0]_j$ on $[H]_j$ for $j = 1, \ldots, k - 1$. It follows that $[H_1]_j \sim [H_2]_j$ for any two $(k - 1)$-dimensional parts of X. The mapping just mentioned maps the configuration $\mathfrak{C}_{k-1}$ on a configuration $\mathfrak{C}_{k-1}$ in H, whence each $(k - 1)$-dimensional part of H is spacelike. For each k-dimensional spacelike element X it thus has been established that

(1) $[H_1]_j \sim [H_2]_j$ for any two $(k - 1)$-dimensional parts H_1 and H_2 of X;
(2) each $(k - 1)$-dimensional part of X is spacelike.

For example, if U is the spacelike n-dimensional universe of an n-dimensional space, $[H_1]_j \sim [H_2]_j$ $(j = 0, 1, \ldots, n + 2)$ for any two hyperplanes H_1 and H_2, and each hyperplane is spacelike.

In order to establish the homogeneity of an n-dimensional space one must demonstrate for $m = n - 1, n - 2, \ldots, 2, 1$ that

(1_m) $[Y_1]_j \sim [Y_2]_j$ $(j = 0, 1, \ldots, m + 1)$ for any two m-dimensional elements Y_1 and Y_2 ;
(2_m) each m-dimensional element is spacelike.

Since (1_{n-1}) and (2_{n-1}) have just been proved it is sufficient to prove (1_{m-1}) and (2_{m-1}) under the assumptions (1_m) and (2_m).

Suppose dim $Y = \dim Y' = m - 1$ and assume

$$\dim(Y \cap Y') = t - 1, \qquad Y = (Y \cap Y') \cup R_1 \cup \cdots \cup R_{m-t},$$
$$Y' = (Y \cap Y') \cup S_1 \cup \cdots \cup S_{m-t}.$$

If one sets

$$Z_i = (Y \cap Y') \cup R_1 \cup \cdots \cup R_{m-t-i} \cup S_1 \cup \cdots \cup S_{m-t} \ (i = 1, \ldots, m - t),$$

then each Z_i is $(m - 1)$-dimensional and each $W_i = Z_i \cup Z_{i+1}$ is m-dimensional. Consequently, by (1_m) each W_i is spacelike and by (2_m) $[Z_i]_j \sim [Z_{i+1}]_j$. Hence $[Y]_j \sim [Y']_j$. Moreover, each $(m - 1)$-dimensional element, being a part of an m-dimensional element, is spacelike. Thus (1_{m-1}) and (2_{m-1}) are established. This concludes the proof of

THEOREM 4.38   *Each finite-dimensional projective space is homogeneous.*

Since one of the lines in the configuration $\mathfrak{C}_2$ in any plane, as well as each one-dimensional projective space, has at least three points as parts, the preceding theorem has the following

COROLLARY 1   *In a finite-dimensional space, each line has at least three points as parts.*

This (not self-dual) statement about *all* lines in the space (which has just been derived from the existence of *one single* fundamental configuration) is one of the traditional postulates of projective geometry. In view of Theorem 4.35, one further obtains

COROLLARY 2   *Each finite-dimensional space is simple.*

A structure is called *multi-complemented* if each nondistinguished element has more than one complement.

THEOREM 4.39   *Each finite-dimensional projective space is multicomplemented.*

In a finite-dimensional projective structure $\mathfrak{S}$, let X be an element that has exactly one complement, X'. First one can show that each point P is a part of either X or X'. Suppose $P \nsubseteq X$, so that $P \cap X = V$. One must prove that $P \subseteq X'$. Let Y be a complement of the element $X \cup P$, so that

$$(X \cup P) \cap Y = V, \quad (X \cup P) \cup Y = U \quad \text{and hence} \quad (P \cup Y) \cup X = U.$$

Then $(P \cup Y) \cap X = V$. Indeed, by the First Projectivity Law,

$$P \cup ((P \cup Y) \cap X) = P \cup ((P \cup X) \cap Y) = P \cup V = P$$

and thus $(P \cup Y) \cap X \subseteq P$; and since also $(P \cup Y) \cap X = X$ it follows that $(P \cup Y) \cap X \subseteq P \cap X = V$, as claimed. Hence $P \cup Y$ and X are complementary. Since X has only the complement X' it follows that $P \cup Y = X'$ and hence $P \subseteq X'$.

Now let P and P' be points of X and X', respectively. The line $P \cup P'$ does not contain a third point since such a point P'' would be a part of either X or X', say of X. This would make $P \cup P''$ a part of X; and $P' \subseteq P \cup P'' \subseteq X$, contrary to the assumption that $X \cap X' = V$.

Since the structure $\mathfrak{S}$ includes a line with exactly two points, $\mathfrak{S}$ is not a space. This concludes the proof of the theorem.

Postulates I, II, III* thus constitute a complete foundation of the geometry of projective spaces of finite dimensions > 1.

The concepts of projective structures and modular lattices overlap. There are nonassociative projective structures as well as modular lattices without distinguished elements. A general concept comprising both projective structures and (even nonmodular) lattices has recently been developed by Dr. Helen Skala. A *trellis* is a set with two operations, $\cup$ and $\cap$, which are commutative and absorptive and satisfy the implications

$$\text{if} \quad a \cup b = a \quad \text{and} \quad a \cup c = a, \quad \text{then} \quad a \cup (b \cup c) = a,$$
$$\text{if} \quad a \cap b = a \quad \text{and} \quad a \cap c = a, \quad \text{then} \quad a \cap (b \cap c) = a.$$

The operations are easily seen to be idempotent and alternative. The roles of associativity, the projectivity laws, modularity, distributivity and complementarity are studied in an extensive, forthcoming trellis theory.

EXERCISE

1. Show that if the nondistinguished element X in a finite-dimensional projective structure $\mathfrak{S}$ has only one complement X', then $\mathfrak{S}$ is isomorphic with $\mathfrak{X} \times \mathfrak{X}'$, where $\mathfrak{X}$ and $\mathfrak{X}'$ are the structures consisting of all parts of X and X', respectively.

REFERENCE

K. Menger, F. Alt, and O. Schreiber, "New foundations of projective and affine geometry," Annals of Math. 37: 456–482 (1936). In this reference projective geometry is based on a set of postulates the most important of which is the following restriction of the fundamental property of points and hyperplanes:

$$\text{if} \quad X \subseteq Y \subseteq X \cup P \neq U, \quad \text{then} \quad Y = X \quad \text{or} \quad Y = X \cup P,$$
$$\text{if} \quad X \supseteq Y \supseteq X \cap H \neq V, \quad \text{then} \quad Y = X \quad \text{or} \quad Y = X \cap H.$$

By adjoining the implication in both cases $X \cup P = U$ and $X \cap H = V$, one obtains a self-dual foundation of projective geometry. By adjoining it only in the case $X \cup P = U$ and assuming Euclid's Parallel Postulate, one arrives at affine geometry.

CHAPTER 5

Projective and Related Planes and Three-Dimensional Spaces

5.1 THE PROJECTIVE PLANE

The power of the theory of projective structures as a foundation of projective geometry becomes manifest in operations with flats of high dimension—for example, in determining the dimension of the join of a 35-dimensional and a 19-dimensional flat whose meet is 7-dimensional. In that theory, the vacuum V and the universe U are indispensable if satisfactory general laws are to be formulated. In the geometry of the plane, however, where the only interesting flats are points and lines, it is preferable to disregard V and U and to restrict the operations to the joining of distinct points and the intersecting of distinct lines—in other words, to revoke the five stipulations formulated in Section 4.1. One can even go a step further and define the two operations in terms of a single (undefined) binary relation existing between some points and some lines, called *incidence*.

It seems natural to demand that the theory be based on a set of postulates that satisfy the principle of duality, as does the entire theory; that is to say, the set should also include, for each postulate, the dual proposition resulting from an interchange of the terms *point* and *line*. From this point of view, the traditional postulate "there exist three points that are noncollinear (that is, not incident with one and the same line)" is not acceptable. A set of postulates including this proposition can, of course, be made self-dual by adjoining the dual proposition "there exist three lines that are non-concurrent (that is, not incident with one and the same point)." But in the presence of the other traditional assumptions, the existence of nonconcurrent lines is a consequence of the existence of noncollinear points, so that the price for obtaining a self-dual set of postulates in this way is redundancy of the system. The following self-dual set of postulates for the projective plane consists of assumptions that are independent.

The undefined terms are *points* (denoted by capital letters), *lines* (denoted by lower case letters) and a binary relation called *incidence*. If the point P and the line l are incident we say that each *is on* the other and write $P \circ l$ or, synonymously, $l \circ P$. We also write $P \circ l \circ Q$ for $P \circ l$ and $l \circ Q$; $P \circ l \neq m$ for $P \circ l$ and $l \neq m$; and the like.

POSTULATE I

(1) *If $P \neq Q$, then there exists at least one line l such that $P \circ l \circ Q$.*
(2) *If $m \neq n$, then there exists at least one point R such that $m \circ R \circ n$.*

POSTULATE II   *There do not exist two distinct points P, Q and two distinct lines l, m such that*

$$
\begin{array}{ccc}
P & \circ & l \\
\circ & & \circ \ ; \\
m & \circ & Q
\end{array}
$$

that is, $m \circ P \circ l$ and $m \circ Q \circ l$.

It readily follows that
(1) *if $P \neq Q$ then there exists exactly one line* (called the *join* of P and Q and denoted by PQ or synonymously by QP) *that is on both P and Q*;
(2) *if $l \neq m$ then there exists exactly one point* (called the *meet* of l and m and denoted by lm or synonymously by ml) *that is on both l and m.*

When we write PQ or lm it will be understood that $P \neq Q$ and $l \neq m$; the symbols PP and ll are undefined. In order to save parentheses, we write PQl for $(PQ)l$, which is the intersection of the join of P and Q with l; and $PQlR$ for $(PQl)R$ or $((PQ)l)R$. More generally,

$$P_0 P_1 l_1 P_2 \text{ is the line } (P_0 P_1 l_1)P_2;$$
$$P_0 P_1 l_1 P_2 l_2 \text{ is the point } (P_0 P_1 l_1 P_2)l_2;$$
$$P_0 P_1 l_1 P_2 l_2 P_3 \text{ is the line } (P_0 P_1 l_1 P_2 l_2)P_3,$$

and so on. Similarly, $l_0 l_1 P_1 l_2 \ldots l_m P_m$ denotes a line, $l_0 l_1 P_1 l_2 \ldots l_m P_m l_{m+1}$, a point. But parentheses are omitted only in strings beginning with the symbols of two points or two lines and then alternating. They are not omitted in $l(PQ)$, which is the same flat as PQl, or in $P(lm) = lmP$, or in $PQ(RS)$. For the sake of greater clarity we often even write $(PQ)(RS)$. Clearly the two operations and the incidence relation are connected as follows.

THEOREM 5.1 (BASIC LAW) $PQl = Q$ *if and only if* $Q \circ l$, *and* $mnR = n$ *if and only if* $n \circ R$.

The use of the symbol PQl implies that $P \neq Q$ and $PQ \neq l$. Because of $PQ = QP$ and $lm = ml$, it follows from the Basic Law that

$PQl = P$ *if and only if* $P \circ l$; *and* $mnR = m$ *if and only if* $m \circ R$.

For $m = PQ$, $n = l$, $R = Q$, because $PQ \circ Q$, one obtains the following

COROLLARY $PQlQ = PQ = QPlQ$ *and* $mnRm = mn = nmRm$.

Since the set of postulates is self-dual it is sufficient, here and in the sequel, to prove only one of any two dual assertions.

The corollary makes it possible to shorten certain strings, for example, $P_0 P_1 l_1 P_2 l_2 P_2 l_1$. Indeed, $P_0 P_1 l_1 P_2 \circ P_2$ implies $P_0 P_1 l_1 P_2 l_2 P_2 = P_0 P_1 l_1 P_2$, whence $P_0 P_1 l_1 P_2 l_2 P_2 l_1 = P_0 P_1 l_1 P_2 l_1 = P_0 P_1 l_1$.

LEMMA 5.1 *If* $PQ \circ R \neq P$, *then* $Q \circ PR$ *and* $PQ = PR$;
$\qquad\qquad$ *if* $lm \circ n \neq l$, *then* $m \circ ln$ *and* $lm = ln$.

Since $Q \circ PQ$, if $Q \not\circ PR$ then $PQ \neq PR$. (We write $\not\circ$ for *is not on*.) Since both PQ and PR are on both P and R this contradicts Postulate II.

If $Q \neq P \neq R \neq Q$ and $PQ \circ R$, then $PQ = PR = QR$ and $\{P, Q, R\}$ is said to be a *collinear triple*. Dually, if $lm = ln = mn$, then $\{l, m, n\}$ is said to be a *concurrent triple*.

From Postulates I and II one cannot conclude that any points or lines exist. The following self-dual postulate implies the traditional assumption that there exist three noncollinear points as well as its dual.

POSTULATE III   *There exist two points P_0, Q_0 and two lines l_0, m_0 such that*

$$
\begin{array}{cc}
P_0 & \circ & l_0 \\
\varnothing & & \varnothing \,. \\
m_0 & \circ & Q_0
\end{array}
$$

The points P_0, Q_0, $l_0 m_0$ are not collinear. For if they were on a line n, then n and l_0 would have P_0 and $l_0 m_0$ in common without being identical, since $l_0 \varnothing Q_0 \circ n$. Similarly, l_0, m_0, $P_0 Q_0$ are not concurrent.

LEMMA 5.2   *For each line k, there exist at least two points on k, and at least one point not on k.*

If $k = P_0 Q_0$, then $P_0 \circ k \circ Q_0$ and $l_0 m_0 \varnothing k$. If $k \neq P_0 Q_0$, then from the nonconcurrence of l_0, m_0, $P_0 Q_0$ it follows that at least two of the points $l_0 k$, $m_0 k$, $P_0 Q_0 k$ are distinct.

Postulates I, II, III do not imply the existence of more than three points and three lines, since they are satisfied by the set

$$\{P_0, Q_0, l_0 m_0, l_0, m_0, P_0 Q_0\}.$$

They are also satisfied by one line k that is on more than two points and one point Q that is on more than two lines, whereas all other lines (points) are on exactly two points (lines); for example, $P_1, P_2, P_3, \ldots$ are on k, and $l_1, l_2, l_3, \ldots$ are on Q, with $P_i = kl_i$ and $l_i = QP_i$. (Upon the adjunction of distinguished elements V and U, this example corresponds to a composite two-dimensional projective structure in the sense of Section 4.11.) In order to rule out such degenerate systems another assumption is needed.

POSTULATE IV   *There exist two points P', Q' and two lines l', m' such that*

$$P' \varnothing l'$$
$$\varnothing \quad \varnothing \qquad \text{and} \qquad l'm' \circ P'Q'.$$
$$m' \varnothing Q'$$

(The two points and two lines constitute a fundamental configuration in the sense of Section 4.12.) A set of points and lines satisfying the self-dual set of Postulates I to IV is called a *projective plane*. A projective plane satisfies the following statement, whose first (not self-dual) half has traditionally served as a postulate.

LEMMA 5.3   *Each line is on at least three points. Each point is on at least three lines.*

Let n be a line in the plane containing P', Q', l', m'. If $n \varnothing l'm'$, then n is on the points $l'n$, $m'n$, $P'Q'n$. The line $P'Q'$ is on P', Q', $l'm'$. If $P'Q' \neq n \circ l'm'$, then n is distinct from at least one of the lines l', m', say $n \neq l'$. By Lemma 5.2 there exists a point R not on $P'Q'$. Then n, which is on $l'm'$, is on R and $RP'l'Q'_n$ or on $RP'n$ and $RQ'n$ according as R is or is not on n. The second half of the lemma is dual to the first.

LEMMA 5.4   *For any two distinct lines (points), there exists a point (line) that is on neither.*

Let k and n be two distinct lines. Since each line contains at least three points, k and n contain two points K_1, K_2 and N_1, N_2 which are distinct from the point kn. The point $(K_1 N_1)(K_2 N_2)$ is on neither line.

For any line l, the set of all points on l is denoted by $[l]$ and called a *range*; more specifically, the range on l. For any point P, the set $[P]$ of all lines on P is called a *pencil*. The line l and the point P are called the *carriers* of $[l]$ and $[P]$, respectively.

If $l \varnothing P \varnothing m \neq l$ and one associates with each point X on l the point XPm on m, one obtains a mapping of $[l]$ onto $[m]$ which is called a *perspectivity*; more specifically, the perspectivity of $[l]$ onto $[m]$ with the *center* P. It is, first of all, a mapping *onto* m; for any point Y on m is the image of the point YPl on l, since $YPlPm = YPm = Y$. If X_1 and X_2 are distinct points on l, then $X_1 Pm \neq X_2 Pm$, and so the mapping is one-to-one. We denote the mapping by π_{lPm} and, accordingly, set $\pi_{lPm} X = XPm$ for each $X \circ l$.

Dually, if $Q \varnothing n \varnothing R \neq Q$, we set $\pi_{QnR} x = xnR$ for each $x \circ Q$, and call

π_{QnR} the perspectivity of $[Q]$ onto $[R]$ with the axis n. Since, by Lemma 5.4, for any two distinct lines l and m there exists a point P that is on neither l nor m, a projective plane has the following important homogeneity property.

THEOREM 5.2   *For any two ranges and any two pencils there exists a perspectivity, and hence a one-to-one mapping, of one onto the other.*

Since the points P', Q', $l'm'$ of Postulate IV are collinear, this theorem has the following consequence whose first half is traditionally formulated as a postulate:

COROLLARY   *Each line is on at least three points; each point is on at least three lines.*

LEMMA 5.5   *A perspectivity of one range (one pencil) onto another leaves the meet (the join) of the carriers fixed.*

By the Corollary of Theorem 5.1, $lmPm = lm$.

Postulates I to IV imply that there exist at least seven points—namely, the collinear triple P' Q', $l'm'$—and, besides $l'm'$, at least two points on l' and two points on m', since each line is on at least three points. Dually, each point is on at least three lines and so there exist at least seven lines. But the postulates do not imply that there are necessarily more than seven points and seven lines, since they are satisfied by the points P, Q, R, S, T, U, V and the lines $p, q, \ldots, v$ with the following incidence relation table:

	P	Q	R	S	T	U	V
p		○	○	○			
q	○		○		○		
r	○	○				○	
s	○			○			○
t		○			○		○
u			○			○	○
v			○	○	○		

If one line is on at least four points, then so is each line, and each point is then on at least four lines. It is clear that in this case there exist at least thirteen points and thirteen lines, since each of the four lines on a point P is on at least three points distinct from P, and no two of these twelve points can be identical.

While there are no projective planes containing exactly k points if $7 < k < 13$, the postulates are indeed satisfied by the set of the thirteen points $A, B, \ldots, L, M$ and the lines $a, b, \ldots, l, m$ with the following incidence relation table:

	A	B	C	D	E	F	G	H	I	J	K	L	M
a	O	O	O	O									
b	O				O			O			O		
c	O					O			O				O
d	O						O			O	O		
e		O				O	O	O					
f			O		O	O						O	
g				O				O	O			O	
h		O			O		O		O				
i			O					O	O				O
j				O	O					O			O
k		O									O	O	O
l				O		O	O			O			
m			O					O		O	O		

Other finite projective planes (that is, finite sets of points and lines satisfying the postulates) are discussed in Exercises 6 and 7 of this section and, more generally, in Exercise 3 of Section 5.3.

EXERCISES

1. Prove that Postulate I (1) is independent of I (2), II, III, and IV.

2. Show that Postulate III is not a consequence of I, II, and IV.

3. If P, Q, R and P, Q, S are collinear triples, under what condition is Q, R, S a collinear triple?

4. If P, Q, R are noncollinear then $PQl = RQl$ if and only if $l \circ Q$.

5. Prove that in a projective plane there is a one-to-one mapping of each pencil onto each range. (*Hint:* First consider the case in which the carriers are not incident.)

6. Show that if one line in a projective plane is on at least n points, then there are at least $n^2 - n + 1$ points.

7. A *model* of the projective plane, with seven points and seven lines, consists of seven points (labeled 1, 2, ..., 7) and seven triangles (labeled *I*, *II*, ..., *VII*) on a torus (anchor ring) such that a point on the torus is a vertex of a triangle if and only if the corresponding point and line in the projective plane are incident. The points 1, 2, ..., 7 lie on the equator of the torus. By cutting the torus along the equator and then along a perpendicular circle, one can flatten the torus into a rectangle whose upper and lower sides stem from the equator. Figure 5 shows points 1, 2, . . . , 7 on the upper and lower

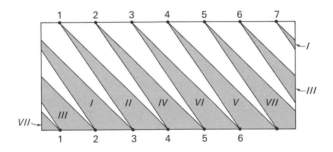

Figure 5

sides of the rectangle. Each upper point i is connected with the lower points $i + 2$ and $i + 3$; for example, upper 1 with lower 3 and 4; of course, 5 with 7 and 1; 6 with 1 and 2; 7 with 2 and 3, where the joining lines begin in the right half of the rectangle and end in its left half. Verify that the points 1, 2, . . . , 7 and the shaded triangles *I*, *II*, . . . , *VII* are indeed a model in the sense explained, by letting correspond to each other the points

    1, 2, 3, 4, 5, 6, 7   in the rectangle
    P, R, Q, U, V, T, S   in the projective plane,

and, similarly, the triangles *I*, *II*, . . . , *VII* and the lines p, r, . . . , s.

8. Set up a table for the incidence relation in a projective plane in which each line is on exactly five points.

5.2   TRIANGLES AND COMPLETE QUADRUPLES

A *complete point triple* is a set of three noncollinear points and their three (nonconcurrent) joins. If the points are denoted by P_1, P_2, P_3, then an appropriate symbol for the join of P_i and P_j would be $l_{\{i,j\}}$, instead of which it is customary, however, to write l_{ij} and/or l_{ji}. The complete point triple is thus the set

$$\{P_1, P_2, P_3, l_{23}, l_{13}, l_{12}\}.$$

It may also be symbolized by the symmetric 3-by-3 matrix

$$\begin{matrix} P_1 & l_{12} & l_{13} \\ l_{21} & P_2 & l_{23} \\ l_{31} & l_{32} & P_3 \end{matrix} \qquad (l_{ij} = l_{ji}).$$

The dual, called a *complete line triple*, is a set

$$\{l_1, l_2, l_3, P_{23}, P_{13}, P_{12}\},$$

where the l_i are nonconcurrent and $P_{ij} = l_i l_j$. By setting $P_i = P_{jk}$ and $l_i = l_{jk}$ for each permutation i, j, k of $1, 2, 3$ (that is, whenever $\{i, j, k\} = \{1, 2, 3\}$), one sees that the two complete triples are the same set, which will be called, briefly, a *triangle*.

If $l \varnothing P_i$ and $P \varnothing l_i$ ($i = 1, 2, 3$), then the following triples of distinct elements are said to be *generated* by the triangle $\mathfrak{T}$ on l and P, respectively:

$$\mathfrak{T}l = \{l_{23}l, l_{13}l, l_{12}l\} \qquad \text{and} \qquad \mathfrak{T}P = \{P_{23}P, P_{13}P, P_{12}P\}.$$

If P is not on l and not collinear with any two points of $\mathfrak{T}$, then a triple of distinct points can be defined as follows:

$$\mathfrak{T}Pl = \{P_1 Pl, P_2 Pl, P_3 Pl\}.$$

Setting $P = P_4$ and $P_i P_j l = Q_{ij}$ one can write the triples $\mathfrak{T}l$ and $\mathfrak{T}Pl$ as the rows of the matrix

$$\begin{matrix} Q_{12} & Q_{13} & Q_{23} \\ Q_{34} & Q_{24} & Q_{14} \end{matrix}$$

in which elements in different columns are distinct (while elements in the same column may be identical). A 2-by-3 matrix satisfying this condition will be called a *matroid*. Two matroids are considered as *equal* if the matrices can be obtained from one another by interchanging the rows and/or permuting the columns (see Exercise 5). Any five indexed elements of a matroid are said to constitute a *matroid fragment*. Elements in the same column of a matrix or a matroid fragment are said to be *opposite*.

A matroid must not be confused with a set of six points. In a matroid, opposite elements may be identical, whereas any two elements of a set are distinct. Except for permutations of rows and columns, the arrangement of the elements of the matroid is significant, whereas there is no distinguished arrangement of the elements of a set.

A *complete point quadruple*, or *quadrangle*, is a set consisting of four points, no three of which are collinear, and their six joins, called the *sides* of the quadrangle. The elements of a quadrangle $\mathfrak{Q}$ with the points $Q_1, \ldots, Q_4$ and the sides $l_{ij} = Q_i Q_j$ can be arranged in a symmetric 4-by-4 matrix; and the sides also constitute a matroid with six distinct elements:

$$
\begin{matrix}
Q_1 & l_{12} & l_{13} & l_{14} \\
l_{21} & Q_2 & l_{23} & l_{24} \\
l_{31} & l_{32} & Q_3 & l_{34} \\
l_{41} & l_{42} & l_{43} & Q_4
\end{matrix}
\qquad (l_{ij} = l_{ji}) \qquad
\begin{matrix}
l_{12} & l_{13} & l_{14} \\
l_{34} & l_{24} & l_{23}
\end{matrix}
$$

In order that six lines l_{ij} be the sides of a quadrangle it is necessary and sufficient that they be distinct and that l_{hi}, l_{hj}, l_{hk} be concurrent if $\{h, i, j, k\} = \{1, 2, 3, 4\}$. The meets of these four triples of lines are the points of the quadrangle. The meets of opposite sides—that is, the points $l_{12} l_{34}, l_{23} l_{14}, l_{31} l_{24}$—are called the *diagonal points* of $\mathfrak{Q}$; the joins of any two diagonal points are referred to as the *diagonal sides* of $\mathfrak{Q}$. In the projective plane with seven points, $\{P, Q, S, T, p, q, r, s, t, v\}$ is a quadrangle. Its diagonal points, R, U, V, are collinear.

If l is a line that is not on any of the points of $\mathfrak{Q}$, then the matroid $\mathfrak{Q}l$ of the points $Q_{ij} = Q_i Q_j l = Q_{ji}$ is called a *quadrangular* matroid on l. If in two columns the elements are identical (that is, if l is on two diagonal points of $\mathfrak{Q}$), then the matroid is said to be *harmonic*.

Let P_{12} and P_{13}, P_{24} and P_{23}, P_{14} be a matroid fragment of points on a line l. For any two points Q_1 and Q_2 not on l but collinear with P_{12} one may define

$$P_{13} Q_1 = l_{13}, P_{23} Q_2 = l_{23}, l_{13} l_{23} = Q_3;$$
$$P_{14} Q_1 = l_{14}, P_{24} Q_2 = l_{24}, l_{14} l_{24} = Q_4.$$

No three of the points Q_1, Q_2, Q_3, Q_4 are collinear. The quadrangle of these points generates on l a matroid consisting of the given fragment and the point $P_{34} = Q_3 Q_4 l$. Hence the following

Remark. A matroid fragment of points on a line l in conjunction with two points not on l but collinear with a point of the fragment can be completed in one and only one way to a quadrangular matroid generated by a quadrangle including the two given points.

The dual of a quadrangle is called a *complete line* quadruple or a *quadrilateral*.

EXERCISES

1. Give examples of a quadrangle and a quadrilateral in the planes with seven and thirteen points. In the former, find a quadrangle different from the one mentioned in the text.

2. A quadrangle $\mathfrak{P}$ contains four triangles $\mathfrak{T}_k$ with the points P_h, P_i, P_j, where $\{h, i, j, k\} = \{1, 2, 3, 4\}$. Show that if l is on none of the points of $\mathfrak{P}$, then for each k the points of $\mathfrak{T}_k l$ and $\mathfrak{T}_k P_k l$ determine one and the same quadrangular matroid on l.

3. Set up the incidence relation and sketch a set of ten points and ten lines including a complete point quadruple $\mathfrak{P}$ with points P_1, P_2, P_3, P_4 and a complete line quadruple $\mathfrak{L}$ with lines l_1, l_2, l_3, l_4 and the six incidences $P_h P_i \circ l_j l_k$ if $\{h, i, j, k\} = \{1, 2, 3, 4\}$.

4. How many different matrices determine the same matroid? (Remember the plane with seven points!)

5.3   THE AFFINE PLANE

If from a projective plane a line, which will be denoted by l_∞, and all points on l_∞ are deleted, the remaining set of points and lines is called an *affine plane* derived from the given projective plane. In an affine plane, the principle of duality is not valid; for although there is a line on each pair of points, there is not a point on each pair of lines. Two lines concurrent with l_∞ in the projective plane do not meet in the affine plane, since their meet has been deleted.

One may describe an affine plane in terms of points, lines, and incidence without reference to the projective plane from which it has been derived, namely by a set of (not self-dual) postulates. In an affine plane, Postulates I (1) and II are valid, whereas I (2) is not. The conjunction of I (1) and II can be replaced by

POSTULATE A *For any two distinct points, there is exactly one line that is on both points.*

It follows that for any two distinct lines, l and m, there is at most one point that is on both lines. If there is no such point, then l and m are said to be *parallel*—in symbols, $l \parallel m$, or synonymously, $m \parallel l$. The affine plane further satisfies the following version of Euclid's famous axiom:

POSTULATE B (PARALLEL POSTULATE) *If $P \varnothing l$, then there exists exactly one line l' such that $P \circ l'$ and $l' \parallel l$.*

If one writes $l \# m$ for the alternative $l \parallel m$ or $l = m$, then it is easy to show that the reflexive and symmetric relation $\#$ is transitive.

Postulates A and B as well as Postulate III for the projective plane are satisfied by the set consisting of two points P, Q and three lines PQ, l, m, where $l \parallel m$. The following stronger assumption excludes this set.

POSTULATE C *There exist two points P_0, Q_0 and two nonparallel lines l_0, m_0 such that*

$$\begin{array}{ccc} P_0 & \circ & l_0 \\ \varnothing & & \varnothing \, . \\ m_0 & \circ & Q_0 \end{array}$$

One obtains a projective plane (see Exercise 3) by adjoining to an affine plane (1) *ideal points* (or *points at infinity*)—namely, one such point for each family consisting of a line l and all lines m such that $m \# l$; (2) one *ideal line* (or *line at infinity*, l_∞), which is on each of the ideal points. If k is a line different from l_∞ and $P \varnothing k$, then $kl_\infty P$ is the (unique) line on P that is parallel to k.

Postulates A, B, C yield

THEOREM 5.3 *There exist four points no three of which are collinear and six lines no four of which are concurrent.*

Let k be the line on P_0 parallel to m_0, and n the line on $l_0 m_0$ parallel to $P_0 Q_0$. If k and n were parallel, then, contrary to Postulate B, the lines k and $P_0 Q_0$ on P_0 would be parallel to n. Consequently, there is a point

$R = kn$ which is readily seen to be noncollinear with any two of the points P_0, Q_0, and $l_0 m_0$. No four of the lines $l_0, m_0, k, n, P_0 Q_0, Q_0 R$ are concurrent.

If h is the line on Q_0 parallel to l_0, there exists a point hn. But Postulates A, B, C do not imply that $kn \neq hn$. In fact, they are satisfied by the set of four points $P_0, Q_0, l_0 m_0, kn$ and the six lines $l_0, m_0, P_0 Q_0, n, h, k$ (resembling the vertices and edges of a tetrahedron). In order to rule out this example on may assume a Postulate C' identical with Postulate IV for a projective plane (see Exercise 7).

Important examples of affine planes can be defined arithmetically. Let $\mathfrak{F}$ be the set of all real numbers or, more generally, any field. A *coordinatized affine plane* or, more precisely, the *affine plane over* $\mathfrak{F}$, is defined as follows: *Point* is a sequence (p, q) of two elements of $\mathfrak{F}$; *line* is a sequence denoted by $\langle p, q \rangle$. (To each pair of elements p, q of $\mathfrak{F}$, there corresponds a point as well as a line.) Two points (two lines) associated with unequal sequences are considered unequal; for example, $\langle p, q \rangle \neq \langle q, p \rangle$. Moreover, there corresponds a line denoted by $\langle r \rangle$ to each element r of $\mathfrak{F}$, and $r \neq s$ implies $\langle r \rangle \neq \langle s \rangle \neq \langle p, q \rangle$ for any p and q. Finally, *incidence* is defined as follows:

$$(x, y) \circ \langle a, b \rangle \text{ if and only if } y = ax + b,$$

$$(x, y) \circ \langle c \rangle \text{ if and only if } x = c.$$

(In elementary analytic geometry, $\langle m, c \rangle$ is defined as the set of all points (x, y) such that $y = mx + c$; and $\langle v \rangle$, as the set of all points (x, y) such that $x = v$.)

For a field $\mathfrak{F}$ to contain, for some number n, exactly n elements it is necessary and sufficient that $n = p^k$ for some prime number p and some natural number k. It follows that (coordinatized) affine planes exist with exactly n points on each line for $n = 2, 3, 4, 5, 7, 8, 9, 11, 13, 16, 17, \ldots$. But no affine planes over fields exist with exactly $6, 10, 12, 14, 15, 18, \ldots$, points on each line. There are, however, affine planes that are not isomorphic with planes over fields. Whether such planes with exactly 10 and/or 12 points on each line exist are open questions. There certainly are no affine planes whatsoever with exactly 6, 14, 22 points on each line or, more generally, $8n + 6$ points if $4n + 3$ is a prime number. (It follows that there are projective planes with exactly $p^k + 1$ but not with exactly $7, 15, 23, \ldots$ points on each line. The questions as to the existence of projective planes with 11 and 13 points on each line are undecided.)

EXERCISES

1. Prove that each coordinatized affine plane satisfies Postulates A, B, C.

2. Prove that if one line of an affine plane is on at least three points, then the plane contains at least nine points and twelve lines. Derive an affine plane with exactly nine points and twelve lines from the projective plane with exactly four points on each line.

3. If ideal points and an ideal line are adjoined to an affine plane, how must the incidence relation be defined in order that a projective plane results?

4. Show that the affine plane over the field of the residues 0,1 modulo 2 consists of four points and six lines; and that the projective plane obtained by adjoining ideal elements is isomorphic with the projective plane with seven points and seven lines.

5. Set up incidence tables for the affine plane over (1) the field of residues $0, 1, 2 \bmod 3$; (2) the field consisting of four elements $0, 1, \omega, \omega'$ for which addition and multiplication are defined as follows:

$+$	0	1	ω	ω'
0	0	1	ω	ω'
1	1	0	ω'	ω
ω	ω	ω'	0	1
ω'	ω'	ω	1	0

$\cdot$	0	1	ω	ω'
0	0	0	0	0
1	0	1	ω	ω'
ω	0	ω	ω'	1
ω'	0	ω'	1	ω

6. Show that coordinatization by the residues 0, 1, 2, 3 mod. 4 (which do not constitute a field) leads to points and lines satisfying neither of the Postulates A and B.

7. Whereas Postulate III is independent of Postulates I, II, IV of a projective plane (see Exercise 2, Section 5.1), Postulate C is a consequence of Postulates A, B, C' ($=$ IV) of an affine plane.

5.4 A SELF-DUAL FRAGMENT OF THE AFFINE PLANE AND ITS APPLICATION TO GALILEAN KINEMATICS

Self-dual systems can be obtained from an affine plane not only by adjoining an ideal line and the points on it (which leads to a projective plane) but also by deleting a line and all lines parallel to it (but, of course, retaining the points on the deleted lines). In the self-dual fragment of the affine plane thus obtained there are not only lines that do not meet in a point but also points that are not joined by a line. Two points that are not on one and the same line of the fragment (because they are on one of the deleted lines of the affine plane) are said to be vertical to one another—briefly, *vertical*. Vertical points are dual to parallel lines.

The motivation for the term "vertical" lies in elementary plane analytic geometry, where it is customary to choose the X-direction as horizontal. A branch of that geometry deals with the lines $\langle m, c \rangle$ (considered as the sets of points (x, y) such that $y = mx + c$)—the nonvertical lines. There are pairs of points that cannot be joined by such lines—for example, (x_0, y) and (x_0, y'). They are the pairs of points that are on one and the same vertical line of the coordinate plane.

The self-dual fragment of the affine plane satisfies the following postulates, the first of which is identical with Postulate II for the projective plane, and the second of which is the conjunction of the Parallel Postulate and its dual.

POSTULATE α *There do not exist two distinct points and two distinct lines such that both points are on both lines.*

POSTULATE β (SELF-DUAL EXTENSION OF EUCLID'S AXIOM)
If $P \oslash l$, then there exists exactly one line on P that is parallel to l, and exactly one point on l that is vertical to P.

In elementary analytic geometry, if (x_0, y_0) is a point that is not on the line $\langle m_0, c_0 \rangle$ (that is to say, such that $m_0 x_0 + c_0 \neq y_0$), then there exist: (1) exactly one line on (x_0, y_0) that is parallel to $\langle m_0, c_0 \rangle$, namely, the line $\langle m_0, y_0 + m_0 c_0 \rangle$; and (2) exactly one point on $\langle m_0, c_0 \rangle$ that is vertical to (x_0, y_0), namely, the point $(x_0, c_0 + x_0 m_0)$. Thus the set of the points and the nonvertical lines of the coordinatized affine plane satisfies the Self-dual Extension of Euclid's Axiom. A symmetric arithmetical model of the fragment results from defining the incidence of (x, y) and $\langle a, b \rangle$ by $ax = b + y$. In this way, b is the *negative* Y-intercept of the line $\langle a, b \rangle$. In order to rule out degenerate self-dual fragments one may assume

POSTULATE γ *There exist two nonvertical points P_0, Q_0 and two nonparallel lines l_0, m_0 such that*

$$
\begin{array}{cc}
P_0 \circ l_0 \\
\oslash \quad \oslash \; . \\
m_0 \circ Q_0
\end{array}
$$

A proof similar to that of Theorem 5.3 yields

THEOREM 5.5   *There exist at least nine points and nine lines.*

Postulates α, β, γ are satisfied by what is called the *Pappus configuration*, consisting of nine points $P, Q, \ldots, X$ and lines $p, q, \ldots, x$ (see Fig. 6) with the following incidence relation table:

	P	Q	R	S	T	U	V	W	X
p	o	o	o						
q	o			o					o
r	o				o		o		
s		o			o	o			
t			o	o			o		
u							o	o	o
v				o	o	o			
w		o				o	o		
x		o				o			o

The triples of parallel lines and vertical points are

p, u, v; q, s, w; r, t, x and P, U, V; Q, S, W; R, T, X.

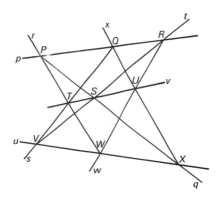

Figure 6

The kinematics of the uniform motions in a one-dimensional space can be represented in a two-dimensional space-time world. In analytic kinematics, each *world point* (sometimes referred to as *event*) can be characterized by a pair of numbers (t, u) indicating the instant t in time and the position u in space. Each uniform motion can be characterized by a pair of numbers $\langle v, w \rangle$ indicating the velocity v and the location w at the instant 0. The motion $\langle v, w \rangle$ passes through the world point (t, u) if and only if $u = v \cdot t + w$.

In Galilean kinematics (in contrast to Einstein's), simultaneity is an absolute relation between events in the sense that two world points that are simultaneous for one observer are also simultaneous for any observer uniformly moving relative to the first. At any instant t_0, the space is the set consisting of the world points (t_0, u) for all u, that is, a set consisting of a world point and all simultaneous world points.

Geometrically, each world point can be represented by a point in a self-dual fragment of an affine plane, each motion, by a (nonvertical) line of the fragment. Postulational kinematics, therefore, may start with two undefined concepts: *world point* and *motion*, satisfying Postulates α, β, γ. The analogue of vertical points are *simultaneous* world points, which cannot be reached from one another by any motion. The analogue of parallel lines are *isokinetic* motions, which do not pass through any one and the same world point. Thus in the Galilean kinematics of a one-dimensional space, *there is a complete duality between world points and motions*.

In Galilean kinematics (as in Einstein's) there is no absolute *rest*. Of two observers in uniform motion relative to one another, either one may consider himself as being at rest. Newton's kinematics results from Galileo's partial relativity theory by distinguishing a motion and all isokinetic motions as representing rest relative to an absolute space. Geometrically, it is obtained from the self-dual fragment by distinguishing (though not deleting) a family consisting of a line and all parallel lines; and from the projective plane by deleting a line l_∞ and all its points, as well as all lines on a certain point V_∞ of l_∞, and by distinguishing a point R ($\neq V_\infty$) on l_∞ and all lines on R (corresponding to rest motions). Because of this last feature, Newton's kinematic, unlike Galileo's, is not self-dual.

EXERCISES

1. Verify Postulate β for the Pappus configuration.

2. From the projective plane with thirteen points and thirteen lines, omit one line and all points on it, as well as all lines on one of those deleted points. By properly labeling the elements of the fragment, obtain the Pappus configuration.

3. Show that by deleting from a projective plane a line l_∞ and all points on it, as well as a point P_0 *not* on l_∞ and all lines on P_0, one obtains another self-dual fragment of the affine plane studied in elementary analytic geometry —namely, the points $\neq (0, 0)$ and the lines not passing through $(0, 0)$ which have equations of the form $x/a + y/b = 1$. In this fragment *"proportional"* points (that is, points collinear with $(0, 0)$) are the dual of parallel lines.

4. Consider the torus obtained from the square in Figure 7 by identifying the sides AB and DC as well as AD and BC (so that A, B, C, D become a single point). Verify that the points $P, \ldots, X$ and the lines $p, \ldots, x$ in the figure (which are simple closed curves on the torus) constitute a model of the Pappus configuration.

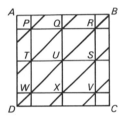

Figure 7

5. Formulate Postulates α, β, γ in purely kinematic terms.

6. What is the dual of Newton's absolute rest?

5.5 THE THREE-DIMENSIONAL PROJECTIVE SPACE

Even in describing the three-dimensional space it is sometimes advantageous to disregard V and U and to restrict the operations to forming (1) the join of two distinct points or of a point and a nonincident line; (2) the meet of two distinct planes or of a plane and a nonincident line; (3) the join as well as the meet of two nonskew (that is, coplanar) lines.

Points and lines will be denoted by capital and lower case letters, respectively, planes by capital script letters. It is then clear, for example, that $Pl\mathscr{P}Qm$ designates the point obtained by joining P with the line l, intersecting with the plane $\mathscr{P}$, joining with the point Q, and intersecting with the line m. But a complete description of geometric constructions by mere chains of symbols for points, lines, and planes is unfortunately impossible.

If l and m are two nonskew lines, then the symbol lm would be ambiguous, and one must therefore indicate what is intended—the point $l \cap m$ or the plane $l \cup m$.

A self-dual set of postulates in terms of points, lines, planes, and the incidence relation follows. (It is assumed that distinct symbols denote distinct elements.)

POSTULATE 1   *If* $P \circ l \circ \mathscr{P}$, *then* $P \circ \mathscr{P}$.

POSTULATE 2   *If P and Q are on both $\mathscr{P}$ and l, then $l \circ \mathscr{P}$; and if $\mathscr{P}$ and $\mathscr{Q}$ are on both P and l, then $l \circ P$.*

POSTULATE 3   *If $P \neq Q$, then there is at least one line on both P and Q. If $\mathscr{P} \neq \mathscr{Q}$, then there is at least one line on both $\mathscr{P}$ and $\mathscr{Q}$.*

POSTULATE 4   *If $P \,\emptyset\, l$, then there exists at least one plane on both P and l; if $\mathscr{P} \,\emptyset\, l$, then there is at least one point on both $\mathscr{P}$ and l.*

POSTULATE 5   *There do not exist distinct points and lines or planes and lines such that*

$$
\begin{array}{ccc}
P \circ l & & \mathscr{P} \circ l \\
\circ \quad \circ & \text{or} & \circ \quad \circ. \\
m \circ Q & & m \circ \mathscr{Q}
\end{array}
$$

POSTULATE 6   *Each line is on at least two points and on at least two planes.*

From Postulates 3 and 5, one readily proves the uniqueness of the joins PQ of points and the meets $\mathscr{P}\mathscr{Q}$ of planes. From Postulates 4 and 5 it follows that there is exactly one plane Pl if $P \,\emptyset\, l$, and exactly one point $\mathscr{P}l$ if $\mathscr{P} \,\emptyset\, l$. For if there were two planes $\mathscr{P}$ and $\mathscr{Q}$ on both P and l, then one could let Q be a point on l (which, by Postulate 6, certainly exists), and $\mathscr{P}$ and $\mathscr{Q}$ would be both on l and PQ, contrary to Postulate 5. For any noncollinear points P, Q, R, there is a unique plane PQR that is on all three points. Dually, there is a unique point $\mathscr{P}\mathscr{Q}\mathscr{R}$ on any three noncollinear planes $\mathscr{P}$, $\mathscr{Q}$, $\mathscr{R}$. Of great importance is

LEMMA 5.6   *If two distinct lines l and m meet, that is, are on one and the same point, then they are coplanar, that is, on one and the same plane, and conversely.*

Suppose $l \circ P \circ m$. By Postulate 6, there is a second point Q on m, and consequently the plane Ql is on both l and m. The converse follows from duality.

Lines that meet and are coplanar are also said to be *dependent*. Independent lines are called *skew*.

   Postulates 1 through 6 are satisfied by a system consisting of one point and one plane. Even if one postulates further the existence of a line l_0, the postulates are satisfied by a set consisting of l_0 and two points and two planes on l_0. The following assumption is stronger.

POSTULATE 7   *There exist three noncollinear points P^*, Q^*, R^* and three noncollinear planes $\mathscr{P}^*$, $\mathscr{Q}^*$, $\mathscr{R}^*$ such that*

$$P^* \varnothing \mathscr{P}^*, \quad Q^* \varnothing \mathscr{Q}^*, \quad R^* \varnothing \mathscr{R}^* \qquad and \qquad P^*Q^*R^* \varnothing \mathscr{P}^*\mathscr{Q}^*\mathscr{R}^*.$$

It then follows that there exist four noncoplanar points and four noncopunctual planes (in particular, no three of them collinear) and six lines. More cannot be proved, since, as one readily sees, Postulates 1 through 7 are satisfied by a *tetrahedron*—that is, a set $\mathfrak{T}$ consisting of (1) four noncoplanar points P_1, P_2, P_3, P_4; (2) four noncopunctual planes $\mathscr{P}_1, \mathscr{P}_2, \mathscr{P}_3, \mathscr{P}_4$ such that $P_i \circ \mathscr{P}_j$ if and only if $i \neq j$; (3) six lines $P_h P_i = \mathscr{P}_j \mathscr{P}_k$ for $\{h, i, j, k\} = \{1, 2, 3, 4\}$.

   Spaces reducing to tetrahedra are ruled out by the following additional assumption of the existence of a fundamental configuration in the sense of Section 4.2.

POSTULATE 8   *There exist three noncollinear points P^0, Q^0, R^0 and three noncollinear planes $\mathscr{P}^0$, $\mathscr{Q}^0$, $\mathscr{R}^0$ such that*

$$P^0 \varnothing \mathscr{P}^0, \quad Q^0 \varnothing \mathscr{Q}^0, \quad R^0 \varnothing \mathscr{R}^0 \qquad and \qquad P^0 Q^0 R^0 \circ \mathscr{P}^0 \mathscr{Q}^0 \mathscr{R}^0$$

and that no three of the points P^0, Q^0, R^0, $\mathscr{P}^0 \mathscr{Q}^0 \mathscr{R}^0$ and the planes $\mathscr{P}^0$, $\mathscr{Q}^0$, $\mathscr{R}^0$, $P^0 Q^0 R^0$ are collinear.

From Postulates 1 through 8 one readily infers the following assertions.

THEOREM 5.5   *For each tetrahedron, there exists a point that is on none of its planes, and a plane that is on none of its points.*

For each plane $\mathscr{P}$, let $[\mathscr{P}]$ denote the set of all points and lines on $\mathscr{P}$. Then one has

THEOREM 5.6   *If $\mathscr{S}^0 = P_1^0 P_2^0 P_3^0$, then $[\mathscr{S}^0]$ satisfies Postulates I through IV of a projective plane. $[\mathscr{S}^0]$ contains at least seven points, at least three on each line.*

In analogy to Lemma 5.3, one proves that *if $\mathscr{P}_1 \neq \mathscr{P}_2$, then there exists a point Q such that $\mathscr{P}_1 \varnothing Q \varnothing \mathscr{P}_2$.* In analogy to Theorem 5.2, one further obtains

THEOREM 5.7   *If $\mathscr{P}_1 \varnothing Q \varnothing \mathscr{P}_2$, then the perspectivity that associates with each point P_1 on $\mathscr{P}_1$ the point $P_1 Q \mathscr{P}_2$ on $\mathscr{P}_2$ is a one-to-one mapping of $[\mathscr{P}_1]$ onto $[\mathscr{P}_2]$ which preserves incidence; that is, if P_1 and P_1' are mapped on P_2 and P_2' then the range $[P_1 P_1']$ is mapped onto the range $[P_2 P_2']$.*

An immediate consequence of the last two theorems is the following

COROLLARY   *Each plane is on at least seven points; and each line is on at least three points. For each plane $\mathscr{P}$, the set $[\mathscr{P}]$ is a projective plane.*

In the remainder of this chapter, *space* always means *three-dimensional space*.

EXERCISES

1. Dualize the theorems stated in this section, and carry out their proofs in detail.

2. Prove that if l and m are independent (skew) lines and $l \varnothing P \varnothing m$, then there exists one and only one line n such that $n \circ P$ and that n and l as well as n and m are dependent. Dualize this theorem.

3. Prove that if l, m, l', m' are lines such that the four pairs l, m; l, m'; l', m; l', m' are dependent while the two pairs l, l' and m, m' are independent, then

$$(l \cup m') \cap (l' \cup m) = (l \cap m) \cup (l' \cap m').$$

Note that in these formulas the symbols $\cup$ and $\cap$ for the operations are indispensable.

4. If, from a projective space, one deletes a plane, $\mathscr{P}_\infty$, and all points and lines on $\mathscr{P}_\infty$, then the remaining set of points, lines, and planes is called an *affine space* derived from the projective space. In the affine space, two lines as well as two planes as well as a line and a plane are said to be *parallel* if and only if their intersection in the projective space is on $\mathscr{P}_\infty$. Formulate postulates for the affine space (analogous to Postulates A through D for the affine plane) in terms of points, lines, planes, and incidence without reference to $\mathscr{P}_\infty$. (In definining parallelism, distinguish parallel and skew lines.)

5. The affine space over the field $\mathfrak{F}$ is defined as follows: *Points* are the sequences of three elements of $\mathfrak{F}$, denoted by (x, y, z). *Planes* are of three kinds: triples, pairs, or singles of elements of $\mathfrak{F}$, denoted by $\langle u, v, w \rangle$, $\langle v, w \rangle$, and $\langle w \rangle$. Incidence is defined by $z = ux + vy + w$ or $y = vx + w$ or $x = w$, respectively. *Lines* are of four kinds:

   1. lines defined by *point-line coordinates*, that is, quadruples denoted either by $[(a, b, c_1, c_2)]$ $(a \neq 0 \neq b)$ to indicate that the line is on the points $(a, 0, c_1)$ and $(0, b, c_2)$, or by *plane-line coordinates*, that is, quadruples $[\langle p, q, r_1, r_2 \rangle]$ $(p \neq 0 \neq q)$ to indicate that the line is on the planes $\langle p, 0, r_1 \rangle$ and $\langle 0, q, r_2 \rangle$. The point-line and the plane-line coordinates of each line are related as follows:

   $$c_1 = r_2, \qquad c_2 = r_1, \qquad pa + qb = 0, \qquad pa - qb = 2(r_2 - r_1).$$

   The incidences $(x, y, z) \circ [\langle p, q, r_1, r_2 \rangle]$ and $\langle u, v, w \rangle \circ [(a, b, c_1, c_2)]$ are defined, respectively, by

   $$z = px + r_1 = qy + r_2 \qquad \text{and} \qquad w = au + c_1 = bv + c_2 \, ;$$

   2. lines parallel to the *YZ*-plane, defined as quadruples $[(a, 0, c, q)]$ to indicate that the line is on the point $(a, 0, c)$ and on the plane $\langle q, c \rangle$;

   3. lines parallel to the *XZ*-plane;

   4. vertical lines defined as pairs $[a, b]$ to indicate that they are on the point (a, b).

   Give arithmetical definitions of the lines of type 3 and define the incidence of lines of the types 2, 3, 4 with planes of the three possible types and with points.

6. Define a self-dual fragment of the affine space by deleting from the projective space a plane $\mathscr{P}_\infty$ and all points and lines on it as well as a point P_∞ (on $\mathscr{P}_\infty$) and all planes and lines on it. Describe this fragment by postulates in terms of points, planes, lines, and incidence analogous to Postulates α, β, γ for the fragment of the affine plane. Define such a self-dual fragment over a field $\mathfrak{F}$. (There is only one kind of point, one kind of plane, and one kind of line, hence the definitions are much simpler than those in Exercise 5.)

7. In the affine space over the field of the real numbers, consider the line l_0 which is on the points $(t, 0, 0)$ for all numbers t, and the lines l_1, l_2, m_0, m_1, m_2 which are on the points $(t, 1, t)$, $(t, 2, 2t)$, $(0, t, 0)$, $(1, t, t)$, $(t, t - 2, 3t - 6)$ for all numbers t, respectively. Find the point-line coordinates of these lines and show that l_i and m_j are dependent for exactly eight of the nine pairs (i, j).

8. In the affine plane over the noncommutative field of the quaternions, let l_0, l_1, l_2, l_3, m_0, m_1, m_2, m_3 be the lines that are on the points

$$(t, 0, 0),\ (t, 1, t),\ (t, i, ti),\ (t, j, jt),\ (0, t, 0),\ (1, t, t),\ (i, t, ti),\ (k, t, kt),$$

for all quaternions t, respectively. Find the point-line coordinates of these lines and show that l_r and m_s are dependent for exactly fifteen of the sixteen pairs (r, s).

9. Show that the fragment discussed in Exercise 6 does not describe the Galilean kinematics of a two-dimensional space in time. In order to describe the latter, delete from the projective space a plane $\mathscr{P}_\infty$ and all its points and lines as well as all ("vertical") planes on a certain line s_∞ on $\mathscr{P}_\infty$ and all lines meeting s_∞; in other words, delete from an affine space a plane and all parallel planes (planes of *simultaneity*) as well as all lines in such planes. Only the remaining lines represent uniform motions.

5.6   DESARGUES' THEOREM

In a projective space, the analogue of a triangle in the plane is a tetrahedron $\mathfrak{T}$ consisting of four points, P_i, four planes $\mathscr{P}_i$, and six lines $P_h P_i = \mathscr{P}_j \mathscr{P}_k$.

Let $\mathscr{P}$ and P be such that $\mathscr{P} \varnothing P_i$ and $P \varnothing \mathscr{P}_i$ ($i = 1, 2, 3, 4$). The meets of $\mathscr{P}$ with the planes and lines of $\mathfrak{T}$ constitute a quadrilateral in $\mathscr{P}$, namely

$$\mathfrak{T}\mathscr{P} = \{n_1, \dots; R_{12}, \dots\}, \text{ where } n_h = \mathscr{P}_h \mathscr{P} \text{ and } R_{hi} = \mathscr{P}_h \mathscr{P}_i \mathscr{P} = n_h n_i.$$

Dually, the joins of P with the points and lines of $\mathfrak{T}$ constitute what may be called a complete line quadruple on P,

$$\mathfrak{T}P = \{P_1 P, \dots; P_1 P_2 P, \dots\}.$$

The meets of $\mathscr{P}$ with the lines and planes of $\mathfrak{T}P$ constitute a quadrangle in P, namely

$$\mathfrak{T}P\mathscr{P} = \{Q_1, \dots; m_{12}, \dots\}, \text{ where } Q_j = P_j P\mathscr{P} \text{ and } m_{jk} = P_j P_k P\mathscr{P} = Q_j Q_k.$$
Q_i and n_i as well as m_{hi} and R_{hi} are said to be *conjugate* with respect to P (see Fig. 8).

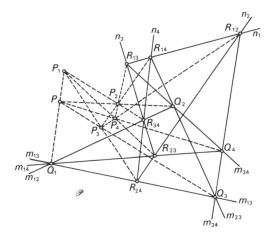

Figure 8

None of the points Q_i is identical with any of the vertices of $\mathfrak{T}\mathscr{P}$; for example, $R_{23} \neq Q_1 \neq R_{12}$. For in view of $P_1 \circ P_2 \circ R_{12}$ from $Q_1 \circ n_2$ it would follow that $P_1 Q_1 \circ P_2$ and hence $P \circ P_2$, contrary to the assumption. Hence $Q_1 \varnothing n_2$, whereas $R_{23} \circ n_2 \circ R_{12}$. Dually, none of the lines n_i is identical with any of the sides of $\mathfrak{T}P\mathscr{P}$.

The union of the complete line and point quadruples $\mathfrak{T}\mathscr{P}$ and $\mathfrak{T}P\mathscr{P}$ is therefore a set consisting of ten points Q_i and $R_{hi} = R_{ih}$ and ten lines n_i and $m_{hi} = m_{ih}$ with the following incidences: If $\{h, i, j, k\} = \{1, 2, 3, 4\}$, then

$$Q_i \varnothing n_j, \qquad Q_i \circ m_{ij}, \qquad Q_i \varnothing m_{jk},$$

$(+)$

$$R_{hi} \circ n_h, \qquad R_{hi} \varnothing n_j, \qquad R_{hi} \varnothing m_{ij}, \qquad R_{hi} \circ m_{jk}.$$

It has been shown above that $Q_1 \varnothing n_2$. From $Q_1 \circ PP_1P_2$ it follows that $Q_1 \circ m_{12}$. If Q_1 were on m_{23}, then Q_1 would be on PP_2P_3 as well as on PP_1P_2 and PP_1P_3, which would imply $Q_1 = P$, whereas $Q_1 \neq P$. Since Q_1, Q_2, and R_{34} are collinear on the plane PP_1P_3, it follows that $R_{34} \circ Q_1 Q_2 = m_{12}$. Similarly $R_{12} \circ m_{34} = Q_3 Q_4$. Hence from $R_{12} \circ m_{13} = Q_1 Q_3$ it would follow that $R_{12} = Q_3$, which is not true. The proof of the other incidences $(+)$ is left to the reader.

The only pairs consisting of one of the ten points and one of the ten lines whose incidence is not described in the formulas $(+)$ are the pairs of conjugate elements: n_i and Q_i as well as R_{jk} and m_{jk}. Indicating these pairs, which may or may not be incident, by question marks and leaving the places of nonincident pairs blank one obtains the following incidence table:

	Q_1	Q_2	Q_3	Q_4	R_{12}	R_{13}	R_{23}	R_{14}	R_{24}	R_{34}
n_1	?				o	o		o		
n_2		?			o		o		o	
n_3			?			o	o			o
n_4				?				o	o	o
m_{12}	o	o			?					o
m_{13}	o		o			?			o	
m_{23}		o	o				?	o		
m_{14}	o			o			o	?		
m_{24}		o		o		o			?	o
m_{34}			o	o	o				o	?

Now consider any quadrangle and any quadrilateral in a plane of a projective space. If certain incidences in the preceding table hold, then a tetrahedron and a point P can be constructed which generate the two quadruples, whence the entire table is valid. Consequently certain of those incidences imply the others. This fact has been expressed by H.S.M. Coxeter in the following

THEOREM 5.8 *If in a plane $\mathscr{P}$ of a projective space, $\{Q_1, \ldots, m_{34}\}$ is a quadrangle and $\{n_1, \ldots, R_{34}\}$ is a quadrilateral such that $R_{hi} \circ m_{jk}$ for five of the six pairs (h, i), then the incidence holds also for the sixth pair.*

Assuming $R_{hi} \circ m_{jk}$ for all pairs except possibly for $(h, i) = (3, 4)$ one can prove that also $R_{34} \circ m_{12}$. Let $\mathscr{P}_1$, $\mathscr{P}_2$, $\mathscr{P}_3$, $\mathscr{P}_4$ be noncopunctual planes such that $\mathscr{P}_i \circ n_i$ $(i = 1, 2, 3, 4)$ and set $\mathscr{P}_h \mathscr{P}_i \mathscr{P}_j = P_k$. From $n_1 n_2 = R_{12} \circ m_{34} = Q_3 Q_4$ it follows that $\mathscr{P}_1 \mathscr{P}_2 (= P_3 P_4)$ and $Q_3 Q_4$ are dependent. Hence so are $P_3 Q_3$ and $P_4 Q_4$. Let P be their meet. Because of the other four assumed incidences, $R_{hi} \circ m_{jk}$, $P_1 Q_1$ and $P_2 Q_2$ both meet $P_3 Q_3$ and $P_4 Q_4$. Since no three $\mathscr{P}_i$ are collinear, no three of the four lines $P_i Q_i$ are coplanar. Hence all four are on P. It follows that $P_1 Q_1$ and $P_2 Q_2$ intersect, whence $\mathscr{P}_3 \mathscr{P}_4 P = n_3 n_4 = R_{34} \circ m_{12}$.

If $n_i \neq Q_j Q_k$ and $Q_i \neq n_j n_k$ for $\{i, j, k\} = \{1, 2, 3\}$, then four of the six incidences $R_{hi} \circ m_{jk}$ can be stated as follows:

$$Q_1 Q_2 n_3 \circ n_4 \circ Q_1 Q_3 n_2 \qquad \text{and} \qquad n_1 n_2 Q_3 \circ Q_4 \circ n_1 n_3 Q_2 .$$

That under these conditions the fifth incidence implies the sixth, and vice versa, is equivalent to the statement

$$Q_2 Q_3 n_1 \circ n_4 \qquad \text{if and only if} \qquad n_2 n_3 Q_1 \circ Q_4 .$$

One thus obtains the famous theorem due to the French geometer Giraud Desargues (1593–1662) in a form that is self-dual as an assertion about points and lines on a plane. (A dual assertion in space deals with planes and lines on a point.)

THEOREM 5.9 (SELF-DUAL FORM OF DESARGUES' THEOREM)
If in a plane of a projective space, Q_1, Q_2, Q_3 are noncollinear, and n_1, n_2, n_3 are nonconcurrent such that

$$Q_i \neq n_j n_k \text{ and } n_i \neq Q_j Q_k \text{ for } \{i, j, k\} = \{1, 2, 3\},$$

then the following assertions are equivalent:

the lines of the triple $\mathfrak{N}: n_1 n_2 Q_3, n_1 n_3 Q_2, n_2 n_3 Q_1$ are concurrent;

the points of the triple $\mathfrak{Q}: Q_1 Q_2 n_3, Q_1 Q_3 n_2, Q_2 Q_3 n_1$ are collinear.

This theorem is the conjunction of the following two statements:

5.9.A   *The concurrence of the lines $\mathfrak{N}$ implies the collinearity of the points $\mathfrak{Q}$.*

5.9.B   *The collinearity of the points $\mathfrak{Q}$ implies the concurrence of the lines $\mathfrak{N}$.*

Either half implies the other. Assuming A, one proves B by considering

$$\text{the points } S_1 = Q_1, \ S_2 = n_2 n_3, \ S_3 = Q_1 Q_3 n_2;$$
$$\text{the lines } r_1 = n_1, \ r_2 = Q_2 Q_3, \ r_3 = n_1 n_3 Q_2 .$$

If the points $\mathfrak{Q}$ are collinear, then

$$\text{the lines } r_1 r_2 S_3, \, r_1 r_3 S_2, \, r_2 r_3 S_1 \text{ are concurrent.}$$

By virtue of A,

$$\text{the points } S_1 S_2 r_3, \, S_1 S_2 r_3, \, S_2 S_3 r_1 \text{ are collinear,}$$

whence the lines $\mathfrak{N}$ are readily seen to be concurrent.

Another self-dual form of Desargues' assertion about planes in a projective space is as follows.

THEOREM 5.10   *For two indexed triangles in a plane, the joins of corresponding points are concurrent if and only if the meets of corresponding lines are collinear.*

The theorem is often expressed and used in the following form, which is not self-dual.

THEOREM 5.11   *If P_1, P_2, P_3 and P'_1, P'_2, P'_3 are two noncollinear triples such that the three lines $P_i P'_i$ are concurrent, then the three points $(P_i P_j)(P'_i P'_j)$ are collinear and vice versa.*

The equivalence of the two halves of this theorem with the statements A and B is obvious. Hence the two halves imply one another.

EXERCISES

1. Dualize Conditions $(+)$ in the projective space, and derive the dual in space of Theorems 5.8 and 5.9.

2. In the ordinary plane, construct complete quadruples of lines and points satisfying Conditions $(+)$ and some of the following conditions: $Q_i \circ n_i$ and/or $R_{hi} \circ m_{hi}$. How many of these "diagonal" conditions can hold simultaneously?

3. What can be said about (a) two noncoplanar triples P_i and P'_i $(i=1, 2, 3)$ and (b) about two quadruples P_i and P'_i $(i=1, 2, 3, 4)$ each of which is noncoplanar, provided that all lines $P_i P_i$ are concurrent?

4. It is necessary to assume in Theorem 5.11 that the two triples be noncollinear?

5.7 ARGUESIAN PROJECTIVE AND AFFINE PLANES

Desargues' assertion, though valid in every plane of a projective space, is not a consequence of Postulates I through IV. In fact, there are projective planes for which the assertion is not valid, among them projective planes with exactly ten points on each line. Affine planes (and even planes with perpendicularity and congruence) that do not satisfy Desargues' assertion were first discovered by Hilbert. A simple example has been constructed by F. R. Moulton as follows: In the affine plane over the field of the real numbers, vertical lines and lines with nonpositive slopes are retained, whereas lines with positive slope are refracted in the X-axis. Thus points are pairs of numbers; lines are defined as the sets of points (x, y) such that (1) $x = c$ for some number c; or (2) $y = mx + c$ for some c and some $m \leqslant 0$; (3) for some c and some $m > 0$, $y = mx + c$ if $x \leqslant -c/m$ and $y = \frac{1}{2}(mx + c)$ if $x > -c/m$. In this plane, consider the two triples of points

$$(-1, 0), (-2, -1), (-\tfrac{2}{3}, -\tfrac{1}{3}) \quad \text{and} \quad (1, 0), (3, -2), (\tfrac{2}{3}, -\tfrac{2}{3}).$$

The joins of corresponding points meet at the point $(-7, 0)$. Two pairs of corresponding lines meet at $(0, -1)$ and $(0, 0)$, thus on the Y-axis. So does the third pair in the ordinary plane. In the modified plane, however, the third pair meets at $(\tfrac{1}{3}, \tfrac{2}{3})$, which is not collinear with the other two meets.

It is not surprising that the simple Postulates I through IV fail to yield a statement as complicated as Desargues'. Nor is any fifth assumption concerning projective planes known that is essentially simpler than Desargues' assertion and which implies it. But Desargues' assertion has many consequences of great importance for the development of plane geometry. One therefore adjoins it to Postulates I through IV by assuming

POSTULATE V. *The points and lines satisfy the assertion of Theorem* 5.10.

The postulate amounts to attributing to projective planes a property that is common to all planes in projective spaces. Projective planes that satisfy Postulate V are called *Desargues planes* or said to be *Arguesian*.

One of the most important consequences of Postulate V is that the Remark at the end of Section 5.2 about the extension of matroid fragments of points on a line to quadrangular matroids can be sharpened. In a Desargues plane, the completed matroid is the same in conjunction with any two points that are collinear with one point of the fragment; and hence each matroid fragment can be extended in one and only one way to a quadrangular matroid. This is an immediate consequence of

THEOREM 5.12   *In a Desargues plane, two quadrangles that generate equal matroid fragments on a line l generate the same matroid of points.*

If the quadrangles with the points T_i and T_i' ($i = 1, 2, 3, 4$) generate on l the same points Q_{ij} ($i, j \neq 3, 4$), then, since Q_{12}, Q_{13}, Q_{23} are collinear, by assertion 5.9.B the lines $T_1 T_1', T_2 T_2', T_3 T_3'$ are concurrent. So are $T_1 T_1', T_2 T_2', T_4 T_4'$. Hence $T_2 T_2', T_3 T_3', T_4 T_4'$ are concurrent and, by assertion 5.9.A, the points Q_{23}, Q_{24} (which are on l) and $(T_3 T_4)(T_3' T_4')$ are concurrent. Hence $T_3 T_4, T_3' T_4'$, and l are concurrent in a point Q_{34}.

In an affine plane, Desargues' assertion is adjoined to Postulates A, B, C IV in the following form:

POSTULATE D   *If P_1, P_2, P_3 are noncollinear points, and l_1, l_2, l_3 are nonconcurrent lines such that $l_i l_j \neq P_k$ and $P_i P_j \neq l_k$ for $\{i, j, k\} = \{1, 2, 3\}$, then*

$$l_1 l_2 P_3, l_1 l_3 P_2, l_2 l_3 P_1 \text{ are concurrent or parallel}$$

if and only if

$$\text{either } P_1 P_2 l_3, P_1 P_3 l_2, P_2 P_3 l_1 \text{ are collinear}$$

$$\text{or } P_i P_j \parallel l_k \parallel (P_k P_i l_j)(P_k P_j l_i) \text{ for some permutation } i, j, k \text{ of } 1, 2, 3$$

$$\text{or } P_1 P_2 \parallel l_3, P_1 P_3 \parallel l_2, P_2 P_3 \parallel l_1.$$

In a Desargues affine plane, a quadrangle $\mathfrak{Q}$ does not generate a point on a line l parallel to one of its sides. If two (opposite) sides of $\mathfrak{Q}$, say l_{12} and l_{34}, are parallel, then all that $\mathfrak{Q}$ generates on a line parallel to l_{12} and l_{34} are the points $Q_{13}, Q_{24}, Q_{23}, Q_{14}$. Of particular importance is the case in which $Q_{13} = Q_{24}$ (Q_{13} is called the midpoint of Q_{23} and Q_{14}).

DEFINITION 5.1   On a line l of a Desargues affine plane, M is called the *midpoint* of the points X and Z if there exists a quadrangle such that

$$l_{12} \parallel l_{34} \parallel l, \qquad X = l_{23} l, \qquad Z = l_{14} l, \qquad M = l_{13} l = l_{24} l.$$

In the plane of a blackboard or the affine plane over the field of the real numbers one readily verifies (by construction or by computations of analytic geometry, respectively) that this definition yields what is ordinarily called midpoints. In a projective plane—that is, in terms of points, lines, and incidence satisfying Postulates I through V—midpoints cannot be defined. Such a definition becomes possible only if one introduces a nonprojective concept—namely, that of a distinguished line l_∞. In this case, if any two lines concurrent with l_∞ are said to be parallel, then on all lines $\neq l_\infty$ midpoints can be defined just as in affine geometry. But the definition is not in terms of the basic projective concepts of point, line, and incidence. Moreover, the midpoint of two points P and Q depends upon the choice of the line l_∞.

But even in Arguesian planes, where midpoints can be defined, they possess only some but not all of the properties of midpoints on blackboards or the affine plane over the field of the real numbers. For example, from Theorem 5.12 it follows that for *each* quadrangle with two sides parallel to l, generating X and Z and having a diagonal point on l, this latter point is M. Hence in each Arguesian affine plane (just as on a blackboard) two points have *at most one midpoint*. On the other hand, in the affine plane with nine points and twelve lines (which is Arguesian, see Exercise 4), if M is the midpoint of X and Z, then (in blatant contradiction to blackboard geometry) Z is also the midpoint of X and M.

Using the concept of midpoint the Italian mathematician Mario Pieri (1860–1913) defined a ternary relation of betweenness for collinear points of an Arguesian affine plane.

DEFINITION 5.2   The point Y is said to be *between* X and Z if and only if $X \neq Y \neq Z$ and either $Y = M$ or there exists a pair of points R, S which is conjugate to both X, Z and M, Y; that is, such that both

$$
\begin{array}{ccc}
X & Z & R \\
X & Z & S
\end{array}
\qquad \text{and} \qquad
\begin{array}{ccc}
M & Y & R \\
M & Y & S
\end{array}
$$

are harmonic matroids.

On a blackboard as well as in the affine plane over the field of the real numbers, Y is between X and Z in this sense if and only if Y belongs to the interval determined by X and Z. For example, on the line $\langle 0, 0 \rangle$ of such a plane the midpoint of $X = (-1, 0)$ and $Z = (1, 0)$ is $M = (0, 0)$. The points $R = (r, 0)$ and $S = (s, 0)$ are readily seen to be conjugate to X and Z if and only if $s = 1/r$ $(r \neq 0 \neq s)$. For R and S to be conjugate to $M = (0, 0)$ and

$Y = (y, 0)$ it is, as one easily verifies, necessary and sufficient that

(*)
$$r^2 - \frac{2}{y}r - 1 = 0.$$

This equation has two (mutually reciprocal) real roots if and only if $0 < y^2 \leqslant 1$. Hence Y is between X and Z if and only if $-1 \leqslant y \leqslant 1$. In this plane, of any three collinear points exactly one is seen to be between the other two.

In general, however—that is, on the basis of Postulates A, B, C, D—betweenness has only some but not all of the properties that this relation possesses on a blackboard. For example, in the plane over the field of the complex numbers, equation (*) has roots for each number $y \neq 0$, whence in the complex affine plane each point of the line $\langle 0, 0 \rangle$ is between X and Z, and, more generally, of any three collinear points each is between the other two. On the other hand, in the plane over the field of the rational real numbers, none of the points $(-1, 0)$, $(\frac{1}{2}, 0)$, $(1, 0)$ is between the other two; for example, $(\frac{1}{2}, 0)$ is not between X and Z, because equation (*) has no rational root for $y = \frac{1}{2}$.

Summarizing, one can say that in contrast to projective planes, in Arguesian affine planes midpoints and betweenness can be defined, but in general they lack some of the most important properties that these concepts possess when defined on a blackboard or in the plane over the field of the real numbers (unless of course one introduces special postulates to the effect that these properties hold).

EXERCISES

1. By methods of elementary analytic geometry prove that the affine planes (1) over the field of the real numbers, (2) over any commutative field, (3) over the noncommutative field of the quaternions are Arguesian.

2. By analytic geometry show that in the planes of Exercise 1 the midpoint of $(x, 0)$ and $(z, 0)$ in the sense of Definition 5.1 is $(\frac{1}{2}(x + z), 0)$.

3. By analytic geometry prove that in the plane over the field of the real numbers the point $(y, 0)$ is between $(-1, 0)$ and $(1, 0)$ in the sense of Definition 5.2 if and only if there exist points $(r, 0)$ and $(s, 0)$ such that r and s are the roots of the equation (*). Which points are between $(a, 0)$ and $(b, 0)$?

4. Show that the affine plane with nine points and twelve lines is Arguesian.

5. On a sheet of paper, let l and l' be two lines that have no common point on the sheet (without necessarily being parallel), and let P be a point that

is on neither line. Using the fact that the sheet is a part of an Arguesian affine plane, show that the line joining P to the meet of l and l' (outside of the sheet) or the line on P parallel to l and l' (if $l \parallel l'$) can be obtained by the following construction on the sheet. Three concurrent lines m_1, m_2, m_3 can be chosen in such a way that the six points

$$Q_1 = lm_1,\ Q_2 = lm_2,\ Q_3 = Q_1 Pm_3\ ;\ Q_1' = l'm_1,\ Q_2' = l'm_2,\ Q_3' = Q_1' Pm_3'$$

as well as the point $R = (Q_2 Q_3')(Q_2' Q_3)$ lie on the sheet. Then PR is the desired line.

5.8 PROJECTIVITIES

The concept of *projectivity* in projective geometry is wider than that of *projection* in other branches of mathematics. The former includes perspectivities such as π_{lPm} (which is called the projection of $[l]$ on $[m]$ from P) and, in addition, all mappings that result from chains of perspectivities; and such mappings are not in general called projections.

Suppose, for example, that $l \varnothing P \varnothing m \varnothing Q \varnothing n$ and that $[l]$ is mapped onto $[m]$ by π_{lPm}, and then $[m]$ is mapped on $[n]$ by π_{mQn}. The resulting mapping of $[l]$ onto $[n]$ is a projectivity. We denote it by π_{lPmQn} and write

$$\pi_{lPmQn} X = \pi_{mQn}(XPm) = XPmQn \text{ for each } X \circ l.$$

More generally, a projectivity of $[l]$ onto $[m]$ is defined as a mapping ρ such that for some natural number k, for k points, and for $k-1$ lines that satisfy the condition

(∗) $$l \varnothing P_1 \varnothing l_1 \varnothing P_2 \varnothing \dots \varnothing l_{k-1} \varnothing P_k \varnothing m,$$

one has for each $X \circ l$

$$\rho X = XP_1 l_1 P_2 \cdots l_{k-1} P_k m.$$

A more descriptive symbol for this mapping ρ is $\pi_{lP_1 l_1 P_2 \dots l_{k-1} P_k m}$. Whenever this symbol is written it is understood that Conditions (∗) are satisfied, as well as corresponding conditions for similar symbols. Clearly,

$$\pi_{lP_1 l_1 \dots l_{k-1} P_k m P_k l_{k-1} \dots l_1 P_1 l}$$

is the identical mapping of l onto itself. Hence $\pi_{mP_k l_{k-1} \dots l_1 P_1 l}$ is called the *inverse* of $\pi_{lP_1 l_1 \dots l_{k-1} P_k m}$.

In a dual way, one defines projectivities of the pencil $[P]$ on the pencil $[Q]$.

In a Desargues plane, the definition of projectivities can be essentially simplified because of the following three lemmas.

LEMMA 5.7 *If l, l_1, and m are distinct concurrent lines, then $\pi_{lP_1l_1P_2m} = \pi_{lP*m}$ for some point $P*$ on P_1P_2 such that $l \varnothing P* \varnothing m$.*

Choose A on l such that $ll_1 \neq A \varnothing P_1P_2$, and set

$$P* = (AP_1\, l_1\, P_2\, mA)(P_1P_2).$$

Using Postulate V one readily proves

$$\pi_{lP_1l_1P_2m}\, X = \pi_{lP*m}\, X \qquad \text{for each } X \text{ on } l,$$

which is the assertion.

Lemma 5.7 replaces a chain of two perspectivities from $[l]$ onto $[l_1]$ onto $[m]$ by a single perspectivity in case l, l_1, m are concurrent. In case they are not, Lemma 5.8 allows us to replace l_1 by almost any given line l'.

LEMMA 5.8 *If l, l_1, and m are nonconcurrent, and if l' is almost any line (precisely, any line such that $lm \varnothing l' \neq (ll_1\, P_2\, m)(l_1\, mP_1\, l)$), then*

$$\pi_{lP_1l_1P_2m} = \pi_{lP'_1l'P'_2m} \qquad \text{for two points } P'_1, P'_2 \text{ on } P_1P_2\,.$$

Clearly, $l' \varnothing ll_1\, P_2\, m$ and/or $l' \varnothing l_1mP_1\, l$. One may assume the first, since the second case can be reduced to the first by letting l and m change roles. Setting $l* = (ll_1)(l'm)$ one has $P_2 \varnothing l* \neq l$. If A is any point on l such that $A \varnothing P_1P_2$ and $ll_1 \neq A \neq ml'P_2\, l_1\, P_1\, l$ and $B = AP_1\, l_1\, P_2\, m$, then set

$$P'_1 = (BP_2\, l*A)(P_1P_2) \qquad \text{and} \qquad P'_2 = P_2 \qquad \text{or} \qquad P'_2 = (AP'_1\, l'B)(P_1P_2)$$

according as $l' \varnothing ll_1$ or $l' \varnothing ll_1$. Using Postulate V one readily demonstrates the assertion.

LEMMA 5.9 $\pi_{lP_1mP_2lP_3m} = \pi_{lP*m}$ *for some point $P*$.*

If $P_1P_2\, l = A$, $AP_3\, m = A'$, $P_2P_3\, m = B'$, $B'P_1\, l = B$ then $P* = (AA')(BB')$ satisfies the assertion as a consequence of Postulate V.

Although a projectivity π of $[l]$ onto $[m]$ is by definition the result of a chain of *some finite number* of perspectivities, the preceding lemmas allow one to obtain π as the result of *two* perspectivities.

THEOREM 5.13 *If π is any projectivity from $[l]$ onto $[m]$, where $l \neq m$, then*

$$\pi = \pi_{lQ_1kQ_2m} \qquad \textit{for some } Q_1, k, Q_2 \textit{ such that } l \oslash Q_1 \oslash k \oslash Q_2 \oslash m.$$

By induction, the assertion obviously can be reduced to the case where

$$\pi = \pi_{lP_1l_1P_2l_2P_3m}.$$

If l, l_1, m are nonconcurrent, then by Lemma 5.7, if l^* is almost any line concurrent with l and l_1, one has

$$\pi_{l_1P_2l_2P_3m} = \pi_{l_1P_2'l^*P_3'm},$$

whence by Lemma 5.7 $\pi_{lP_1l_1P_2'l^*} = \pi_{lP*l^*}$ for some P^*, and $\pi = \pi_{lP*l^*P_3'm}$. The case where $l, l_2,$ and m are nonconcurrent is treated similarly. This leaves the case where l, l_1, l_2, m are concurrent, which can be handled by Lemma 5.7 or 5.9.

EXERCISES

1. State conditions that are sufficient to make $\pi_{lP_1l_1P_2l_2P_3m}$ reducible to a single perspectivity.

2. Prove that each projectivity of $[l]$ onto $[l]$ can be obtained by a chain of two or three perspectivities. Find properties of those projectivities of $[l]$ onto $[l]$ which are the results of two perspectivities.

3. If ρ is a projectivity of $[P]$ onto $[Q]$, and l is a line such that $P \oslash l \oslash Q$, then the mapping ρ^* of $[l]$ onto $[l]$ such that $\rho^*X = l\rho\,(XP)$ for each $X \circ l$ is a projectivity. This often used mapping is called the projectivity of $[l]$ onto $[l]$ *induced by* ρ. If $\rho = \pi_{PmCnQ}$, what is ρ^*?

4. The projectivity π_{lPmQl} leaves the point PQl fixed.

5.9   DANDELIN SPACES

In a three-dimensional affine space over the field of the real numbers, one can prove (see Exercise 3) the following theorem discovered by the French mathematician G. P. Dandelin (1794–1847):

THEOREM 5.14 *If* l_i *and* $m_i (i = 0, 1, 2, 3)$ *are two quadruples of skew lines such that for fifteen of the sixteen pairs,* i, j *the lines* l_i *and* m_j *are dependent, then so are the lines of the sixteenth pair.*

By completing that affine space to a projective space one can extend Dandelin's Theorem to the latter. But the theorem is not a consequence of Postulates 1–8, since it does not hold in a three-dimensional affine space over the (non-commutative) field of the quaternions (see Exercise 8, Section 5.5); nor, consequently, does it hold in the projective space obtained by completing that affine space, even though this projective space satisfies Postulates 1–8. Nor does the theorem, which has important consequences, seem to follow from any assumption that is essentially simpler than the theorem itself. Just as one adjoins Desargues' Theorem to Postulates I–IV, one adjoins Dandelin's Theorem to Postulates 1–8.

POSTULATE 9 *The lines satisfy the assertion of Theorem 5.14.*

Projective spaces satisfying Postulate 9 are called *Dandelin spaces*. Important consequences of Postulate 9 concern planes in Dandelin spaces.

A cycle of six points in a projective plane

$$
\begin{array}{ccc}
 & P_1 \quad Q_2 & \\
Q_3 & & P_3 \\
 & P_2 \quad Q_1 &
\end{array}
$$

is called a *Pappus hexagon* if P_1, P_2, P_3 as well as Q_1, Q_2, Q_3 are collinear and the meet of the two lines is not among the six points. Each of the six lines $P_i Q_{i+1}, Q_i P_{i+1}$ (where $3 + 1 = 1$) is called a *side* of the hexagon; $P_i Q_{i+1}$ and $Q_i P_{i+1}$ are said to be *opposite* sides. The Greek geometer Pappus (in the early part of the fourth century A.D.) discovered that *the three meets of opposite sides of any Pappus hexagon are collinear* (Pappus Law). A projective plane in which this law holds is called a *Pappus plane*. If the hexagon mentioned above lies in a Pappus plane, then the following points are collinear:

$$(P_1 Q_2)(P_2 Q_1) = R_3, \qquad (P_2 Q_3)(P_3 Q_2) = R_1, \qquad (P_3 Q_1)(P_1 Q_3) = R_2.$$

The nine points $P_1, P_2, P_3, Q_1, Q_2, Q_3, R_1, R_2, R_3$ can readily be identified with the points $P, Q, R, V, W, X, U, S, T$, respectively, in the Pappus configuration discussed in Section 5.4. Such a configuration also includes a cycle of six lines—two triples of concurrent lines (not including the join of the two meets of the triples), which determine a third concurrent triple. This is brought out in the following self-dual

DEFINITION A projective plane is a *Pappus plane* provided that, *if A, B, C are noncollinear, and a, b, c are nonconcurrent, and if*

$$A \, \emptyset \, a$$
$$\emptyset \quad \emptyset, \qquad ab \circ AB, \qquad C \circ c, \qquad a \, \emptyset \, C \, \emptyset \, b, \qquad A \, \emptyset \, c \, \emptyset \, B,$$
$$b \, \emptyset \, B$$

then ACb, BCa, (acA)(bcB) are collinear
and acA, bcB, (ACb)(BCa) are concurrent.

THEOREM 5.15 *Each plane in a Dandelin space is a Pappus plane.*

Let $\mathscr{P}_0$ be a plane in a Dandelin space. If $P_1, \ldots, Q_3$ is a Pappus hexagon in $\mathscr{P}_0$ let l_0 and m_0 be the lines on P_1, P_2, P_3 and Q_1, Q_2, Q_3, respectively, $l_0 m_0$ being distinct from the six points. There exist six lines $l_1, l_2, l_3, m_1, m_2, m_3$ such that

(1) $l_i \circ Q_i$ and $m_i \circ P_i$ $(i = 1, 2, 3)$;

(2) any two lines l_i and l_j as well as m_i and m_j $(i = 0, 1, 2, 3)$ are independent;

(3) any two lines l_i and m_j, except possibly l_3 and m_3, are dependent.

In order to obtain such lines one may first choose lines l_1 on Q_1 and l_2 on Q_2 such that l_0, l_1, l_2 are independent. Then let m_1 be the line (uniquely determined according to the theorem in Exercise 2, Section 5.5) which is on P_1 and such that m_1 and l_1 as well as m_1 and l_2 are dependent. Clearly, m_0 and m_1 are independent. Similarly, let m_2 and m_3 be the lines on P_2 and P_3, respectively, such that m_2 as well as m_3 and l_1 as well as l_2 are dependent. Then m_0, m_1, m_2, m_3 are pairwise independent. Finally, let l_3 be the line on Q_3 which is dependent of m_1 and m_2. Clearly, l_0, l_1, l_2, l_3 are independent. Because of Postulate 9, l_3 and m_3 are dependent.

According to the theorem in Exercise 3, Section 5.5,

$$n_k = (l_i \cup m_j)(l_j \cup m_i) = (l_i \cap m_j)(l_j \cap m_i) \text{ if } \{i, j, k\} = \{1, 2, 3\}.$$

Clearly, the lines n_1, n_2, n_3 are pairwise dependent (for example, both n_2 and n_3 are on $l_1 \cap m_1$). Consequently, there exists a plane $\mathscr{P}$ which is on n_k

($k = 1, 2, 3$). Since $Q_i \circ l_i$ and $P_j \circ m_j$ it follows that $Q_i P_j \circ (l_i \cup m_j)$ and

$$R_k = (Q_i P_j)(Q_j P_i) \circ n_k \circ \mathscr{P} \quad (k = 1, 2, 3).$$

But $R_k \circ \mathscr{P}_0$, whence $R_k \circ \mathscr{P}\mathscr{P}_0$ ($k = 1, 2, 3$); that is to say, R_1, R_2, R_3 are collinear, and $[\mathscr{P}_0]$ is a Pappus plane.

EXERCISES

1. How many Pappus hexagons can be extracted from (a) two collinear triples of points $\{A, B, C\}$ on m, and $\{D, E, F\}$ on n, neither triple including the point mn; (b) a Pappus configuration?

2. Construct a Pappus hexagon such that the three lines on the P_i, the Q_j, and the R_k are concurrent. Dualize the configuration thus obtained.

3. Prove Dandelin's Theorem in the affine space over the field of the real numbers. Without any loss of generality, one can assume that l_0 is the X-axis, and m_0 is the Y-axis, and that l_1 and m_1 meet at a point $(c_1, d_1, 1)$. Let (c_2, d_2, e_2) be the meet of l_2 and m_2; and $(c_3, d_3, 0)$, the meet of the projections of l_3 and m_3 in the XY-plane. Let (c_3, d_3, f) and (c_3, d_3, g) be the points on l_3 and m_3 with the projection $(c_3, d_3, 0)$. Under the assumption that l_i and m_j meet for each of the pairs (i, j) $(i, j = 1, 2, 3)$ except possibly $(3, 3)$, prove that $f = g$; that is to say, prove that l_3 and m_3 also meet.

   Note that in an affine space two lines are skew if and only if they are not coplanar. Skew lines are not copunctual (that is, do not meet), but not vice versa. Nonskew lines are coplanar and hence either copunctual or parallel.

4. From the Pappus plane delete a line, l_∞, and the points on it. In the remaining affine plane, define parallelism of lines as their concurrence with l_∞ and consider the fragments of Pappus configurations, for example, Pappus hexagons with parallel opposite sides. Define Pappus affine planes by an assumption about Pappus configurations and fragments of theirs without any reference to l_∞.

5.10   PAPPUS PLANES

The affine plane over the field of the quaternions is Arguesian but not a Pappus plane; and the same is true for the projective plane obtained by its completion. On the other hand, the German mathematician Gerhard Hessenberg proved in 1905

THEOREM 5.16   *Each Pappus plane is Arguesian.*

In a Pappus plane, let P, Q, R and P', Q', R' be two triples of points such that the lines PP', QQ', RR' are concurrent, say, in the point O. It must be proved that the points (see Fig. 9):

$$A = (QR)(Q'R'), \qquad B = (PR)(P'R'), \qquad C = (PQ)(P'Q')$$

are collinear. This is done by introducing the following four auxiliary points (see Fig. 9):

$S = (PR)(R'Q')$. It follows that $SQ' = Q'R'$ and $PS = PR$.
$T = (RR')(PQ') = (RO)(PQ') = (R'O)(PQ')$.
$U = (OS)(QP)$. It follows that $UP = QP$.
$V = (OS)(Q'P')$. It follows that $Q'V = Q'P'$.

The following three hexagons are easily seen to be Pappus hexagons.

Q	R	$(QR)(SQ')$	$= (QR)(Q'R') = A,$
P	O	$(RO)(Q'P)$	$= T,$
Q'	S	$(OS)(PQ)$	$- U.$ Hence A is on TU.

P'	R'	$(P'R')(SP)$	$= (P'R')(PR) = B,$
Q'	O	$(R'O)(PQ')$	$= T,$
P	S	$(OS)(Q'P')$	$= V.$ Hence B is on VT.

S	Q'	$(SQ')(TU)$	$= A$, since $A \circ TU$ and $A \circ Q'R' = SQ'$
P	V	$(Q'V)(UP)$	$= (Q'P')(QP) = C,$
U	T	$(VT)(PS)$	$= B$, since $B \circ VT$ and $B \circ PS = PR$.

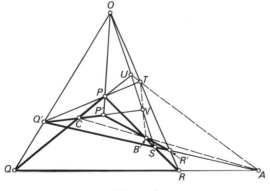

Figure 9

Pappus' Law applied to the third hexagon yields the asserted collinearity of A, B, C.

Pappus planes are characterized among the Desargues planes by certain properties of the projectivities. The most important of these properties are the following three, which are equivalent to Pappus' Law and to one another.

THEOREM 5.17   *In order that a Desargues plane be a Pappus plane each of the following conditions is necessary and sufficient:*

CONDITION I (ON FIXED ELEMENTS)   *Each projectivity of a range or a pencil on itself that is not the identity maps at most two elements on themselves.*

CONDITION II (ON THE DETERMINATION OF PROJECTIVITIES)   *Each projectivity is determined by any three elements and their images; in other words, any two projectivities which map three distinct elements on the same images are identical (that is, either projectivity maps each element on the same image as the other).*

CONDITION III (CHARACTERIZATION OF PERSPECTIVITIES)   *In order that a projectivity of a range on another range or of a pencil on another pencil be a perspectivity it is necessary and sufficient that the element common to the two ranges or pencils remain fixed (that is, be mapped on itself).*

The theorem will be demonstrated by proving that Condition I holds in any Pappus plane and then deriving Condition II from I, Condition III from II, and Pappus' Law from Condition III.

(a) *Pappus' Law implies Condition I.*   In a Pappus plane, let ρ be a projectivity of $[l]$ onto $[l]$ such that $\rho F_i = F_i$ for three distinct points F_1, F_2, F_3. It must be shown that ρ is the identity; that is to say, $\rho X = X$ for each X on l. Since the Pappus plane is a Desargues plane, Theorem 5.13 and Exercise 2 in Section 5.8 can be applied; that is, it may be assumed that

$$\rho = \pi_{lSl_1 Tl_2 Ul} \qquad \text{for some } S, l_1, T, l_2, U.$$

At least one of the points F_1, F_2 is distinct from ll_2, say $F_1 \neq ll_2$. Hence by Lemma 5.8,

$$\pi_{lSl_1 Tl_2} = \pi_{lQ_1 m_1 Q' l_2} \qquad \text{for some } Q_1, Q' \text{ and } m_1, \text{ where } m_1 \circ F_1.$$

Consequently, $F_2 \neq lm_1 \neq F_3$. By virtue of the same lemma,

$$\pi_{m_1 Q' l_2 U l} = \pi_{m_1 Q_2 m_2 Q_3 l}$$

for any line $m_2 \varnothing lm_1 = F_1$ with one exception. At least one of the lines $Q_1 F_3 m_1 F_2$, $Q_1 F_2 m_1 F_3$ is nonexceptional, say the first, and can be chosen as m_2. Thus

$$\rho = \pi_{lQ_1 m_1 Q_2 m_2 Q_3 l}, \qquad \text{where } m_1 \circ F_1, \, m_2 \circ F_2, \text{ and } m_1 m_2 \circ Q_1 F_3.$$

It follows that

$$\rho F_1 = F_1 Q_1 m_1 Q_2 m_2 Q_3 l = F_1 Q_2 m_2 Q_3 l.$$

From $\rho F_1 = F_1$, it follows that F_1, Q_2, Q_3 are collinear. Similarly,

$$\rho F_2 = F_2 Q_1 m_1 Q_2 m_2 Q_3 l = F_2.$$

whence F_2, Q_1, Q_2 are collinear; and

$$\rho F_3 = F_3 Q_1 m_1 Q_2 m_2 Q_3 l = F_3.$$

Hence F_3, Q_3, $m_1 m_2$ and thus F_3, Q_1, Q_3 are collinear. Now let X be any point on l. In order to prove $\rho X = X$, set

$$X' = X Q_1 m_1, \qquad X'' = X' Q_2 m_2, \qquad X''' = X'' Q_3 l = \rho X.$$

Since F_1, Q_2, Q_3 as well as F_2, $m_1 m_2$, X'' are collinear

$$
\begin{array}{ccc}
 & F_1 \quad F_2 & \\
m_1 m_2 & & Q_2 \\
 & Q_3 \quad X'' &
\end{array}
$$

is a Pappus hexagon. Consequently,

$$(F_2 Q_2)(m_1 m_2 Q_3) = Q_1, \quad (m_1 m_2 F_1)(Q_2 X'') = X', \quad (F_1 F_2)(Q_3 X'') = X'''$$

are collinear. It follows that $X''' = Q_1 X' l = \rho X$, as asserted.

(b) *Condition 1 implies II.* Let ρ and τ be two projectivities of $[l]$ onto $[m]$ such that $\rho P_i = \tau P_i$ for three distinct points P_1, P_2, P_3 on l. If σ is the inverse of ρ, then applying σ after τ one obtains a projectivity of $[l]$ onto $[l]$ that leaves the three points P_i fixed and thus, by Condition I, is the identity. Consequently, σ is also the inverse of τ, which therefore is identical with ρ.

(c) *Condition II implies III.* That a perspectivity of $[l]$ on $[m]$ leaves lm fixed has been stated in Lemma 5.4. Conversely, let ρ be a projectivity of $[l]$ onto $[m]$ $(m \neq l)$ that leaves lm fixed. Let A and B be any two distinct points $\neq lm$ on l and set $\rho A = A'$, $\rho B = B'$, $(AA')(BB') = C$. Then the perspectivity π_{lCm} maps A, B, and lm obviously on the same points on m as ρ does. By Condition II, this perspectivity thus is identical with ρ.

(d) *Condition III implies Pappus Law.* Let P, Q, R; P', Q', R' be distinct points on lines l and l', respectively, neither triple including ll'. In order to prove Pappus' Law it must be demonstrated that

$$(PQ')(P'Q) = R'', \qquad (PR')(P'R) = Q'', \qquad (QR')(Q'R) = P''$$

are collinear. Setting $(Q'R)(Q''R'') = S$ one must show that $P'' = S$. For this purpose one may consider the projectivity

$$\rho = \pi_{mPnQ'l''}, \qquad \text{where } m = P'Q, \ n = P'R, \ l'' = Q''R''.$$

Being collinear with P and Q' the point R'' is a fixed point of this projectivity from $[m]$ onto $[n]$. Since $R'' = ml''$, by Condition III, ρ is a perspectivity. Now clearly $\rho P' = l'l''$, $\rho Q = S$, and $\rho(PR'm) = Q''$. Since ρ is a perspectivity, the joins of the three points P' Q, and $PR'm$ with their respective images— that is, the lines l', QS, and PR'—must be concurrent. Since $PR'l' = R'$ it follows that $QS \circ R'$ and $S \circ QR'$. By definition, $S \circ QR'$. Hence $S = P''$. This concludes the proof of the equivalence of the three properties of projectivities with one another and with Pappus Law.

In a non-Pappus plane, Conditions I, II, III are not in general satisfied. For example, in an affine plane over the field of the quaternions, there is a projectivity ρ which maps the X-axis onto itself in such a way that $\rho(x, 0) = (-ixi, 0)$ for each quaternion x. This mapping has infinitely many fixed points—namely, all points $(x, 0)$ for which x is a quaternion of the form $a + 0i + 0j + 0k$. But ρ is not the identity, since $\rho(j, 0) = (-iji, 0) = (-j, 0)$.

5.11   CONICS

If ρ is a projectivity of $[l]$ onto $[m]$, then the set of all lines that join a point X on l and its image ρX (on m) is called a *line conic*, more specifically, the line conic $\Gamma(l, \rho, m)$ determined by l, ρ, m. Dually, the *point conic* $\Gamma(P, \rho, Q)$ is defined as the set of all points at which a line on P meets its image ρP (belonging to the pencil $[Q]$) under the projectivity ρ of $[P]$ on $[Q]$.

A conic determined by a perspectivity is called *degenerate*. If $\rho = \pi_{PnQ}$, then $\Gamma(P, \rho, Q)$ is the set of the points $x(\rho x) = x(xnQ) = xnQx = xn$ for all lines x on P with the single exception of the line $x = PQ$. Its image being $\rho x = x$, the symbol $x(\rho x)$ is undefined. All points of the range $[n]$ (except PQn) belong to $\Gamma(P, \pi_{PnQ}, Q)$. It is customary, however, to define this set as the union of the ranges $[n]$ and $[PQ]$. *All conics considered in the sequel are assumed to be proper*—that is, determined by a projectivity that is not a perspectivity.

If σ is the inverse of ρ, then clearly $\Gamma(P, \rho, Q) = \Gamma(Q, \sigma, P)$. By Theorem 5.13, and its dual, each projectivity of $[P]$ onto $[Q]$ is equal to

$$\pi_{PlCmQ} \text{ for some } l, C, m \text{ such that } P \not\in l \not\in C \not\in m \not\in Q,$$

whence $\Gamma(P, \rho, Q) = \{xlCmQx \,|\, x \circ P\}$, that is, the set of the points $xlCmQx$ for all lines x on P.

Obviously, $PQ(\rho(PQ)) = Q$. Hence Q belongs to the set $\Gamma(P, \rho, Q)$; and the line $\rho(PQ)$ is on no point of the conic but Q. A line that is on exactly one point, X, of a point conic Γ is said to be a *tangent* of Γ, more specifically, *tangential to Γ at X*. Thus $\rho(PQ)$ is tangential to $\Gamma(P, \rho, Q)$ at Q. If $Q \circ y \neq \rho(PQ)$, then y is also on a point $\neq Q$ of the conic, namely on $y(\sigma y)$, where σ is the inverse of ρ. Similarly, $PQ\sigma(PQ) = P$, whence P also belongs to $\Gamma(P, \rho, Q)$; and $\sigma(PQ)$ is easily seen to be the tangent of $\Gamma(P, \rho, Q)$ at P.

THEOREM 5.18   *In a Pappus plane, a range $[l]$ and a point conic have at most two common points.*

Consider $\Gamma(P, \rho, Q)$. If $l \circ P$ and/or $l \circ Q$, then the assertion is evident. If $P \not\in l \not\in Q$, let ρ^* be the projectivity of $[l]$ onto $[l]$ which is induced by ρ (see Exercise 3, Section 5.8). Clearly, for a point X of C to belong to $[l]$ it is necessary and sufficient that X be a fixed point ρ^*. In a Pappus plane, a projectivity of a range onto itself has at most two fixed points unless it is the identity, in which case ρ would be the perspectivity π_{PlQ}, and the conic, improper.

LEMMA 5.10   *In a Pappus plane, if R_1, R_2, R_3 are points of $\Gamma(P, \rho, Q)$ distinct from P and Q, then*

$$\rho = \pi_{PlCmQ} \qquad \text{if } l = R_i R_j, \ C = (PR_k)(QR_j), \ m = R_i R_k$$

for each of the six permutations i, j, k of $1, 2, 3$.

In view of Theorem 5.17, Condition II, if $\pi_{PlCmQ} = \tau$, then it is sufficient to prove that $\rho x = \tau x$ for three lines x on P. If ρ^* and τ^* are the projectivities of $[l]$ onto $[l]$ induced by ρ and τ, respectively, then it clearly suffices to show that $\rho^* X = \tau^* X$ for three points X on l. Such points are R_i, R_j, and $PR_k l$. Indeed, the first two points on l belong to the conic and hence, according to the proof of Theorem 5.17, are fixed points of the projectivity ρ^*. They are also fixed points of τ^* since $R_i = lm$ and R_j is collinear with P and C. Finally,

$$\tau^*(PR_k l) = PR_k lCmQl = PR_k l((PR_k)(QR_j))(R_i R_k)Ql = (PR_k)(R_i R_k)Ql$$
$$= R_k Ql = \rho^*(PR_k l).$$

This completes the proof.

LEMMA 5.11   *In order that X be a point of the conic $\Gamma(P, \rho, Q)$ containing R_1, R_2, R_3, it is necessary that the three intersections of opposite sides of the hexagons*

$$
\begin{array}{cc}
X & \\
P & Q \\
R_k & R_j, \\
 & R_i
\end{array}
\quad \text{that is, } A = (PX)(R_i R_j),\ B = (R_i R_k)(QX),\ C = (PR_k)(QR_j),
$$

be collinear for the six permutations i, j, k of $1, 2, 3$; and it is sufficient that these points be collinear for any single one of these hexagons.

Let X belong to the conic. Setting $QX = n$ one has $B = mn$. By Lemma 5.10

$$\pi_{PlCmQ}(PX) = Q \qquad \text{if } l = R_i R_j,\ m = R_i R_k,\ C = (PR_k)(QR_j)$$

for any permutation i, j, k of $1, 2, 3$; that is to say

$$(PX)(R_i R_j)CmQ = n \quad \text{or} \quad ACmQ = n, \quad \text{whence } ACmQm = nm = B.$$

It follows that $ACm = B$ and $B \circ AC$. If X does not belong to the conic, then $\pi_{PlCmQ}(PX) \neq QX$ for each of the six permutations i, j, k.

Now let the points R_t be denoted by R, S, T. A rotation of the hexagon mentioned in Lemma 5.11 about its center leaves opposite sides opposite. Hence the sufficiency of the condition in Lemma 5.11 has numerous consequences. For example, counterclockwise rotation of that hexagon by $60°$ and $180°$ transforms

$$
\begin{array}{cc}
X & \\
P \quad Q & \\
T \quad R & \\
S &
\end{array}
\quad \text{into} \quad
\begin{array}{cc}
Q & \\
X \quad R & \\
P \quad S & \\
T &
\end{array}
\quad \text{and} \quad
\begin{array}{cc}
S & \\
R \quad T, & \\
Q \quad P & \\
X &
\end{array}
$$

respectively. Hence, if X belongs to $\Gamma(P, \rho, Q)$ containing R, S, T, then Q belongs to $\Gamma(X, \rho', R)$ containing P, S, T; and S belongs to $\Gamma(R, \rho'', T)$ containing P, Q, X. (The projectivities ρ, ρ', ρ'' can be described by means of Lemma 5.10.) These examples demonstrate that the points P and Q, which occur in the definition of $\Gamma(P, \rho, Q)$, do not really play an exceptional role in that conic. The fact that $\Gamma(P, \rho, Q)$ has exactly one tangent at P can therefore be extended.

THEOREM 5.19   *A point conic has exactly one tangent at each of its points.*

The examples following Lemma 5.11 can easily be generalized to

THEOREM 5.20   *If $P_1, \ldots, P_5$ are distinct points no three of which are collinear, and P_6 belongs to the conic $\Gamma(P_1, \rho, P_2)$ containing P_3, P_4, P_5, then P_h belongs to the conic $\Gamma(P_i, \rho', P_j)$ containing P_k, P_m, P_n for each permutation h, i, j, k, m, n of $1, 2, 3, 4, 5, 6$. Here, ρ' is determined by Lemma 5.10.*

In order to show, for example, that if X belongs to the conic $\Gamma(P, \rho, Q)$ containing R, S, T, then R belongs to the conic $\Gamma(P, \rho', Q)$ containing S, T, X, one rotates the hexagon, as in the preceding example, first by $60°$, then interchanges P and S (which is permissible, since Lemma 5.10 is symmetric in R_1, R_2, R_3), and then rotates again by $60°$.

THEOREM 5.21   *If R and S are any two points of a point conic, if t is the tangent at R, and if ρ denotes the mapping of $[R]$ on $[S]$ such that $\rho t = PQ$ and if ρx is the line joining S to the other point on x and the conic for each $x \neq t$ on R, then ρ is a projectivity.*

Lemma 5.11 can now be extended to a theorem named after the French mathematician Blaise Pascal (1623–1662), who discovered a special case in 1639.

THEOREM 5.22   *In order that six points no three of which are collinear belong to one and the same conic it is necessary that, for each of the sixty different hexagons in which the points can be arranged, the three meets of opposite sides be collinear; and it is sufficient that these points be collinear for any single one of these hexagons.*

EXERCISES

1. Complete the proof of Theorem 5.20.

2. For line conics, the dual of tangents to a point conic is called *point of contact on a line*. Dualize the theorems in this section and their proofs. The dual of Pascal's Theorem was discovered by the French geometer C. J. Brianchon in 1806.

5.12   REGIONS AND HYPERBOLIC PLANES

Under proper assumptions, the undefined concepts *point, line,* and *on* are sufficient for the development not only of projective and affine plane geometry but of a class of other geometries including that of the hyperbolic plane, first studied in Bolyai's and Lobachevsky's noneuclidean geometry.

Models for those other geometries are supplied by strictly convex regions in an affine plane—briefly, *regions*—as represented by a sheet of paper or the interior of a conic. In such a model, the term *point* means an interior point of the region, and *line* means a segment joining two points of the rim. (Because of the strict convexity of the region, all other points of the segment are in the interior of the region.) Two lines are said to *meet* if there exists a point (that is, a point of the region !) which is on both lines. Any two nonmeeting lines, *l* and *m*, are said to be *parallel*, in symbols, *l* ∥ *m*. (This concept will later be sharpened.) Parallel lines are not regarded as meeting in a point outside the region (contrary to the point of view taken in Exercise 5 in Section 5.7, where a region is considered as a part of an affine plane). In the following theory, regions are considered in and by themselves.

The fundamental difference between the geometry of regions and affine geometry is that in the former *each point is on more than one line parallel to a given line*. This assumption, in contrast to Euclid's parallel postulate, makes it not only possible to define betweenness but, in conjunction with very simple postulates (which also hold in affine planes), yields all the properties of the betweenness relation in the affine plane over a field of real numbers. But even more can be said. In regions, the relations of congruency and perpendicularity can be defined, which is impossible in affine planes. These facts (discovered by Menger in 1938) led to a systematic development of hyperbolic geometry in terms of *points, lines,* and *on* by the Notre Dame school of geometry (1938–1946).

Most convenient for the study of betweenness is the following way of introducing the relation (due to F. P. Jenks):

DEFINITION 5.3   The point Q is said to be *between* P and R, written PQR, if for any two meeting lines p and r on P and R, respectively, each line q on Q meets p and/or r. Thus Q is *not* between P and R if there exist three lines p, q, r on P, Q, R, respectively, such that $p \parallel q \parallel r \not\parallel p$. (By $r \not\parallel p$ we indicate that r and p meet.)

This definition may also be formulated in an affine or a projective plane But on the basis of Postulates A–D or I–V, the relation so defined lacks the basic properties of betweenness, since each of any three collinear points is readily seen to be between the other two. In the geometry of regions, however, the following assumptions, due to F. P. Jenks and H. F. De Baggis, yield a complete theory of betweenness. They can easily be verified on a sheet of paper. Hence by a (plane) *region* we mean a set of (undefined) points and lines with an incidence relation satisfying Postulates A–E.

POSTULATE A   *For any two distinct points there is exactly one line that is on both points.*

POSTULATE B   *If l and m meet, and P is on neither line, then there exists a line on P that is parallel to l but not to m.*

POSTULATE Γ   *If P_1, P_2, P_3 are three points and l_{ij} are nine lines such that*

$$P_i \circ l_{ij} \ (j = 1, 2, 3), \quad l_{ii} \parallel l_{jk} \qquad \text{if } \{i, j, k\} = \{1, 2, 3\} \text{ and } l_{ij} \not\parallel l_{ji},$$

then P_1, P_2, P_3 are noncollinear.

In order to obtain an example of three points and nine lines in these relations, consider in the euclidean plane a circular region of radius 2. Let C be the concentric circle of radius 1, and $Q_1, Q_2, \ldots, Q_6, Q_7 = Q_1$ the vertices of a regular hexagon inscribed in C. The three points $P_i = Q_{2i}$ ($i = 1, 2, 3$), the tangents l_{ii} to C at the points Q_{2i}, and the six lines l_{ii+1} and l_{ii-1} joining Q_{2i} to Q_{2i+1} and Q_{2i-1}, respectively, constitute the desired example.

Clearly, Postulates **A, B, Γ** are also satisfied in each affine plane—Postulate Γ vacuously, since a line l_{11} that is parallel to two meeting lines l_{23} and l_{32}

(as in the premise of Postulate Γ) does not exist. Moreover, the following existence postulates are also valid in the affine plane.

POSTULATE Δ   (1) *Each line is on at least one point.* (2) *There exist three collinear points.* (3) *There exist three noncollinear points.*

The only assumption that contradicts those of affine geometry is

POSTULATE E   *For each line l, each point that is not on l is on at least two lines that are parallel to l.*

Under these assumptions, betweenness has the ordinary properties of betweenness for points of a line. (Betweenness in metric spaces and lattices has been studied in Part 1 of this book.)

Property 1.   *If PQR then P, Q, R are collinear.* If P, Q, R are noncollinear points, let q be a line on Q that is parallel to the line p joining P and R. By Postulate E, there exists a line r on R that is distinct from p and parallel to q. Clearly, p and r are meeting lines neither of which meets q. Since p, q, r are on P, Q, R, respectively, it follows that Q is not between P and R.

Property 2.   *If PQR then RQP.* This is an immediate consequence of Definition 5.3.

Property 3.   *If PQR then not PRQ.* Since in an affine plane the relation introduced by Definition 5.3 lacks Property 3, the proof of this property in a region must utilize Postulate E, the only nonaffine assumption about regions. If PQR, let q be a line on Q. By Postulate E there exist two lines r_1, r_2 on R such that $r_1 \parallel q \parallel r_2$. By B there exists a line p on P such that $r_1 \parallel p \nparallel r_2$. Since PQR and $q \parallel r_2 \nparallel p$, it follows that $q \nparallel p$. Hence $R \circ r_1 \parallel q \nparallel p \parallel r_1$, in contradiction to PRQ.

Property 4.   *If PQR and PRS, then QRS.*

Property 5.   *Of any three distinct collinear points, at least one is between the other two.* If for three collinear points P_1, P_2, P_3 neither P_1 nor P_2 is between the other two, then there exist six lines such that

$$l_{11} \circ P_1, \; l_{23} \circ P_2, \; l_{32} \circ P_3, \; l_{23} \parallel l_{11} \parallel l_{32} \nparallel l_{23};$$
$$l_{22} \circ P_2, \; l_{13} \circ P_1, \; l_{31} \circ P_3, \; l_{13} \parallel l_{22} \parallel l_{31} \nparallel l_{13}.$$

Hence if

$$l_{33} \circ P_3,\, l_{12} \circ P_1,\, l_{21} \circ P_2,\text{ and } l_{12} \nparallel l_{21},$$

then in view of the collinearity of P_1, P_2, P_3, Postulate Γ implies that $l_{12} \parallel l_{33} \parallel l_{21}$ is impossible, whence P_3 is between P_1 and P_2.

By Postulate **B**, there is a line p'' on P such that $s \nparallel p'' \parallel q$. Since $p'' \parallel q \parallel r$, from PQR it follows that $p'' \parallel r$. Hence $p'' \parallel r \parallel s$ in contradiction to PRS. This completes the proof of Property 4. Assuming PQR, PRS, and non QRS one derives a contradiction. Because of non QRS there exist lines q, r, s on Q, R, S, respectively, such that $q \parallel r \parallel s \nparallel q$.

Property 6. *If PQR and PRS, then PQS.* From Property 5, it follows that either PSQ or SPQ or PQS. Now PSQ and PRS imply RSQ in contradiction to QRS, which follows because of Property 4; and SPQ and SRP, because of Property 4, imply RPQ in contradiction to PQR.

Property 6 can be proved without using Property 5 or Postulate Γ on the basis of the following assumption: *If l and m are parallel lines and P is a point on neither l nor m, then there exists a line on P that is parallel to both l and m.* Indeed, if Q is not between P and S, then there exist lines p, q, s on P, Q, S such that $p \parallel q \parallel s \nparallel p$. By the assumption mentioned above, there exists a line r on R that is parallel to both q and s. From PRS it follows that p and r meet. Since $p \parallel q \parallel r$, this contradicts PQR.

COROLLARY   *If P, Q, R, S are collinear, then no point is between each of the three pairs of other points.*

For example, PSQ and PSR and QSR are incompatible. Indeed, each of the assumptions PQR, QRP, RPQ leads to a contradiction such as this: PSQ and PQR imply SQR, in contradiction to QSR.

Property 7. *If PQR and QRS, then PQS and PRS.*   E. H. Moore proved that a ternary relation with the Properties 1–5 in a set containing at least five elements also has Property 6. Indeed, PQR, QRS, and non PQS imply RSP and SPQ. Assuming the existence of a fifth point T such that PQT or PTQ or QPT, one readily sees that each case leads to a contradiction.

Hence PQS and, similarly, PRS. Thus all that remains to be demonstrated is the following assertion, which can be considerably sharpened (see Exercise 3).

In a region satisfying Postulates **A**–**E** *each line is on at least five points.*

First of all, if x and w are lines and X, Y, Z are points on x such that XYZ and $wx = X$, then w is on at least four points. For by Postulate Δ (3), there is a point V not on x; and by Postulate **B** there exists a line on V meeting w but not x. Hence there is a point $W \neq X$ on w. By Postulate **E**, there exist two lines y_1, y_2 on Y and parallel to ZW. Because XYZ, these lines meet $XW = w$ in two points $\neq X$ and $\neq W$.

Secondly, if the line x on at least four points, then so is any line y meeting x. For if $xy = X$, then by the Corollary to Property 5 among the points $\neq X$ on x there are two points such that X is not between them, whence say XYZ. By the first remark, y is on at least four points.

Finally, any line z not meeting x is on at least four points, for if one lets y be the join of a point on x and a point on z, then by the second remark, y is on at least four points and so is z because z meets y.

Hence *each* line is on at least four points if *one* line is. But by Δ (2) and Property 2, there exists a line l_0 on at least three points, which may be so labeled that PQR. Then by the first remark, any line joining P to a point not on l_0 is on at least four points.

In order to show that any line l is on at least five points, consider three points on l such that XYZ, and let W be a fourth point. By the corollary of Property 5, it is impossible that XYZ and XYW and ZYW. Hence either XYW or ZYW is impossible. In either case, the four points can be so labeled P, Q, R, S that PQR and QRS. Let $l' \neq l$ be a line on P. By the corollary of Property 5, among the at least three points $\neq P$ on l' there are two points, P' and P'', such that $PP'P''$. By Postulate **E**, there are two lines, m_1 and m_2, on P' that are parallel to $P''Q$. Because $PP'P''$ they both meet l in two points T_1, T_2 which obviously are $\neq P$, $\neq Q$, and $\neq R$. Thus at least one of the points T_1, T_2 is a fifth point on l.

The results can be summarized in the following theorem due to De Baggis, which, on the basis of stronger postulates, had been proved by Jenks.

THEOREM 5.23 *Under the assumptions* **A–E**, *about a plane region, betweenness as introduced in Definition 3 has all the properties of that ternary relation in the set of all real numbers.*

For further theorems about linear and planar order in a region, see Exercises 1–4.

Certain pairs of nonmeeting lines in a region can be said to be mutually parallel in a strict sense. Let *l* be a line, and *P*, *Q* points such that $PQ = m$ and *l* do not intersect. Then we say that *m* and *l* are *strictly parallel*, more specifically, that they are strictly parallel *on the Q-side of P* (and we write $P \to Q \parallel l$) if, for every line *p* on *P* that intersects *l*, each line on *Q* intersects at least one of the lines *l*, *p*. It is in this strict sense of parallelism that Bolyai and Lobachevsky, in their development of the hyperbolic noneuclidean geometry, replaced Euclid's parallel postulate by the following assumption, which is more stringent than Postulate **E**.

POSTULATE $\mathbf{E}_{BL}$ *If l is a line and P is a point not on l, then there exist exactly two lines on P that are strictly parallel to l.*

Postulates **A**–**Δ** and assumption $\mathbf{E}_{BL}$ yield, besides the theory of order, a theory of strict parallelism, developed by Jenks and De Baggis.

For any two strictly parallel lines there exists exactly one line that is strictly parallel to both lines—briefly, a *common strict parallel*—the three lines constituting what may be called an *asymptotic triangle*. Any two lines that are not strictly parallel have exactly four common strict parallels, which come in two pairs such that the lines of each pair are not strictly parallel. If *l* and *l'* are *separate*, that is, neither meeting nor strictly parallel, then the lines of exactly one of these two pairs meet. Their meet $C_{ll'}$, is called the *center* of *l* and *l'*.

An affine plane can be completed by introducing families of mutually parallel lines as ideal points on an ideal line l_∞ in a projective plane, each line $\neq l_\infty$ being on exactly one ideal point. In a similar way, a region can be completed by introducing families of mutually strictly parallel lines as ideal points on an ideal rim of the region in an affine plane, each line being on exactly two ideal points, called its *ends*. If the ends of *l* are *E* and *F*, then it is convenient to refer to *l* as *EF*, and to the lines on *P* that are strictly parallel to *l* as *PE* and *PF* (even though *E* and *F* are not points of the region).

If l and l' are separate lines, then their ends can be labeled E, F and E', F' in such a way that the meet of EF' and FE' is the center $C_{ll'}$. With this *proper* labeling, for any points $P \circ l$ and $P' \circ l'$ the pairs PE' and $P'E$ as well as PF' and $P'F$ meet.

It is possible to define *congruence* of two pairs of points. First assume that the pairs are on two separate lines l and l', and let E, F and E', F' be a proper labeling of their ends. Then the pairs P, Q on l and P', Q' on l' are congruent if and only if there exists a projectivity of l on l' that maps E, P, Q, F on E', P', Q', F', respectively. If Pappus' Law holds in the region, then it is easy to show that the following condition is necessary and sufficient for the existence of such a projectivity: the three points at which the pairs of lines

$$PE' \text{ and } P'E, \qquad QF' \text{ and } Q'F, \qquad EF' \text{ and } E'F$$

meet (the third point being $C_{ll'}$) are collinear. Hence the collinearity of these three points may serve as a definition of the congruence. If l and l' are not separate, then there exists a line l^* that is separate from both, and P, Q and $P'Q'$ are congruent if and only if there exists a pair P^*, Q^* on l^* that is congruent to both pairs.

For this relation to have the ordinary properties of congruence it is necessary that the rim of the region be a conic, in which case the region is Felix Klein's model of the hyperbolic plane of Bolyai and Lobachevsky. In terms of elements of the region, the conic character of the rim can be expressed by postulating that Pascal's Law holds asymptotically: *If* $l_1, l_2, l_3, l_4, l_5, l_6, l_7 = l_1$ *constitute an asymptotic hexagon—that is, if each pair of consecutive lines, but no other pair, is strictly parallel, and if the three pairs of opposite sides* l_1, l_4; l_2, l_5; l_3, l_6 *meet, then their three meets are collinear.*

One can develop postulational theories of *unbounded* convex regions in the affine plane, for example, of a halfplane (whose theory seems to be applicable to the visual plane) or of a region whose rim is an unbounded conic—in particular, a degenerate conic consisting of two lines. If these lines are parallel one obtains what is called a *strip* in the affine plane. Of particular significance are fragments of strips in which the lines parallel to the rim are deleted; in other words, strips in a self-dual fragment of an affine plane. The importance of such strips with verticality lies in their applicability to the kinematics of the special theory of relativity. They will be studied in the following section.

EXERCISES

1. Show that for any two points P and R there exist points Q and S such that PQR and PRS.

2. On the basis of Properties 1–7 of betweenness, introduce a binary order relation for the points on each line such that PQR if and only if $P < Q$ and $Q < R$ or $P > Q$ and $Q > R$.

3. Using the result of Exercise 1 prove the theorem of B. G. Topel: *Each line in a region is on infinitely many points.*

4. Assuming that X, Y, Z are noncollinear and that XUY, let u be a line on U parallel to XZ and consequently meeting YZ in a point U'. Prove non $U'ZY$. (*Hint:* Consider a line z on Z such that $XZ \neq z \parallel u$. Let y be a line $\neq YZ$ on Y such that $z \parallel y \nparallel XZ$ and show that $U'ZY$ contradicts XUY.

5. Using Properties 1–7 of betweenness and especially Exercise 4, prove the following theorem due to Jenks and De Baggis: *If P_1, P_2, P_3 are noncollinear points then each line meeting one of the three lines $P_i P_j$ in a point between P_i and P_j is either on P_k or meets one of the other two other lines, say $P_i P_k$, in a point that is between P_i and P_k and meets the third line either not at all or in a point that is not between P_j and P_k.* In Euclidean geometry, this statement is the most famous of the postulates introduced by Moriz Pasch in 1882 in order to fill the gaps in Euclid's Elements concerning linear and planar order. The assertion is a consequence of the simple assumptions **A–E**, valid in any region.

6. Formulate conditions concerning an asymptotic hexagon under which the three pairs of opposite sides meet.

The following three problems are unsolved on the basis of Postulates **A–E**.

7. Can Postulate be replaced by the following simpler assumption? If *for three points P_1, P_2, P_3 six lines l_h^i (h, $i = 1, 2, 3$, $h \neq i$) exist such that*

$$P_1 \circ l_1^2 \nparallel l_2^1 \circ P_2 \circ l_2^3 \nparallel l_3^2 \circ P_3 \circ l_3^1 \nparallel l_1^3 \circ P_1$$

$$l_1^2 \parallel l_3^i \parallel l_2^1, \quad l_2^3 \parallel l_1^i \parallel l_1^2, \quad l_3^1 \parallel l_2^k \parallel l_1^3$$

for some i in $\{1, 2\}$, j in $\{2, 3\}$, k in $\{3, 1\}$,

then P_1, P_2, P_3 are not collinear.

8. Can Postulate Γ be replaced by the following assumption (which is a consequence of the statement in Problem 7)? *If P, Q, R, S, T are five points such that no two consecutive lines*

$$QR, \quad ST, \quad PQ, \quad RS, \quad TP$$

intersect, then the lines PQ and RT do not intersect.

5.13   STRIPS WITH VERTICALITY
AND THEIR APPLICATION TO
EINSTEIN-MINKOWSKI KINEMATICS

By a strip with verticality—briefly, *strip*—we mean a set consisting of the following elements of a self-dual fragment of an affine plane (see Section 5.4): (1) all lines; (2) roughly speaking, one half of the points; more precisely, after choosing two nonvertical points P_1 and P_2 of the fragment we include a point P in the strip if and only if there exist four points Q_1, R_1 (vertical to P_1) and Q_2, R_2 (vertical to P_2) such that

$$Q_1 Q_2 \parallel R_1 R_2 \qquad \text{and} \qquad P = (Q_1 R_2)(R_1 Q_2).$$

In the affine plane, a strip is a set consisting of (1) all lines except a pencil of parallel ("vertical") lines; (2) all points between two vertical lines, v_1 and v_2. For each line l, the range $[l]$ is the set of all points between lv_1 and lv_2.

From a projective plane one obtains a strip as follows: (a) by deleting a point V_∞ and a line v_∞ on V_∞ as well as all elements incident with V_∞ and v_∞; (b) by choosing two distinct lines v_1 and v_2 on V_∞ and $\neq v_\infty$, and deleting all points P such that the two pairs of lines PV_∞, v_∞ and v_1, v_2 do not separate one another.

In a strip (just as in a self-dual fragment) a pair of points not joined by a line is said to be *vertical*; and (just as in a region) a pair of lines not meeting in a point is said to be *parallel*. In the projective plane, vertical points are collinear with V_∞, and parallel lines meet in a deleted point. Lines meeting in a point of v_∞ do not play any special role. (The line v_∞ only serves in stipulation (b) to describe the half of the points that belong to the strip.)

It is easy to see that a strip satisfies Postulates **B**, Γ, Δ for a region. Postulates **A** and **E** must be replaced by the following

POSTULATE A^V   *For two distinct points, there is at most one line that is on both points.*

POSTULATE E^V   *If l is a line and P a point not on l, then there exist* (1^V) *exactly one point on l that is vertical to P;* (2^V) *at least two lines on P that are parallel to l.*

The theory of betweenness, except for minor modifications, can be developed for triples of collinear or vertical points, just as in a region. For three concurrent lines, m can be said to be between l and n if for three vertical points L, M, N on l, m, n, respectively, M is between L and N. (This holds for each vertical triple if it holds for one.) Verticality is a transitive relation (as in a self-dual fragment), parallelism is not.

Strictly parallel or, as we shall also say, *almost meeting* are those pairs of lines which, in the affine or projective plane, meet on a point of v_1 or v_2. In terms of elements of the strip, strict parallelism can be defined precisely in the same way as in regions. More stringent than $\mathbf{E}^V$ is the following analogue of Postulate $\mathbf{E}_{BL}$ for a region.

POSTULATE $\mathbf{E}_{BL}^V$ obtained from Postulate $\mathbf{E}^V$ if (2^V) is replaced by (2_{BL}^V)
There exist exactly two lines on P that are strictly parallel to l.

In a strip, two strictly parallel lines have no common strict parallel, and asymptotic triangles do not exist. Two lines, l and l', that are not strictly parallel have exactly two common strict parallels that meet if and only if l and l' are separate.

In an affine plane, two pairs of strictly parallel lines of a strip may be said to be parallel on opposite sides or on the same side according as one pair meets in a point of v_1, and the other in a point of v_2 or both pairs are concurrent with the same v_i. In terms of elements of the strip itself this relation, which is insignificant in regions, can be defined as follows: Two pairs of strictly parallel lines are *parallel on opposite sides* if there exists at least one line that is strictly parallel to exactly one line of each pair. If l' and l'' are meeting lines that are strictly parallel to a third line l, then they are parallel on opposite sides.

In a region, two pairs of points P, Q and P', Q' (or, as one might say, two *segments*) on separate lines have been called congruent if the center of the lines $l = PQ = EF$ and $l' = P'Q' = E'F'$ is collinear with the points $(PE')(P'E)$ and $(QF')(Q'F)$ for proper labeling of the ends; and for pairs on nonseparate lines, an intermediate pair P^*, Q^* has been used. Exactly in the same way one can define congruence of two nonvertical pairs of points of a strip. Proper labeling may be replaced by stipulating that the pairs PE', l' and $P'E$, l be parallel on the same side and that this be true also for QF', l' and $Q'F$, l.

Congruence can also be defined for two pairs of not strictly parallel lines (or, as one might say, for two *angles*). For example, if m and n are separate lines, then the pairs l, m and l, n are said to be congruent if l is on the center C_{mn}. Each such line l meets either both m and n or neither.

In developing the geometry of a strip one needs the law of Pappus for hexagons and asymptotic hexagons. (Since the rim consists of two lines, the asymptotic hexagons are Pascal hexagons.)

We are now going to consider the dual of a strip, that is, a set consisting of all points and, roughly speaking, one half of the lines of a self-dual fragment of an affine plane. More precisely, a *dual strip* is obtained from a projective plane as follows: (a) by deleting a line l_∞ and a point L_∞ on l_∞ as well as all elements incident with l_∞ or L_∞; (b) by choosing two distinct points L_1 and L_2 on l_∞ and $\neq L_\infty$, and deleting all lines l such that the two pairs of points ll_∞, L_∞ and L_1, L_2 do not separate one another. In the affine plane obtained by deleting the line l_∞ and its points, each point P of a dual strip is on what may be called a *cross* consisting of two end lines (not belonging to the dual strip)—namely, the lines $e = PL_1$ and $f = PL_2$ in the projective plane, which are the dual of the ends E and F of a line in a strip. For any two points, P and Q, these crosses are parallel, since PL_1 and QL_1 as well as PL_2 and QL_2 meet on l_∞. The dual strip consists of all points and all lines in two quadrants of the crosses.

In a euclidean plane, one obtains a dual strip, for example, by choosing L_∞ as the vertical direction, and L_1 and L_2 as directions symmetric to the vertical, say with the slopes 1 and -1. The model of the dual strip then consists of all points and all lines with slopes between 1 and -1.

Lines that do not meet (the dual of vertical points in a strip) will be said to be *parallel*. The dual of two nonmeeting (that is, parallel) and almost meeting (that is, strictly parallel) lines are *nonjoined* and *almost joined* points. In the euclidean model, almost joined points lie on lines of slope 1 or -1.

It is easy to dualize Postulates $\mathbf{A}^V$, $\mathbf{B}$, Γ, and Δ. The dual of $\mathbf{E}^V_{BL}$ reads as follows.

POSTULATE $\mathbf{E}^l_{BL}$ *If P is a point and l a line not on P, then there exist* (1) *exactly one line of P that is parallel to l,* (2) *exactly two points on l that are not joined to P.*

All definitions and assertions concerning strips can be dualized. In particular, one can develop the theory of betweenness for concurrent and parallel triples of lines and, by means of parallel lines, define betweenness for collinear points. The dual of the two common strict parallels of two not strictly parallel lines are the two points common to the two crosses on two not almost joined points, P and Q. If and only if P and Q are separate, then those two points are joined. Their join (the dual of the center of two lines in a strip) is called the *axis* of P and Q and denoted by a_{PQ}. By dualizing the definition of two lines that are strictly parallel on the same side one introduces two points that are *almost joined in the same direction*. Dualizing the congruence of two pairs of nonvertical points one obtains the *congruence of two pairs of nonparallel lines*.

Congruence can also be defined for pairs of not almost joined points (or, as one might say, *segments*). For example, if P and Q are separate points to which the point O is not almost joined, then the pairs O, P and O, Q are congruent if and only if O is on a_{PQ}. If P and Q are not separate, then a_{PQ} does not exist. Points that are almost joined to O are ruled out since for any two such points P, Q the axis a_{PQ} is on O. Hence if Q' is any point that is separate from P and almost joined to O, then O, P would be congruent to O, Q as well as to O, Q'.

If P and Q are separate and O, P and O, Q are congruent according to this definition, then O is either joined to both P and Q or separate from both. In the former case, there is exactly one other point, Q_0, on the line OQ such that O, P and O, Q_0 are congruent, namely the meet of OQ and the line on P parallel to a_{PQ}. The point Q_0 is called the *reflection* of Q in O.

For any two not almost joined points M and N there exist exactly two points that are almost joined to both M and N. They are the dual of the common strict parallels of two lines in a strip and will be denoted by K'_{MN} and K''_{MN}. In terms of these points one can define two points, Q'_0 and Q''_0 which, in the euclidean model of the dual strip, are the *reflections of Q in the cross on O*. First assume that Q is joined to O. Then K'_{OQ} is on exactly one line parallel to OQ, and that line is on exactly one other point, R', that is almost joined to O. There are exactly two points that are almost joined to R' and K'_{QR}. One of them is O. The other is one of the desired reflections and will be denoted by Q'_0. In a similar way one defines Q''_0, which is the reflection of Q'_0 in O. Secondly, assume that the point S is separated from O. There is exactly one line on O that is parallel to a_{OS}. The two points on that line that are almost joined to S are the reflections S'_0 and S''_0 of S in the cross on O. The points O, Q'_0 are separate or joined according as O, Q are joined or separate. Even so, extending the previous definition, one calls the two pairs congruent. The points Q'_0 and Q''_0 are also referred to as the *conjugates* of Q with regard to O.

The assumptions from which the geometry of a dual strip can be developed must include the dual of Pappus' Law and its asymptotic form: *If in the hexagon $P_1, P_2, P_3, P_4, P_5, P_6, P_7 = P_1$ each pair of consecutive points, but no other pair, is almost joined (P_1, P_2 and P_3, P_4 and P_5, P_6 in the same direction, the other three pairs in the opposite direction), and the pairs P_1, P_3 and P_2, P_4 and P_3, P_5 are joined, then the joins of these three pairs are concurrent.*

Just as a self-dual fragment of an affine plane represents the Galilean kinematics of the uniform motions in a 1-dimensional space, the geometry of a dual strip represents the Einstein-Minkowski kinematics, points and lines representing world points or events and motions, respectively.

In both theories, isokinetic motions are represented by parallel lines; but the two kinematics profoundly differ with regard to the connection of world points by motions. In the Galilean kinematics, any two world points can be connected except simultaneous events, which correspond to vertical points in a self-dual fragment of an affine plane. In the relativistic world, there is no absolute simultaneity, just as there is no verticality in a dual strip. A pair of connected events corresponds to a pair of joined points in a dual strip and is referred to as a *timelike* pair. Pairs of events corresponding to separate and almost-joined pairs of points are said to be *spacelike* and *lightlike*, respectively. The events of a lightlike pair can be connected by a light ray but not by an ordinary motion. If A, B is a spacelike pair, then an observer in the world point A considers the event B as simultaneous if and only if he is engaged in a motion passing through the world point B'_A, conjugate to B with regard to A and defined in terms of events and motions, just as the point Q'_O in a dual strip is defined in terms of points and lines. Equivalently one can say that if an observer at the world point A is engaged in the motion m, then the events that are simultaneous to him are world points conjugate with regard to A to the world points on m.

In this way, the entire kinematics of the special relativity theory can be developed in terms of three undefined concepts, *world point*, *motion*, and *incidence*, without any analytic definitions, assumptions about coordinatization, or the like. The theory can be based on a few simple assumptions about those three undefined concepts—namely, translations of postulates for a dual strip. For example, it is easy to demonstrate the Lorentz-FitzGerald contractions qualitatively. But on the basis of those assumptions it is also possible to introduce coordinates and to develop the quantitative details of those contractions as well as the other results of the analytical kinematics of relativity.

REFERENCES

For the self-dual postulates for the projective plane, the self-dual fragment of the affine plane, and the development of hyperbolic geometry in terms of points, lines, and incidence, see the papers by J. C. Abbott, H. F. De Baggis, F. P. Jenks, J. Landin, K. Menger, and B. J. Topel in the Reports of a Mathematical Colloquium (II) 1-8, Notre Dame, 1939–1948. (The short proof of Property 4 in Section 5.12 is due to Mr. Robert R. Jensen.) For self-dual postulates for the 3-dimensional projective space, see K. Menger, "The projective space," *Duke Jouranl* 17 (1950), 1-14.

Projectivities and conics are more extensively treated in most traditional texts on projective geometry—for example, in Veblen and Young, *Projective Geometry*, vol. I, Waltham, Mass, Blaisdell, 1965.

In Section 5.10, Hessenberg's Theorem 5.16 is proved following H. S. M. Coxeter, *The Real Projective Plane*, Cambridge University Press, 1955, from which Figure 9 is taken. The models shown in Figures 5 and 7 are due to Sr. M. Petronia van Straten, "The topology of the configurations of Desargues and Pappius," *Reports of a Mathematical Colloquium (II)* 8 (1948) 1-17.

In connection with Chapters 4 and 5, see also K. Menger, "On algebra of geometry and recent progress in non-Euclidean geometry," *The Rice Institute Pamphlets* 27 (1940) 41-79.

PART 3

METRIC GEOMETRY

CHAPTER 6

Selected Metric Properties of Banach Spaces.

Metric Segments and Lines

6.1 THE METRIC CHARACTERIZATION PROGRAM

The reader should refer to Sections 1.7 and 1.8 for the definitions of metric space and normed linear space. It was shown that every normed linear space is a metric space when the distance xy of elements x, y of the space is defined to be the norm of the element $x - y$; that is, $xy = \|x - y\|$. Thus, while "distance" is a primitive concept in the definition of a metric space, it is a *derived* notion in a normed linear space, being defined in terms of the three primitive notions of the space. In view of its genesis, it might be suspected that in a normed linear space distance has properties that it does not possess in a general (free) metric space. The determination of those distinctive properties for (complete) normed linear spaces is the object of this chapter; their specification is an important part of the *metric characterization program* applied to this class of spaces.

Stated in another way, we seek to ascertain those properties of the distance function of a metric space **M** that are necessary and sufficient in order (1) to define in **M** a binary operation (addition), and a (real) scalar multiplication, in conformity with the seven conventions (postulates) listed in Example 3 of Section 1.8, and (2) to associate with each element x of **M** a nonnegative real number $\|x\|$, its norm, that satisfies the three requirements (norm postulates) 8, 9, 10 of that section. Those distance properties, stated wholly and explicitly in terms of the metric, are then interpreted as a set of metric postulates for complete normed linear spaces, in terms of just *one* primitive concept "distance" instead of the *three* primitive notions "addition," "scalar multiplication," and "norm" customarily used to define those spaces. From another point of view, the desired properties provide a *metric characterization* of the class of complete normed linear spaces with respect to the class of metric spaces, since they present the metric features that are possessed by those and only those metric spaces that are complete normed, linear spaces.

6.2 BANACH SPACES WITH UNIQUE METRIC LINES

If a normed linear space is not complete, then by a procedure quite similar to that employed for normed lattices in Section 2.11 it may be extended to a complete normed linear space. Such spaces are called Banach spaces. We shall denote them by the German letter $\mathfrak{B}$.

DEFINITION   A subset G of a metric space is called a *metric line* provided there is a one-to-one correspondence f between the elements of G and the elements of the set R of real numbers such that if $a, b \in G$ and $x = f(a)$, $y = f(b)$ $(x, y \in R)$, then $ab = |f(x) - f(y)|$.

Such a one-to-one, distance-preserving correspondence we have called a congruence (Section 1.9), and so G is a metric line provided it is congruent with the euclidean one-dimensional space $\mathbf{E}_1$.

Now each two distinct elements x_0, y_0 of a Banach space belong to at least one metric line G of the space. For the subset of all elements of $\mathfrak{B}$ that may be written $\lambda x_0 + (1 - \lambda)y_0$, $-\infty < \lambda < \infty$, contains the two elements x_0, y_0, and may be put in a one-to-one, distance-preserving correspondence f with the points of the E_1 by writing $f(\lambda x_0 + (1 - \lambda)y_0) = \lambda \cdot \|x_0 - y_0\|$ (Exercise 1). Hence that subset of $\mathfrak{B}$ is a metric line containing the elements x_0, y_0. Let us denote it by $G(x_0, y_0)$.

In those Banach spaces usually considered the most important, the subset

$$[\lambda x_0 + (1 - \lambda)y_0 \mid x_0, y_0 \in \mathfrak{B}, x_0 \neq y_0, -\infty < \lambda < \infty]$$

is the *only* metric line containing the elements x_0, y_0. In such spaces metric lines are said to be unique. But Banach spaces exist in which metric lines are not unique. An example of such a space is the following.

Consider the set of all ordered pairs (x, y) of real numbers x, y, and define

$$(x, y) + (u, v) = (x + u, y + v),$$
$$\lambda \cdot (x, y) = (\lambda \cdot x, \lambda \cdot y), \qquad \lambda \text{ real,}$$
$$\|(x, y)\| = |x| + |y|.$$

Then

$$\text{dist}[(x, y), (u, v)] = \|(x - u, y - v)\| = |x - u| + |y - v|.$$

It is easily shown that the resulting space is a Banach space, and the elements $(0, 0)$, $(1, 1)$ are contained in infinitely many metric lines (Exercise 2).

In what follows we shall restrict ourselves to the consideration of Banach spaces in which each two distinct points are joined by a unique metric line. Besides being the most interesting, our metric characterization of those Banach spaces leads more readily to the metric axiomatization of euclidean spaces, which is one of the principal objectives of this Part of our study. We denote such spaces by the symbol $\mathfrak{B}^*$.

EXERCISES

1. If x_0, y_0 are any two distinct elements of a Banach space, show that the subset $[\lambda x_0 + (1 - \lambda)y_0 \mid -\infty < \lambda < \infty]$ is a metric line.

2. Show that the space defined in Section 6.2 is a Banach space and that the elements $(0, 0)$, $(1,1)$ are contained in infinitely many metric lines.

3. Prove that the space referred to in Exercise 2 is a two-dimensional Minkowski space $\mathbf{M}(\mathbf{\Gamma})$ (see Section 1.8, Example 2), where $\mathbf{\Gamma}$ is the perimeter of the square bounded by the four lines $\pm x \pm y = 1$. Denote this space by $\mathbf{M}_2^1$.

4. Let addition and scalar multiplication be defined in the set of all ordered pairs (x, y) of real numbers x, y as in Section 6.2, but define $\|(x, y)\| = \max(|x|, |y|)$. Show that the resulting space is a Banach space, and that it is the two-dimensional Minkowski space $\mathbf{M}(\mathbf{\Gamma})$, where $\mathbf{\Gamma}$ is the perimeter of the square bounded by the four lines $x = \pm 1$, $y = \pm 1$. Denote this space by $\mathbf{M}_2^\infty$.

5. Prove that $\mathbf{M}_2^\infty \approx \mathbf{M}_2^1$.

6. Show that any metric $(n+1)$-tuple is congruently contained in $\mathbf{M}_n^\infty$, where $\mathbf{M}_n^\infty$ is the space of all ordered n-tuples of real numbers $x = (x_1, x_2, \ldots, x_n)$, with $\|x\| = \max(|x_1|, |x_2|, \ldots, |x_n|)$.

7. Is $\mathbf{M}_n^\infty \approx \mathbf{M}_n^1$ (where $\mathbf{M}_n^1$ is the same *linear* space as $\mathbf{M}_n^\infty$ but $\|x\| = |x_1| + |x_2| + \cdots + |x_n|$, in $\mathbf{M}_n^1$) for $n > 2$?

6.3 SIX BASIC METRIC PROPERTIES OF $\mathfrak{B}^*$ SPACES

In Section 2.10 a metric space was defined to be metrically convex provided it contains for each two of its distinct points a, c at least one point b such that $a \neq b \neq c$ and $ab + bc = ac$. The point b is called a between-point of a, c and the relation is symbolized throughout this Part by writing abc. A metric space is *metrically externally convex* if and only if it contains for each two of its distinct points a, b at least one point c such that abc subsists.

Clearly each $\mathfrak{B}^*$ space is (1) metric, (2) complete, (3) metrically convex, and (4) metrically externally convex. The metricity of $\mathfrak{B}^*$ was shown in Section 1.8 and the completeness is part of the definition of $\mathfrak{B}^*$ spaces. If $x, z \in \mathfrak{B}^*$, $x \neq z$, then $y = (\frac{1}{2})(x + z)$ is an element of $\mathfrak{B}^*$ such that $xy = (\frac{1}{2}) \cdot \|x - z\| = yz$, and so xyz subsists. Finally, if $x, y \in \mathfrak{B}^*$, $x \neq y$, then for $z = 2y - x$ we have $yz = \|x - y\| = xy$, $xz = 2\|x - y\|$, and so xyz holds.

Thus, the class $\{\mathfrak{B}^*\}$ of spaces is a subclass of the class $\{\mathbf{M}\}$ of all complete, metrically convex, and externally convex metric spaces. To establish two additional metric properties of $\mathfrak{B}^*$ spaces, we prove two lemmas concerning metric spaces (which are also useful in later developments) and begin by redeeming a promise made in Section 2.8.

LEMMA 6.1   *In every metric space the betweenness relation has the following transitive property: pqr and prs imply pqs and qrs, and conversely.*

Proof.   From the relations pqr, prs we infer that p, q, r, s are pairwise distinct, and addition of the equalities implied by those betweennesses gives $pq + qr + rs = ps \leqq pq + qs$. Hence $qr + rs \leqq qs$, and from the triangle inequality applied to q, r, s we conclude that $qr + rs = qs$, and consequently also $pq + qs = ps$. In a similar way, the converse is established.

DEFINITION   A subset of S of a metric space is a *metric segment* provided it is congruent with a straight line segment. If S is a metric segment and

$a, b \in S$ that correspond by the congruence to the end points a', b', respectively, of the corresponding line segment, then a, b are called the end points of the metric segment and we write $S = S_a^b$. Occasionally the end points a, b will be omitted from the symbol for a metric segment.

LEMMA 6.2   *If S_a^p, S_p^b are metric segments of a metric space and apb subsists, then $S_a^p \cup S_p^b$ is a metric segment with end points a, b.*

Proof.   The set products $S_a^p \cap S_p^b = \{p\}$, for otherwise a point x exists such that axp and pxb subsist. But by the preceding lemma, apb and axp imply xpb, which contradicts pxb.

Since apb subsists, points a', p', b' of $\mathbf{E}_1$ exist (with p' between a' and b') such that $ap = a'p'$, $pb = p'b'$, $ab = a'b'$; that is, the metric triple a, p, b is congruent with the euclidean (linear) triple a', p', b', with a, p, b corresponding to a', p', b', respectively. We symbolize this by writing

$$a, p, b \approx a', p', b'.$$

Congruences f_1, f_2 exist such that $f_1(S_a^p) = \text{seg}[a', p']$, $(f_1(a) = a')$, $f_2(S_p^b) = \text{seg}[p', b']$, $(f_2(b) = b')$. Let f denote the mapping of $S_a^p \cup S_p^b$ onto $\text{seg}[a', b']$ which coincides with f_1 on S_a^p and with f_2 on S_p^b. The proof is completed when it is shown that f is a congruence.

It suffices to consider $x \in S_a^p$ $(a \neq x \neq p)$, $y \in S_p^b$ $(p \neq y \neq b)$, $f(x) = x'$, $f(y) = y'$, $x' \in \text{seg}[a', p']$, $y' \in \text{seg}[p', b']$. Clearly $x'y' = x'p' + p'y' = xp + py = xy$, where the second equality results directly from the definition of f, and the last equality follows from xpy, which is a consequence of apb, axp, and pyb (Lemma 6.1). If the conditions $a \neq x \neq p$, $p \neq y \neq b$ are not satisfied, the equality $xp + py = xy$ is obtained with less argument.

LEMMA 6.3   *Each two distinct points x, y of a $\mathfrak{B}^*$ space are end points of exactly one metric segment.*

Proof. We have seen that there exists a metric line $G(x, y)$ of $\mathfrak{B}^*$. If f is a congruence of $G(x, y)$ with the $\mathbf{E}_1$ $(f(\mathbf{E}_1) = G(x, y))$ and x', $y' \in \mathbf{E}_1$ such that $f(x') = x$, $f(y') = y$, then $f(\text{seg}[x', y']) = S_x^y$, a metric segment with end points x, y. Hence x, y are end points of at least one metric segment.

Let S', S^* be two distinct metric segments of $\mathfrak{B}^*$ *with end points* x, y. From the congruence of $G(x, y)$ with the $\mathbf{E}_1$ it follows that

$$G(x, y) = \overleftarrow{G}(x, y) \cup S_x^y \cup \overrightarrow{G}(x, y),$$

where

$$\overleftarrow{G}(x, y) = [z \in G(x, y) \mid zxy \text{ or } z = x],$$
$$S_x^y = [z \in G(x, y) \mid xzy \text{ or } z = x \text{ or } z = y],$$
$$\overrightarrow{G}(x, y) = [z \in G(x, y) \mid xyz \text{ or } z = y].$$

Clearly $\overleftarrow{G}(x, y) \cap S_x^y = \{x\}$, $\overleftarrow{G}(x, y) \cap \overrightarrow{G}(x, y) = \varnothing$, $\overrightarrow{G}(x, y) \cap S_x^y = \{y\}$.

Since the metric segments S^*, S', and S are each congruent with $\text{seg}[x', y']$ of $\mathbf{E}_1$, put

$$f^*(S^*) = \text{seg}[x', y'] = f'(S'),$$

and $f(G(x, \ddot{y})) = \mathbf{E}_1$, where f, f', and f^* are congruences, with $f(x) = f'(x) = f^*(x) = x'$, $f(y) = f'(y) = f^*(y) = y'$.

A mapping g of $\overleftarrow{G}(x, y) \cup S' \cup \overrightarrow{G}(x, y)$ onto $\mathbf{E}_1$ is defined by $g(z) = f(z)$, $z \in \overleftarrow{G}(x, y) \cup \overrightarrow{G}(x, y)$, and $g(z) = f'(z)$, $z \in S'$. Note that $\overleftarrow{G}(x, y) \cap S' = \{x\}$ and $\overrightarrow{G}(x, y) \cap S = \{y\}$. To show that this mapping is a congruence, let p, q be elements of $\overleftarrow{G}(x, y) \cup S' \cup \overrightarrow{G}(x, y)$, and suppose $p \in \overleftarrow{G}(x, y)$, $p \neq x$, $q \in S'$, $x \neq q \neq y$. From pxy and xqy we conclude pxq, and so

$$pq = px + xq = f(p)f(x) + f'(x)f'(q) = g(p)g(x) + g(x)g(q) = g(p)g(q).$$

Similarly, in case $p \in \overrightarrow{G}(x, y)$, $q \in S$; and if p, q belong to the same summand, $pq = g(p)g(q)$ by congruence f' or f.

Hence, the set $\overleftarrow{G}(x, y) \cup S' \cup \overrightarrow{G}(x, y)$ is a metric line containing x, y, and in like manner it is shown that $\overleftarrow{G}(x, y) \cup S^* \cup \overrightarrow{G}(x, y)$ is also a metric line that contains x, y. Since $S' \neq S^*$, these two metric lines are distinct, contradicting the definition of $\mathfrak{B}^*$.

DEFINITION   Three points of a metric space are *linear* provided they are congruent with three points of $\mathbf{E}_1$.

COROLLARY 1   *Every linear triple of points of a $\mathfrak{B}^*$ space is contained in a metric segment which is unique, unless the three points coincide.*

Proof. If x, y, z are a linear triple of a $\mathfrak{B}^*$ space and $x \neq y \neq z \neq x$, the labeling may be assumed so that xyz holds. Then unique metric segments S_x^y, S_y^z exist (Lemma 6.3) and the set sum $S_x^y \cup S_y^z$ is a metric segment (Lemma 6.2) containing x, y, z, and it is the only such segment, by the preceding lemma.

If x, y, z are not pairwise distinct, the proof of the corollary is immediate.

Remark. If $x, y \in \mathfrak{B}^*$, the unique metric segment S_x^y is part of the metric line $G(x, y)$. Hence each linear triple is contained in a unique metric line (unless the points coincide).

DEFINITION   If p, r are points of a metric space $\mathbf{M}$ and $q \in \mathbf{M}$ such that $pq = qr = (\frac{1}{2})pr$, then q is called a *middle point* of p, r.

COROLLARY 2   *A $\mathfrak{B}^*$ space contains for each two of its distinct points exactly one middle point.*

Proof. If $x, z \in \mathfrak{B}^*$, $x \neq z$, and y, y^* are middle points of x, z, then the (unique) metric segment S_x^z contains both y and y^* (Corollary 1, since xyz and xy^*z subsist), and from the congruence of S_x^z with a straight line segment it follows that $y = y^*$.

We conclude this section by establishing two more metric properties of a $\mathfrak{B}^*$ space.

THEOREM 6.1   *Each $\mathfrak{B}^*$ space has*

(i) *the two-triple property (that is, if four pairwise distinct points of $\mathfrak{B}^*$ contain two linear triples, then all four of the triples are linear) and,*

(ii) *the Young property (that is, if x, y, z are a nonlinear triple of $\mathfrak{B}^*$ and $m(x, y)$, $m(x, z)$ are middle points of x, y and x, z, respectively, then $\mathrm{dist}[m(x, y), m(x, z)] = (\frac{1}{2})yz$.**

Proof. Let x, y, z, u be pairwise distinct points of $\mathfrak{B}^*$, with the linear triples x, y, z and x, y, u. Then, by a preceding Remark, each of these triples is contained in a unique metric line; that is $y, z \in G(x, y)$ and hence the triples u, x, z and u, y, z are linear.

---

* In 1910 W. H. Young showed that the property stated in (ii) could be used as a substitute for Euclid's parallel postulate in the foundations of euclidean geometry (*Quarterly Journal of Pure and Applied Mathematics*, vol. XLI (1910), pp. 353–363). Later he used the inequality $\mathrm{dist}[m(x, y), m(x, z)] < (\frac{1}{2})yz$ as a *characteristic* property of hyperbolic geometry (*American Journal of Mathematics*, vol. 33 (1911), pp. 249–286). For these reasons it seems appropriate to refer to (ii) as the Young property.

To prove (ii), we first observe that since $(\frac{1}{2})(x + y)$ and $(\frac{1}{2})(x + z)$ are middle points of x, y and x, z, respectively, then, by Corollary 2 of the preceding lemma, $m(x, y) = (\frac{1}{2})(x + y)$ and $m(x, z) = (\frac{1}{2})(x + z)$. Then

$$\text{dist}[m(x, y), m(x, z)] = (\tfrac{1}{2}) \|x + y - (x + z)\| = (\tfrac{1}{2}) \|y - z\| = (\tfrac{1}{2})yz.$$

6.4 METRIC SEGMENTS

In the preceding section attention has been focused on six metric properties possessed by every $\mathfrak{B}^*$ space: (1) metricity, (2) completeness, (3) metric convexity, (4) metric external convexity, (5) the two-triple property, and (6) the Young property. These six properties of $\mathfrak{B}^*$ spaces were selected because *they are characteristic metric properties of those spaces*; that is, not only does every $\mathfrak{B}^*$ space possess them (as we have seen in Section 6.3), but, as we shall show, any space that enjoys those six properties is a $\mathfrak{B}^*$ space! To accomplish this, we begin in this section by establishing the very important property of any complete and metrically convex metric space stated in the following theorem.

THEOREM 6.2   *Each two distinct points of a complete, metrically convex metric space* **M** *are end points of a metric segment.*

Proof.   A function f is called isometric in **M** provided it defines a *congruent mapping* of its domain $D(f)$ onto its range $R(f)$, where $D(f) \subset \mathbf{M}$, $R(f) \subset \mathbf{E}_1$. Thus f is a real-valued function, and $D(f) \underset{f}{\approx} R(f)$, where the symbol $\underset{f}{\approx}$ denotes that the congruence of $D(f)$ with $R(f)$ is established by f. Any function g that is isometric in **M**, with $D(g) \supset D(f)$ and $g(x) = f(x)$, $x \in D(f)$, is called an *isometric extension* of f, and we write $f \subset g$.

For $p, q \in \mathbf{M}$, $p \neq q$, put $D_1 = \{p, q\}$. The function f_1, defined by $f_1(p) = 0$, $f_1(q) = d = pq$ is clearly isometric in **M**. Denote by Σ_2 the set of all isometric extensions f of f_1 such that $R(f) \subset [0, d]$, let I_2 denote the set of four non-overlapping equal subsegments of $[0, d]$ with length $d/4$, and denote by $N_2(f)$ the number of members of I_2 with points in common with $R(f)$. Now select any member f_2 of Σ_2 for which $N_2(f_2)$ is maximal (that is, $N_2(f_2) \geqq N_2(f), f \in \Sigma_2$).

Supposing that n real functions $f_1 \subset f_2 \subset \cdots \subset f_n$ have been selected, an $(n + 1)$-st function f_{n+1} is chosen from the set Σ_{n+1} of all isometric extensions f of f_n with $R(f) \subset [0, d]$ by the property $N_{n+1}(f_{n+1}) \geqq N_{n+1}(f)$, where $N_{n+1}(f)$ is the number of members of I_{n+1} that have points in common with $R(f), f \in \Sigma_{n+1}$, and I_{n+1} is the set of nonoverlapping equal subsegments of $[0, d]$ with length $d/2^{n+1}$. Thus an infinite sequence

$$f_1 \subset f_2 \subset \cdots \subset f_n \subset \cdots$$

of real functions, isometric in $\mathbf{M}$, is obtained (with ranges in $[0, d]$), each of which is an isometric extension of each function with smaller index.

If $v \in \bigcup_{i=1}^{\infty} D(f_i)$, put $f(v) = f_k(v)$, where k is the smallest index n for which $v \in D(f_n)$. Then $f(v) = f_i(v)$ for every $i \geq k$, $R(f) = \bigcup_{i=1}^{\infty} R(f_i)$, and $\bigcup_{i=1}^{\infty} D(f_i) = D(f) \underset{f}{\approx} R(f) \subset [0, d]$. Hence f is an isometric extension of f_i $(i = 1, 2, \ldots)$. We show next that the congruence $D(f) \underset{f}{\approx} R(f)$ can be extended in a unique way to the closures $\mathrm{cl}[D(f)]$, $\mathrm{cl}[R(f)]$.

If $v \in D'(f)$, $v \notin D(f)$, there exists an infinite sequence $\{v_n\}$ of pairwise different points of $D(f)$ with $\lim v_n = v$ (Theorem 1.3). Since $\{v_n\}$ has a limit and $\mathbf{M}$ is metric, it is a Cauchy sequence (Remark, Section 2.11) and so the sequence $\{f(v_n)\}$ of real numbers is also a Cauchy sequence $(v_i v_j = f(v_i)f(v_j)$ $(i, j = 1, 2, \ldots))$ and consequently has a limit v'. It is clear that $v' \in R'(f) \subset [0, d]$. If $v'' = \lim \mathrm{f}(\bar{v}_n)$, where $\lim \bar{v}_n = v$, $\bar{v}_n \in D(f)$, $n = 1, 2, \ldots$, then the staggered sequence $v_1, \bar{v}_1, v_2, \bar{v}_2, \ldots, v_n, \bar{v}_n, \ldots$ also has limit v. Denoting by v^* the limit of the sequence $f(v_1), f(\bar{v}_1), f(v_2), f(\bar{v}_2), \ldots, f(v_n), f(\bar{v}_n), \ldots$, then $v^* = v'' = v'$, since v' and v'' are limits of subsequences of a sequence with limit v^* (Remark, Section 1.9). Hence the point $v' = \lim f(v_n)$ is uniquely determined by $v \in D'(f)$, *independently of the choice of the sequence* $\{v_n\}$ of points of $D(f)$ with limit v. Putting $g(v) = f(v)$, $v \in D(f)$, and $g(v) = v'$, $v \in D'(f)$, we see that $D(g) = D(f) \cup D'(f)$, $R(g) \subset [0, d]$, and the continuity of the metric (Theorem 1.2) yields

$$D(g) \underset{g}{\approx} R(g).$$

Hence g is an isometric extension of f, and consequently $f_n \subset g$, $n = 1, 2, \ldots$.

It is important to observe that $R(g) = \mathrm{cl}[R(f)] = R(f) \cup R'(f)$; for it is clear that $R(g) \subset \mathrm{cl}[R(f)]$, and if $v' \in R'(f)$ and $\{f(v_n)\}$ is an infinite sequence of pairwise distinct points of $R(f)$ with limit v', then the Cauchy character of $\{f(v_n)\}$ implies that $\{v_n\}$ is a Cauchy sequence, and *since M is complete* it contains a point $v = \lim v_n$. Hence $v \in D'(f)$ and $v' = g(v)$; that is, $v' \in R(g)$, and so $R(g) = \mathrm{cl}[R(f)]$.

The proof is completed when it is shown that $R(g) = [0, d]$, for then the function g establishes a congruence between the subset $D(g)$ of M and the line segment $[0, d]$. Consequently $D(g)$ is a metric segment, and since

$p, q \in D(g)$ and $g(p) = 0$, $g(q) = d$, the metric segment $D(g)$ has end points p, q.

If we now assume that $R(g) \neq [0, d]$, then, since $R(g) = \mathrm{cl}[R(f)]$, a closed set, and 0, $d \in R(g)$, there exists an open interval (a, c) of $[0, d]$, $a < c$, that contains no points of $R(g)$ and whose end points a, $c \in R(g)$. Denoting by r, $t \in D(g)$ such that $a = g(r)$, $c = g(t)$, then $rt = c - a$, where we make the labels of points of $[0, d]$ serve also as coordinates.

The reader might have observed that so far the assumed metric convexity of $\mathbf{M}$ has not entered into the rather lengthy argument. All of the functions $f_1, f_2, \ldots, f_n, \ldots$ as well as f and g might be identical with f_1, $D(g) = \{p, q\}$, $R(g) = \{0, d\}$, $a = 0$, and $c = d$! It is time to bring into play the convexity hypothesis.

Since r, $t \in \mathbf{M}$, $r \neq t$, the metric convexity of $\mathbf{M}$ assures the existence of a point s of $\mathbf{M}$ such that rst subsists. Let b denote the point of (a, c) such that $rs = b - a$, $st = c - b$. Since $b \notin R(g)$, then $s \notin D(g)$. Let $D^* = D(g) \cup \{s\}$, and define $g^*(x) = g(x)$, $x \in D(g)$, and $g^*(s) = b$. To show that g^* is isometric in $\mathbf{M}$ (and hence is an isometric extension of g) it evidently suffices to prove that $sx = g^*(s)g^*(x) = bg(x) = |b - g(x)|$ for every $x \in D(g)$, $r \neq x \neq t$.

Since $a = g(r)$, $c = g(t)$, $g(x)$ are pairwise distinct points of $[0, d]$, one of them is between the two others, and since the interval (a, c) contains no points of $R(g)$, either $g(x)g(r)g(t)$ or $g(r)g(t)g(x)$ subsists. Because g is isometric in $\mathbf{M}$, it follows that either xrt or rtx holds. Now xrt and rst imply xrs, and so $sx = xr + rs = a - g(x) + b - a = b - g(x)$; and rtx and rst imply stx, and then $sx = st + tx = c - b + g(x) - c = g(x) - b$. Hence in any event $sx = |b - g(x)|$, and g^* is an isometric extension of g and of every function f_n, $n = 1, 2, \ldots$.

Let n be any positive integer such that $(d/2^{n+1}) < \min(b - a, c - b)$. Then at least one of the segments of I_{n+1} *is contained in the interval* (a, c) *and contains* b, and so $R(g^*)$ has a point in common with at least one interval of I_{n+1} whose product with $R(f_{n+1})$ is null. On the other hand, since $R(f_{n+1}) \subset R(g^*)$, then $R(g^*)$ has points in common with each segment of I_{n+1} that has a nonempty product with $R(f_{n+1})$. Hence $N_{n+1}(g^*) > N_{n+1}(f_{n+1})$, which, since g^* is an isometric extension of f_n and consequently is a member of Σ_{n+1}, contradicts the definition of f_{n+1}, and the proof of the theorem is complete.

COROLLARY   *Each linear triple of distinct points of a complete, metrically convex metric space is contained in a metric segment of the space.*

The corollary follows from the preceding theorem and Lemma 6.2.

EXERCISES

1. Give an example of a complete metrically convex, semimetric space with continuous distance function for which it is not the case that each two distinct points are joined by a metric segment.

2. Show (without using Theorem 6.2) that a compact, metrically convex, metric space contains for each two of its distinct points p, q, a point m such that $pm = mq = \frac{1}{2}pq$ (m is called a *middle point* of p, q). Show that m is not necessarily unique.

3. Use the fact proved in Exercise 2 to construct a proof that each two distinct points of a compact, metrically convex, metric space are joined by a metric segment.

6.5 METRIC LINES

We have defined a subset of a space to be a metric line provided it is congruent with the euclidean straight line. The object of this section is to establish the existence of lines for the class of spaces described above, but first we prove the following lemma.

LEMMA 6.4 *If a, b are distinct points of a complete, metrically externally convex metric space* $\mathbf{M}$, *the set of numbers* $[ap \mid p \in \vec{B}(a, b)]$ *is not bounded, where* $\vec{B}(a, b) = [p \in M \mid abp \text{ or } p = b]$.

Proof. If the contrary be assumed, and $r_0 = $ l.u.b. ap, $p \in \vec{B}(a, b)$, let p_1 denote a point of $\vec{B}(a, b)$, $p_1 \neq b$, such that $ap_1 > r_0 - 1$. From the transitive property of metric betweenness, $\vec{B}(a, b) \supset \vec{B}(a, p_1)$, and it follows that the distances ap, $p \in \vec{B}(a, p_1)$ are bounded also. Putting $r_1 = $ l.u.b. ap, $p \in \vec{B}(a, p_1)$, it is clear that $r_0 \geq r_1 > ab$, where the last inequality is insured by the external convexity of $\mathbf{M}$.

By induction, an infinite sequence $\{p_n\}$ of pairwise distinct points of $\vec{B}(a, b)$, and a bounded, monotone decreasing sequence $\{r_n\}$ of real numbers are obtained (putting $b = p_0$) such that

(i) $r_{n-1} = $ l.u.b. ap, $p \in \vec{B}(a, p_{n-1})$,

(ii) $p_n \in \vec{B}(a, p_{n-1})$, $r_n > ap_n > r_{n-1} - 1/n$,

(iii) $r_0 \geqq r_1 \geqq r_2 \geqq \cdots \geqq r_n \geqq \cdots > ab,$

(iv) $\vec{B}(a, p_0) \supset \vec{B}(a, p_1) \supset \vec{B}(a, p_2) \supset \cdots \supset \vec{B}(a, p_n) \supset \cdots,$

with (i), (ii) valid for $n = 1, 2, \ldots$.

Putting $\lim r_n = r$, we obtain from (ii) $\lim ap_n = r$. Use of (ii), the transitivity of metric betweenness, and induction yield $ap_n\, p_{n+k}$ $(n, k = 1, 2, \ldots)$; that is, $ap_n + p_n\, p_{n+k} = ap_{n+k}$, and consequently $\lim_{n,\, k \to \infty} p_n\, p_{n+k} = 0$. Hence the sequence $\{p_n\}$ is a Cauchy sequence of points of the complete metric space **M**, and so $\lim p_n = p$, $p \in$ **M**. Continuity of the metric gives $r = \lim ap_n = ap$.

Turning again to the relations $ap_n + p_n\, p_{n+k} = ap_{n+k}$, for n fixed and $k = 1, 2, \ldots$, we have $ap_n + p_n p = ap$. Since the elements of the sequence $\{p_n\}$ are pairwise distinct, each $p_n \neq a$, and at most for one index i is $p_i = p$, then we may suppose $ap_n\, p$ holds for $n = 1, 2, \ldots$. Now the external convexity of **M** insures the existence of a point p^* such that app^* subsists. This relation, together with $ap_n\, p$, gives $ap_n\, p^*$, and so $p^* \in \vec{B}(a, p_n)$. It follows that $ap^* \leqq r_n$, $ap + pp^* = ap^* \leqq r_n$ and so $r + pp^* \leqq r_n$, $n = 1, 2, \ldots$. But this clearly contradicts $r = \lim r_n$, and establishes the lemma.

THEOREM 6.3 *Each two distinct points of a complete, metrically convex and externally convex metric space **M** are on a metric line of the space.*

Proof. If $a, b \in$ **M**, $a \neq b$, let $S_0 = S_a^b$ denote a metric segment of **M** with end points a, b (Theorem 6.2). By the preceding lemma, **M** contains a point p_1 such that abp_1 holds and $ap_1 > 2ab$. If $S_b^{p_1}$ is a metric segment with end points b, p_1 then $S_a^b \cup S_b^{p_1}$ is a metric segment with end points a, p_1 (Lemma 6.2). Denoting this segment by S_1, we say that S_0 *has been prolonged through b to a segment of more than twice its length.* In a similar way, S_1 may be prolonged through a to a segment S_2 of more than twice its length.

By induction, an infinite sequence $\{S_n\}$, $n = 0, 1, 2, \ldots$ of metric segments is obtained, with S_n a prolongation of S_{n-1} to more than twice its length. The set sum $G = \bigcup_{n=0}^{\infty} S_n$ is mapped congruently onto the euclidean line $\mathbf{E}_1$ by associating with $p \in G$ that point p' of $\mathbf{E}_1$ such that $ap = a'p'$, $bp = b'p'$, where a', b' are two fixed points of $\mathbf{E}_1$ with $ab = a'b'$. Then, by definition, G is a metric line.

COROLLARY *Each metric segment of a complete, metrically convex and external convex metric space can be prolonged to a metric line.*

Remark.   It is interesting to observe that Theorem 6.3 is not valid if "metric space" is replaced by "semimetric space with continuous distance function." The space whose points are all the ordered pairs (r_1, r_2) of real numbers, with distance

$$xy = (x_1 - y_1)^2 + (x_2 - y_2)^2,$$

$x = (x_1, x_2)$, $y = (y_1, y_2)$, is easily seen to be complete, metrically convex and externally convex, semimetric with continuous distance function, but there are no metric lines in this space. There are no metric segments either, so the corollary of Theorem 6.3 is (vacuously) satisfied. Note also that Lemma 6.4 is valid in this example, but not, of course, Theorem 6.2.

6.6 UNIQUENESS OF METRIC LINES

We recall that a metric space has the *two-triple property* provided the linearity of two of the triples of any quadruple of pairwise distinct points of the space implies the linearity of all triples of the quadruple. Note that this does *not* imply the linearity of the quadruple, for the metric quadruple p_1, p_2, p_3, p_4 with $p_1 p_2 = p_2 p_3 = p_3 p_4 = p_4 p_1 = 1$, $p_1 p_3 = p_2 p_4 = 2$ is *not* congruent with four points of the E_1, but all four of its triples are linear. The important connection between the two-triple property and uniqueness of metric lines is exhibited in the following theorem.

THEOREM 6.4   *Let* **M** *denote any complete, metrically convex and externally convex metric space. Each two distinct points of* **M** *are contained in a unique metric line if and only if* **M** *has the two-triple property.*

Proof.   Let a, b, c, d denote pairwise distinct elements of M, with a, b, c and a, b, d linear triples. By the corollary of Theorem 6.2 there exists a metric segment S^* that contains a, b, c, and a metric segment S^{**} that contains a, b, d, and S^*, S^{**} can be prolonged to metric lines G^*, G^{**}, respectively (Corollary, Theorem 6.3), each of which contains the distinct points a, b. If there is exactly one metric line of **M** that contains a, b, then $G^* = G^{**} = G$ and $a, b, c, d \in G$. Then the two triples a, c, d and b, d, c are linear. Hence uniqueness of lines implies the two-triple property.

Now let **M** have the two-triple property and suppose that there are distinct elements a, b of **M** contained in *distinct* metric lines $G(a, b)$, $G^*(a, b)$. We may assume the labeling so that $c \in G(a, b)$ and $c \notin G^*(a, b)$, and distinguish two cases.

Case 1.   *The point c is contained in the metric segment S_a^b of $G(a, b)$.*

Let c^* denote the unique point on $S_{a,b}^*$, the metric segment of $G^*(a, b)$ with end points a, b, such that $ac^* = ac$ and $bc^* = bc$. Then $c^* \neq c$ and a, b, c, c^* are four pairwise distinct points with two linear triples a, b, c and a, b, c^*. It follows that a, c, c^* and b, c, c^* are also linear triples, and that cac^*, cbc^* subsist.

If $d^* \in S_{a,b}^*$ with ad^*c^* holding, that relation together with ac^*b implies ad^*b and d^*c^*b, and so the quadruple of pairwise distinct points b, c, c^*, d^* contains the linear triples b, c, c^* and b, c^*d^*. Consequently the remaining two triples c, c^*, d^* and b, c, d^* are linear also.

Now neither cc^*d^* nor c^*cd^* can subsist, for the results of combining each with cac^* contradict ad^*c^*. Hence c^*d^*c holds. Now bd^*c and bc^*d^* imply bc^*c, contradicting cbc^*; bcd^* and cad^* (which follows from cac^* and ad^*c^*) yield bad^*, contrary to ad^*b; and finally, cbd^* and bc^*d^* give cc^*d^*, contradicting c^*d^*c.

Hence the assumption of Case 1 is untenable.

Case 2.   *The point c does not belong to S_a^b.*

With proper labeling of points a, b, the relation abc holds. Let c^* denote that point of $G^*(a, b)$ such that abc^* subsists, $ac^* = ac$ (and consequently $bc^* = bc$).

The four pairwise distinct points a, b, c, c^* have triples a, b, c and a, b, c^* linear, and hence a, c, c^* and b, c, c^* are linear also. It follows readily that cac^* and cbc^* must subsist. From $ab + bc = ac$ and $ab + bc^* = ac^*$ we get $2ab + bc + bc^* = ac + ac^*$, from which $ab = 0$. Hence Case 2 is impossible also, and the proof is complete.

Remark.   It is interesting to observe that in a complete, metrically convex metric space, the two-triple property implies uniqueness of metric segments (and hence uniqueness of middle points) but *not conversely.* For if three pairwise distinct rays of the E_2 (euclidean plane) with common initial point p be metrized convexly (that is, if q, r are points of any one of the rays, then $\text{dist}(q, r) = qr$, the euclidean distance of points q, r, while if q, r belong to different rays, $\text{dist}(q, r) = qp + pr$) the complete, metrically convex (and externally convex) metric space so obtained has unique segments, but it does not have the two-triple property (since it does not have unique lines). On the other hand, if a, b are joined by two metric segments, the reader may easily show that the two-triple property does not hold.

EXERCISES

1. Let $S = [x \in \mathfrak{B}| \ \|x\| = 1]$. If S does not contain a metric segment, is $\mathfrak{B}$ a $\mathfrak{B}^*$ space? Is the converse valid?

2. Give an example (other than the one given in the text) of a metric space with unique segments, but not unique lines. Can this happen in a Banach space?

REFERENCE

Section 6.4   N. Aronszajn, *Neuer Beweis der Streckenverbundenheit vollständiger konvexe Räume*, Ergebnisse eines math. Kolloquiums (Wien), Heft 6 (1935), pp. 45–46.

CHAPTER 7

Metric Postulates for Banach Spaces with Unique Metric Lines

7.1 EXTENSION OF THE YOUNG PROPERTY

From now on we shall understand by a *space* $\mathbf{M}_Y$ any complete, metrically convex and externally convex metric space with the two-triple and Young properties. The object of this section is to prove that if p, q, r are nonlinear points of a space $\mathbf{M}_Y$ and $q' \in S_p^q$, $r' \in S_p^r$ such that

$$\frac{pq'}{pq} = \lambda = \frac{pr'}{pr} \qquad (0 \leq \lambda \leq 1),$$

then $q'r' = \lambda \cdot qr$. Observe that since p, q, r are nonlinear, they are pairwise distinct, the metric segments S_p^q, S_p^r exist by Theorem 6.2, and have only point p in common (see Remark following Theorem 6.4). This extension of the Young property is trivial for $\lambda = 0$ (then $q' = p = r'$) and for $\lambda = 1$ (then $q' = q$ and $r' = r$). In the following we suppose $0 < \lambda < 1$, and hence $q' \neq r'$.

We set $T(p, q, r) = S_p^q \cup S_q^r \cup S_p^r$, and refer to $T(p, q, r)$ as *the triangle determined by* p, q, r. The *sides* of $T(p, q, r)$ are the *metric lines* $G(p, q)$, $G(q, r)$, $G(p, r)$ which, by Theorem 6.4, exist and are unique. They are the unique prolongations of the metric segments S_p^q, S_q^r, S_p^r, respectively (Corollary, Theorem 6.3).

If $x, y \in \mathbf{M}_Y$, $(x \neq y)$, we denote by $m(x, y)$ the unique middle point of x, y.

LEMMA 7.1 *If* $t \in G(q, r)$, *a side of* $T(p, q, r)$, *the middle point* $m(p, t)$ *of* p *and* t *is on the line* $G(m(p, q), m(p, r))$ *and* $m(p, t) \in seg[m(p, q), m(p,r)]$ *if and only if* $t \in S_q^r$.

Proof. The lemma is trivial if $t = q$ or $t = r$. Assuming $q \neq t \neq r$ and applying the Young property to $T(p, t, q)$ gives $dist[m(p, q), m(p, t)] = (\frac{1}{2})qt$, and applying that property to $T(p, t, r)$ yields $dist[m(p, t), m(p, r)] = (\frac{1}{2})rt$. Since $dist[m(p, q), m(p, r)] = (\frac{1}{2})qr$, the linearity of the point-triple $m(p, t)$, $m(p, q)$, $m(p, r)$ follows from the linearity of the point-triple q, r, t. Moreover, $m(p, t)$ is between $m(p, q)$ and $m(p, r)$ if and only if qtr holds.

Use of the corollaries to Theorems 6.2 and 6.3 and of Theorem 6.4 justifies the conclusion that $m(p, t) \in G(m(p, q), m(p, r))$.

COROLLARY *For each* $t \in G(q, r)$, *the lines* $G(p, t)$, $G(m(p, q), m(p, r))$ *intersect.*

LEMMA 7.2 *To each point* $t' \in G(m(p, q), m(p, r))$ *there corresponds a point* $t \in G(q, r)$ *such that* $t' = m(p, t)$, *and* $t \in S_q^r$ *if and only if* $t' \in seg[m(p, q), m(p, r)]$.

Proof. If $t' = m(p, q)$, then $t = q$, and if $t' = m(p, r)$, then $t = r$. Suppose $m(p, q) \neq t' \neq m(p, r)$ and consider the point t such that $pt't$ holds and $pt = 2pt'$. Then $t' = m(p, t)$ and the Young property applied to $T(p, t, q)$ and $T(p, t, r)$ yields $qt = 2 \cdot t'm(p, q)$, $rt = 2 \cdot t'm(p, r)$. These relations, together with $qr = 2 \; dist[m(p, q), m(p, r)]$ and the linearity of $m(p, q)$, $m(p, r)$, t', show that q, r, t are linear, and consequently $t \in G(q, r)$. Moreover, qtr holds if and only if t' is between $m(p, q)$ and $m(p, r)$.

COROLLARY *For each* $t' \in G(m(p, q), m(p, r))$, *the lines* $G(p, t')$, $G(q, r)$ *intersect.*

LEMMA 7.3 *If G denotes any metric line of* $\mathbf{M}_Y$ *and p any point of* $\mathbf{M}_Y$, *then there exists at least one foot of p on G.*

Proof. If $p \in G$, then p is itself a foot of p on G.

If $p \notin G$, select any point q of G and two points a, b of G such that aqb holds and $qa = ab = 2pq$. Since the metric segment S_a^b of G is closed and compact, there is a foot f_p of p on S_a^b (Theorem 1.16); that is, $pf_p \leqq px$, $x \in S_a^b$. Now if $y \in G$, $y \notin S_a^b$, then either yaq or qby holds. In the first case

$$py + pq \geqq qy = aq + ay = 2pq + ay,$$

and so $py \geqq pq + ay > pq \geqq pf_p$. The same result is obtained if qby subsists, and hence $pf_p \leqq py$ for every point y of G; that is, f_p is a foot of p on G.

Remark. Only the metricity of $\mathbf{M}_Y$ and the congruence of G with the euclidean straight line are needed to establish the preceding lemma.

LEMMA 7.4 *Let* f_p *denote a foot of p on* $G(q, r)$, *a side of* $T(p, q, r)$. *Then lines* $G(p, f_p)$, $G(m(p, q), m(p, r))$ *intersect, and their intersection* f' *is a foot of p on* $G(m(p, q), m(p, r))$.

Proof. If f' is not a foot of p on $G(m(p, q), m(p, r))$, this line contains a point g' such that $pg' < pf'$. Then by Lemma 7.2, line $G(q, r)$ contains a point g such that $g' = m(p, g)$. Hence $pg = 2pg' < 2pf' = pf_p$, and f_p is not a foot of p on $G(q, r)$, contrary to the hypothesis.

The existence of f' is ensured by the corollary of Lemma 7.1.

LEMMA 7.5 *Let* f_p' *denote a foot of p on* $G(m(p, q), m(p, r))$. *Then lines* $G(p, f_p')$, $G(q, r)$ *intersect, and their intersection f is a foot of p on* $G(q, r)$.

The proof, which makes use of Lemma 7.1, is similar to that of the preceding lemma.

LEMMA 7.6 *Let* q', r' *denote points of* $T(p, q, r)$ *such that* $q' \in S_p^q$, $r' \in S_p^r$. *If* $pq'/pq = \frac{3}{4} = pr'/pr$, *then* $q'r' = (\frac{3}{4})qr$.

Proof. Let p^*, q^*, r^* denote the middle points of q and r, p and r, and p and q, respectively, and $s^* = m(p^*, r^*)$ and $t^* = m(p^*, q^*)$ (Fig. 10). Clearly

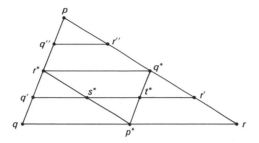

Figure 10

$q' = m(q, r*), r' = m(r, q*)$, and the Young property applied to the appropriate triangles gives

$$q's* = (\tfrac{1}{2})qp* = (\tfrac{1}{4})qr; \quad s*t* = (\tfrac{1}{2})q*r* = (\tfrac{1}{4})qr; \quad t*r' = (\tfrac{1}{2})rp* = (\tfrac{1}{4})qr.$$

According to Lemma 7.1, $G(r, r*)$ intersects $G(p*, \dot{q}*)$ in the point $x = m(r, r*)$. Applying the Young property to $T(p, r, r*)$ and $T(q, r, r*)$ gives $xq* = (\tfrac{1}{2})pr*$ and $xp* = (\tfrac{1}{2})qr*$. Since $pr* = qr*$, then $xp* = xq*$, and so $x = t*$. Hence $t* = m(r, r*)$, and consideration of $T(p*, r*, q)$, $T(p*, r*, r)$, and $T(q, r, r*)$ leads to the conclusion that $s* = m(q', t*)$.

It is shown similarly that $s* = m(q, q*)$, and $t* = m(s*, r')$. Now $s* = m(q', t*)$ implies $q's*t*$, and $t* = m(s*, r')$ implies $s*t*r'$. Hence $q', s*, t*, r' \in G(s*, t*)$, *in that order*, and so

$$q'r' = q's* + s*t* + t*r' = (\tfrac{3}{4})qr.$$

We observe that if $q'' \in S_p^q$, $r'' \in S_p^r$, with $pq''/pq = \tfrac{1}{4} = pr''/pr$, then the Young property applied to $T(p, q*, r*)$ gives $q''r'' = (\tfrac{1}{2})q*r* = (\tfrac{1}{4})qr$.

Hence for every number of the form $k/2^n$ $(n = 1, 2; k = 0, 1, \ldots, 2^n)$, $a \in S_p^q$, $b \in S_p^r$, $pa = (k/2^n)pq$, $pb = (k/2^n)pr$ implies $ab = (k/2^n)qr$, and so the Young property has a valid extension for the "ratios of division" $\tfrac{1}{4}$ and $\tfrac{3}{4}$.

Numbers of the form $k/2^n$, where n is a natural number and $k = 0, 1, 2, \ldots$ 2^n, are called *dyadically rational*. We wish to show now that the Young property can be extended to *any* dyadically rational ratio of division.

It is instructive to illustrate the inductive procedure we shall use by consideration of the ratio of division $\tfrac{5}{8} = (\tfrac{1}{2})(\tfrac{2}{4} + \tfrac{3}{4})$.

Let $a*$, a, a' be points of S_p^q, and $b*$, b, b' points of S_p^r such that (Fig. 11)

$$\frac{pa*}{pq} = \frac{1}{4} = \frac{pb*}{pr}; \quad \frac{pa}{pq} = \frac{2}{4} = \frac{pb}{pr}; \quad \frac{pa'}{pq} = \frac{3}{4} = \frac{pb'}{pq}.$$

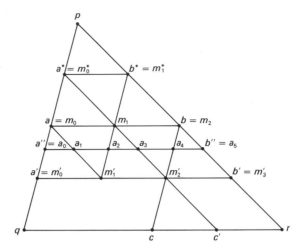

Figure 11

Let m_1 divide S_a^b into two equal subsegments, and let m_1', m_2' divide $S_{a'}^{b'}$ into three equal subsegments, $S_{a'}^{m_1'}$, $S_{m_1'}^{m_2'}$, $S_{m_2'}^{b'}$. If $c = m(q, r)$ it is easily seen that $m_2' = m(b, c)$ and the points a^*, m_1, m_2', c' of Figure 9 (with $c' = m(c, r)$) are constructible in the same way as the points q', s^*, t^*, r' of Figure 8. Hence, by the proof of Lemma 7.6, the points a^*, m_1, m_2', c' are collinear, in that order, and $m_1 = m(a^*, m_2')$.

Now if $a'' \in S_p^q$, $b'' \in S_p^r$ with $pa''/pq = \frac{5}{8} = pb''/pr$ and $a'' = a_0 = m(a, m_0')$, $(m_0' = a')$, $a_1 = m(a, m_1')$, $a_2 = m(m_1, m_1')$, $a_3 = m(m_1, m_2')$, $a_4 = m(b, m_2')$, $b'' = a_5 = m(b, b')$, the Young property applied to the appropriate triangle gives $a_i a_{i+1} = (\frac{1}{8})qr$, $i = 0, 1, 2, 3, 4$.

Moreover, examination of $T(a^*, a', m_2')$ shows that (since $a_1 = m(a, m_1')$, $a_2 = m(m_1, m_1')$ and $a = m(a^*, a')$, $m_1 = m(a^*, m_2')$) the points a_0, a_1, a_2, a_3 play the same role as the points q', s^*, t^*, r' of Figure 8, and consequently they are collinear, with $a_1 = m(a_0, a_2)$ and $a_2 = m(a_1, a_3)$.

Similarly it is shown that b^*, m_1', m_1 are collinear, with $m_1 = m(b^*, m_1')$, and an examination of $T(b^*, m_1', b')$ shows that a_2, a_3, a_4, a_5 are collinear, since $m_2' = m(m_1', b')$ and $b = m(b^*, b')$.

Consequently, $a_0, a_1, a_2, a_3, a_4, a_5$ are collinear (in that order) and so $a''b'' = a_0 a_5 = (\frac{5}{8})qr$.

Before proceeding to set up the inductive argument for any dyadically rational number, we observe that the Young extension for $1/2^{n+1}$ follows from its validity for $1/2^n$ by one direct application of the Young property to the triangle $T(p, d, d')$, where $d \in S_p^q$, $d' \in S_p^r$, and $pd/pq = 1/2^n = pd'/pr$.

Now let n be an arbitrary but fixed natural number, and denote by a_k, b_k points of S_p^q, S_p^r, respectively, such that

$$\frac{pa_k}{pq} = \frac{k}{2^n} = \frac{pb_k}{pr}, \qquad k = 0, 1, \ldots, 2^n.$$

Denote by $a_k = c_0^k, c_1^k, \ldots, c_{k-1}^k, c_k^k = b_k$, $k + 1$ points that divide $\text{seg}[a_k, b_k]$ into k equal (abutting, nonoverlapping) subsegments $(k = 1, 2, \ldots, 2^n)$ (Fig. 12).

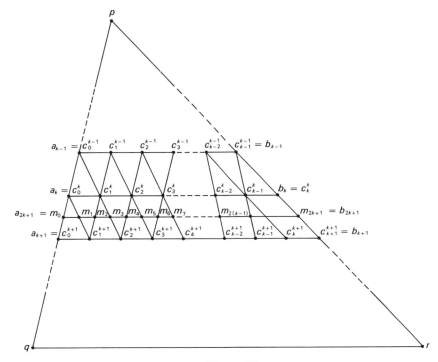

Figure 12

We make the inductive hypothesis that (i) the Young extension is valid for each of the numbers $k/2^n$, $k = 0, 1, \ldots, 2^n$ (which we have verified for $n = 2$), and (ii) $c_i^k = m(c_{i-1}^k, c_{i+1}^k)$ $(i = 1, 2, \ldots, k)$, $c_i^k = m(c_i^{k-1}, c_i^{k+1})$ $(i = 0, 1, \ldots, k)$ for every $k = 1, 2, \ldots, 2^n - 1$ (which is also valid for $n = 2$).

For $k = 2, 3, \ldots, 2^n$, the inductive hypothesis gives

$$\text{dist}[c_i^{k-1}, c_{i+1}^{k-1}] = \text{dist}[c_i^k, c_{i+1}^k] = \text{dist}[c_i^{k+1}, c_{i+1}^{k+1}] = (1/2^n)qr,$$

for every index $i = 0, 1, \ldots, k$ for which the corresponding point exists, and hence $c_i^{k-1} = m(c_{i-1}^{k-1}, c_{i+1}^{k-1})$ $(i = 1, 2, \ldots, k - 2)$; $c_i^k = m(c_{i-1}^k, c_{i+1}^k)$ $(i = 1, 2, \ldots, k - 1)$; and $c_i^{k+1} = m(c_{i-1}^{k+1}, c_{i+1}^{k+1})$ $(i = 1, 2, \ldots, k)$.

If $a_{2k+1} = m(a_k, a_{k+1})$, $b_{2k+1} = m(b_k, b_{k+1})$, then

$$pa_{2k+1}/pq = (2k + 1)/2^{n+1} = pb_{2k+1}/pr,$$

and we wish to show that

$$\text{dist}[a_{2k+1}, b_{2k+1}] = \left[\frac{2k + 1}{2^{n+1}}\right] qr.$$

Letting $m_{2i} = m(c_i^k, c_i^{k+1})$ $(i = 0, 1, \ldots, k)$, and $m_{2i-1} = m(c_{i-1}^k, c_i^{k+1})$ $(i = 1, \ldots, k + 1)$, we observe that since $c_0^k = m(c_0^{k-1}, c_0^{k+1})$, $c_1^k = m(c_0^{k-1}, c_2^{k+1})$ and $c_1^{k+1} = m(c_0^{k+1}, c_2^{k+1})$, the points m_0, m_1, m_2, m_3 play the same role with respect to $T(c_0^{k-1}, c_0^{k+1}, c_2^{k+1})$ that q', s^*, t^*, r' played in Lemma 7.6 with respect to triangle $T(p, q, r)$. It follows that

$$m_0 m_1 = m_1 m_2 = m_2 m_3 = (1/2)(1/2^n)qr,$$

and m_0, m_1, m_2, m_3 are collinear.

The same procedure shows m_2, m_3, m_4, m_5 are collinear, and $m_2 m_3 = m_3 m_4 = m_4 m_5 = (1/2^{n+1})qr$; and repetition establishes that $m_0, m_1, \ldots, m_{2k+1}$ are collinear, with $m_i m_{i+1} = (1/2^{n+1})qr, i = 0, 1, \ldots, 2k$. We conclude that

$$\text{dist}[a_{2k+1}, b_{2k+1}] = \left[\frac{2k + 1}{2^{n+1}}\right] qr.$$

Finally, it is observed that part (ii) of the inductive hypothesis is easily established for the pertinent point-triples introduced in the $(n + 1)$-st stage of the process.

THEOREM 7.1   *Let λ be any real number between zero and 1. If $a \in S_p^q$, $b \in S_p^r$ with $pa/pq = \lambda = pb/pr$, then $ab = \lambda \cdot qr$.*

Proof.   If $\{\lambda_n\}$ is a sequence of dyadically rational numbers with $\lim_{n \to \infty} \lambda_n = \lambda$, and $a_n \in S_p^q$, $b_n \in S_p^r$ such that $pa_n/pq = \lambda_n = pb_n/pr$ $(n = 1, 2, \ldots)$, then $\lim_{n \to \infty} a_n = a$, $\lim_{n \to \infty} b_n = b$, and (Theorem 1.2)

$$ab = \lim_{n \to \infty} a_n b_n = \lim_{n \to \infty} \lambda_n \cdot qr = \lambda \cdot qr,$$

where the second equality is a result of the Young extension to each dyadically rational ratio of division.

LEMMA 7.7   *Let* p, q, r *denote noncollinear points of* $\mathbf{M}_Y$. *If* $q', r' \in \mathbf{M}_Y$ *with* qpq', rpr', *and* $pq'/pq = \lambda = pr'/pr$, $\lambda \geqq 0$, *then* $q'r' = \lambda \cdot qr$.

Proof.   If $\lambda = 0$, then $q' = p = r'$, and the conclusion follows. For $\lambda > 0$ it follows readily from the hypotheses that $qp/qq' = 1/(1 + \lambda) = rp/rr'$. Let s denote a point of $\mathbf{M}_Y$ such that qsr holds and $qs/qr' = 1/(1 + \lambda)$. Applying Theorem 7.1 to $T(q, q', r')$ yields $ps = q'r'/(1 + \lambda)$.

Now $r's/r'q = \lambda/(1 + \lambda) = r'p/r'r$, and the same theorem applied to $T(r', q, r)$ gives $ps = \lambda \cdot qr/(1 + \lambda)$; hence $q'r' = \lambda \cdot qr$.

With the aid of Lemma 7.7 and the foregoing theorem, the following result is easily established (let the reader supply the proof).

THEOREM 7.2   *Let* p, q, r *denote noncollinear points of* $\mathbf{M}_Y$, *and* q', r' *points of* $G(p, q)$, $G(p, r)$, *respectively* $(q \neq q' \neq p, \; r \neq r' \neq p)$ *such that either both of the relations* qpq', rpr' *subsist, or neither does. If* $pq'/pq = pr'/pr$, *then* $t \in G(q, r)$ *implies the existence of* $t' \in G(p, t) \cap G(q', r')$ *such that* $pt'/pt = pq'/pq$. *Moreover, if* f_p *is a foot of* p *on* $G(q, r)$, *then the corresponding point* $f'_p = G(p, f_p) \cap G(q', r')$ *is a foot of* p *on* $G(q', r')$.

THEOREM 7.3   *Let* p, q, r *denote noncollinear points of* $\mathbf{M}_Y$. *If* $pq'q$ *and* $pr'r$ *hold, and* $pq'/pq = pr'/pr$, *then* $G(q, r) \cap G(q', r') = \varnothing$.

Proof.   If $t \in G(q, r) \cap G(q', r')$, then by the preceding theorem there is a point t' common to $G(q', r')$ and $G(p, t)$ with $pt'/pt = pq'/pq \neq 1$ (for since $pq'q$ subsists, $q' \neq q$). Since $G(p, t)$ meets $G(q', r')$ in at most one point t', then $t = t'$ and $pt'/pt = 1$, contrary to the above.

7.2   PASCH AND MENELAUS TRIANGLE THEOREMS

Of fundamental importance in determining the nature of a geometry is the investigation of the behavior of lines with respect to their intersections with the sides of a triangle, first studied systematically by Pasch in 1882. This section is devoted to establishing such intersection properties (in particular, the property stated in Theorem 7.7, which is commonly referred to as the Pasch Theorem). The importance of the theorem of Menelaus (proved in the first century A.D. by the Greek astronomer who, together with Heron and Pappus, witnessed the demise of the great school of Greek geometry) for a theory of transversals was emphasized in a paper by L. N. M. Carnot written in 1806. We establish it in Theorem 7.5

THEOREM 7.4 *If $p, q, r, q', r' \in M_Y$ such that p, q, r are noncollinear, $pq'q$ and $pr'r$ subsist, and $(pq'/pq) \neq (pr'/pr)$, then a point s of $G(q, r)$ exists with $q'r's$ or $sq'r'$ holding.*

Proof. If $pq'/pq = \lambda$ and $pr'/pr = \mu$, then $0 < \lambda < 1$, $0 < \mu < 1$, and the labeling may be chosen so that $\lambda > \mu$. Since $pr'/\lambda < pr'/\mu = pr$, a point t of S_p^r exists such that $pt = pr'/\lambda$. Then $pr'/pt = pq'/pq = \lambda$, and consequently $q'r'/qt = \lambda$.

From

$$rt = rp - pt = rp - \frac{pr'}{\lambda} = rp - \frac{\mu \cdot rp}{\lambda} = \left[\frac{\lambda - \mu}{\lambda}\right] \cdot rp,$$

follows

$$(*) \qquad \qquad \frac{rt}{rp} = \frac{\lambda - \mu}{\lambda}.$$

Now since $0 < (\lambda - \mu)/\lambda < 1$, a point u of $G(q, r)$ exists such that rqu holds and $rq/ru = (\lambda - \mu)/\lambda$, and so $qt = [(\lambda - \mu)/\lambda] \cdot pu$. This equality, together with $q'r' = \lambda \cdot qt$ gives

$$(**) \qquad \qquad q'r' = (\lambda - \mu) \cdot pu.$$

Since $qq'/pq = (pq - pq')/pq = 1 - \lambda$, there exists a point s of S_q^u such that $qs = (1 - \lambda) \cdot qu$. Consequently

$$\frac{qs}{qu} = 1 - \lambda = \frac{qq'}{pq},$$

and Theorem 7.1 applied to $T(q, u, p)$ yields

$$(***) \qquad \qquad q's = (1 - \lambda) \cdot pu.$$

The proof is completed by observing that

$$rs = ru - (qu - qs) = ru - qu + (1 - \lambda) \cdot qu = ru - \lambda \cdot qu,$$

and $qu = ru - qr = 1 - [(\lambda - \mu)/\lambda] \cdot ru = (\mu/\lambda) \cdot ru$; so $rs = (1 - \mu) \cdot ru$. Since $rr' = pr - pr' = (1 - \mu) \cdot pr$, then $rs/ru = rr'/rp$, and Theorem 7.1

applied to $T(r, u, p)$ gives

(†)
$$sr' = (1 - \mu) \cdot pu.$$

Then (**), (***), and (†) yield

$$r'q' + q's = (\lambda - \mu) \cdot pu + (1 - \lambda) \cdot pu = (1 - \mu) \cdot pu = sr',$$

and consequently $s \in G(q', r') \cap G(q, r)$.

Note that $r'q's$ follows from assuming that $(pq'/pq) > (pr'/pr)$.

COROLLARY *If $p, q, r, q', r' \in \mathbf{M}_Y$, such that p, q, r are noncollinear, pqq', prr' hold and $(pq'/pq) \neq (pr'/pr)$, then $G(q', r') \cap G(\ ,r) \neq \varnothing$.*

Apply the preceding theorem to $T(p, q', r')$.

THEOREM 7.5 (Menelaus) *If $p, q, r, q', r', s \in \mathbf{M}_Y$ such that p, q, r are noncollinear, $pq'q$ and $pr'r$ subsist, $s \in G(q, r)$, $q \neq s \neq r$, and q', r', s are collinear, then*

$$\left(\frac{pr'}{rr'}\right) \cdot \left(\frac{qq'}{pq'}\right) \cdot \left(\frac{rs}{qs}\right) = 1.$$

Proof. Since $G(q', r')$ intersects $G(q, r)$, then $(pq'/pq) \neq (pr'/pr)$ (Theorem 7.3), and the labeling may be assumed so that $\lambda = (pq'/pq) > (pr'/pr) = \mu$ $(0 < \lambda < 1, \ 0 < \mu < 1)$. Let t be a point of S_p^r with $pt = pr'/\lambda$, and let u satisfy the relations rqu and $ru = [\lambda/(\lambda - \mu)] \cdot rq$. An easy computation shows that $rs = (1 - \mu) \cdot ru$.

Since $pr' = \mu \cdot pr$ and $rr' = pr - pr' = (1 - \mu) \cdot pr$, then $pr'/rr' = \mu/(1 - \mu)$.

It is observed that the points t, u selected above are the same as those with the same labels that were chosen in the proof of the preceding theorem. In the proof of that theorem it was shown that the point s common to $G(q, r)$ and $G(q', r')$ is the point of S_q^u such that $qs = (1 - \lambda) \cdot qu$. Since rqu holds,

$$qs = (1 - \lambda) \cdot (ru - qr) = (1 - \lambda)\left[1 - \frac{\lambda - \mu}{\lambda}\right] ru$$

$$= \left[\frac{\mu(1 - \lambda)}{\lambda}\right] ru,$$

and so

$$\frac{rs}{qs} = \frac{\lambda(1 - \mu)}{\mu(1 - \lambda)}.$$

From $qq' = pq - pq' = (1 - \lambda) \cdot pq$, $pq' = \lambda \cdot pq$, $qq'/pq' = (1 - \lambda)/\lambda$, and consequently

$$\left(\frac{pr'}{rr'}\right) \cdot \left(\frac{qq'}{pq'}\right)\left(\frac{rs}{qs}\right) = \left[\frac{\mu}{1 - \mu}\right]\left[\frac{1 - \lambda}{\lambda}\right]\left[\frac{\lambda(1 - \mu)}{\mu(1 - \lambda)}\right] = 1.$$

THEOREM 7.6 (Pasch) *If* $p, q, r, q', s \in \mathbf{M}_Y$, *with* p, q, r *noncollinear, pq'q subsisting, while sqr or qrs holds, then* $G(q', s)$ *contains a point* r' *such that pr'r subsists.*

Proof.   Suppose, first, that *sqr* holds, put $sq/sr = v$, $0 < v < 1$, $pq'/pq = \lambda$, $0 < \lambda < 1$, and define μ by the equation

(∗) $$\frac{\mu(1 - \lambda)}{\lambda(1 - \mu)} = v.$$

It is easily seen that $0 < \mu < 1$ and so there exists a point r' of S_p^r such that $pr'/pr = \mu$, $p \neq r' \neq r$. Since, moreover, it follows from (∗) that $\lambda > \mu$, the preceding theorem yields the existence of a point s' with $r'q's'$ and rqs' holding, and

$$\frac{s'q}{s'r} = \frac{\mu(1 - \lambda)}{\lambda(1 - \mu)} = v = \frac{sq}{sr}.$$

Hence $s' = s$, and so r' is the desired point.

To complete the proof, suppose given $r' \in S_p^r$, $p \neq r' \neq r$, and s with *sqr* holding. Writing $sq/sr = v$, $pr'/pr = \mu$, and setting $v = \mu(1 - \lambda)/\lambda(1 - \mu)$, it is seen that $0 < \lambda < 1$. Thus there is a point q' of S_p^q such that $pq'/pq = \lambda$, $p \neq q' \neq q$, and q' is the desired point.

COROLLARY   *If* $p, q, r, r', s \in \mathbf{M}_Y$, *with* p, q, r *noncollinear and sqr, prr' holding, then* $G(p, q)$ *intersects* $G(r', s)$.
   Apply the preceding theorem to $T(s, r, r')$.

THEOREM 7.7   *If* $p, q, r, q', r' \in \mathbf{M}_Y$, *with* p, q, r *noncollinear, qpq', rpr' subsisting, and* $(pq'/pq) \neq (pr'/pr)$, *then* $G(q, r)$ *intersects* $G(q', r')$.

The proof, which is similar to that of Theorem 6.4 and uses Lemma 7.7, is left to the reader.

7.3 DEFINITION AND PROPERTIES OF PLANES OF $\mathbf{M}_Y$

The intersection theorems proved in the preceding section are needed for an acceptable definition of planes of $\mathbf{M}_Y$.

DEFINITION   If $p \in \mathbf{M}_Y$ and G is a line of $\mathbf{M}_Y$, $p \notin G$, *the plane* $\pi(p; G)$ is the topological closure of the set of all points of $\mathbf{M}_Y$ collinear with p and a point of G.

Let us observe that in hyperbolic space the point set defined above would not be intuitively acceptable as a "plane," since the set would not have the desired property of containing with each two of its distinct points the line containing them. Our first task is to show that for a space $\mathbf{M}_Y$, the set so defined *does* have this important property—that is, that $\pi(p; G)$ is a *linear space*. To do this, we need the following lemma.

LEMMA 7.8   *If $p \in \mathbf{M}_Y$ and $\{q_i\}$ is an infinite sequence of points of $\mathbf{M}_Y$ with limit q ($q_i \neq p \neq q$,) then $t \in G(p, q)$ implies the existence of an infinite sequence $\{t_i\}$, $t_i \in G(p, q_i)$ ($i = 1, 2, \ldots$), such that $\lim_{i \to \infty} t_i = t$.*

Proof.   Let t be any point of $G(p, q)$. Select a point t_i on $G(p, q_i)$ such that p, q_i, t_i satisfy the same betweenness relation as p, q, t (for example, $pq_i t_i$ holds if pqt does, and so on) and such that $pt_i/pq_i = pt/pq$. Then clearly $\lim_{i \to \infty} pq_i = pq$, $\lim_{i \to \infty} pt_i = pt$, and linearity of p, q_i, t_i yields $\lim_{i \to \infty} q_i t_i = qt$.

Since the sequence $\{q_i\}$ converges, it is a Cauchy sequence; from Theorem 7.1, $t_i t_j = (pt/pq) \cdot q_i q_j$ ($i, j = 1, 2, \ldots$), and consequently the sequence $\{t_i\}$ is a Cauchy sequence also. Let $t^* = \lim_{i \to \infty} t_i$ ($\mathbf{M}_Y$ is complete). Continuity of the metric yields

$$\lim_{i \to \infty} pt_i = pt^*, \qquad \lim_{i \to \infty} q_i t_i = qt^*,$$

and hence p, q, t^* are linear. But since $pt = pt^*$ and $qt = qt^*$, then $t^* = t$, and the lemma is proved.

THEOREM 7.8   *If $p \in \mathbf{M}_Y$ and G is a line of $\mathbf{M}_Y$, $p \notin G$, the plane $\pi(p; G)$ is linear.*

Proof. If $x, y \in \pi(p; G)$, $x \neq y$, and $G(x, y)$ denotes the unique line containing them, it is desired to show that $G(x, y) \subset \pi(p; G)$.

First, suppose that points x', y' of G exist such that both triples p, x, x' and p, y, y' are linear, with $x' \neq x$ and $y' \neq y$. (If $x' = x$ and $y' = y$, then $G(x, y) = G$ and then clearly $G(x, y) \subset \pi(p; G)$.) Five possibilities present themselves for consideration.

Case I. *The relations pxx' and pyy' subsist.* If $px/px' \neq py/py'$, then $G(x, y)$ intersects $G(x'\ y')$ in a point s (Theorem 7.4), with $x' \neq s \neq y,'$ and the labeling may be chosen so that yxs holds.

Let q be any point of $G(x, y)$, $q \neq x, y, s$. If either yqs or ysq is valid, a point q' of G exists which is linear with p and q (apply Theorem 7.6 to $T(y, s, y')$ if yqs holds and its corollary to $T(x, s, x')$ in case ysq does, noting that yxs and ysq imply xsq). Consequently, in those two cases q is a point of $\pi(p; G)$. If syq holds, and $yp/yy' \neq yq/ys$, then $G(p, q)$ intersects $G(s, y') = G$, by Theorem 7.7, and again $q \in \pi(p; G)$. On the other hand, if syq subsists and $yp/yy' = yq/ys$, let $\{q_i\}$ be any infinite sequence of points of $S_y^q - \{q\}$ with limit q; that is, yq_iq holds ($i = 1, 2, \ldots$), and $\lim_{i \to \infty} q_i = q$. Then $yq_i < yq$ and so $(yq_i/ys) \neq (yp/yy')$ ($i = 1, 2, \ldots$). Hence to each q_i there corresponds a point q_i' of G such that p, q_i, q_i' are linear, and so q is an accumulation point of the set $\{q_1, q_2, \ldots, q_n, \ldots\}$, each point of which is linear with p and a point of G. It follows from the definition of $\pi(p; G)$ that q is one of its points.

To complete the examination of Case I, we observe that if $px/px' = py/py'$ then $q \in G(x, y)$ implies the existence of $q' \in G(x', y')$ such that p, q, q' are linear (Theorem 7.2) and so $q \in \pi(p; G)$.

Case II.   *The relations xpx', ypy' hold.*
Case III.   *The relations px'x, py'y hold.*
Case IV.   *The relations pxx', py'y hold.*

These cases present no novelties. They are treated in all respects in the manner employed in Case I.

Case V.   *The relation xpx', together with one of the relations pyy' or py'y, subsists.*

In this case, $G(x, y)$ intersects G, by the Pasch theorem or its Corollary. Let $\{s\} = G(x, y) \cap G$, and suppose $q \in G(x, y)$. If neither pyy' and qys nor $py'y$ and qsy holds, application of the Pasch theorem or its Corollary to

$T(y, y', s)$ yields the existence of a point q' of G that is linear with p and q, and so $q \in \pi(p; G)$. If, on the other hand, one of the above pair of betweenness relations does hold and $yp/yy' \neq yq/ys$, then $G(p, q)$ intersects G by one of the preceding theorems, and $q \in \pi(p; G)$; and in case $yp/yy' = yq/ys$, the procedure applied at a similar juncture in the proof of Case I yields the desired result here.

It remains to consider the possibility that the points x and/or y selected in $\pi(p; G)$ are not linear with p and points of G. In this situation sequences $\{x_i\}, \{y_i\}$ of points of $\pi(p; G)$ exist with $\lim_{i \to \infty} x_i = x$, $\lim_{i \to \infty} y_i = y$ $(x_i \neq y_i)$, and such that x_i is linear with p and a point x_i' of G and y' is linear with p and a point y_i' of G $(i = 1, 2, \ldots)$. Then by the previous argument, the lines $G(x_i, y_j) \subset \pi(p; G)$, whenever $x_i \neq y_j$ $(i, j = 1, 2, \ldots)$.

Now by Lemma 7.8 each point t of $G(x_i, y)$ (the index i is fixed) is the limit of a sequence $\{t_j\}$, where $t_j \in G(x_i, y_j)$ $(j = 1, 2, \ldots)$, and consequently $G(x_i, y) \subset \pi(p; G)$. The same lemma ensures that each point of $G(x, y)$ is the limit of a sequence $\{s_i\}$ of points of $G(x_i, y)$, and since $\pi(p; G)$ is closed, it follows that $G(x, y)$ is contained in the plane $\pi(p; G)$, and the proof is complete.

THEOREM 7.9   *If p, q, r are noncollinear points of $\mathbf{M_Y}$, then $\pi(p; G(q, r)) = \pi(q; G(p, r)) = \pi(r; G(p, q))$.*

Proof.   It suffices to establish that

$$\pi(p; G(q, r)) \subset \pi(q; G(p, r)),$$

for the inclusion relation may be reversed by applying the same argument to a relabeling of the points, and another relabeling will establish the second equality stated in the theorem.

If $x \in \pi(p; G(q, r))$, $x \neq p$, and $x \notin G(q, r)$ (trivial, otherwise), suppose, first, that a point x' of $G(q, r)$ exists such that p, x, x' are linear. If $x' = q$ or $x' = r$, then x is linear with q and a point of $G(p, r)$ and so $x \in \pi(q; G(p, r))$. Assume $x' \neq q, r$, and consider the following three cases.

Case I. *The relation xpx' holds.* If either $x'qr$ or $qx'r$ holds, the Pasch theorem applied to $T(x', q, x)$ shows that $G(q, x)$ intersects $G(p, r)$; that is, x is linear with q and a point of $G(p, r)$, and so $x \in \pi(q; G(p, r))$.

If qrx' holds and $x'r/x'q \neq x'p/x'x$, then, again, $G(p, r)$ intersects $G(q, x)$ and the desired conclusion is obtained; and if $x'r/x'q = x'p/x'x$, there exists a sequence $\{x_i\}$ with limit x and $px_i x$ holding $(i = 1, 2, \dots)$. Then $x'r/x'q \neq x'p/x'x_i$ and consequently for each $i = 1, 2, \dots$, there is a point x_i'' of $G(p, r)$ which is linear with q and x_i. Then $x_i \in \pi(q; G(p, r))$, $i = 1, 2, \dots$, and consequently x belongs to $\pi(q, G(p, r))$.

Case II.   *The relation pxx' holds.* If $x'qr$ holds and $x'x/x'p \neq x'q/x'r$, then $G(q, x)$ intersects $G(p, r)$, and so x is linear with q and a point of $G(p, r)$; and if $(x'x/x'p) = (x'q/x'r)$ then the "sequence method" used before shows that $x = \lim_{i \to \infty} x_i$, with $x_i \in \pi(q; G(p, r))$, $i = 1, 2, \dots$, and so $x \in \pi(q; G(p, r))$. If either $qx'r$ or qrx' holds, the desired result follows by applying the Pasch theorem to $T(x', p, r)$.

Case III.   *The relation px'x holds.* The treatment of this case is in all respects the same as that of the two preceding cases. (Apply the Pasch theorem to r, x', p, q, x in case $x'qr$ holds.)

Finally, if $x \in \pi(p; G(q, r))$ and x is not linear with p and a point of $G(q, r)$, then by the now familiar argument, it is the limit of a sequence $\{x_i\}$ with $x_i \in \pi(q; G(p, r))$, $i = 1, 2, \dots$, and so $x \in \pi(q; G(p, r))$. The proof of the theorem is complete.

Notation

If p, q, r are noncollinear points of $\mathbf{M}_Y$, denote the coincident planes $\pi(p; G(q, r)) = \pi(q; G(p, r)) = \pi(r; G(p, q))$ by $\pi(p, q, r)$.

In several of the preceding proofs, different arguments are needed according as a point x of $\pi(p; G)$ is or is not linear with p and a point of G. It is convenient to write

$$\pi(p; G) = \pi_I(p; G) \cup \pi_B(p; G),$$

where $\pi_I(p; G)$ denotes those points of $\pi(p; G)$ that are linear with p and a point of G, and $\pi_B(p; G)$ denotes the set of points of $\pi(p; G)$ that are not linear with p and a point of G. Of course, this decomposition of the points of a plane into two (nonnull), mutually exclusive subsets is relative to point p and line G. Note that $p \in \pi_I(p; G)$.

7.4   FURTHER PROPERTIES OF PLANES

The object of the theorems of this section is to show that the plane $\pi(p; G)$ is determined by *any three of its noncollinear points*, and hence that there is a *unique plane* containing three noncollinear points.

THEOREM 7.10   *If* $p^* \in \pi_I(p; G)$ *and* $p^* \notin G$, *then the planes* $\pi(p^*; G)$ *and* $\pi(p; G)$ *are identical.*

Proof.   Clearly, we may assume that $p^* \neq p$. Since $p^* \in \pi_I(p; G)$, then $p \in \pi_I(p^*; G)$ and consequently it suffices to prove that $\pi(p; G) \subset \pi(p^*; G)$ because the reverse relation $\pi(p^*; G) \subset \pi(p; G)$ follows by change of labeling.

Now let x be a point of $\pi(p; G)$ and suppose that $p' \in G$ which is linear with p and p^*. The following cases arise.

Case I.   *The point* $x \in \pi_I(p; G)$. Then there exists a point x' of G which is linear with p and x. If pxx' holds, application of previous theorems to $T(p, p', x')$ shows that $G(x, p^*)$ intersects G (and hence $x \in \pi(p^*; G)$) except in the case that $pp^*/pp' = px/px'$, and pp^*p' holds. The "sequence method" is then used to establish that $x \in \pi(p^*; G)$. A similar argument is employed if xpx' or $px'x$ holds. Clearly $p \in \pi(p^*; G)$, and so the theorem is proved in this case.

Case II.   *The point* $x \in \pi_B(p; G)$. Then there is a sequence $\{x_i\}$ of points of $\pi_I(p; G)$ that converges to x, and by Case I, each $x_i \in \pi(p^*; G)$, $i = 1, 2, \ldots$. Since $\pi(p^*; G)$ is a closed set, it follows that $x \in \pi(p^*; G)$.

LEMMA 7.9   *Let p be a point of $\mathbf{M}_Y$, and $\{q_i\}$ an infinite sequence of points of $\mathbf{M}_Y$ with* $\lim_{i \to \infty} q_i = q$ $(q \neq p \neq q_i, i = 1, 2, \ldots)$. *If a line G of $\mathbf{M}_Y$, distinct from $G(p, q)$, intersects $G(p, q)$ in a point r, and $G(p, q_i)$ in a point r_i $(i = 1, 2, \ldots)$, then* $\lim_{i \to \infty} r_i = r$.

Proof.   The assertion is trivial if $p = r$. Suppose, then, $p \neq r$, and on each line $G(p, q_i)$ select the point t_i such that p, q_i, t_i are in the same betweenness relation as p, q, r (for example, $pq_i t_i$ holds provided pqr does, and so on) and $pt_i/pq_i = pr/pq$. Since $\lim_{i \to \infty} q_i = q$ and, by Theorem 7.1, $rt_i = (pr/pq) \cdot qq_i$ $(i = 1, 2, \ldots)$, it follows that $\lim_{i \to \infty} t_i = r$. Hence the lemma is proved in case $r_i = t_i$ for infinitely many values of the index i. In the contrary case, delete from $\{t_i\}$ each $t_i = r_i$. If $\varepsilon > 0$, points r', r^* of G exist such that $r'rr^*$ holds and $rr' = rr^* < \varepsilon$, and since r is not on $G(p, r')$ or $G(p, r^*)$, there exists a $\delta > 0$ such that $\delta < \min[\text{dist}(r, G(p, r')), \text{dist}(r, G(p, r^*))]$. Let N be a natural number such that for each $i > N$, $rt_i < \delta$ (such an N exists since $\lim_{i \to \infty} t_i = r$). For each $i > N$, consider the point r_i common to G and $G(p, t_i)$ $(i = 1, 2, \ldots)$. The following cases arise.

Case I.   *Either* $r_i t_i p$ *or* $t_i r_i p$ *holds.* Now r_i is in the segment with end points r', r^*, for if this is not the case, the Pasch theorem or its corollary ensures the existence of a point t' at which $G(p, r')$ intersects $G(r, t_i)$, or a point t^* at which $G(p, r^*)$ intersects $G(r, t_i)$, according as $r_i t_i p$ or $t_i r_i p$ holds, respectively, and $rt' t_i$ or $rt^* t_i$ holds. In the first case $rt_i > rt'$, which is impossible, since $rt_i < \delta$ and $rt' > \delta$ (since t' is a point of $G(p, r')$). A similar contradiction is encountered in the second case. Hence $rr_i < \varepsilon$ for each $i > N$.

Case II.   *The relation* $t_i p r_i$ *holds.* If r_i is not a point of the segment determined by r', r^*, the labeling may be chosen so that $r_i r r^*$ holds. The Pasch theorem applied to $T(r, r_i, t_i)$ shows that $G(p, r^*)$ intersects $G(r, t_i)$ in a point t^*, and $rt^* + t^* t_i = rt_i$. This is impossible since $rt_i < \delta < rt^*$. Hence we may conclude that r_i is a point of the segment with end points r', r^*, and so again $rr_i < \varepsilon$ for each $i > N$.

Thus, in every case, $\varepsilon > 0$ implies the existence of a natural number N such that whenever $i > N$, $rr_i < \varepsilon$. It follows that $\lim_{i \to \infty} r_i = r$, and the lemma is proved.

We establish now several theorems that lead to the important property of the uniqueness of parallels, defining two lines to be *mutually parallel* provided (a) they lie in the same plane, and (b) they do not intersect each other.

THEOREM 7.11   *If* $q \in \pi_1(p; G)$, $q \notin G$, *there is a unique line* G^* *of* $\pi(p; G)$ *that contains* q *and does not intersect* G.

Proof.   Suppose, first, that $q \neq p$, and let q' denote that point of G that is linear with p and q. Let t^* be any point of G ($t^* \neq q'$), and let q^* be the point of $\pi(p; G)$ such that p, q^*, t^* are in the same betweenness relation as p, q, q', and $pq^*/pt^* = pq/pq'$. Then the line $G(q, q^*)$ contains q, is contained in $\pi(p; G)$, by Theorem 7.8, and does not intersect G, by Theorem 7.3. Hence there is at least one line $G^* = G(q, q^*)$ of $\pi(p; G)$ which contains q and is parallel to G.

The line G^* is independent of the construction used to obtain it. Suppose, for example, that $t^{**} \in G$ with $q' t^{**} t^*$ holding, and let q^{**} be that point of $\pi(p; G)$ such that p, q^{**}, t^{**} are in the same betweenness relation as p, q, q', with $pq^{**}/pt^{**} = pq^*/pt^*$. The Pasch theorem or its corollary applied to $T(p, t^*, t^{**})$ shows that $G(q', q^{**})$ meets $G(p, t^*)$ in a point s, and the same theorem applied to $T(p, q', s)$ shows that $G(q^*, q^{**})$ meets $G(p, q')$ in a point x. Since $G(q^*, q^{**})$ does not intersect G, then $px/pq' = pq^*/pt^* = pq/pq'$, and consequently $q = x$. Thus $q^{**} \in G(q, q^*)$. Similar arguments yield the desired result in case $t^{**}q't^*$ or $q't^*t^{**}$ holds, and we conclude that G^* is independent of the arbitrariness in the choice of t^*.

Suppose, now, that $\pi(p; G)$ contains *another* line G'' on q which does not intersect G.

Assertion.   The line G'' contains at least one point different from q which belongs to $\pi_I(p; G)$.

If the contrary be assumed, then $G'' - \{q\} \subset \pi_B(p; G)$ and there is an infinite sequence $\{q_i\}$ of points of $\pi_B(p; G)$ with $\lim_{i \to \infty} q_i = q$, and for each $i = 1, 2, \ldots$, there is an infinite sequence $\{q_{ij}\}$ of points of $\pi_I(p; G)$ with $\lim_{j \to \infty} q_{ij} = q_i$. Corresponding to each point q_{ij} is a point q'_{ij} of G that is linear with p and q_{ij}. Now for each $i = 1, 2, \ldots$, the sequence $\{q'_{ij}\}$ is unbounded, for otherwise some point q_i would be in $\pi_I(p; G)$. (Let the reader verify this.) Thus for each $i = 1, 2, \ldots$, a subsequence of $\{q_{ij}\}$ exists whose corresponding sequence $\{q'_{ij}\}$ has its points receding monotonically from q' in one direction. Taking a subsequence $\{q_{i_k}\}$ of $\{q_i\}$, sequences $\{q_{i_k j}\}$ are obtained whose corresponding sequences $\{q'_{i_k j}\}$ have their points all receding from q' in the *same direction for all* $k = 1, 2, \ldots$. Renumber the elements of $\{q_{i_k}\}, q_1, q_2, \ldots$.

Hence for any $K > 0$ and any index $i = 1, 2, \ldots$, there exists a positive number N_i such that if $n > N_i$, then the distance $q'_{in} q' > K$. The sequences $\{q'_{ij}\}$ may be renumbered so that

$$q'_{ij} q' > K \qquad (i, j = 1, 2, \ldots).$$

A final selection of subsequences ensures that the sequence $\{q_{ii}\}$ converges to q.

Then $\lim_{i \to \infty} q_{ii} = q$, while $q'_{ii} q' > K$ $(i = 1, 2, \ldots)$; consequently, the sequence $\{q'_{ii}\}$ does *not* converge to q', which contradicts Lemma 7.9, and establishes the assertion.

Let r denote any point of G'', which is distinct from q and which is linear with p and a point r' of G. Since G'' does not intersect G, it follows from the Pasch theorem that the triples p, q, q' and p, r, r' are in the same betweenness relation, and then Theorem 7.4 or Theorem 7.7 may be invoked to conclude that $pr/pr' = pq/pq'$. Hence G'' coincides with G^*.

Finally we remark that $q \in \pi_I(p; G)$, $(q \notin G)$ implies that $p \in \pi_I(q; G)$, and since by Theorem 7.10, $\pi(p; G) = \pi(q; G)$, then there is exactly one line of $\pi(p; G)$ on p that does not intersect G. The proof of the theorem is complete.

Theorem 7.10 can now be extended to the following more inclusive one.

THEOREM 7.12 *If p^* is any point of $\pi(p; G)$ not belonging to G, then $\pi(p^*; G) = \pi(p; G)$.*

Proof. If $p^* \in \pi_I(p; G)$, then Theorem 7.10 gives the desired result.

Suppose $p^* \in \pi_B(p; G)$ and let q be an arbitrary point of $\pi_I(p; G)$ which is not on G $(q \neq p)$. Then by Theorem 7.10 $\pi(q; G) = \pi(p; G)$, and by Theorem 7.11 (since $p \in \pi_I(q; G)$) $G(p, p^*)$ is the unique line of $\pi(q; G)$ that contains p and does not intersect G.

If x is a point with p^*xq holding, then $x \notin G(p, p^*)$ and so $G(p, x)$ intersects G in a point, say s'. If t is a point with qpt holding, then consideration of $T(p, q, x)$ shows that $G(p^*, t)$ intersects $G(p, x)$ in a point, say s, and consideration of $T(p, q', s')$ shows that $G(p^*, t)$ intersects G in a point v'. Finally, considering $T(p^*, s, x)$, we see that line G meets $G(p^*, q)$ in a point q^*. Hence $p^* \in \pi_I(q; G)$ and so $\pi(p^*; G) = \pi(q; G) = \pi(p; G)$.

The result of the preceding theorem permits a more inclusive form of Theorem 7.11.

THEOREM 7.13 *If $q \in \pi(p; G)$, $q \notin G$, there is a unique line of $\pi(p; G)$ through q that does not meet G.*

Proof. If $q \in \pi_I(p; G)$ then the desired result was proved in Theorem 7.11.

If $q \in \pi_B(p; G)$ then, by the preceding theorem, $\pi(q; G) = \pi(p; G)$, and since $q \in \pi_I(q; G)$, the existence and uniqueness of the line through q that does not meet G follows from the earlier theorem.

THEOREM 7.14 *If p^*, q^*, r^* are noncollinear points of $\mathbf{M}_Y$, there is one and only one plane of $\mathbf{M}_Y$ that contains them.*

Proof. Clearly, the plane $\pi(p^*, q^*, r^*)$ contains p^*, q^*, and r^*.

Let $\pi(p; G)$ be any plane that contains p^*, q^*, r^*, and suppose $q, r \in G$ $(q \neq r)$. Then $\pi(p; G(q, r)) = \pi(p, q, r)$ (see Section 7.3), and one of the points p^*, q^*, r^* is not on $G(q, r)$. Assuming the labeling so that $p^* \notin G(q, r)$, then, by Theorem 7.12,

$$\pi(p; G(q, r)) = \pi(p^*; G(q, r)) = \pi(p^*, q, r) = \pi(q; G(p^*, r)).$$

Similarly, one of the points q^*, r^* is not on $G(p^*, r)$. With the labeling chosen so that $q^* \notin G(p^*, r)$,

$$\pi(q; G(p^*, r)) = \pi(q^*; G(p^*, r)) = \pi(r; G(p^*, q^*)),$$

and since $r^* \notin G(p^*, q^*)$,

$$\pi(r; G(p^*, q^*) = \pi(r^*; G(p^*, q^*) = \pi(p^*, q^*, r^*).$$

Hence $\pi(p^*, q^*, r^*) = \pi(p; G)$, and the theorem is proved.

COROLLARY   *Any plane of* $\mathbf{M}_Y$ *is determined by any three of its non-collinear points. Any plane of* $\mathbf{M}_Y$ *is determined by a point of* $\mathbf{M}_Y$ *and a line of* $\mathbf{M}_Y$ *not containing the point.*

THEOREM 7.15   *If p is a point and G is a line of any plane π of* $\mathbf{M}_Y$, *with $p \notin G$, then π contains exactly one line on p that is parallel to G.*

Proof.   By the above corollary, $\pi = \pi(p; G)$, and the theorem follows from Theorem 7.11.

COROLLARY   *If a line of* $\mathbf{M}_Y$ *intersects one of two mutually parallel lines of* $\mathbf{M}_Y$, *it intersects the other also.*

We have frequently referred to Theorem 7.6 as the Pasch theorem. This section is concluded by establishing what is usually called the Pasch axiom.

THEOREM 7.16 (Pasch Axiom)   *Let p, q, r denote noncollinear points of* $\mathbf{M}_Y$ *and G a line of $\pi(p, q, r)$, distinct from $G(p, q)$. If G intersects $G(p, q)$ in a point between p and q, then G contains a point between p and r, or a point between q and r, or G contains the point r.*

Proof.   Suppose that G contains no point between q and r, and $r \notin G$. If q' denotes the point of G between p and q, then $\pi(p, q, r) = \pi(q'; G(q, r))$, by the corollary to Theorem 7.14. If G intersects $G(q, r)$ in a point p', then qrp' or $p'qr$ holds, and in either case G contains a point r' between p and r (Pasch Theorem).

If, on the other hand, G does not intersect $G(q, r)$, let r' be chosen such that $pr'r$ holds, and $pr'/pr = pq'/pq$. Then $G(q', r')$ is parallel to $G(q, r)$, and hence must coincide with G (since the plane contains only one line through q' that is parallel to G).

7.5   THE THEORY OF PARALLELS IN A SPACE M_Y

In this section we develop that part of the theory of parallels in a space M_Y that will be useful in achieving our objective—the metric characterization of complete normed linear spaces with unique metric lines.

THEOREM 7.17   *If G_1, G_2, G_3 are pairwise distinct lines of M_Y, with G_1 parallel to G_2, and G_2 parallel to G_3, then G_1 is parallel to G_3.*

Proof.   Let q, r denote distinct points of G_1, and q' a point of G_2. If $p \in G(q, q')$ with $q'qp$ holding, then clearly $G(p, r)$ is not parallel to G_2 (since G_1 is the unique parallel to G_2 through r, and $G_1 \neq G(p, r)$) and since $G(p, r)$ and G_2 are clearly coplanar, then $G(p, r)$ intersects G_2 in a point r' with $pq/pq' = pr/pr'$.

Select a point q'' of G_3 and a point p' of $G(q', q'')$ with $p'q''q'$ holding and $p'q''/p'q' < pq/pq'$. Since the coplanar lines $G(p', r')$ and G_3 are not parallel, then they intersect in a point r'' with

$$\frac{p'r''}{p'r'} = \frac{p'q''}{p'q'}.$$

From $pq/pq' > p'q''/p'q'$ and pqq', $p'q''q'$ it follows that

$$(*) \qquad\qquad \frac{q'q}{q'p} < \frac{q'q''}{q'p'},$$

so the coplanar lines $G(q, q'')$ and $G(p, p')$ are not parallel. Their point of intersection s is such that $pp's$ holds (see Theorem 7.4 and its proof, which shows that the inequality $(*)$ implies $pp's$ rather than $p'ps$).

Use of the Menelaus theorem yields

$$\frac{sp}{sp'} = \frac{1}{\dfrac{q''p'}{q''q'} \cdot \dfrac{qq'}{qp}},$$

and in a similar manner it is seen that $G(r, r'')$ meets $G(p, p')$ in a point s' with

$$\frac{s'p}{s'p'} = \frac{1}{\dfrac{r''p'}{r''r'} \cdot \dfrac{rr'}{rp}}.$$

But $r''p'/r''r' = q''p'/q''q'$ and $rr'/rp = qq'/qp$; consequently $ps/p's = ps'/p's'$, and $s = s'$. It follows that lines G_1 and G_3 are coplanar.

Applying the Menelaus theorem to $T(s, q, p)$ and $T(s, r, p)$ gives

$$\frac{sq''}{q''q} = \frac{1}{\dfrac{qq'}{pq'} \cdot \dfrac{pp'}{sp'}} = \frac{1}{\dfrac{rr'}{pr'} \cdot \dfrac{pp'}{sp'}} = \frac{sr''}{r''r},$$

from which follows

$$\frac{sq''}{sq} = \frac{sr''}{sr}.$$

Hence G_1 is parallel to G_3.

It should be observed that the lines G_1, G_2, G_3 are not necessarily coplanar.

THEOREM 7.18 *If G_1, G_2, G_3 are three pairwise distinct and mutually parallel lines which are cut by two intersecting lines G and G' in the points q_1, q_2, q_3 and r_1, r_2, r_3, respectively, then*

$$\frac{q_1 q_2}{r_1 r_2} = \frac{q_2 q_3}{r_2 r_3}.$$

We leave the proof to the reader.

THEOREM 7.19 *If G_1, G_2, G_3 are three pairwise distinct and mutually parallel lines which are cut by mutually parallel lines G and G' in points q_1, q_2, q_3 and r_1, r_2, r_3, respectively, then*

$$\frac{q_1 q_2}{r_1 r_2} = \frac{q_2 q_3}{r_2 r_3}.$$

Proof.   On line G select a point p $(p \neq q_1, q_2, q_3)$, on line G_3 select a point s_3 $(q_3 \neq s_3 \neq r_3)$, and draw $G(p, s_3)$. That line intersects G_1, G_2, and G' in points s_1, s_2, and p', respectively. (Let the reader show this.)

A twofold application of the preceding theorem yields

$$\frac{q_1 q_2}{q_2 q_3} = \frac{s_1 s_2}{s_2 s_3} = \frac{r_1 r_2}{r_2 r_3},$$

and the desired result follows immediately.

Defining a *parallelogram* as a quadrilateral whose opposite sides are segments of mutually parallel lines, the following two important theorems are established.

THEOREM 7.20 *The diagonals of a parallelogram bisect each other.*

Proof. Let the pairs of opposite sides of the parallelogram be segments of the mutually parallel lines G_1 and G_2, and of the mutually parallel lines G and G', with G_1 and G_2 intersecting G and G' in the points p_1, p_2 and q_1, q_2, respectively. Put $p' = m(p_1, p_2)$, $q' = m(q_1, q_2)$, and $t = m(p_2, q_1)$. In $T(q_1, p_2, q_2)$, $q_1 t/q_1 p_2 = q_1 q'/q_1 q_2 = 2$, and so $G(q', t)$ is parallel to G_2; similar consideration of $T(p_2, p_1, q_1)$ shows that $G(p', t)$ is parallel to G_1 and hence parallel to G_2 (Theorem 7.17). Since $G(q', t)$ and $G(p', t)$ both contain point t, it follows from the uniqueness theorem for parallels that $G(p', t) = G(q', t)$, so p', q', t are on a line G^*.

If $r_1 = m(p_1, q_1)$, $r_2 = m(p_2, q_2)$, and $s = m(p_1, q_2)$ the above argument may be applied to show that r_1, r_2, s are on a line G^{**}.

Consideration of $T(q_2, p_1, p_2)$ shows that G^{**} is parallel to G, and examination of $T(q_1, p_1, p_2)$ shows that $G(r_1, t)$ is parallel to G. Since G^{**} and $G(r_1, t)$ both contain point r_1, then these two lines are identical and consequently $s = t$, which proves the theorem.

THEOREM 7.21 *The opposite sides of a parallelogram are equal in length.*

Proof. Labeling as in the preceding theorem, it is clear that $p_2 q_2 = 2 \cdot p't = p_1 q_1$, and $q_1 q_2 = 2 \cdot r_1 t = p_1 p_2$.

7.6 METRIC CHARACTERIZATION OF BANACH SPACES WITH UNIQUE METRIC LINES

We are now in position to accomplish one of the major objectives of our study of metric geometry. In Chapter 6 it was shown that each Banach space with unique metric lines is a complete, metrically convex and externally convex metric space which has the two-triple and the Young properties; that is, each Banach space with unique metric lines is an $\mathbf{M_Y}$ space. By means of the properties of $\mathbf{M_Y}$ spaces established in the foregoing sections of this chapter, we are now able to prove the converse proposition; namely, that each $\mathbf{M_Y}$ space is a Banach space with unique metric lines. Thus we lay bare the metric structure of such Banach spaces and exhibit those metric properties which characterize Banach spaces among the class of all metric spaces.

To avoid trivialities, we assume that an $\mathbf{M_Y}$ space contains at least two distinct points.

The reader should consult Example 3 of Section 1.8 for the postulates of a normed linear space, and recall that a Banach space is a (metrically) complete, normed linear space.

The definition of addition for elements of a space $\mathbf{M_Y}$ is based upon the notion of *reflection in a point*.

DEFINITION   If p, q are points of a metric space, a point p' of the space is the *reflection of p in q* provided $pq = qp' = (\frac{1}{2}) \cdot pp'$; that is, *$q$ is a middle point of p, p'*.

Remarks. Since each two distinct points of $\mathbf{M_Y}$ are on a unique metric line (which is congruent to the euclidean straight line), the reflection p' of p in q is *uniquely* determined for each two distinct points p, q of $\mathbf{M_Y}$, and if $p = q$, then clearly $p' = p$. If p' is the reflection of p in q, then p is the reflection of p' in q.

Now let o denote any arbitrarily selected point of $\mathbf{M_Y}$, which is kept fixed throughout the remainder of this section.

DEFINITION   If $p, q \in \mathbf{M_Y}$, the point $p + q$ is defined as that point of $\mathbf{M_Y}$ which is the reflection of point o in $m(p, q)$, the middle point of p, q.

Remark. Since for each two elements p, q of $\mathbf{M_Y}$, distinct or not, $m(p, q)$ is unique, and since (as remarked above) each point of $\mathbf{M_Y}$ has a unique reflection in any point of $\mathbf{M_Y}$, then for each two points p, q of $\mathbf{M_Y}$, the point $p + q$ of $\mathbf{M_Y}$ is uniquely determined. Clearly, $o + p = p$ for every point p of $\mathbf{M_Y}$.

DEFINITION   If $p \in \mathbf{M_Y}$, $(p \neq o)$, and $\lambda > 0$, the point $\lambda \cdot p$ is defined as that point p' of the line $G(o, p)$ such that $p'op$ does *not* hold, and $op' = \lambda \cdot op$.

If $\lambda = 0$, then $\lambda \cdot p = o$; if $\lambda < 0$, and $p \neq o$, then $\lambda \cdot p$ is the reflection of $(-\lambda) \cdot p$ in o. For every real λ, $\lambda \cdot o = o$.

Remark. For each point p of $\mathbf{M}_Y$, and each real number λ, the point $\lambda \cdot p$ is uniquely determined.

DEFINITION   If $p, q \in \mathbf{M}_Y$, the binary operation $p + q$ is called *addition*, and if λ is any real number, the operation $\lambda \cdot p$ is called *scalar multiplication*.

THEOREM 7.22   *An $\mathbf{M}_Y$ space is a linear space with respect to the operations of addition and scalar multiplication.*

Proof. Since $m(x, y) = m(y, x)$, then $x + y = y + x$, and property (i) is valid.

To establish property (iii), note that the equality $x + y = x + z$ implies that $m(x, y)$ and $m(x, z)$ are both middle points of o and $x + y$, and consequently $m(x, y) = m(x, z)$. The latter equality clearly implies that $y = z$.

Property (iv) is proved by observing that the points o, $\lambda \cdot x$, $\lambda \cdot y$, and $\lambda \cdot (x + y)$ are the vertices of a parallelogram, and $\lambda \cdot x + \lambda \cdot y$ is the reflection of the vertex o in the middle point $m(\lambda \cdot x, \lambda \cdot y)$ of the diagonal whose end points are $\lambda \cdot x$ and $\lambda \cdot y$. It follows from Theorem 7.19 that $\lambda \cdot (x + y) = \lambda \cdot x + \lambda \cdot y$.

Since the points o, $\lambda_1 \cdot x$, $\lambda_2 \cdot x$, $(\lambda_1 + \lambda_2) \cdot x$ are collinear, property (v) follows from the congruence of lines of $\mathbf{M}_Y$ with euclidean straight lines. The same consideration establishes property (vi), and property (vii) is an immediate consequence of the definition of scalar multiplication.

To establish property (ii), the associativity of addition, it suffices to show that the middle points $m(x, y + z)$ and $m(x + y, z)$ coincide.

From $T(y, x, z)$, $xz = 2 \cdot \text{dist}[m(x, y), m(y, z)]$,   and   consideration of $T(o, x + y, y + z)$ gives

$$\text{dist}[x + y, y + z] = 2 \cdot \text{dist}[m(x, y), m(y, z)].$$

Consequently $\text{dist}[x + y, y + z] = xz$.

Also, $G(x, x + y)$ and $G(y + z, z)$ are both parallel to $G(o, y)$ and hence are parallel to each other. Similarly, $G(x, z)$ and $G(x + y, y + z)$ are mutually parallel, since each of these lines is parallel to $G(m(x, y), m(y, z))$. Hence x, $x + y$, $y + z$, and z are the vertices of a parallelogram, the diagonals of which bisect each other; that is, $m(x, y + z) = m(x + y, z)$, and the proof of the theorem is complete.*

DEFINITION   If $p \in \mathbf{M}_Y$, the *norm* $\|p\|$ of p is defined to be the distance op.

THEOREM 7.23   *An $\mathbf{M}_Y$ space is a Banach space with unique metric lines.*

Proof.   Clearly $\|p\| \geqq 0$ for every point p of $\mathbf{M}_Y$ and $\|p\| = 0$ if and only if $p = o$. Also, $\|\lambda \cdot p\| = \|p'\| = op' = |\lambda| \cdot op = |\lambda| \cdot \|p\|$, and $\|p + q\| \leqq \|p\| + \|q\|$, since $o, p, q, p + q$ are vertices of a parallelogram whose opposite sides are equal in length (Theorem 7.21).

We state the principal result of this chapter in the following form.

THEOREM 7.24   *A Banach space with unique metric lines is characterized metrically among all metric spaces by being complete, metrically convex and externally convex, and having the two-triple and the Young properties.*

EXERCISES

Let M^* denote a complete, metrically convex and externally convex metric space with the following property:
(*) if p, q, r denote any three nonlinear elements of M^*, and q' is any middle point of p, q, and r' is any middle point of p, r, then $q'r' = (\frac{1}{2})qr$.

1. Show that M^* contains for each two distinct points exactly one middle point.

2. Show that each two distinct points of M^* are on exactly one metric line of M^*.

3. Prove that if p, q, r are *any* elements of M^*, with q' a middle point of p, q and r' a middle point of p, r, then $q'r' = (\frac{1}{2})qr$.

4. Use Theorem 7.24 and Exercise 2 to conclude that the class of Banach spaces with unique lines is characterized metrically among metric spaces by being complete, metrically convex and externally convex, and having property (*).

---

* The proof has been given for the "general" case in which the points x, y, z are not collinear and no two are collinear with o. The argument is simplified in any of the several special cases.

Let $M\dagger$ denote a complete, metrically externally convex metric space with the following property:

($\dagger$) if p, q, r denote any three elements of $M\dagger$ $(q \neq p \neq r)$, there exists a middle point q' of p, q and a middle point r' of p, r such that $q'r' = (\frac{1}{2})qr$.

5. Show that every Banach space is an $M\dagger$ space.

6. Is the class of all Banach spaces metrically characterized among metric spaces by completeness, external convexity, and property ($\dagger$)?

REFERENCES

Sections 7.1 to 7.6   E. Z. Andalafte and L. M. Blumenthal, "Metric characterizations of Banach and euclidean spaces," *Fund. Math.*, **55**: 23–55, (1964).

CHAPTER 8

Metric Postulates for Euclidean Space

8.1 THE WEAK PYTHAGOREAN PROPERTY

In Hilbert's famous axiomatization of euclidean geometry a postulate is formulated to the effect that whenever two triangles have two sides and the included angle of one congruent, respectively, to two sides and the included angle of the other, then a second angle of the one is congruent to the corresponding angle of the other. From that postulate it follows readily that the two triangles have their corresponding angles and sides congruent, and hence are, by definition, congruent triangles.

The postulate that we wish to introduce in this section, and which will play a very important role in the developments of this chapter, can be thought of as a very weakened form of Hilbert's triangle congruence axiom. Instead of asserting a property of *any* two triangles with two sides and included angle of one congruent, respectively, to two sides and the included angle of the other, our postulate is restricted to two *right triangles* T_1, T_2 *which have a leg in common*; and instead of concluding that a second angle of T_1 is congruent to a second angle of T_2, it is asserted that the remaining side (the hypotenuse) of T_1 is congruent to the remaining side (hypotenuse) of T_2.

Since such an assertion is obviously valid if the theorem of Pythagoras concerning right triangles holds, it seems reasonable to refer to the assumption, formulated below for metric spaces, as the weak pythagorean postulate. We shall see later that the right triangle theorem of Pythagoras is a consequence of that postulate.

The weak pythagorean property. Let p denote a point, and G a line of a metric space **M**, *with* $p \notin G$. *If* f_p *is any foot of p on G, and* $q, r \in G$ *such that* $qf_p = rf_p$, *then* $sq = sr$ *for every point s of* $G(p, f_p)$.

We proceed to a study of metric spaces, denoted by $\mathbf{M}_P$ (containing at least two points) which are (1) complete, (2) metrically convex and externally convex, have (3) the two-triple property, and (4) the weak pythagorean property.

EXERCISES

1. If p is a point and G is a metric line of any metric space M, show that for each real number $k \geqq pf_p$, where f_p denotes a foot of p on G, each half-line of G determined by and containing f_p has a point s such that $ps = k$.

2. In any metric space **M**, show that if f_p is any foot of p on a metric line G of **M** and s is any point of G with $sf_p > 2 \cdot pf_p$, then $ps > pf_p$, and so s is not a foot of p on G.

8.2   PROPERTIES OF FOOT OF A POINT ON A LINE. PERPENDICULARITY

It was observed in the remark following Lemma 7.3 that if p is a point and G is a metric line of any metric space, then there exists at least one foot of p on G. Since $f_p = p$ if $p \in G$, the foot is unique in this case.

Let us suppose that $p \notin G$ and f'_p is a foot of p on G, with $f'_p \neq f_p$. Put $a = pf_p > 0$ and $b = f_p f'_p > 0$, and let n be a positive integer such that

$n \cdot b > 2a$. If f'' denotes that point of G with $f''f'_p = b$ and $f_p f'_p f''$ holding, then by the weak pythagorean property, $pf_p = pf''$, and so $f'' = f''_p$ is another foot of p on G.

Make the inductive assumption that $f_p, f'_p, f''_p, \ldots, f_p^{(k)}$ are feet of p on G, with $f_p f'_p = f'_p f''_p = \cdots = f_p^{(k-1)} f_p^{(k)} = b$ and $f_p, f'_p, f''_p, \ldots, f_p^{(k)}$ encountered in that order in traversing that half-line of G determined by f_p that contains them. Denote by $f^{(k+1)}$ the point of G with $f_p^{(k)} f^{(k+1)} = b$ and $f_p^{(k-1)} f_p^{(k)} f^{(k+1)}$ holding. Then it follows from the weak pythagorean property that

$$pf_p^{(k-1)} = pf^{(k+1)},$$

and so $f^{(k+1)}$ is a foot of p on G, which may be denoted by $f_p^{(k+1)}$.

Proceeding in this manner, a foot $f_p^{(n)}$ of p on G is obtained with $f_p f_p^{(n)} = n \cdot b > 2a$, contradicting the result of Exercise 2 of the preceding section.

We have thus established the following theorem.

THEOREM 8.1   *In an* $\mathbf{M}_P$ *space each point has a unique foot on each line.*

Now if p is a point and G is a line of an $\mathbf{M}_P$ space, $p \notin G$, let s denote any point of the line $G(p, f_p)$, where, as usual, f_p denotes the unique foot of p on G, and suppose $f_s \neq f_p$, with f_s the foot of s on G. Let t denote the point of G with $f_s f_p = tf_p$ and $tf_p f_s$ holding. Then $sf_s = st$ and so t is a foot of s on G distinct from f_s. Since this is impossible we have proved the following result.

THEOREM 8.2   *If, in an* $\mathbf{M}_P$ *space,* f_p *is the foot of a point* p *on a line* G $(p \notin G)$, *then the point* f_p *is the foot on* G *of every point of* $G(p, f_p)$.

The next theorem is the basis for the important *symmetric property of perpendicularity* (to be defined later).

THEOREM 8.3   *If, in an* $\mathbf{M}_P$ *space,* f_p *is the foot of a point* p *on a line* G $(p \notin G)$, *then* f_p *is the foot on* $G(p, f_p)$ *of each point of the line* G.

Proof.   Let q be any point of G $(q \neq f_p)$ and denote by f_q the unique foot of q on the line $G(p, f_p)$. If $q' \in G$ with $q'f_p = qf_p$ and $qf_p q'$ holding, then (since f_p is the foot of f_q on G) $qf_q = q'f(p)$. Now $qf_q + q'f_q \geq qq' = qf_p + q'f_p$, and it follows that $qf_q \geq qf_p$. Hence $f_q = f_p$, and since $q = f_p$ obviously implies $f_q = f_p$, the proof is complete.

DEFINITION   A line G^* is *perpendicular* to a line G provided there exists a point p of G^* $(p \notin G)$, whose foot f_p on G is the intersection of G and G^*.

Remark. It follows from Theorem 8.3 that if G^* is perpendicular to G, then G is perpendicular to G^*. The relation of perpendicularity being symmetric, we say that G and G^* are mutually perpendicular.

THEOREM 8.4   *If p, q are distinct points and G is a line of an $\mathbf{M}_P$ space $(p, q \notin G)$, let f_p, f_q denote the feet of p, q, respectively, on G. If $G(p, q)$ intersects G in the point r, then $f_p = f_q$ implies $f_p = f_q = r$.*

Proof.   Let $f = f_p = f_q$ and denote by r' the reflection of r in f (see Section 7.6). Then the weak pythagorean property yields $pr = pr'$ and $qr = qr'$.

Now one of the relations rpq, rqp, prq holds. If rpq, then $rp + pq = qr$ and consequently $pr' + pq = qr'$; if rqp, then $qr + pq = pr$, and so $qr' + pq = pr'$; and if prq, then $pr + rq = pq$, and hence $pr' + r'q = pq$. In each case, p, q, r' are linear and so $r' \in G(p, q)$. Since r' is also a point of G, then $r' = r$ and consequently $r = f$.

THEOREM 8.5   *If f_p denotes the foot on a line G of a point p, then f_p is a continuous function of p in $\mathbf{M}_P$.*

Proof.   Let $\{p_k\}$ denote any infinite sequence of points of $\mathbf{M}_P$ with $\lim_{k \to \infty} p_k = p_0$, $p_0 \in \mathbf{M}_P$. We wish to show that $\lim_{k \to \infty} f_k = f_0$, where f_k denotes the foot on G of p_k $(k = 0, 1, 2, \ldots)$.

Since $f_i f_0 \leqq f_0 p_0 + p_0 p_i + p_i f_i \leqq f_0 p_0 + p_0 p_i + p_i f_0$ $(i = 1, 2, \ldots)$, it is clear that the sequence $\{f_k\}$ of points of G is bounded.

If f_0 is not the limit of that sequence, then a positive number K and a subsequence $\{f_{i_n}\}$ exist such that $f_0 f_{i_n} > K$ $(n = 1, 2, \ldots)$. This subsequence contains a subsequence $\{f_k'\}$ with a limit g and $g \neq f_0$. Denote by $\{p_k'\}$ the subsequence of $\{p_k\}$ corresponding to $\{f_k'\}$. Then $\lim_{k \to \infty} p_k' = p_0$, and since $p_k' f_0 \geqq p_k' f_k'$ $(k = 1, 2, \ldots)$, then $p_0 f_0 = \lim_{k \to \infty} p_k' f_0 \geqq \lim_{k \to \infty} p_k' f_k' = p_0 g$. Since f_0 is the unique foot of p_0 on G, and $g \neq f_0$, a contradiction to the supposition that f_0 is not the limit of $\{f_k\}$ is obtained.

LEMMA 8.1   *If q, r are distinct points of G, and $p \in \mathbf{M}_P$ with $pq = pr$, then the foot f_s on G of each point of $S_p^q \cup S_p^r$ lies in S_q^r.*

Proof. If $t \in S_p^q$ and $f_t \notin S_q^r$, then either $f_t\,qr$ or qrf_t holds. Suppose the first alternative.

Since $f_r = r$ and f_x is a continuous function of x, $x \in S_p^t \cup S_p^r$, that set contains a point s with $f_s = q$. Now if $s \in S_p^t$, then by Theorem 8.2 $f_t = q$, contrary to $f_t\,qr$ holding, but if $s \in S_p^r$, then, since $sq < sr$,

$$sq + ps < sr + ps = pr = pq,$$

in violation of the triangle inequality.

If the second alternative holds, then S_q^t contains a point s such that $f_s = r$, and a contradiction is obtained as above.

A similar argument is applied if it is assumed that $t \in S_p^r$, to complete the proof of the theorem.

THEOREM 8.6 *For each point p and line G of $\mathbf{M}_P$, the function px is monotone increasing without bound as x recedes from f_p (the foot of p on G) along either half-line of G determined by f_p.*

Proof. It follows from Exercise 1 of Section 7.1 that px is not bounded on either of the half-lines considered.

If px is not monotone increasing on a half-line of G determined by f_p, then points q, r of G exist with $f_p\,qr$ holding and $pq = pr$. But then, by the preceding lemma, $f_p \in S_q^r$, which contradicts the relation $f_p\,qr$.

The isosceles triangle property. If p is a point and G is a line of a metric space $(p \notin G)$, let q, r be distinct points of G with pq = pr. Then the middle point $m = m(q, r)$ of q, r on G is a foot on G of every point of the line G(p, m).

In euclidean geometry the isosceles triangle property is valid, due to the theorem that the line joining the middle point of the base of an isosceles triangle to the opposite vertex is perpendicular to the base. The following theorem shows that every $\mathbf{M}_P$ space has the isosceles triangle property, and we will establish later the *equivalence* of that property and the weak pythagorean property in any complete, metrically convex and externally convex metric space with the two-triple property.

THEOREM 8.7 *Every $\mathbf{M}_P$ space has the isosceles triangle property.*

Proof. If $\mathbf{M}_P$ does not have the isosceles triangle property, then a point p and distinct points q, r of a line G exist with $pq = pr$, but the middle point

$m = m(q, r)$ of q, r is *not* the foot f_p of p on G. By Lemma 8.1, the foot f_p of p on G lies in S_q^r. If q' denotes the reflection of q in f_p then $pq' = pq = pr$, and since $m \neq f_p$, then $q' \neq r$. Either $f_p\, rq'$ or $f_p\, q'r$ holds, but in either event the equality $pq' = pr$ contradicts the monotone property established in Theorem 8.6.

8.3   SPACES WITH THE TWO-TRIPLE AND THE ISOSCELES TRIANGLE PROPERTIES

Let $\{\mathbf{M}_I\}$ denote the class of complete, metrically convex, and externally convex metric spaces $\mathbf{M}_I$ with the properties listed in the heading of this section. It was shown in Theorem 8.7 that $\{\mathbf{M}_P\} \subset \{\mathbf{M}_I\}$, where $\{\mathbf{M}_P\}$ denotes the class of all $\mathbf{M}_P$ spaces. The objective of this section is to prove that $\{\mathbf{M}_I\} \subset \{\mathbf{M}_P\}$, and consequently that the weak pythagorean property and the isosceles triangle property are logically equivalent in the class of complete, metrically convex and externally convex metric spaces with the two-triple property. We begin by establishing properties of a foot of a point on a line.

LEMMA 8.2   *If, in a space* $\mathbf{M}_I$, *q and r are feet of a point p on a line G ($q \neq r$), then every point of the segment S_q^r is a foot of p on G.*

Proof.   If q, r are feet of p on G ($q \neq r$), then $pq = pr$ and so, by the isosceles triangle property, their middle point $m = m(q, r)$ is also a foot of p on G. Hence, the middle point of each two distinct feet of p on G is itself a foot of p on G, and consequently the set of feet of p on line G is dense in S_q^r. The continuity of the metric assures that each point of S_q^r is a foot of p on G.

THEOREM 8.8   *A point p of $\mathbf{M}_I$ either has a unique foot on a line G of $\mathbf{M}_I$ or the set of all feet of p on G constitutes a segment.*

Proof.   If p does not have a unique foot on G, let f, f' denote distinct feet of p on G. Then by the preceding lemma, each point of $S_f^{f'}$ is a foot of p on G.

Let $F(p)$ denote the set of all feet of p on G. Then by Exercise 2 of Section 8.2, $F(p)$ is a bounded subset of G, and so $F(p)$ has "rightmost" and "leftmost" bounds q, r which, from the continuity of the metric, evidently belong to $F(p)$, and $q \neq r$. Then by the preceding lemma, each point of the segment S_q^r is a foot of p on G and from the definitions of q, r, no point of G not in S_q^r is a foot of p on G.

THEOREM 8.9   *If each point of a segment S_q^r is a foot of a point p on a line G, each point f of G with qfr holding is the unique foot on G of each point t of $G(p, f)$, distinct from p.*

Proof. Since qfr holds, S_q^r contains two distinct points a, b such that $af = fb = (\frac{1}{2}) \cdot ab$, and $pa = pb$. Then by the isosceles triangle property, f is a foot on G of each point t of $G(p, f)$. We show that if $t \neq p$, then f is the only foot of t on G. This is obvious in case $t = f$, so we assume $t \neq f$.

Suppose t has another foot f' on G and $f' \in S_q^r$. Then f' is a foot of p on G, and we have $pf = pf'$, and $tf = tf'$. Now since p, t, f are pairwise distinct points of $G(p, f)$, then fpt or ftp or tfp holds.

In the first case, $ft = fp + pt$ and hence $tf' = pf' + pt$; that is, $f' \in G(p, t)$, and since also $f' \in G$, then $f' = f$, contrary to assumption. The other two cases yield, in analogous manner, the same contradiction.

The above argument depends on t having a foot on G, distinct from f, which is a point of S_q^r. But this occurs whenever t has a foot on G distinct from f. For if f'' is such a foot and $f'' \notin S_q^r$, then each point of the segment determined by f, f'' is a foot of t on G, by Lemma 8.2, and this segment clearly has a point distinct from f in S_q^r, since f is an interior point of S_q^r.

THEOREM 8.10   *If S_q^r is the set of all feet on G of a point p, and t is any point of S_p^q, distinct from p, then q is the unique foot of t on G.*

Proof.   First, it is clear that q is a foot of t on G; for if s is a point of G with $ts < tq$, then

$$pq = pt + tq > pt + ts \geqq ps.$$

But $pq > ps$ contradicts the assumption that q is a foot of p on G.

Moreover, q is the unique foot of t on G, for suppose t has another foot q^* on G. If $q^* \in S_q^r$ then q^* is also a foot of p on G, and a contradiction is obtained by use of the same argument employed in the proof of the preceding theorem; if $q^* \notin S_q^r$, then

$$pq^* > pq = pt + tq = pt + tq^*,$$

in violation of the triangle inequality.

THEOREM 8.11   *For each point p and each line G of an M_I space, the foot of p on G is unique.*

Proof. If p has more than one foot on G, the set of all feet of p on G constitutes a segment S_q^r. Let $m = m(q, r)$, and let t be any point with tpm holding. Then by Theorem 8.9, m is the unique foot of t on G.

Now $tm = tp + pm = tp + pq > tq$, for since $G(p, t)$ intersects G in m, and $m \neq q$, then q is not linear with p and t, and consequently $tp + pq \neq tq$. But $tm > tq$ contradicts the statement that m is a foot of t on G.

The property of unique feet, established in the above theorem, is the only result of the weak pythagorean property used in the proof of Theorem 8.5. Hence the important continuity property enunciated in that theorem is valid also in an $\mathbf{M}_I$ space. We restate it in the following form.

THEOREM 8.12 *In any $\mathbf{M}_I$ space, the foot f_x on a given line G of an arbitrary point x is a continuous function of x.*

THEOREM 8.13 *If, in any $\mathbf{M}_I$ space, f_p denotes the foot on a given line G of a given point p of the space, then the function px is monotone increasing without bound as x recedes along either half-line of G determined by f_p.*

Proof. If px is not monotone increasing, then points t and x_2 of G exist with $f_p t x_2$ holding, and $pt > px_2$. The continuity of px on the segment determined by f_p, t implies the existence of a point x_1 of this segment with $px_1 = px_2$; and the isosceles triangle property implies that $m = m(x_1, x_2)$ is a foot of p on G. Since p has the unique foot f_p on G, then $m = f_p$, which is impossible since m is an interior point of the segment with end points x_1, x_2 and f_p is not.

The unbounded property of px follows from Exercise 1 of Section 8.1.

We proceed toward our objective of proving the equivalence of the weak pythagorean property and the isosceles triangle property by first establishing the validity in every $\mathbf{M}_I$ space of a restricted form of the weak pythagorean property.

LEMMA 8.3 (RESTRICTED FORM OF THE WEAK PYTHAGOREAN PROPERTY). *If, in any $\mathbf{M}_I$ space, f_p is the foot of p on the line G, and q, r are points of G with $qf_p = f_p r$, then $pq = pr$.*

Proof. The lemma is trivially valid if $q = r$. Let us suppose $q \neq r$ and assume $pq \neq pr$, with the labeling selected so that $pq > pr$. The continuity of px implies that a point q^* of G exists with qq^*f_p holding, and $pq^* = pr$. If $m = m(q^*, r)$ denotes the middle point of q^* and r, then, by the isosceles triangle property, m is a foot of p on G. But since p has only one foot on G, then $m = f_p$, and consequently $q = q^*$. This contradicts qq^*f_p, and proves the lemma.

LEMMA 8.4 *If, in any $\mathbf{M}_I$ space, f_p is the foot of the point p on the line G $(p \notin G)$, then f_p is the foot on G of each point of the line $G(p, f_p)$.*

Proof. Let q, r denote distinct points of G with $qf_p - f_p r$. Then by the preceding lemma, $pq = pr$ and the isosceles triangle property asserts that f_p is the foot on G of each point of the line $G(p, f_p)$.

THEOREM 8.14 *In any $\mathbf{M}_I$ space, the weak pythagorean property is valid.*

Proof. Let p be any point and G any line of an $\mathbf{M}_I$ space $(p \notin G)$, and let f_p denote the foot of p on G. If s is any point of $G(p, f_p)$, then f_p is a foot of s on G (Lemma 8.4), and if q, r are points of G with $qf_p = f_p r$, then $sq = sr$ (Lemma 8.3).

We close this section with the following theorem that summarizes some of the results of this and the preceding section.

THEOREM 8.15 *In any complete, metrically convex and externally convex metric space with the two-triple property, the weak pythagorean property and the isosceles triangle property are mutually equivalent.*

8.4   EQUIDISTANT LOCI IN THE PLANE

The results established in the preceding sections of this Part are all valid in every complete, metrically convex and externally convex metric space with the two-triple, Young, and isosceles triangle properties. In such spaces the weak pythagorean property is valid also (since, as we have seen, the weak pythagorean property and the isosceles triangle property are mutually equivalent in any complete, metrically convex and externally convex metric space with the two-triple property). Denote the class of these spaces by $\{\mathbf{M}_{PY}\}$. Clearly

$$\{\mathbf{M}_{PY}\} = \{\mathbf{M}_Y\} \cap \{\mathbf{M}_P\} = \{\mathbf{M}_Y\} \cap \{\mathbf{M}_I\}.$$

In this section the theory of parallels in $\mathbf{M}_{PY}$ spaces is developed further, with the objective of establishing the linearity of the equidistant locus in the plane.

THEOREM 8.16 *If distinct lines G_1, G_2 of a plane π of an $\mathbf{M}_{PY}$ space are each perpendicular to a line G, then G_1 and G_2 are mutually parallel.*

Proof. If G_1 and G_2 are perpendicular to G at points q and r, respectively, and $p \in G_1 \cap G_2$, then p has two distinct feet q, r on G, which contradicts the uniqueness of feet (for example, Theorem 8.1).

THEOREM 8.17 *If, in a plane π of an $\mathbf{M}_{PY}$ space, a line G is perpendicular to one of two mutually parallel lines G_1, G_2, then G is perpendicular to the other also.*

Proof. Suppose G is perpendicular to G_1 at the point q. Then G intersects G_2 (since G_1 is the unique line through q that is parallel to G_2, and $G_1 \neq G$). Let r denote the intersection of G with G_2, and let p denote any point of G_2, distinct from r. If f_p is the unique foot of p on G, then $G(p, f_p)$ is perpendicular to G, and, according to the preceding theorem, $G(p, f_p)$ is parallel to G_1 (since clearly $G(p, f_p) \neq G_1$). Consequently $G(p, f_p) = G_2$, and hence G is perpendicular to G_2.

DEFINITION The *equidistant locus in the plane* π of two distinct points q, r of π, denoted by $\mathscr{E}(q, r)$, is the set of points p of π such that $pq = pr$.

THEOREM 8.18 *Let q, r be distinct points of a plane π of an $\mathbf{M}_{PY}$ space. Then $\mathscr{E}(q, r)$ is the set of all points p of π such that $f_p = m(q, r)$, where f_p is the foot of p on $G(q, r)$, and $m(q, r)$ is the middle point of q, r.*

Proof. If $p \in \mathscr{E}(q, r)$, then $pq = pr$ and $f_p = m(q, r)$ by the isosceles triangle property, and if $p \in \pi$ and $f_p = m(q, r)$, then $pq = pr$, by the weak pythagorean property, and so $p \in \mathscr{E}(q, r)$.

LEMMA 8.5   *In a plane π of an $\mathbf{M}_{PY}$ space, let p, q be distinct points of $\mathscr{E}(r, r')$, $r, r' \in \pi$ $(r \neq r')$. If $G(p, q)$ intersects $G(r, r')$, then $G(p, q) \subset \mathscr{E}(r, r')$*

Proof. If $G = G(r, r')$, $G^* = G(p, q)$, and $t \in G \cap G^*$, let t' denote the reflection of t in $m = m(r, r')$.

By Theorem 8.18, m is the foot on G of p and of q, so by the weak pythagorean property, $pt = pt'$ and $qt = qt'$.

If $p = t$, then $rt = tr'$; that is, $p = m(r, r')$ and p is the foot on G of q. Consequently, each point of $G(p, q)$ has the middle point of r, r' for its foot on G, and it follows from the preceding theorem that $G(p, q) \subset \mathscr{E}(r, r')$. The same conclusion is obtained in case $q = t$.

If $p \neq t \neq q$, then pqt or qpt or ptq holds. In case pqt subsists, then

$$pq + qt' = pq + qt = pt = pt',$$

and it follows that p, q, t' are collinear. Hence $t' = t = m$, and the intersection m of G and G^* is the foot on G of every point of G^*. Again, we have $G(p, q) \subset \mathscr{E}(r, r')$.

The two remaining possibilities (qpt and ptq) are treated similarly to complete the proof of the theorem.

THEOREM 8.19   *If in a plane π of an $\mathbf{M}_{PY}$ space, p, q are distinct points of $\mathscr{E}(r, r')$, $r, r' \in \pi$ $(r \neq r')$, then $G(p, q)$ intersects $G(r, r')$ at $m = m(r, r')$ and $G(p, q) \subset \mathscr{E}(r, r')$.*

Proof.   The desired conclusion is obtained from the preceding lemma when it is shown that $G(p, q)$ intersects $G(r, r')$.

Suppose $G(p, q)$ is parallel to $G(r, r')$. Since $pr = pr'$ and $qr = qr'$, the isosceles triangle property insures that the middle point m of r, r' is the foot of p on $G(r, r')$, as well as the foot of q on $G(r, r')$. Hence each of the lines $G(p, m)$, $G(q, m)$ is perpendicular to $G(r, r')$, for since $G(p, q)$ is assumed to be parallel to $G(r, r')$, then $p \neq m \neq q$.

Then, by Theorem 8.17, each line $G(p, m)$, $G(q, m)$ is perpendicular to $G(p, q)$, and consequently each of the distinct points p, q is the foot on $G(p, q)$ of the point m. This contradiction proves that $G(p, q)$ intersects $G(r, r')$ and hence is contained in $\mathscr{E}(r, r')$. It follows that the point in which $G(p, q)$ intersects $G(r, r')$ is the middle point of r, r'.

THEOREM 8.20   *If in a plane π of an $\mathbf{M}_{PY}$ space, $p, q \in \mathscr{E}(r, r')$, with $p \neq q$, $r \neq r'$ $(p, q, r, r' \in \pi)$, then $\mathscr{E}(r, r') = G(p, q)$.*

Proof.   According to the preceding theorem, $G(p, q) \subset \mathscr{E}(r, r')$, and $G(p, q)$ intersects $G(r, r')$ at $m = m(r, r')$. If $p^* \in \mathscr{E}(r, r')$ and $p^* \notin G(p, q)$, then $G(p^*, p) \subset \mathscr{E}(r, r')$ and also intersects $G(r, r')$ at m. But then it coincides with $G(p, q)$.

COROLLARY   *In a plane π of an $\mathbf{M}_{PY}$ space, the perpendicular bisector of a segment is the locus of points of π equidistant from the end points of the segment.*

Remark.   The results we have obtained concerning planes of an $\mathbf{M}_{PY}$ space (in particular, the important property of the equidistant locus proved in Theorem 8.20) might be combined with theorems established by Busemann to show that each plane π of a *finitely compact* $\mathbf{M}_{PY}$ space is euclidean. From this it would not be difficult to show that the whole space is euclidean. We shall, however, pursue a different route to that end.

8.5   REFLECTIONS AND THE PYTHAGOREAN PROPERTY

To exhibit the euclidean nature of $\mathbf{M}_{PY}$ spaces we define in this section a class of mappings (*reflections in a line*) of a plane onto itself, which we show to be one-to-one and distance-preserving (that is, congruences). These mappings are used to prove the validity of the pythagorean right triangle theorem.

DEFINITION   A point p' of an $\mathbf{M}_{PY}$ space is a reflection of a point p in a line G of the space provided the foot f_p of p on G is the middle point of p and p'. The *reflection mapping* R_G of a plane π in a line G of π is the mapping which maps each point of π onto its reflection in G; that is, $p \in \pi$ implies $p' = R_G(p)$, the reflection of p in G.

THEOREM 8.21   *The reflection R_G of a plane onto itself is a congruence.*

Proof.   The reflection R_G is clearly a one-to-one "onto" mapping. To show it is a congruence, first let $p' = R_G(p)$, $q' = R_G(q)$, where p, q are distinct points of π, not on G. We wish to prove $pq = p'q'$.

Case I.   *The line $G(p, q)$ intersects G.* Let r denote the point of intersection of G and $G(p, q)$. Since $pf_p p'$ holds, with $pf_p = f_p p'$, and (by a previous theorem) f_p is also the foot of r on $G(p, p')$, the weak pythagorean property gives $rp = rp'$. Similarly, it is seen that $rq = rq'$.

Let q'' be that point of $G(r, p')$ such that $r, p', q'' \approx r, p, q$; that is, $rq'' = rq$ and $p'q'' = pq$. Applying Theorem 7.2 to the triples r, p, p' and r, q, q'', it is seen that G intersects $G(q, q'')$ in the foot of r on $G(q, q'')$. But that intersection point is then the foot of q on G, and consequently $q'' = q'$, so $pq = p'q'' = p'q'$.

(In case r is the foot of p on G, then each of the triples r, p, p' and r, p', q' is linear and the desired equality $pq = p'q'$ follows at once from the equalities $rp = rp'$ and $rq = rq'$.)

Case II.   *The line $G(p, q)$ does not intersect G.* Each of the lines $G(p, p')$, $G(q, q')$ is perpendicular to G, and hence the two lines are mutually parallel (Theorem 8.16). Also, each of the lines $G(p, q)$, $G(p', q')$ is parallel to G; for $G(p, q)$ is parallel to G by the hypothesis of Case II, and applying Theorem 7.21 to the parallelograms with vertices p, q, f_p, f_q, we conclude that $pf_p = qf_q$, where f_p, f_q are the feet on G of p, q, respectively. The parallel to G through p' intersects $G(q', f_q)$ in a point q'', with $f_q q'' = f_p p' = f_p p = f_q q = f_q q'$, and consequently $q'' = q'$.

Thus the quadrilateral with vertices p, q, p', q' is a parallelogram and hence $pq = p'q'$, completing the proof of the theorem.

(The case in which both points $p, q \in G$ is trivial, and if exactly one of the points, say p, belongs to G, then $p = p'$ and $pq = pq' = p'q'$ by a direct application of the weak pythagorean property.)

Remark.   The following are easy consequences of Theorem 8.21:

(i) if $p, q \in \pi$ $(p \neq q)$ and $G \subset \pi$, $p' = R_G(p)$, $q' = R_G(q)$, then $G(p', q') = R_G(G(p, q))$,

(ii) if r is a point of π, $r \notin G(p, q)$, and f denotes the foot of r on $G(p, q)$, then $f' = R_G(f)$ is the foot of $r' = R_G(r)$ on $G'(p, q) = R_G(G(p, q))$.

DEFINITION   A space has the *pythagorean property* provided for each three of its points p, q, r, with p a foot of q on the metric line $G(p, r)$, $(pq)^2 + (pr)^2 = (qr)^2$.

LEMMA 8.6   *Let p, q, r be points of an* $\mathbf{M}_{PY}$ *space, with $p \neq q \neq r \neq p$. If p is the foot of q on $G(p, r)$, and f_p is the foot of p on $G(q, r)$, then the relation $qf_p r$ holds.*

Proof. The points q, r, f_p are pairwise distinct; for if $r = f_p$, then r is the foot of q on $G(p, r)$ and so $p = r$; and if $q = f_p$, then $G(p, q)$ is perpendicular to $G(p, r)$ and to $G(q, r)$. This implies that $G(p, r)$ and $G(q, r)$ are mutually parallel, which is clearly not the case.

If $pq = pr$, then by the isosceles triangle property, the foot f_p of p on $G(q, r)$ is the middle point $m = m(q, r)$, and consequently $qf_p r$ surely subsists.

If $pq \neq pr$, let the labeling be so chosen that $pr > pq$. Then, by the monotone property (for example, Theorem 8.13), qrf_p cannot hold, and so rqf_p or $qf_p r$ holds.

Now f_p is the foot of r on $G(p, f_p)$, and so $rf_p < rp$. But since p is the foot of r on $G(p, q)$, then $rp < qr$. Hence $rf_p < qr$, which eliminates rqf_p and establishes $qf_p r$.

THEOREM 8.22   *Each* $\mathbf{M}_{PY}$ *space has the pythagorean property.*

Proof.   Let p denote the foot of q on $G(p, r)$ and f the foot of p on $G(q, r)$. By the preceding lemma, qfr holds. We shall show that

$$\frac{qr}{rp} = \frac{rp}{rf} \quad \text{and} \quad \frac{qr}{qp} = \frac{qp}{qf}.$$

Since f is the foot of p on $G(q, r)$, then f is the foot of r on $G(p, f)$, and so $rf < rp$. Hence S_r^p contains a point f' with $rf' = rf$. According to the isosceles triangle property, the middle point $m = m(f, f')$ is the foot of r on $G(f, f')$.

By reflection in $G(m, r)$, the points f, p, r map onto f', p', r, respectively, where $p' \in G(q, r)$ with $rp' = rp$. By (ii) of the Remark following Theorem 8.21, f' is the foot on $G(r, f') = G(p, r)$ of p' (since f is the foot on $G(f, r) = G(q, r)$ of p). Hence the lines $G(p, q)$, $G(p', f')$ are mutually parallel and $(rf'/rp) = (rp'/rq)$. Since $rp' = rp$ and $rf' = rf$, we obtain $(pr)^2 = rf \cdot qr$.

Interchanging the labels of q and r yields $(pq)^2 = qf \cdot qr$, and addition gives

$$(pq)^2 + (pr)^2 = qr \cdot (qf + fr) = (qr)^2,$$

completing the proof of the theorem.

8.6   METRIC AXIOMATIZATION OF EUCLIDEAN SPACES OF FINITE OR INFINITE DIMENSIONS

In this section we establish the euclidean nature of an $\mathbf{M}_{PY}$ space, and accomplish our objective of providing a new, purely metric axiomatization of euclidean geometry.

DEFINITION   A metric space **M** is a *euclidean number line* **R** provided there is a one-to-one correspondence γ between the elements of **M** and the set R of all real numbers such that if p, $q \in$ **M** and $\gamma(p)$, $\gamma(q) \in R$, then $pq = |\gamma(p) - \gamma(q)|$.

Remark.   Since each (metric) line of an **M**$_{PY}$ space is congruent with the euclidean straight line **E**$_1$, each line of an **M**$_{PY}$ space is a euclidean number line.

DEFINITION   A metric space **M** is a *euclidean number plane* provided there is a one-to-one correspondence Γ between the points of **M** and the elements of the set of all ordered pairs of real numbers such that if p, $q \in$ **M** and $\Gamma(p) = (x_p, y_p)$, $\Gamma(q) = (x_q, y_q)$, then

$$pq = [(x_p - x_q)^2 + (y_p - y_q)^2]^{1/2}.$$

THEOREM 8.23   *Each plane π of an* **M**$_{PY}$ *space is a euclidean number plane.*

Proof.   Let the plane π of **M**$_{PY}$ be determined by the point p and the line X ($p \notin X$), let o denote the foot of p on X, and put $Y = G(o, p)$. Since X and Y are each euclidean number lines, we may suppose that their common point o corresponds to the number 0 in each of the correspondences γ, γ^* between their points and the set of real numbers; that is, if $\gamma(X) = \mathbf{R}$ and $\gamma^*(Y) = \mathbf{R}$, then $\gamma(o) = \gamma^*(o) = 0$.

If $q \in X$ with $oq = 1$, then, since $oq = |0 - \gamma(q)|$, it follows that $\gamma(q) = 1$ or $\gamma(q) = -1$. If $\gamma(q) = 1$, then all points x of X such that oxq or oqx holds (or $x = q$) have $\gamma(x) > 0$, and if xoq holds, then $\gamma(x) < 0$. Similar remarks are applicable to the line Y, and so each of these lines has a nonnegative and a nonpositive "half," say X_+, X_- and Y_+, Y_-. Now if $q \in X_+$ and $q^* \in Y_+$ with $oq = oq^*$, we may suppose that $\gamma(q) = \gamma^*(q^*)$.

Let s be any point of π and let f_s, f_s^* denote the feet of s on X and Y, respectively. Associate with point s the ordered pair of real numbers (x_s, y_s), where $x_s = \gamma(f_s)$ and $y_s = \gamma^*(f_s^*)$. Since a point of π has a unique foot on a line, the *coordinates* (x_s, y_s) are uniquely determined by point s. Further, each pair of real numbers (x, y) are the coordinates of a unique point of π; for if f, f^* are the points of X, Y, respectively, with $x = \gamma(f)$, $y = \gamma^*(f^*)$, the lines perpendicular to X and Y at f and f^*, respectively, intersect (why?), the point of intersection has coordinates (x, y) and is, moreover, the only point with that property (why?). (Let the reader show how to construct a line perpendicular to a given line and through a given point of that line.)

If s, $t \in \pi$ and (x_s, y_s), (x_t, y_t) are their coordinates, respectively, we show that

$$st = [(x_s - x_t)^2 + (y_s - y_t)^2]^{1/2}.$$

The equality obviously holds if $s = t$ (for then $x_s = x_t$ and $y_s = y_t$); it also holds if $x_s = x_t \neq 0$, for then $f_s = f_t$ and s, t, f_s^*, f_t^* are vertices of a parallelogram whose opposite sides are equal (Theorem 7.21), and so $st = f_s^* f_t^* = |y_s - y_t|$, by the congruence γ^*; if $x_s = x_t = 0$, then $s = f_s^*$, $t = f_t^*$ and $st = |y_s - y_t|$. The desired equality is established by a similar argument in case $y_s = y_t$.

Suppose now that $x_s \neq x_t$ and $y_s \neq y_t$. If neither s nor t are points of X or Y, and g_s denotes the foot of s on $G(t, f_t)$, with g_t the foot of t on $G(s, f_s)$, then g_s is the foot of t on $G(s, g_s)$, and the pythagorean property gives $(st)^2 = (tg_s)^2 + (sg_s)^2$. But $tg_s = f_t^* f_s^* = |y_t - y_s|$ and $sg_s = f_s f_t = |x_s - x_t|$. Hence

$$st = [(x_s - x_t)^2 + (y_t - y_s)^2]^{1/2}.$$

Finally, if s or t belongs to X or Y, the desired equality is obtained at once from the pythagorean property. The proof of the theorem is complete.

We have now established that each $\mathbf{M}_{PY}$ space is a complete, normed, linear space (Banach space) with unique metric lines, and every plane of $\mathbf{M}_{PY}$ is a euclidean number plane. A variety of methods are now available to prove that every k-dimensional subspace of $\mathbf{M}_{PY}$ is a euclidean k-dimensional number space, and hence that $\mathbf{M}_{PY}$ is *generalized euclidean* (that is, euclidean of finite or infinite dimension). The reader will observe that none of the postulates of an $\mathbf{M}_{PY}$ space implies finite dimensionality of the space.

DEFINITION   A *k-dimensional subspace* of an $\mathbf{M}_{PY}$ space is the topological closure of the set of all points of $\mathbf{M}_{PY}$ that are collinear with a point of a $(k-1)$-dimensional subspace of $\mathbf{M}_{PY}$ and a point not belonging to it. A point is a zero-dimensional subspace.

Remark.   The lines and planes of an $\mathbf{M}_{PY}$ space are its one-dimensional and two-dimensional subspaces, respectively.

THEOREM 8.24   *For every positive integer n, each n-dimensional subspace of an* $\mathbf{M}_{PY}$ *space is congruent with euclidean space* $\mathbf{E}_n$ *of n dimensions.*

Proof. The theorem has been established for $n = 1, 2$. We assume its validity for $n = k$.

Let π_k denote any k-dimensional subspace, p_0 a point of $\mathbf{M}_{PY}$ not belonging to it, and $\pi_{k+1} = \{p_0, \pi_k\}$ the $(k + 1)$-dimensional subspace they determine. If f_0 is the foot of p_0 on π_k (let the reader prove the existence of f_0), let p_0' be a point and $\mathbf{E}_k$ a k-dimensional subspace of a euclidean $(k + 1)$-dimensional space $\mathbf{E}_{k+1}$, with $p_0 f_0 = p_0' f_0'$, where f_0' denotes the foot of p_0' on $\mathbf{E}_k$.

Consider now a congruence γ of π_k with $\mathbf{E}_k$ such that $\gamma(f_0) = f_0'$. This congruence, together with the association of p_0 with p_0', is extended to

$$\{p_0\} + \pi_k \approx \{p_0'\} + \mathbf{E}_k$$

by direct application of the pythagorean property; for if $q \in \pi_k$ $(q \neq f_0)$, and $q' = \gamma(q)$, then f_0 is the foot of p_0 on $G(f_0, q)$, f_0' is the foot of p_0' on $G(f_0', q')$, and, since $p_0 f_0 = p_0' f_0'$, $f_0 q = f_0' q'$ and the pythagorean property is valid in both π_{k+1} and $\mathbf{E}_{k+1}$, then $p_0 q = p_0' q'$.

If $x \in \pi_{k+1}$, $x \notin \pi_k$, $x \neq p_0$, then either x is collinear with p_0 and a point q of π_k or $x = \lim x_n$, where x_n is collinear with p_0 and a point q_n of π_k, $n = 1, 2, \ldots$. If the first alternative holds, associate with x that unique point x' of $\mathbf{E}_{k+1}$ such that $p_0, q, x \approx p_0', q', x'$, where $q' = \gamma(q)$; that is, x' has the same distances from p_0', q' that x has from p_0, q, respectively, and if the second alternative holds, then define x_n' by $p_0, q_n, x_n \approx p_0', q_n', x_n'$, where $q_n' = \gamma(q_n)$, $n = 1, 2, \ldots$. The planes $\pi(p_0, q_i, q_j)$, $\mathbf{E}_2(p_0', q_i', q_j')$ are both euclidean and since $p_0, q_i, q_j \approx p_0', q_i', q_j'$; $p_0, q_i, x_i \approx p_0', q_i', x_i'$; $p_0, q_j, x_j \approx p_0', q_j', x_j'$, it follows that $x_i x_j = x_i' x_j'$ $(i, j = 1, 2, \ldots, i \neq j)$.

Hence $\{x_n'\}$ $(n = 1, 2, \ldots)$ is a Cauchy sequence of points of $\mathbf{E}_{k+1}$ and consequently has a limit x', which we associate with the point x of π_{k+1}. This point x' is easily seen to be independent of the choice of the sequence $\{x_n\}$ with $x = \lim x_n$.

The mapping of π_{k+1} into $\mathbf{E}_{k+1}$ so defined is a *congruence* (that is, one-to-one, and distance-preserving) and *onto*, which establishes the theorem by complete induction. (Let the reader verify this.)

The possibility of infinite dimensionality of $\mathbf{M}_{PY}$ can be eliminated by replacing the completeness condition of $\mathbf{M}_{PY}$ by the stronger one of finite compactness, which assumes that every *bounded* infinite subset of $\mathbf{M}_{PY}$ has a nonnull derived set. It is easily shown that finite compactness implies completeness. The preceding developments are summarized as follows.

Metric Postulates for Banach Spaces with Unique Lines

An abstract set S is a Banach space with unique metric lines provided the following postulates are satisfied.

POSTULATE 1   *A metric space* $\mathbf{M}$ *is defined over* S; that is, there is a mapping μ of ordered pairs of elements of S into the set of nonnegative real numbers such that (i) if $a, b \in S$, $\mu(a, b) = 0$ if and only if $a = b$, (ii) $\mu(a, b) = \mu(b, a)$, and (iii) if $a, b, c \in S$,

$$\mu(a, b) + \mu(b, c) \geq \mu(a, c).$$

POSTULATE 2   *The metric space* $\mathbf{M}$ *is metrically complete;* that is, if $p_1, p_2, \ldots, p_n, \ldots$ is any infinite sequence of elements of S such that $\lim_{i, j \to \infty} \mu(p_i, p_j) = 0$, then S contains an element p such that $\lim_{i \to \infty} \mu(p, p_i) = 0$.

POSTULATE 3   *The metric space* $\mathbf{M}$ *is metrically convex and externally convex;* that is, if $a, c \in S$, $a \neq c$, then $b \in S$ exists such that $\mu(a, b) + \mu(b, c) = \mu(a, c)$, with $a \neq b \neq c$, and S contains at least one element d such that $\mu(a, c) + \mu(c, d) = \mu(a, d)$, with $c \neq d$.

POSTULATE 4   *The metric space* $\mathbf{M}$ *has the two-triple property;* that is, if any four pairwise distinct elements of S contain two triples that are congruent with triples of the euclidean straight line, then the remaining two triples of the quadruple have that property also.

POSTULATE 5   *The metric space* $\mathbf{M}$ *has the Young property;* that is, if p, q, r are any three elements of S which are not congruent with points of a euclidean straight line, and $m(p, q)$, $m(p, r)$ are middle points of p, q and p, r, respectively; that is,

$$\mu(p, m(p, q)) = \mu(m(p, q), q) = (\tfrac{1}{2}) \cdot \mu(p, q),$$
$$\mu(p, m(p, r)) = \mu(m(p, r), r) = (\tfrac{1}{2}) \cdot \mu(p, r),$$

then

$$\mu(m(p, q), m(p, r)) = (\tfrac{1}{2}) \cdot \mu(q, r).$$

Metric Postulates for Euclidean Spaces of Finite or Infinite Dimensions

Adjoin to the preceding five postulates 1, 2, 3, 4, 5, *either* of the following two postulates.

POSTULATE 6 *The metric space* **M** *has the weak pythagorean property;* that is, if f_p is a foot of p on a metric line G of **M** $(p \notin G)$, and $q, r \in G$ with $\mu(q, f_p) = \mu(r, f_p)$, then $\mu(s, q) = \mu(s, r)$ for each element s of the metric line $G(p, f_p)$ joining p and f_p. (An element f_p is a *foot* of an element p on a line G provided $\mu(p, f_p) \leq \mu(p, x)$ for every element x of G.)

POSTULATE 6* *The metric space* **M** *has the isosceles triangle property;* that is, if $p, q, r \in S$ $(q \neq r)$, with $\mu(p, q) = \mu(p, r) > 0$, and $p \notin G(q, r)$, then the middle point $m(q, r)$ of q, r is the foot on $G(q, r)$ of every point of the line $G(p, m(q, r))$.

A set of metric postulates for euclidean spaces of *finite dimensions* is formed by Postulates 1, 2*, 3, 4, 5, 6, *or* 6*, where Postulate 2* is the following strengthening of Postulate 2.

POSTULATE 2* *The metric space* **M** *is finitely compact;* that is, each bounded, infinite subset of **M** has a nonnull derived set.

A system of metric postulates for a euclidean space of *given dimension n* is given by Postulates 1, 2, 3, 4, 5, 6 (*or* 6*), and 7.

POSTULATE 7 *The metric space* **M** *contains an equilateral set of* $n + 1$ *points, but does not contain an equilateral set of* $n + 2$ *points.*

For if **M** does not contain an equilateral $(n + 2)$-tuple, then **M** does not contain an $(n + 1)$-dimensional subspace (since by Theorem 7.24 such a subspace is congruent to $\mathbf{E}_{n+1}$, which contains an equilateral $(n + 2)$-tuple).

On the other hand, if **M** contains an equilateral $(n + 1)$-tuple, then **M** contains an n-dimensional subspace, for in the contrary case each subspace of **M** has dimension *less than n*, and, being congruent with a euclidean space of the same dimension, does not contain an equilateral $(n + 1)$-tuple.

It follows that Postulate 7 assures that **M** is n-dimensional and hence congruent with $\mathbf{E}_n$.

REFERENCES

Sections 8.1 to 8.5 E. Z. Andalafte and L. M. Blumenthal, "Metric characterizations of Banach and euclidean spaces," *Fund. Math.*, **55**: 23–55 (1964).

Integral Geometry of Metric Arcs

9.1 INTRODUCTORY REMARKS.
THE n-LATTICE THEOREM

An important equivalence class of figures of a metric space $\mathbf{M}$ (with respect to homeomorphisms) is the class $\{A\}$ of *arcs* of the space. An arc A is any subset of M that is homeomorphic with a closed interval or line segment $I = [r_1, r_2]$, $r_1 < r_2$; that is, I is the metric space of all real numbers x such that $r_1 \leqq x \leqq r_2$, with $xy = |x - y|$. Since a homeomorphism between any two line segments $I = [r_1, r_2]$, $r_1 < r_2$, and $I^* = [r_1^*, r_2^*]$, $r_1^* < r_2^*$, is easily established, each arc is homeomorphic with the line segment $I_0 = [0, 1]$; that is, A is an arc if and only if there exists a mapping f, biuniform and bicontinuous in I_0, such that $A = f(I_0)$.

The two points $f(0) = a$, $f(1) = b$ are called the *end points* of $A = A(a, b)$, and a finite subset $P = \{p_1, p_2, \ldots, p_n\}$ of A is called *normally ordered* (the points of P are said to be encountered in the order of their subscripts when the arc is traversed from a to b) provided $f^{-1}(p_i) < f^{-1}(p_{i+1})$ for $(i = 1, 2, \ldots, n - 1)$. The utility of this concept comes from its being independent of the particular homeomorphism f; for it is not difficult to show that if $p_1, p_2, \ldots, p_n$ are normally ordered points of A when $A = f(I_0)$, then they are also normally ordered points of A if $A = g(I_0)$, where g is any homeomorphism with $g(0) = a = f(0)$ and $g(1) = b = f(1)$, since the natural order of the points of the closed interval $[0, 1]$ is preserved by any homeomorphism of that interval with itself that keeps each end point fixed.

While the definition of an arc as a homeomorph of a line segment is useful whenever it is desirable to parameterize the arc, the property of being an arc is an *intrinsic* (topological) property and hence the property should be expressible without the use of an intermediary, extraneous, line segment. The question arises: what topological properties of a subset of a metric space are necessary and sufficient in order that the subset be an arc?

It is clear that each line segment is a continuum, and each point of the segment, different from the two end points, is a cut-point (see Section 12.6). Since these properties are topological, it follows that every arc possesses them; moreover, it turns out that these rather obvious necessary properties are also sufficient: that is, *a subset of a metric space is an arc if and only if it is a continuum with every point a cut-point except two*. A proof of this interesting topological characterization of an arc is given in Section 12.6. In that characterization, the end points of an arc are defined as the two points of the arc that are not cut-points.

A normally ordered subset $P = \{p_0, p_1, \ldots, p_n\}$ of a metric arc $A(a, b)$ forms an *n-lattice* L_n of the arc provided $a = p_0$, $b = p_n$, and $p_0 p_1 = p_1 p_2 = \cdots = p_{n-1} p_n = \lambda(n)$. The number $\lambda(n)$ is called the *side* of the lattice. Although it is, of course, uniquely determined by the points or vertices $p_0, p_1, \ldots, p_n$ of L_n, it is not a single-valued function of the natural number n. An *n*-lattice is said to be *homogeneous* provided $p_i p_j \geq \lambda(n)$ whenever $|i - j| > 1$ $(i, j = 0, 1, \ldots, n)$.

THEOREM 9.1 (Schoenberg)   *If $A(a, b)$ is any arc of a metric space, and n is any natural number, $A(a, b)$ contains an n-lattice L_n.*

Proof.   Let $A(a, b) = f(I_0)$, where $I_0 = [0, 1]$, and f is a homeomorphism, with $f(0) = a, f(1) = b$.

Putting $p_s = f(s)$, $p_t = f(t)$ $(s, t \in I_0, p_s, p_t \in A(a, b))$, the function $F(s, t) = p_s p_t$ is nonnegative and continuous on I_0, vanishes if and only if $s = t$, and $F(0, 1) = p_0 p_1 = f(0)f(1) = ab$.

Corresponding to each n-tuple $(x_1, x_2, \ldots, x_n)$ of positive real numbers x_i $(i = 1, 2, \ldots, n)$, with $\sum_{i=1}^{n} x_i = 1$ there is (1) a decomposition of the segment $[0, 1]$ into a sum of n abutting (nonoverlapping) subsegments $S_1, S_2, \ldots, S_n$ (numbered in the order in which they are encountered in traversing the segment $[0, 1]$ from 0 to 1) with lengths $x_1, x_2, \ldots, x_n$ respectively, and (2) a point X of n-dimensional euclidean space $\mathbf{E}_n$, with cartesian coordinates $(x_1, x_2, \ldots, x_n)$. Since the coordinates of X are all positive, and their sum is 1, it is clear that the locus of all such points is the interior of the $(n - 1)$-dimensional simplex Σ_{n-1} that is cut from the hyperplane $X_1 + X_2 + \cdots + X_n = 1$ by the n coordinate hyperplanes $X_i = 0$, $i = 1, 2, \ldots, n$.

We associate, now, with each point $Y = (y_1, y_2, \ldots, y_n)$ of Σ_{n-1} the point $Y' = (y_1', y_2', \ldots, y_n')$, where

$$y_1' = \frac{F(0, y_1)}{N},$$

$$y_2' = \frac{F(y_1, y_1 + y_2)}{N}, \qquad y_3' = \frac{F(y_1 + y_2, y_1 + y_2 + y_3)}{N}, \ldots,$$

$$y_n' = \frac{F(y_1 + y_2 + \cdots + y_{n-1}, 1)}{N},$$

where

$$N = F(0, y_1) + F(y_1, y_1 + y_2) + \cdots + F(y_1 + y_2 + \cdots + y_{n-1}, 1).$$

It is easily seen that the association just defined is a continuous mapping of Σ_{n-1} into itself that maps each k-dimensional "face" of Σ_{n-1} into itself, $k = 0, 1, \ldots, n - 2$, and hence maps each vertex onto itself. (Let the reader verify these properties of the mapping.)

It follows that the mapping is *onto*, and so each point of Σ_{n-1} is the image by the mapping of a point of Σ_{n-1}. Applying this to $(1/n, 1/n, \ldots 1/n)$, an *interior* point of Σ_{n-1}, there exists an *interior* point $(z_1, z_2, \ldots, z_n)$ of Σ_{n-1} such that

$$\frac{F(0, z_1)}{N} = \frac{F(z_1, z_1 + z_2)}{N} = \cdots = \frac{F(z_1 + z_2 + \cdots + z_{n-1}, 1)}{N} = \frac{1}{n},$$

and consequently

(*) $F(0, z_1) = F(z_1, z_1 + z_2) = \cdots = F(z_1 + z_2 + \cdots + z_{n-1}, 1).$

Since $(z_1, z_2, \ldots, z_n)$ is an interior point of Σ_{n-1}, $z_1 > 0$, $z_2 > 0$, $\ldots$, $z_n > 0$, $z_1 + z_2 + \cdots + z_n = 1$, and so there is a decomposition of $[0, 1]$ into the sum of n abutting (nonoverlapping) subsegments with lengths $z_1, z_2, \ldots, z_n$ (the subsegments ordered in the usual manner). Then by (∗) the points $a = p_0 = f(0)$, $p_1 = f(z_1)$, $p_2 = f(z_1 + z_2)$, $\ldots$, $p_{n-1} = f(z_1 + z_2 + \cdots + z_{n-1})$, $p_n = f(1) = b$, form an n-lattice of $A(a, b)$.

Remark. In the ingenious proof of the n-lattice theorem presented above we have purposely refrained from giving an argument to show that the mapping of Σ_{n-1} into itself, defined in the proof, is actually an *onto* mapping. Let the reader investigate this matter.

9.2   SPREAD OF A HOMEOMORPHISM. ARC LENGTH AS A RIEMANN INTEGRAL. RATIO OF ARC TO CHORD

The length $\ell(A)$ of a metric arc A is defined to be the *least upper bound* of the numbers

$$l(P) = \sum_{i=1}^{n-1} p_i p_{i+1}$$

for all normally ordered subsets $P = \{p_1, p_2, \ldots, p_n\}$ of A. The arc is *rectifiable* in case the least upper bound is finite, and *nonrectifiable* otherwise.

We show in this section, with the aid of the concept of the *spread of a homeomorphism*, that the lengths of the members of a large class of metric arcs can be expressed as Riemann integrals.

Let f denote a homeomorphism of $I_0 = [0, 1]$ with the metric arc A. If $x, y, t \in I_0$,

$$f^*(t) = \lim_{x, y \to t} \frac{f(x)f(y)}{xy}$$

is called the *spread* of f at t, for each point t of I_0 at which the (finite) limit exists.

Remark 1. If $f^*(t)$ exists for every $t \in I_0$, then the distance quotient $f(x)f(y)/xy$ is bounded in I_0.

For if the contrary be assumed, corresponding to each positive integer n, points x_n, y_n of I_0 exist such that $f(x_n)f(y_n)/x_n y_n > n$. Since $f(x_n)f(y_n) \leqq \operatorname{diam} A$, the distance quotient cannot be made arbitrarily large if a subsequence $\{x_{i_n} y_{i_n}\}$ of $\{x_n y_n\}$ is bounded away from zero, and so we may conclude that $\lim_{n \to \infty} x_n y_n = 0$. From the compactness of I_0, a point t_0 of I_0 exists with $\lim x_{i_n} = t_0 = \lim y_{i_n}$, and

$$f^*(t_0) = \lim_{n \to \infty} \frac{f(x_{i_n})f(y_{i_n})}{x_{i_n} y_{i_n}},$$

which is impossible, since the limit indicated in the right member of the equality does not exist, and $f^*(t_0)$ does exist, by hypothesis.

Remark 2.   If $f^*(t)$ exists for every $t \in I_0$, then the arc $A = f(I_0)$ is rectifiable.

By the preceding remark, there exists a positive constant K such that $f(x)f(y)/xy < K$ for all $x, y \in I_0$. Now if $p_1, p_2, \ldots, p_n$ is any normally ordered subset of A, and $x_1 < x_2 < \cdots < x_n$ are the points of I_0 with $p_i = f(x_i)$ $(i = 1, 2, \ldots, n)$, then $p_i p_{i+1}/x_i x_{i+1} < K$ $(i = 1, 2, \ldots, n-1)$, and

$$0 < \sum_{i=1}^{n-1} p_i p_{i+1} < K \cdot \sum_{i=1}^{n-1} x_i x_{i+1} \leqq K.$$

Hence l.u.b. $\ell(P)$, for all normally ordered subsets P of arc A exists and is less than or equal to K; that is $\ell(A) \leqq K$.

Remark 3.   The spread f^* is continuous wherever it exists.

Let E denote a subset of I_0, and suppose $f^*(t)$ exists at each point t of E. From the existence of f^* at t $(t \in E)$, it follows that to each $\varepsilon > 0$ there corresponds a $\delta(t, \varepsilon) > 0$ such that $xt < \delta(t, \varepsilon)$, $yt < \delta(t, \varepsilon)$ $(x, y \in E)$, imply

$$\left| \frac{f(x)f(y)}{xy} - f^*(t) \right| < \frac{\varepsilon}{2}.$$

To prove f^* continuous at t_0 $(t_0 \in E)$, let t' denote any point of E with $t_0 t' < \delta(t_0, \varepsilon)$, and select points x, y such that $xt_0 < \delta(t_0, \varepsilon)$, $yt_0 < \delta(t_0, \varepsilon)$, and $xt' < \delta(t', \varepsilon)$, $yt' < \delta(t', \varepsilon)$. If E does not contain points x, y, t', then

t_0 is an isolated point of E, and f^* is obviously continuous at t_0. Clearly

$$|f^*(t_0) - f^*(t')| = \left| f^*(t_0) - \frac{f(x)f(y)}{xy} + \frac{f(x)f(y)}{xy} - f^*(t') \right|$$

$$\leq \left| f^*(t_0) - \frac{f(x)f(y)}{xy} \right| + \left| \frac{f(x)f(y)}{xy} - f^*(t') \right|$$

$$< \varepsilon,$$

and f^* is continuous at t_0.

Remark 4. If $f^*(t)$ exists at each point t of a closed subset E of I_0, then the distance quotient $f(x)f(y)/xy$ converges uniformly to $f^*(t)$ as $x, y \to t$.

For, if the contrary be assumed, then an $\varepsilon_0 > 0$ exists, corresponding to which there is a sequence $\{t_n\}$ of points of E and sequences $\{x_n\}$, $\{y_n\}$ of points of I_0 with $x_n t_n < 1/n$, $y_n t_n < 1/n$ $(n = 1, 2, \ldots)$, and

$$\left| f^*(t_n) - \frac{f(x_n)f(y_n)}{x_n y_n} \right| > \varepsilon_0.$$

If $\{t_{i_n}\}$, $\{x_{i_n}\}$, $\{y_{i_n}\}$ are subsequences with $\lim_{n\to\infty} x_{i_n} = \lim_{n\to\infty} y_{i_n} = \lim_{n\to\infty} t_{i_n} = t_0$ $(t_0 \in E)$, then $\lim_{n\to\infty} f^*(t_{i_n}) = f^*(t_0)$, by Remark 3, but on the other hand,

$$f^*(t_0) = \lim_{x_{i_n}, y_{i_n} \to t_0} \left[\frac{f(x_{i_n})f(y_{i_n})}{x_{i_n} y_{i_n}} \right] \neq \lim_{n \to \infty} f^*(t_{i_n}),$$

a contradiction which establishes this remark.

Remark 5. Let f and g be two homeomorphisms of I_0 onto arc A, and suppose $f^*(t_0)$ exists, $t_0 \in I_0$. Then $g^*(t_0)$ exists if $\lim_{x, y \to t_0} g(x)g(x)/f(x)f(y)$ does, and if $f^*(t_0) \neq 0$ and $g^*(t_0)$ exists, then so does

$$\lim_{x, y \to t_0} \frac{g(x)g(y)}{f(x)f(y)}.$$

The proof of this Remark is immediate upon writing

$$\frac{g(x)g(y)}{xy} = \left[\frac{g(x)g(y)}{f(x)f(y)} \right] \cdot \left[\frac{f(x)f(y)}{xy} \right],$$

and $\lim_{x, y \to t_0} f(x)f(y)/xy = f^*(t_0)$.

THEOREM 9.2 *Let $A = f(I_0)$ be a metric arc. If $f^*(t)$ exists (finite) for every $t \in I_0$, then*

$$\ell(A) = \int_0^1 f^*(t)\, dt.$$

Proof. Let $P = \{p_0, p_1, \ldots, p_k\}$ be a normally ordered subset of A such that

$$\ell(P) = \sum_{i=1}^{k} p_{i-1} p_i > \ell(A) - \varepsilon \qquad (\varepsilon > 0),$$

and put $f^{-1}(p_i) = x_i$ $(i = 0, 1, \ldots, k)$. Then $x_0 < x_1 < \cdots < x_{k-1} < x_k$.

Clearly a normally ordered subset $Q = (q_0, q_1, \ldots, q_m)$ of A exists such that (1) $P \subset Q$, and (2)

$$\left| \sum_{i=1}^{m} f^*(y_{i-i}) \cdot (y_i - y_{i-1}) - \int_0^1 f^*(t)\, dt \right| < \varepsilon,$$

where $f^{-1}(q_i) = y_i$ $(i = 0, 1, \ldots, m)$, and $y_0 < y_1 < \cdots < y_{m-1} < y_m$. Since Q is a refinement of P, we have

$$\sum_{i=1}^{m} q_{i-1} q_i \geq \sum_{i=1}^{k} p_{i-1} p_i > \ell(A) - \varepsilon.$$

Now by Remark 4, there exists $\gamma > 0$ such that $x, y \in I_0$, and $xy < \gamma$ implies

$$\left| f^*(x) - \frac{f(x)f(y)}{xy} \right| < \varepsilon,$$

and since by Remark 3, f^* is continuous (and hence *uniformly* continuous) on I_0, there exists $\eta > 0$ such that $t_1, t_2 \in I_0$ and $t_1 t_2 < \eta$ implies

$$|f^*(t_1) - f^*(t_2)| < \varepsilon.$$

Select $\delta = \min(\gamma, \eta)$, and choose a normally ordered subset $R = \{r_1, r_2, \ldots, r_n\}$ of A such that (1) $Q \subset R$ and (2) $z_i z_{i+1} < \delta$ $(i = 0, 1, \ldots, n - 1)$, where $z_i = f^{-1}(r_i)$ $(i = 0, 1, \ldots, n)$. (Let the reader show that such a selection of R is possible.) Since $Q \subset R$,

$$(*) \qquad \sum_{i=1}^{n} r_{i-1} r_i > \ell(A) - \varepsilon,$$

and also

$$(**) \qquad \left| \sum_{i=1}^{n} f^*(z_{i-1}) \cdot (z_i - z_{i-1}) - \int_0^1 f^*(t)\, dt \right| < \varepsilon.$$

It follows easily that

(***)
$$\left| \sum_{i=1}^{n} f^*(z_{i-1}) \cdot (z_i - z_{i-1}) - \sum_{i=1}^{n} r_{i-1} r_i \right| < \varepsilon,$$

and hence (using (**) and (***)),

$$\left| \sum_{i=1}^{n} r_{i-1} r_i - \int_0^1 f^*(t)\, dt \right| < \varepsilon + \varepsilon.$$

Combining the preceding inequality with (*) gives

$$\left| \ell(A) - \int_0^1 f^*(t)\, dt \right| < 2\varepsilon + \varepsilon,$$

and since ε may be selected arbitrarily small (but positive), we conclude that

$$\ell(A) = \int_0^1 f^*(t)\, dt.$$

The arcs considered in classical differential geometry have the important property that as two points q, r of the arc approach a third point p of the arc (along the arc) the limit of the ratio of the length of the chord determined by the points (that is, the *distance qr*) to the length $\ell(A_q^r)$ of the subarc with end points q, r, is unity. Easy examples of arcs that do not have this property at a point p are readily found, and one is led to inquire *what metric arcs possess this property at every point*. It is clear that rectifiability of the curve in a neighborhood of each of its points (and hence global rectifiability of the arc) is a necessary condition.

Let A, then, denote a rectifiable metric arc of length s_0, and parameterize A according to arc length; that is, $A = f(I)$, where I denotes the closed interval $[0, s_0]$, and f is a homeomorphism from I to A.

If $s \in [0, s_0]$, the limit under investigation (if it exists) is

$$\lim_{x,\, y \to s} \frac{f(x)f(y)}{xy} = f^*(s).$$

Now, by the preceding theorem,

(†)
$$s_0 = l(A) = \int_0^{s_0} f^*(s)\, ds,$$

and since f^* is continuous wherever it exists and (in this application) $f^*(s) \leq 1$, $s \in [0, s_0]$, then (†) yields $f^*(s) \equiv 1$ in $[0, s_0]$. We have, therefore, established the following theorem.

THEOREM 8.3   *Let A be a metric arc. At each point of A the limit of the ratio of chord to arc is unity if and only if that limit exists at each point of A.*

It should be observed that "chord" refers here to the distance of a pair of points and *not* to a line segment joining the points, since such a segment may not exist. Also, the limit is taken as two distinct points of the arc approach a *third* point of the arc, and *not* as one point approaches another (fixed) point of the arc.

EXERCISES

1. Interpret the spread of a homeomorphism in the case of an arc of euclidean space E_3, with equations $x_i = x_i(t)$ $(i = 1, 2, 3)$, $0 \leq t \leq 1$.

2. If $x, y \in [0, 1]$, put $t = |x - y|$ and define the distance xy by

$$xy = t - \left(\frac{1}{6}\right)t^3 + \left(\frac{1}{24}\right)t^4 \sin\left(\frac{1}{t}\right), \qquad xy = 0 \text{ if } t = 0.$$

The resulting space is a metric arc. Find its length.

9.3   LATTICES AND ARC LENGTH

This section exhibits the length of a metric arc as the *limit* of the lengths of n-lattices inscribed in the arc. The following two lemmas are needed.

LEMMA 9.1   *If A is any metric arc and $\lambda(n)$ is the side of any n-lattice of A, then $\lim_{n \to \infty} \lambda(n) = 0$.*

Proof.   Writing $A = f[I_0]$, it follows from the uniform continuity of f in I_0 that to each $\varepsilon > 0$ there corresponds a $\delta > 0$ such that $p, q \in I_0$ and $pq < \delta$ implies $f(p)f(q) < \varepsilon$. Let N denote a natural number such that $N\delta > 1$. Then for every $n > N$ each n points of I_0 has at least one pair of consecutive points with distance less than δ. It follows that the side of each n-lattice of A $(n > N)$ is less than ε, and consequently $\lim_{n \to \infty} \lambda(n) = 0$.

We remark that if Q denotes a normally ordered subset of a metric arc A, there might not exist any n-lattice of A with length $n \cdot \lambda(n) \geq \ell(Q)$. For if

a, b, c are noncollinear points of the euclidean plane, and ac/bc is irrational, then no n-lattice of the arc

$$A(a, b) = \text{seg}[a, c] + \text{seg}[c, b]$$

contains the point c, and consequently, for every n-lattice of $A(a, b)$, $n \cdot \lambda(n) < ac + cb$, the length $\ell(Q)$ of the normally ordered subset $Q = \{a, c, b\}$ of A. This remark gives added point to the following lemma.

LEMMA 9.2 *Let C denote a constant and Q any normally ordered subset of a metric arc $A(a, b)$, with $\ell(Q) > C$. There exists a natural number N such that for every natural number $n > N$, every n-lattice with side $\lambda(n)$ has its length $n \cdot \lambda(n) > C$.*

Proof. Let $Q = \{q_0, q_1, \ldots, q_m\}$, and consider the point q_i, where i is an arbitrary but fixed one of the indices $0, 1, \ldots, m$.

Assertion. *To each $\sigma > 0$ there corresponds a positive integer N_i^* such that for every integer $n > N_i^*$, each n-lattice of $A(a, b)$ has a point on $A(q_i, b)$ (the subarc of $A(a, b)$ with end points q_i, b) with distance from q_i less than σ.*

If $i = 0$ and $q_0 = a$, then a is itself such a point, and b serves in case $i = m$ and $q_m = b$. In case $a \neq q_i \neq b$, consider the two subsets $A(a, q_i)$ and $A(q_i, b) - U(q_i; \sigma) \cdot A(q_i, b)$ of $A(a, b)$, where $U(q_i; \sigma)$ denotes the spherical neighborhood of q_i with radius σ. If $A(q_i, b) - U(q_i, \sigma) \cdot A(q_i, b) = \varnothing$, then point b again serves as the desired point, and if that set is not null, it has a positive distance δ from the set $A(a, q_i)$, and clearly $\delta \leqq \sigma$.

Now since $\lim_{n \to \infty} \lambda(n) = 0$, there exists a natural number N_i^* such that $\lambda(n) < \delta$ whenever $n > N_i^*$. Clearly, for each n-lattice L_n, each of the sets $L_n \cdot A(a, q_i)$ and $L_n \cdot [A(q_i, b) - U(q_i; \sigma) \cdot A(q_i, b)]$ is nonnull, and it follows that for $n > N_i^*$, every n-lattice of $A(a, b)$ contains a point on $A(q_i, b)$ with distance from q_i less than σ.

Similarly, $\sigma > 0$ implies the existence of a natural number N_i^{**} such that every n-lattice with $n > N_i^{**}$ has a point on $A(a, q_i)$ with distance from q_i less than σ $(i = 0, 1, \ldots, m)$.

Consequently, if $n > N(\sigma) = \max[N_0^*, N_0^{**}, N_1^*, N_1^{**}, \ldots, N_m^*, N_m^{**}]$, then every n-lattice L_n has points on each side of q_i with distances from q_i less than σ $(i = 0, 1, \ldots, m)$. Denote by $q_{j_i}^*$ the *first* point of L_n to the *right* of q_i with $q_i q_{j_i}^* < \sigma$, and by $q_{j_{i+1}}^{**}$ the *last* point of L_n to the *left* of q_{i+1} with

$q_{i+1}q_{j_{i+1}}^{**} < \sigma$. Then

$$q_i q_{i+1} \leqq q_i q_{j_i}^* + (\text{length of } L_n \text{ from } q_{j_i}^* \text{ to } q_{j_{i+1}}^{**}) + q_{j_{i+1}}^{***} q_{i+1},$$

and hence

$$q_i q_{i+1} < 2\sigma + (\text{length of } L_n \text{ from } q_{j_i}^* \text{ to } q_{j_{i+1}}^{**}).$$

It follows that for every $n > N(\sigma)$, and each n-lattice L_n,

$$n \cdot \lambda(n) + 2m\sigma > \ell(Q) > C.$$

Take $\sigma = [\ell(Q) - C]/2m$, and put $N = N(\sigma)$. Then for $n > N$, $n \cdot \lambda(n) + [\ell(Q) - C] > \ell(Q)$; that is, $n \cdot \lambda(n) > C$, and the proof of the lemma is complete.

We are now in position to prove the desired theorem.

THEOREM 9.4   *If A is any metric arc, then $\ell(A) = \lim_{n \to \infty} n \cdot \lambda(n)$.*

Proof. If $\ell(A)$ is finite, then $\varepsilon > 0$ implies the existence of a normally ordered subset Q of A such that $\ell(Q) > \ell(A) - \varepsilon$. Taking $\ell(A) - \varepsilon$ as the constant C of the preceding lemma a positive integer $N(\varepsilon)$ exists such that for every n-lattice L_n with $n > N(\varepsilon)$

$$\ell(A) \geqq n \cdot \lambda(n) > \ell(A) - \varepsilon,$$

and consequently $\ell(A) = \lim_{n \to \infty} n \cdot \lambda(n)$.

If $\ell(A)$ is infinite, then $C > 0$ implies the existence of a normally ordered subset Q of A with $\ell(Q) > C$. Then by the preceding lemma, a positive integer N exists such that for $n > N$, the length of every n-lattice of A exceeds C. If follows that $\lim_{n \to \infty} n \cdot \lambda(n) = \infty = \ell(A)$, completing the proof of the theorem.

9.4   THE PROBLEM OF UNIQUE n-LATTICES. THE OBTUSE ANGLE PROPERTY Ω

The planar arc pictured in Figure 13, the construction of which is evident, is an example of a rectifiable metric arc with the property that for each positive integer N there exists a positive integer n $(n > N)$, such that the arc contains two distinct n-lattices. For example,

$$a = p_0, p_1, p_2, p_3, p_4, p_5, p_6 = b \quad \text{and} \quad a = q_0, q_1, q_2, q_3, q_4, q_5, q_6 = b$$

are two distinct 6-lattices of the arc, and the reader will have no difficulty in showing that for $n = 6(2^k - 1)$ $(k = 1, 2, \ldots)$, the arc contains two distinct n-lattices, with $\lambda(n) = (\tfrac{1}{2})^{k-1}$. The length of the arc is clearly 12.

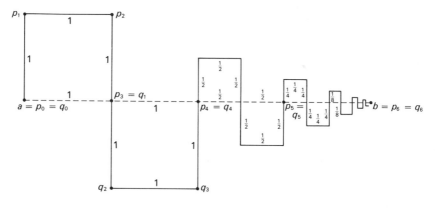

Figure 13

The problem arises: what conditions (*not too restrictive*) are to be imposed on a metric arc in order that "almost" all (that is, all but a finite number) of its n-lattices should be unique?

To solve this problem we define a property which also proves to be very useful in another connection.

A metric triple of pairwise distinct points p, q, r is said to *contain* an angle θ provided θ is one of the angles of a (euclidean) planar triangle whose vertices p', q', r' are congruent to p, q, r. We write $\angle pqr$ to denote the angle at vertex q' of the triangle p', q', r', where $p, q, r \approx p', q', r'$.

The obtuse angle property Ω. A metric continuum K has the obtuse angle property Ω at a point p of K provided a positive number δ_p exists such that each triple of pairwise distinct points of $K \cdot U(p; \delta_p)$ contains an obtuse angle, where

$$U(p; \delta_p) = [q \in \text{space} \mid pq < \delta_p].$$

Denoting a maximum angle contained in a metric triple r, s, t by $\max \angle \{r, s, t\}$, then K has property Ω at p $(p \in K)$ provided $\delta_p > 0$ exists such that $\max \angle \{r, s, t\} > \pi/2$ for every triple of pairwise distinct points r, s, t of K, with $\max\{pr, ps, pt\} < \delta_p$.

A metric continuum K has the obtuse angle property Ω provided K has property Ω at each of its points.

9.5 METRIC CONTINUA WITH THE OBTUSE ANGLE PROPERTY Ω

The uniform nature of the property Ω is established in the following lemma.

LEMMA 9.3 *If a metric continuum K has property Ω, then $\delta_0 > 0$ exists such that for every element p of K, $\max \measuredangle \{r, s, t\} > \pi/2$ for each triple r, s, t of pairwise distinct points of K with $\max\{pr, ps, pt\} < \delta_0$.*

Proof. If the contrary be assumed, then (since K is closed and compact) sequences $\{p_n\}, \{r_n\}, \{s_n\}, \{t_n\}$ of points of K exist, each with limit p_0 $(p_0 \in K)$, such that $r_n \neq s_n \neq t_n \neq r_n$, and $\max \measuredangle \{r_n, s_n, t_n\} \leqq \pi/2$ $(n = 1, 2, \ldots)$. But for n sufficiently large, the triple r_n, s_n, t_n is in the δ_{p_0}-neighborhood of p_0, and consequently $\max \measuredangle \{r_n, s_n, t_n\} > \pi/2$ (since K has the property Ω at p_0).

LEMMA 9.4 *If a metric continuum K has property Ω, then for each point p of K and each triple of its pairwise distinct points, r, s, t with $\max\{pr, ps, pt\} < \delta_0$ (where δ_0 is the constant whose existence is established in the preceding lemma) the equality $rs = st$ implies that $rt > rs$.*

Proof. According to the preceding lemma, $\max \measuredangle \{r, s, t\} > \pi/2$. If $r, s, t \approx r', s', t'$ (points of $\mathbf{E}_2$), then $r's' = s't'$, and so the maximum angle of triangle (r', s', t') has its vertex at s'. Since that angle is obtuse, then $rt = r't' > r's' = rs$.

LEMMA 9.5 *Let p denote an arbitrarily selected point of a metric continuum K with property Ω. For each positive number δ less than δ_0, the set S of all points q of K with $pq = \delta$ is finite.*

Proof. If $q_1, q_2 \in S$ $(q_1 \neq q_2)$, then $(q_1 q_2)^2 = 2 \cdot \delta^2 \cdot (1 - \cos \theta)$, where θ is the angle at the vertex p' of the isosceles planar triangle (q_1', p', q_2') such that $q_1, p, q_2 \approx q_1', p', q_2'$. Since, by Lemma 9.3, $\max \measuredangle \{p, q_1, q_2\} > \pi/2$, then $\theta = \max \measuredangle \{p, q_1, q_2\}$ and so $\cos \theta < 0$. Hence $q_1 q_2 > \delta \sqrt{2}$, and since S is compact (being a subset of K) it follows that S is a finite set (see the proof of Theorem 1.6).

THEOREM 9.5 *A metric continuum K with the obtuse angle property Ω is either an arc or a simple closed curve (that is, a homeomorph of a circle).*

Proof. It follows from the preceding lemma that each point p of K is contained in an open set $U(p)$ of arbitrarily small diameter whose product with K is a connected set (that is, K is a *locally connected* continuum). If the diameter of $U(p)$ is less than δ_0, then $K \cdot U(p)$ does not contain an equilateral triple of points, and it may be concluded that K is *locally* an arc.* The conclusion of the theorem readily follows.

Remark. If the obtuse angle property Ω be strengthened to require that for each point p of K a positive number δ_p^* exists such that for each three pairwise distinct points r, s, t of $K \cdot U(p; \delta_p^*)$, the max $\not\prec \{r, s, t\}$ is obtuse and *bounded away from* $\pi/2$ (that is, max $\not\prec \{r, s, t\} \geq \pi/2 + \alpha_p$ $(0 < \alpha_p \leq \pi/2))$, we say that K has the property Ω^*. Clearly, property Ω^* implies property Ω (but not conversely) and so a metric continuum with property Ω^* is either an arc or a simple closed curve (Theorem 9.5). Arcs with property Ω^* will be studied in a later section.

EXERCISE

1. Prove that if a metric continuum has property Ω or Ω^* at one of its points, then in a neighborhood of that point the continuum is an arc.

9.6   METRIC ARCS WITH PROPERTY Ω.
UNIQUENESS OF n-LATTICES

We recall (Section 9.1) that an n-lattice $p_0, p_1, \ldots, p_n$ of a metric arc A is *homogeneous* provided $p_i p_j \geq \lambda(n)$ whenever $|i - j| > 1$.

THEOREM 9.6   *If a metric arc A has the obtuse angle property Ω, then a positive integer N exists such that if $n > N$, every n-lattice of A is homogeneous.*

Proof. As shown in Lemma 9.1, the side $\lambda(n)$ of an n-lattice approaches zero as n approaches infinity, and so there exists a positive integer N such that for every $n > N$, and each n-lattice

$$L_n = \{p_0^n, p_1^n, \ldots, p_n^n\},$$

---

* A. N. Milgram, "*Some metric topological properties*," *Rep. Math. Colloquium*, Ser. II (Notre Dame), Issue 5, pp. 25– 35 (1944).

$\lambda(n) < \min(ab, \delta_0)$, where a, b are the end points of A and δ_0 is the constant of Lemma 9.3. The theorem is proved by showing that $p_i^n p_{i+j}^n \geqq \lambda(n)$ $(i, j = 0, 1, \ldots, n; 0 < i + j \leqq n)$.

If subscripts i, j exist such that for points p_i^n, p_{i+j}^n, the distance $p_i^n p_{i+j}^n < \lambda(n)$, then clearly $j > 1$ and the subarc $A(p_{i+1}^n, b)$ contains p_{i+j}^n. Let $f(p_i^n)$ denote a foot of p_i^n on that subarc.

Case 1. *The point $f(p_i^n)$ is distinct from b.* Since $p_i^n p_{i+1}^n = \lambda(n) > p_i^n p_{i+j}^n$, then $f(p_i^n) \neq p_{i+1}^n$ and

$$p_i^n f(p_i^n) \leqq p_i^n p_{i+1}^n = \lambda(n) < \min(a\,b, \delta_0).$$

For each positive integer k, choose points x_k, y_k with $x_k \in A(p_{i+1}^n, f(p_i^n))$, $y_k \in A(f(p_i^n), b)$ and $0 < x_k f(p_i^n) = y_k f(p_i^n) < 1/k$. If $x_k p_i^n \leqq y_k p_i^n$, put $x_k' = x_k$ and select $y_k' \in A(f(p_i^n), y_k)$ so that $y_k' p_i^n = x_k' p_i^n$. Similarly, if $x_k p_i^n > y_k p_i^n$, define $y_k' = y_k$ and choose a point x_k' of $A(x_k, f(p_i^n))$ with $x_k' p_i^n = y_k' p_i^n$. (Let the reader show that these selections are possible.) Hence two infinite sequences $\{x_k'\}$, $\{y_k'\}$ of points are obtained with $p_i^n x_k' = p_i^n y_k'$ $(k = 1, 2, \ldots)$, and

$$\lim_{k \to \infty} x_k' = f(p_i^n) = \lim_{k \to \infty} y_k'.$$

Clearly $\lim_{k \to \infty} x_k' y_k' = 0$, and since $p_i^n f(p_i^n) < \delta_0$, then for k sufficiently large, $p_i^n x_k' = p_i^n y_k' < \delta_0$, but $x_k' y_k' < p_i^n x_k'$, in violation of Lemma 9.4.

Case 2. *The point $f(p_i^n)$ coincides with b.* Now $p_i^n b \leqq p_i^n p_{i+j}^n < \lambda(n) < ab$, and so $p_i^n \neq a$ and $i \neq 0$. If $f(b)$ denotes a foot of b on $A(a, p_i^n)$, then

$$bf(p) \leqq bp_i^n \leqq p_i^n p_{i+j}^n < \lambda(n) < ab,$$

and so $f(b) \neq a$.

The procedure employed in Case 1 may be applied to establish the existence of points x, y of $A(a, b)$, on opposite sides of $f(b)$, with $bx = by < \delta_0$ and $xy < bx$, again contradicting Lemma 9.4. This completes the proof of the theorem.

If $x, y \in A(a, b)$, $x \neq y$, let $x \prec y$ denote that x is encountered before y in traversing the arc from a to b (that is, $x \prec y$ denotes that the ordered quadruple a, x, y, b is in normal order). If $p \in A(a, b)$ and $\delta > 0$, let

$$RU_A(p; \delta) = [q \mid q \in A \cdot U(p; \delta), p \prec q]$$

and

$$LU_A(p; \delta) = [q \mid q \in A \cdot U(p; \delta), q \prec p].$$

LEMMA 9.6   *If a metric arc $A(a, b)$ has the obtuse angle property Ω then positive numbers δ_R and δ_L exist such that for every point p of A,*

(i) $x, y \in RU_A(p; \delta_R)$, $x \prec y$, *imply* $px < py$, *and*

(ii) $x, y \in LU_A(p; \delta_L)$, $x \prec y$, *imply* $py < px$.

Proof.   If it be assumed that (i) is false, then an infinite sequence $\{\delta_n\}$ of positive numbers with $\lim_{n \to \infty} \delta_n = 0$ and an infinite sequence $\{p_n\}$ of points of $A(a, b)$ with $\lim_{n \to \infty} p_n = p_0$ ($p_0 \in A(a, b)$) exist such that for every positive integer n, each of the sets $RU_A(p; \delta_n)$ contains points x_n, y_n with $x_n \prec y_n$ but $p_n x_n \geqq p_n y_n$.

If $p_n x_n > p_n y_n$, let y_n' denote a point of $RU_A(p; \delta_n)$ such that $p_n \prec y_n' \prec x_n$ and $p_n y_n' = p_n y_n$; if $p_n x_n = p_n y_n$, put $y_n' = x_n$. Let m_n denote a point on $A(y_n', y_n)$ with maximum distance from p_n. In case $y_n' \neq x_n$, then m_n is an interior point of $A(y_n', y_n)$. If m_n is *not* an interior point of that arc (in which case $m_n = y_n' = x_n$ or $m_n = y_n$), let f_n denote a foot of p_n on $A(y_n', y_n)$. The sequence $\{m_n'\}$ is defined by putting $m_n' = m_n$ if m_n is interior to $A(y_n', y_n)$, and in the contrary case (a) $m_n' = f_n$, if f_n is interior to $A(y_n', y_n)$, and (b) if not, m_n is an arbitrarily chosen interior point of that subarc.

Now since $\delta_n \to 0$, $p_n \to p_0$, and y_n', y_n are points of $RU_A(p_n; \delta_n)$ for every positive integer n, it is clear that $\lim_{n \to \infty} y_n' = p_0 = \lim_{n \to \infty} y_n$. Although the points m_n' are not necessarily points of $RU_A(p_n; \delta_n)$ for *every* value of n, we show that for every open set containing p_0 there exists an index n such that the point m_n' associated with $RU_A(p_n; \delta_n)$ belongs to that open set.

To accomplish this, let $U(p_0)$ denote any open set containing p_0. Since $p_0 \in A$, an *arc*, there exists an open set $V(p_0)$ containing p_0 such that $\text{cl}[V(p_0)] \subset U(p_0)$ and the boundary $\text{Bd}[V(p_0)]$ contains at most one point x of A with $p_0 \prec x$. If, now, n be taken so that $U(p_n; \delta_n) \subset V(p_0)$, then the point m_n' associated with $RU_A(p_n; \delta_n)$ in the manner described above, cannot be exterior to $V(p_0)$ for, if it were, then $\text{Bd}[V(p_0)]$ would contain two points x, y of A with $p_0 \prec x$ and $p_0 \prec y$. Hence $m_n' \in \text{cl}[V(p_0)]$ and consequently $m_n' \in U(p_0)$.

Now choose n so that (i) $p_0 m_n' < \delta_0$, the constant of Lemma 9.3, and (ii) $U(p_n; \delta_n) \subset U(p_0; \delta_0)$. Then the points p_n, y_n, y_n', m_n' associated with $RU_A(p_n; \delta_n)$ belong to $U(p_0; \delta_0)$, and it is clear that distinct points r_n, t_n of $U(p_0; \delta_0)$ exist (arbitrarily close to m_n') such that $r_n \in A(y_n', m_n'), t_n \in A(m_n', y_n)$, and $p_n r_n = p_n t_n > r_n t_n$, in contradiction to Lemma 9.4.

A similar argument is used to establish part (ii) of the lemma.

The preceding lemma leads to the following theorem (let the reader supply the easy argument).

THEOREM 9.7 *If a metric arc A has the obtuse angle property Ω, then a positive constant ρ_0 exists such that*

(i) *each triple of pairwise distinct points of A with diameter less than ρ_0 contains an obtuse angle,*

(ii) *for every point p of A and every positive number $\rho < \rho_0$, the boundary of $U(p; \rho)$ contains at most one point x of A with $x \prec p$ and at most one point y of A with $p \prec y$,*

(iii) *if $p \in A$ and $x, y \in RU_A(p; \rho_0)$, then $x \prec y$ implies $px < py$, and if $x, y \in LU_A(p; \rho_0)$, then $x \prec y$ implies $py < px$,*

(iv) *if $x, y, z \in A$ with $x \prec y \prec z$ and $\operatorname{diam}\{x, y, z\} < \rho_0$, then $\not{\lessdot} xyz = \max \not{\lessdot} \{x, y, z\}$ (that is, the maximum angle contained in the triple of points has its vertex at point y).*

We are now in position to prove the principal theorem of this section.

THEOREM 9.8 *If a metric arc has property Ω, then "almost all" n-lattices are unique (that is, there exists a positive integer N such that if $n > N$, the arc contains exactly one n-lattice).*

Proof. Lemma 9.1 implies the existence of a positive integer N such that if $n > N$ the side $\lambda(n)$ of any n-lattice of $A(a, b)$ is less than the constant ρ_0 of Theorem 9.7. Assume that $A(a, b)$ contains two n-lattices $L_n^{(1)} = \{p_0, p_1, \ldots, p_n\}$ and $L_n^{(2)} = \{q_0, q_1, \ldots, q_n\}$ with $\lambda_1(n) \leqq \lambda_2(n)$, and $n > N$.

Case 1. $\lambda_1(n) < \lambda_2(n)$. Since $p_0 = a = q_0$ and $\lambda_1(n) < \lambda_2(n) < \rho_0$, the points p_1 and q_1 belong to $RU_A(a; \rho_0)$ and $ap_1 < aq_1$ implies $p_1 \prec q_1$ (Theorem 9.7). Now $p_1 q_2 > \lambda_2(n)$, for in the contrary case $p_1, q_1 \in LU_A(q_2; \rho_0)$ and $p_1 \prec q_1$ implies $q_2 p_1 > q_2 q_1 = \lambda_2(n)$. It follows that $p_2 \prec q_2$, for since $q_2 p_1 > \lambda_2(n) > \lambda_1(n)$, then surely $p_2 \neq q_2$, and if $q_2 \prec p_2$, then the interior of $A(p_1, q_2)$ contains a point p_2' such that $p_1 p_2' = \lambda_1(n) = p_1 p_2$. Since $p_2' \prec q_2$ and $q_2 \prec p_2$, then $p_2' \prec p_2$. But $p_2, p_2' \in RU_A(p_1; \rho_0)$, and consequently $p_2' \prec p_2$ implies $p_1 p_2' < p_1 p_2$.

The relation $p_{n-1} \prec q_{n-1}$ now follows by induction. But since $bp_{n-1} = p_n p_{n-1} = \lambda_1(n)$ and $bq_{n-1} = q_n q_{n-1} = \lambda_2(n)$, the points $p_{n-1}, q_{n-1} \in LU_A(b; \rho_0)$ and consequently $p_{n-1} \prec q_{n-1}$ implies $\lambda_1(n) = bp_{n-1} > bq_{n-1} = \lambda_2(n)$, contrary to the assumption of this case.

Case 2. $\lambda_1(n) = \lambda_2(n)$. Application of (iii), Theorem 9.7, gives $p_1 = q_1$, and repeated application yields $p_i = q_i$ $(i = 1, 2, \ldots, n)$; that is, $L_n^{(1)} = L_n^{(2)}$.

See Section 9.1 for the existence of at least one n-lattice for each positive integer n.

9.7 METRIC ARCS WITH PROPERTY Ω^*

We redeem a promise made in Section 9.5 and prove in this section an important consequence of assuming that a metric arc has property Ω^*.

LEMMA 9.7   *If a metric arc A has property Ω^*, positive constants δ_0^* and α_0 $(\alpha_0 \leqq \pi/2)$ exist such that $p \in A$ and $r, s, t \in A \cdot U(p; \delta_0^*)$, $r \neq s \neq t \neq r$, imply*

$$\max \,\, \star \,\, \{r, s, t\} \geqq \frac{\pi}{2} + \alpha_0 .$$

Proof.   The uniformity with respect to the numbers α_p, $p \in A$, is shown first. Assuming the contrary leads to the existence of a point p_0 of A, every spherical neighborhood of which contains pairwise distinct points r_n, s_n, t_n with $\max \,\, \star \,\, \{r_n, s_n, t_n\} \leqq \pi/2 + 1/n$, for n arbitrarily large. But then A does not have property Ω^* at p_0, contrary to the hypothesis.

The uniformity with respect to the numbers δ_p^*, $p \in A$, follows as in the proof of Lemma 9.3.

THEOREM 9.9   *Any metric arc $A(a, b)$ with property Ω^* is rectifiable.*

Proof.   Let $\sigma = (\frac{1}{2})\min\{\delta_0, \delta_0^*, \rho_0\}$, where δ_0, δ_0^*, ρ_0 are the constants of Lemmas 9.3 and 9.7 and of Theorem 9.7, respectively. There exists a finite subset $p_1, p_2, \ldots, p_m$ of A such that

$$A = A \cdot [U(p_1; \sigma) \cup U(p_2; \sigma) \cup \cdots \cup U(p_m; \sigma)].$$

(Let the reader justify this statement.) By (ii) of Theorem 9.7, for each $i = 1, 2, \ldots, m$, the boundary of $U(p_i; \sigma)$ contains at most one point a_i of A with $a_i \prec p_i$, and at most one point b_i of A with $p_i \prec b_i$.

Assertion.   *If p_j is any one of the points $p_1, p_2, \ldots, p_m$ such that b_j exists, the subarc $A(p_j, b_j)$ of $A(a, b)$ is rectifiable.*

If n is any positive integer greater than the integers N of Theorems 9.6 and 9.8, let $L_n = \{q_0, q_1, \ldots, q_n\}$ denote the unique n-lattice of the arc $A(p_j, b_j)$. This lattice is homogeneous. Let $x_k = p_j q_k$ ($k = 1, 2, \ldots, n$), and denote the angle $\not\prec p_j q_k q_{k+1}$ by β_k ($k = 1, 2, \ldots, n-1$).

Now q_k, $q_{k+1} \in U(p_j; \sigma)$ ($k = 0, 1, \ldots, n-2$), and $U(p_j; \sigma) \subset U(p_j; \rho_0)$ imply $x_k < x_{k+1}$, since $p_j \prec q_k \prec q_{k+1}$ if $k \neq 0$, and the inequality is obvious if $k = 0$. Moreover, since the lattice is homogeneous, $x_{k+1} \geq q_k q_{k+1} = \lambda(n)$, the side of L_n. But if $x_{k+1} = q_k q_{k+1}$ for $k \neq 0$, then, since $p_j, q_k, q_{k+1} \in U(p_j; \delta_0)$ and are pairwise distinct, it follows from Lemma 9.4 that $x_k > \lambda(n)$. Hence $x_{k+1} > \lambda(n)$ for $k \neq 0$, and consequently x_{k+1} is the greatest of the three distances determined by the points p_j, q_k, q_{k+1}. Then, since $\sigma < \delta_0^*$, $\beta_k = \not\prec p_j q_k q_{k+1} = \max \not\prec \{p_j, q_k, q_{k+1}\} \geq \pi/2 + \alpha_0$ ($k = 1, 2, \ldots, n-1$), by Lemma 9.7. Now $x_1 = \lambda(n) \geq \lambda(n) \sin \alpha_0$, and we make the inductive assumption that $x_k \geq k \cdot \lambda(n) \sin \alpha_0$, $k < n - 1$. The law of cosines yields

$$x_{k+1}^2 = x_k^2 + \lambda^2(n) - 2x_k \cdot \lambda(n) \cos \beta_k,$$

and since $\beta_k \geq \pi/2 + \alpha_0$, then $\cos \beta_k \leq -\sin \alpha_0$. Hence

$$x_{k+1}^2 \geq \lambda^2(n) \sin^2 \alpha_0 (k^2 + 2k + 1 + \cot^2 \alpha)$$
$$\geq (k+1)^2 \cdot \lambda^2(n) \sin^2 \alpha_0,$$

and so $x_{k+1} \geq (k+1) \cdot \lambda(n) \sin \alpha_0$.

It follows that $p_j b_j = p_j q_n = x_n \geq n \cdot \lambda(n) \sin \alpha_0$, and hence $n \cdot \lambda(n) \leq p_j b_j \csc \alpha_0 = C$, a constant, for every sufficiently large positive integer n. Use of Theorem 9.4 gives $\ell[A(p_j, b_j)] = \lim_{n \to \infty} n \cdot \lambda(n) \leq C$, and so the Assertion is proved.

It is proved in a similar manner that each of the arcs $A(a, p_j)$, $A(a_j, p_j)$, $A(p_m, b)$ is rectifiable, and it follows that $A(a, b)$ is rectifiable.

COROLLARY   *Any metric continuum with property Ω^* is either a rectifiable arc or a rectifiable simple closed curve.*

Proof.   Since property Ω^* implies property Ω, the corollary is a consequence of Theorems 9.5 and 9.9.

Remark.   The preceding development may be *localized* to prove that if a metric continuum has property Ω^* at a point p, then it is a rectifiable arc in a neighborhood of p.

REFERENCES

Section 9.1   I. J. Schoenberg, "On metric arcs of vanishing Menger curvature," *Ann. of Math.*, **41**: 715–726 (1940).

Section 9.2   The spread of a homeomorphism was first defined in W. A. Wilson, "On certain types of continuous transformations of metric spaces," *Amer. J. Math.*, **57**: 62–68 (1935).

Sections 9.4 to 9.7   The obtuse angle properties Ω and Ω^* were introduced in W. R. Abel and L. M. Blumenthal, "Distance geometry of metric arcs II," *Arch. Math.*, **8**: 451–464 (1957).

CHAPTER 10

Differential Geometry
of Metric Arcs

10.1 METRIZATIONS OF CURVATURE

A study of differential properties of metric continua without defining them by means of coordinates or equations was begun in 1930 by K. Menger. Such a program has two objectives: one is to obtain the results of classical geometry in a *purely metric manner* which will exhibit in all clarity those metric properties of euclidean space upon which their validity ultimately rests, and thus, perhaps, to free the classical development from the many geometrically extraneous conditions that the use of analysis imposes; the other objective is to use the metrizations of the basic concepts of the subject, which must be obtained before the metric method is operable, as a basis for extending those concepts to more general spaces. Thus, for example, the metric program applied to differential geometry, has two prongs: a *methodological* one concerned with a fresh (metric) derivation of classical results, in which all of the assumptions needed are expressed directly on the geometric entities involved and not, as in classical differential geometry, on the functions used to represent them; and a second prong that points in the direction of *generalization* and *unification*.

The first concept of classical differential geometry to be metrized is that of (linear) curvature.

If q, r, s are three pairwise distinct points of a metric continuum K, put

$$(*) \qquad k_M(q, r, s) = \frac{2(\sin \measuredangle qrs)}{qs}$$

(see Section 9.4 for definition of $\measuredangle qrs$), and refer to $k_M(q, r, s)$ as the curvature of the point-triple $\{q, r, s\}$. Since $\cos \measuredangle qrs = (qr^2 + rs^2 - qs^2)/2qr \cdot rs$, an easy computation yields

$$k_M(q, r, s) = \frac{[-D(q, r, s)]^{1/2}}{qr \cdot qs \cdot rs},$$

where

$$D(q, r, s) = \begin{vmatrix} 0 & 1 & 1 & 1 \\ 1 & 0 & qr^2 & qs^2 \\ 1 & qr^2 & 0 & rs^2 \\ 1 & qs^2 & rs^2 & 0 \end{vmatrix}.$$

Clearly, $k_M(q, r, s) = 2(\sin \measuredangle rqs)/rs = 2(\sin \measuredangle qsr)/qr$.

Now if $p \in K$, the *curvature* $k_M(p)$ of K at p is defined by

$$k_M(p) = \lim_{q, r, s \to p} k_M(q, r, s).$$

The question arises: what is the nature of a metric continuum K in a neighborhood of a point p_0 at which $k_M(p_0)$ exists? This question is answered in the following theorem.

THEOREM 10.1   *If a metric continuum K has finite curvature $k_M(p_0)$ at a point p_0 of K, then in a neighborhood of p_0, K is a rectifiable arc. If $k_M(p)$ exists (finite) for each point p of K, then K is either a rectifiable arc or a rectifiable simple closed curve.*

Proof.   It suffices to show that K has property Ω^* at p_0. Let $\{r_n\}, \{s_n\}, \{t_n\}$ be infinite sequences of points of K with $\lim_{n \to \infty} r_n = \lim_{n \to \infty} s_n = \lim_{n \to \infty} t_n = p_0$, and let

$$\lim_{n \to \infty} k_M(r_n, s_n, t_n) = k_M(p_0).$$

Then

$$k_M(p_0) = 2 \lim_{n \to \infty} \frac{\sin \angle r_n s_n t_n}{r_n t_n} = 2 \lim_{n \to \infty} \frac{\sin \angle r_n t_n s_n}{r_n s_n}$$

$$= 2 \lim_{n \to \infty} \frac{\sin \angle s_n r_n t_n}{s_n t_n},$$

and consequently (since $\lim_{n \to \infty} r_n s_n = \lim_{n \to \infty} s_n t_n = \lim_{n \to \infty} r_n t_n = 0$),

$$\lim_{n \to \infty} \sin \angle r_n s_n t_n = \lim_{n \to \infty} \sin \angle r_n t_n s_n = \lim_{n \to \infty} \sin \angle s_n r_n t_n = 0.$$

Hence

$$\lim_{n \to \infty} \sin[\max \angle \{r_n, s_n, t_n\}] = 0,$$

and since $\max \angle \{r_n, s_n, t_n\} \geq \pi/3$, then $\lim_{n \to \infty} \max \angle \{r_n, s_n, t_n\} = \pi$.

It follows that positive numbers δ_{p_0}, α_{p_0} exist ($\alpha_{p_0} \leq \pi/2$) such that, if $r, s, t \in K \cdot U(p_0; \delta_{p_0})$, $r \neq s \neq t \neq r$, then $\max \angle \{r, s, t\} \geq \pi/2 + \alpha_{p_0}$; that is, K has property Ω^* at p_0. Both conclusions of the theorem are now justified by the Corollary of Theorem 9.9 and the Remark following it. For the proof of the next theorem it is useful to observe that δ_{p_0} can be chosen so that $\max \angle \{r, s, t\} > 3\pi/4$.

Now let K denote a metric continuum, $p_0 \in K$ such that $k_M(p_0)$ exists, and let $U(p_0; \delta)$ denote a neighborhood of p_0 such that $K \cdot U(p_0; \delta)$ is a rectifiable arc. Since $k_M(p_0)$ exists, then there exists a number δ^* ($0 < \delta^* < \delta$) such that $A = K \cdot \bar{U}(p_0; \delta^*)$ is a rectifiable arc with property Ω^*. Hence, according to Theorem 9.7, a positive constant ρ_0 exists with the properties listed in that theorem. Let $\rho_0^* = (\frac{1}{2})\min(\rho_0, \delta^*)$ and consider the rectifiable arc $A^* = K \cdot \bar{U}(p_0; \rho_0^*)$.

If $q, r \in A^*$, $q \neq r$, put

$$k_H(q, r) = \left\{ \frac{24[\ell(q, r) - qr]}{\ell^3(q, r)} \right\}^{1/2},$$

where $\ell(q, r)$ denotes the length of the subarc of A^* joining q and r, and define

$$k_H(p_0) = \lim_{q, r \to p_0} k_H(q, r).$$

Each of the limits $k_M(p_0)$ and $k_H(p_0)$ metrize the classical notion of curvature of a curve of euclidean three-space, *whenever these limits exist*, and so each provides, a priori, a means of extending the classical concept to more general figures. Though the first metrization $k_M(p_0)$ is meaningful for p_0, a point of *any* metric continuum, and the second metrization $k_H(p_0)$ is applicable only when p_0 is a point of a *rectifiable metric arc*, the apparently wider application of the first metrization is nullified by the result established in the preceding theorem. We shall see, moreover, that the metrization $k_H(p_0)$ is, in fact, the more general one in the sense that if $k_M(p_0)$ exists for any point p_0 of a metric continuum, then so does $k_H(p_0)$, and the two numbers are equal; but the existence of $k_H(p_0)$, for p_0 a point of a rectifiable metric arc, does *not* imply the existence of $k_M(p_0)$.

Whenever it is stated that $k_M(p)$, $k_H(p)$ exist, it is meant that the limits defining them are *finite*, although occasionally that word is added for emphasis.

EXERCISES

1. If $k_M(p)$ exists at each point p of a metric arc A, prove that A is rectifiable.

2. If $k_M(p)$ exists at each point p of a metric arc A, show that $k_M(p)$ is uniformly continuous in A.

3. Show that if $k_M(p)$ exists (finite) at p, $p \in C$, a *euclidean* continuum, then C has a unit tangent vector $\alpha(p)$ at p in the strong sense (that is, as points q, r of C approach p along C, line $L(q, r)$ approaches a unique line L) and $\alpha(p)$ is continuous wherever it exists.

10.2   COMPARISON OF TWO CURVATURE METRIZATIONS

We establish first the following theorem.

THEOREM 10.2   *If $k_M(p)$ exists (finite) for $p \in K$, a metric continuum, then $k_H(p)$ exists and equals $k_M(p)$.*

Proof.   We have seen that the existence of $k_M(p)$ implies that in a spherical neighborhood of p, the continuum K is a rectifiable arc A which has property Ω^*. Moreover, the radius of the neighborhood may be chosen so that (i) each triple of pairwise distinct points of A contains an angle greater than $3\pi/4$, and (ii) if $x, y, z \in A$, with $x \prec y \prec z$, then $\measuredangle xyz = \max \measuredangle \{x, y, z\}$.

Now since $k_M(p)$ exists, $\varepsilon > 0$ implies the existence of $\eta > 0$ such that if $u, v, w \in A \cdot U(p; \eta)$, then

$$k_M(p) - \varepsilon < k_M(u, v, w) < k_M(p) + \varepsilon.$$

We may suppose that $\eta < 1/[k_M(p) + \varepsilon]$, and let $A^* = A \cdot \bar{U}(p; \eta/4)$.

If $x, y, z \in A^*$, with $x \prec y \prec z$, $xz = a$, $xy = b$, $yz = c$ $(a > b \geq c)$, consider the plane triangle x', y', z' with $x'z' = a$, $x'y' = b$, and $y'z' = c$. Clearly

$$a = b \cdot \cos \measuredangle y'x'z' + c \cdot \cos \measuredangle y'z'x'$$
$$= b \cdot [1 - \sin^2 \measuredangle y'x'z']^{1/2} + c \cdot [1 - \sin^2 \measuredangle y'z'x']^{1/2},$$

and so

$(*) \qquad a = b \cdot [1 - (\tfrac{1}{4}) c^2 \cdot k_M^2(x, y, z)]^{1/2} + c \cdot [1 - (\tfrac{1}{4}) b^2 \cdot k_M^2(x, y, z)]^{1/2},$

since

$$k_M(x, y, z) = \frac{2 \sin \measuredangle yxz}{c} = \frac{2 \sin \measuredangle y'x'z'}{c}$$
$$= \frac{2 \sin \measuredangle yzx}{b} = \frac{2 \sin \measuredangle y'z'x'}{b}.$$

Now since $\measuredangle xyz = \max \measuredangle \{x, y, z\} > 3\pi/4$, it is easily seen that

$$0 < (\tfrac{1}{4})b^2 \cdot k_M^2(x, y, z) < \tfrac{1}{2},$$
$$0 < (\tfrac{1}{4})c^2 \cdot k_M^2(x, y, z) < \tfrac{1}{2},$$

and since the inequalities

$$\sqrt{1 - \alpha} \geq 1 - \alpha,$$
$$\sqrt{1 - \alpha} \leq 1 - (\tfrac{1}{2})\alpha,$$
$$\sqrt{1 - \alpha} \geq 1 - (\tfrac{1}{2})\alpha - \alpha^2,$$

are valid for $0 \leq \alpha < \tfrac{1}{2}$, we obtain from $(*)$,

$(\dagger) \qquad\qquad a \geq (b + c)[1 - (\tfrac{1}{4})bck_M^2(x, y, z)],$

$(\dagger\dagger) \qquad\qquad a \leq b + c - (\tfrac{1}{8})b^2ck_M^2(x, y, z),$

$(\dagger\dagger\dagger) \qquad a \geq b + c - (\tfrac{1}{8})k_M^2(x, y, z)bc(b + c) - (\tfrac{1}{8})k_M^4(x, y, z)b^4c.$

To proceed with the proof of the theorem, let q, r denote points of A^*, and $A^*(q, r)$ the subarc of A^* with end points q, r. By Theorem 9.4, $\ell[A^*(q, r)] = \lim_{n \to \infty} n \cdot \lambda(n)$, and since A^* has property Ω^*, there exists a positive integer N such that for every $n > N$ the *unique* n-lattice inscribed in $A^*(q, r)$ is homogeneous (Theorems 9.6 and 9.8). It follows that corresponding to each $\zeta > 0$ there is a natural number N^* such that the unique, homogeneous, n-lattice $p_0, p_1, \ldots, p_n$ ($p_0 = q$ and $p_n = r$) inscribed in $A^*(q, r)$ satisfies the inequalities

$$\ell[A^*(q, r)] \geqq p_0 p_1 + p_1 p_2 + \cdots + p_{n-1} p_n = n \cdot \lambda(n) > \ell[A^*(q, r)] - \zeta,$$

for every $n > N^*$.

Since the points q, p_{k-1}, p_k are such that $q \prec p_{k-1} \prec p_k$ and $qp_k > qp_{k-1} \geqq p_{k-1} p_k = \lambda(n)$ ($k = 2, 3, \ldots, n$), application of (†) gives

$$qp_k > qp_{k-1} + \lambda(n) - (\tfrac{1}{4})\lambda(n)\ell^2[A^*(q, r)] \cdot (k_M(p) + \varepsilon)^2,$$

$k = 2, 3, \ldots, n$. Summation for $k = 2, 3, \ldots, m$ ($m \leqq n$) gives

$$(**) \qquad qp_m > m \cdot \lambda(n)\{1 - (\tfrac{1}{4})\ell^2[A^*(q, r)](k_M(p) + \varepsilon)^2\}.$$

Application of inequality (††) to q, p_{k-1}, p_k yields

$$qp_k < qp_{k-1} + \lambda(n) - (\tfrac{1}{8})\lambda(n)qp_{k-1}^2(k_M(p) - \varepsilon)^2,$$

$k = 2, 3, \ldots, n$, and use of (**) for $m = k - 1$ gives

$$qp_k < qp_{k-1} + \lambda(n) - (\tfrac{1}{8})\lambda^3(n)(k-1)^2(k_M(p) - \varepsilon)^2$$
$$\times \{1 - (\tfrac{1}{4})\ell^2[A^*(q, r)](k_M(p) + \varepsilon)^2\}^2.$$

Summing the last inequality for $k = 2, 3, \ldots, n$, we obtain

$$qr < n \cdot \lambda(n) - (\tfrac{1}{8})\lambda^3(n)(k_M(p) - \varepsilon)^2$$
$$\times \{1 - (\tfrac{1}{2})\ell^2[A^*(q, r)](k_M(p) + \varepsilon)^2\} \sum_{k=2}^{n} (k - 1)^2.$$

Since

$$\sum_{k=2}^{n} (k - 1)^2 = (\tfrac{1}{6})n(n - 1)(2n - 1),$$

and $n \cdot \lambda(n) \leqq \ell[A^*(q, r)]$, it follows that

$$qr < \ell[A^*(q, r)] - (\tfrac{1}{24})(k_M(p) - \varepsilon)^2(\ell[A^*(q, r)] - \zeta)^3[1 - 3/2n]$$
$$\times \{1 - (\tfrac{1}{2})\ell^2[A^*(q, r)](k_M(p) + \varepsilon)^2\},$$

and since ζ is arbitrarily small,

$$(1) \quad qr < \ell[A^*(q, r)] - (\tfrac{1}{24})(k_M(p) - \varepsilon)^2\ell^3[A^*(q, r)]$$
$$+ (\tfrac{1}{40})\ell^5[A^*(q, r)](k_M(p) + \varepsilon)^4.$$

Applying now inequality (†††) to q, p_{k-1}, p_k, and noting that $k_M(q, p_{k-1}, p_k) < k_M(p) + \varepsilon$,

$$qp_k > qp_{k-1} + \lambda(n) - (\tfrac{1}{8})[k_M(p) + \varepsilon]^2 \cdot \lambda(n)[qp_{k-1} + \lambda(n)] \cdot qp_{k-1}$$
$$- (\tfrac{1}{8})[k_M(p) + \varepsilon]^4 \cdot \lambda(n) \cdot qp_{k-1}^4,$$

which, since $qp_{k-1} \leqq (k-1)\lambda(n)$, becomes

$$qp_k > qp_{k-1} + \lambda(n) - (\tfrac{1}{8})[k_M(p) + \varepsilon]^2 k(k-1)\lambda^3(n)$$
$$- (\tfrac{1}{8})[k_M(p) + \varepsilon]^4 \cdot (k-1)^4 \lambda^5(n),$$

$k = 2, 3, \ldots, n$.

Summing these inequalities and noting that

$$\sum_{k=2}^{n} (k-1)^4 = \frac{n(n-1)(2n-1)(3n^2 - 3n - 1)}{30}$$

and

$$\sum_{k=2}^{n} k(k-1) = \frac{n(n-1)(n+1)}{3},$$

we obtain

$$qr > n \cdot \lambda(n) - (\tfrac{1}{24})[k_M(p) + \varepsilon]^2 \cdot \lambda^3(n) \cdot n(n-1)(n+1)$$
$$- (\tfrac{1}{240})[k_M(p) + \varepsilon]^4 \cdot \lambda^5(n) \cdot n(n-1)(2n-1)(3n^2 - 3n - 1).$$

It follows easily that

(2) $$qr > \ell[A^*(q,r)] - (\tfrac{1}{24})[k_M(p) + \varepsilon]^2 \ell^3[A^*(q,r)]$$
$$- (\tfrac{1}{40})[k_M(p) + \varepsilon]^4 \cdot \ell^5[A^*(q,r)].$$

Inequalities (1) and (2) give

$$[k_M(p) - \varepsilon]^2 - (\tfrac{3}{5})[k_M(p) + \varepsilon]^4 \cdot \ell^2[A^*(q,r)]$$
$$< \frac{24\{\ell[A^*(q,r)] - qr\}}{\ell^3[A^*(q,r)]}$$
$$< [k_M(p) + \varepsilon]^2 + (\tfrac{3}{5})[k_M(p) + \varepsilon]^4 \ell^2[A^*(q,r)].$$

Letting q, r approach p, and observing that ε can be chosen arbitrarily small, we obtain

$$\lim_{q,r \to p} \frac{24 \cdot \{\ell[A^*(q,r)] - qr\}}{\ell^3[A^*(q,r)]} = k_M^2(p),$$

and consequently

$$k_H(p) = k_M(p).$$

10.3   FURTHER COMPARISON OF METRIC CURVATURES.
METRIC PTOLEMAIC SPACE

We have seen that if k_M exists at a point of a metric continuum, then so does k_H, and the two curvatures are equal. But the existence of k_H at a point of a rectifiable metric arc does *not* imply the existence of k_M at that point. This is exemplified by the following example, due to J. Haantjes.

Redefine the distance of two points x, y of the unit segment $[0, 1]$ by writing

$$d(x, y) = f(t) = t - (\tfrac{1}{6})t^3 + (\tfrac{1}{24})t^4 \sin(1/t),$$

where $t = |x - y| \neq 0$ and $d(x, x) = f(0) = 0$. Let us show, first of all, that the space A so obtained is metric. Clearly $d(x, y) \geq 0$, $d(x, y) = 0$ if $x = y$, and $d(x, y) = d(y, x)$. It remains to show (i) if $d(x, y) = 0$, then $x = y$, and (ii) the validity of the triangle inequality.

Since, for $0 < t \leq 1$,

$$f'(t) = 1 - (\tfrac{1}{2})t^2 + (\tfrac{1}{6})t^3 \sin(1/t) - (\tfrac{1}{24})t^2 \cos(1/t)$$
$$> 1 - (\tfrac{1}{2}) - (\tfrac{1}{6}) - (\tfrac{1}{24}) > 0,$$

and $f'(0) = 1$, it follows that $f(t)$ increases monotonically, and so $d(x, y) = f(t) = 0$ if and only if $x = y$.

To establish (ii), let $x < y < z$ be three points of $[0, 1]$ and put $y - x = u$, $z - y = v$. Then $d(x, y) = f(u)$, $d(y, z) = f(v)$, $d(x, z) = f(u + v)$, and since the function f increases monotonically on $[0, 1]$, it remains to show that $f(u) + f(v) \geq f(u + v)$.

Since the preceding inequality is symmetric in u, v, we may assume $u \leq v$. Now

(*) $f(u) + f(v) - f(u + v) = (\tfrac{1}{2})uv(u + v) - (\tfrac{1}{24})\{\varphi(u + v) - \varphi(u) - \varphi(v)\}$,

where $\varphi(u) = u^4 \sin(1/u)$, and consequently

$$|\varphi(u + v) - \varphi(u) - \varphi(v)| = |u\varphi'(v + \theta \cdot u) - \varphi(u)|, 0 < \theta < 1$$
$$< 4u(u + v)^3 + u(u + v)^2 + u^4$$
$$< (8 + 2 + 1)uv(u + v).$$

Substitution in (*) yields $f(u) + f(v) > f(u + v)$, and so A is a metric space. From the continuity and monotonicity of $f(t)$ on $[0, 1]$, it follows that the identity mapping of $[0, 1]$ onto A is a homeomorphism and consequently the space A is a metric arc.

Assertion I.   *The metric arc A is rectifiable.*

If $p \in [0, 1]$, the spread $f^*(p)$ at p of the homeomorphism that maps $[0, 1]$ onto A (the identity mapping) is

$$f^*(p) = \lim_{q, r \to p} \frac{\text{distance of } q, r \text{ in } A}{\text{distance of } q, r \text{ in } [0, 1]}$$

$$= \lim_{t \to 0} \frac{t + (\tfrac{1}{6})t^3 - (\tfrac{1}{24})t^4 \sin 1/t}{t},$$

where $t = |q - r|$. Hence $f^*(p) \equiv 1$, $p \in [0, 1]$, and so

$$\ell(A) = \int_0^1 f^*(x)\, dx = 1,$$

by Theorem 9.2.

It is interesting to observe that the length of the arc (segment) $[0, 1]$ is invariant under the change of metric, though the new arc A is not a metric segment since $\ell(A) \neq d(0, 1)$.

Assertion II.   *At each point p of A, $k_H(p) = 2$.*

If $q, r \in A$, $q < r$, then

$$\ell(q, r) = \int_q^r f^*(x)\, dx = r - q = t,$$

since $f^*(x) \equiv 1$, and so

$$24 \cdot \lim_{q, r \to p} \frac{[\ell(q, r) - d(q, r)]}{\ell^3(q, r)}$$

$$= 24 \cdot \lim_{t \to 0} \frac{\{t - [t - (\tfrac{1}{6})t^3 + (\tfrac{1}{24})t^4 \sin(1/t)]\}}{t^3} = 4.$$

Hence $k_H(p) = 2$, for every point p of A.

Assertion III.   *The curvature k_M does not exist at any point p of A.*

Suppose $p \in A$ such that $k_M(p)$ exists $(0 < p < 1)$, and select points q, r of A with $q < p < r$, $p - q = 1/2\pi n$, $r - p = 1/2\pi n^3$, where n is a natural number. Then

$$k_M(p) = \lim_{q, r \to p} k_M(p, q, r),$$

and putting

$$d(p, q) = f(1/2\pi n) = d_1, \; d(p, r) = f(1/2\pi n^3) = d_2,$$
$$d(q, r) = f[(n^2 + 1)/2\pi n^3] = d_3,$$

we have

$$(*) \quad k_M^2(p, q, r) = \frac{(d_1 + d_2 + d_3)(-d_1 + d_2 + d_3)(d_1 - d_2 + d_3)(d_1 + d_2 - d_3)}{d_1^2 d_2^2 d_3^2}.$$

Since $\lim_{t \to 0} f(t)/t = 1$, we may write

$$k_M^2(p) = \lim_{n \to \infty} \frac{(d_1 + d_2 + d_3)(-d_1 + d_2 + d_3)(d_1 - d_2 + d_3)(d_1 + d_2 - d_3)}{(1/2\pi n)^2 \cdot (1/2\pi n^3)^2 \cdot [(n^2 + 1)/2\pi n^3]^2}$$

$$= \lim_{n \to \infty} A \cdot B \cdot C \cdot D,$$

where

$$A = \frac{d_1 + d_2 + d_3}{[(n^2 + 1)/2\pi n^3]}, \qquad C = \frac{d_1 - d_2 + d_3}{(1/2\pi n)},$$

$$B = \frac{-d_1 + d_2 + d_3}{(1/2\pi n^3)}, \qquad D = \frac{d_1 + d_2 - d_3}{(1/2\pi n)(1/2\pi n^3)[(n^2 + 1)/2\pi n^3]}.$$

Now it is easily seen that

$$\lim_{n \to \infty} A = \lim_{n \to \infty} B = \lim_{n \to \infty} C = 2$$

and

$$\lim_{n \to \infty} D = \tfrac{13}{24}.$$

It follows that

$$k_M(p) = (\tfrac{1}{3}) \sqrt{39},$$

which is impossible since, according to Theorem 9.2, the existence of $k_M(p)$ implies $k_M(p) = k_H(p) = 2$.

Hence $k_M'(p)$ does not exist at any point p of A, $p \neq 0$. Now since $k_M(p)$ is a continuous function of p, then, if k_M exists at 0 or 1, clearly $k_M(0) =$

$k_M(1) = (\frac{1}{3})\sqrt{39}$, and the same contradiction is encountered. The proof of the Assertion is complete.

A metric space is called *ptolemaic* provided for each four pairwise distinct points p, q, r, s of the space the inequality $pq \cdot rs + pr \cdot qs \geq ps \cdot qr$ is satisfied; that is, the three products of "opposite" distances $pq \cdot rs$, $pr \cdot qs$, $ps \cdot qr$ satisfy the triangle inequality. Writing

$$C(p, q, r, s) = \begin{vmatrix} 0 & pq^2 & pr^2 & ps^2 \\ pq^2 & 0 & qr^2 & qs^2 \\ pr^2 & qr^2 & 0 & rs^2 \\ ps^2 & qs^2 & rs^2 & 0 \end{vmatrix},$$

it is not difficult to show that

(∗) $C(p, q, r, s) = - (pq \cdot rs + pr \cdot qs + ps \cdot qr)(pq \cdot rs + pr \cdot qs - ps \cdot qr)$
$$\times (pq \cdot rs - pr \cdot qs + ps \cdot qr)(- pq \cdot rs + pr \cdot qs + ps \cdot qr),$$

and hence a metric space is ptolemaic if and only if $C(p, q, r, s) \leq 0$ for each quadruple of pairwise distinct points p, q, r, s (Exercise 1).

The class of metric ptolemaic spaces includes all euclidean and hyperbolic spaces, but spherical and elliptic spaces are not even locally ptolemaic. In the next section we prove that $k_M(p)$ and $k_H(p)$ are equivalent for p a point of an arc in a metric ptolemaic space. It will follow that these two curvatures are equivalent for arcs of any euclidean space.

EXERCISES

1. Establish the expression for $C(p, q, r, s)$ given in (∗) and justify the conclusion that follows (∗).

2. Show that euclidean three-space E_3 is ptolemaic.

3. Prove that the metric arc A defined in this section is not ptolemaic.

10.4   EQUIVALENCE OF $k_M(p)$ AND $k_H(p)$ FOR METRIC PTOLEMAIC ARCS

Let p be an interior point of a metric ptolemaic arc. To show the equivalence of $k_M(p)$ and $k_H(p)$ it suffices, by Theorem 10.2, to prove that the existence of $k_H(p)$ implies the existence of $k_M(p)$. The following proof of this is due to J. Haantjes.

Let A, then, denote a rectifiable metric ptolemaic arc, $p \in A$ (p not an end point of A). Since $k_H(p)$ exists, $\eta > 0$ implies the existence of $\delta > 0$ such that if $q, s \in A \cdot U(p; \delta)$, $q \prec s$, then

$$|d - \ell(q, s) + \sigma \cdot \ell^3(q, s)| < \eta \cdot \ell^3(q, s),$$

where $d = qs$, $\ell(q, s) = \ell[A(q, s)]$, and $\sigma = (\frac{1}{24})k_H^2(p)$.

If $r \in A(q, s)$, $q \prec r \prec s$, put

$$u = \ell(q, r), \qquad v = \ell(r, s), \qquad u + v = \ell(q, s),$$

$$a = qr, \qquad\qquad b = rs, \qquad\qquad c = qs.$$

Then numbers α, β, γ exist such that

$$a = u - \sigma u^3 + \alpha u^3,$$

$$b = v - \sigma v^3 + \beta v^3,$$

$$c = u + v - \sigma(u + v)^3 + \gamma(u + v)^3,$$

where each of the numbers $|\alpha|$, $|\beta|$, $|\gamma|$ is less than η.

Now the existence of $k_H(p)$ implies that

$$\lim_{q, r \to p} \frac{u}{a} = \lim_{r, s \to p} \frac{v}{b} = \lim_{q, s \to p} \frac{u + v}{c} = 1,$$

and consequently $k_M(p)$ exists if and only if

$$\lim_{q, r, s \to p} \frac{[(a + b + c)(a + b - c)(a - b + c)(-a + b + c)]}{u^2 \cdot v^2 \cdot (u + v)^2}$$

exists, and $k_M(p)$ equals the nonnegative square root of this limit. Now

$$\frac{(a + b - c)}{uv(u + v)} = 3\sigma - \frac{[\gamma \cdot (u + v)^3 - \alpha u^3 - \beta v^3]}{uv(u + v)},$$

and if $\lim_{u, v \to 0}(a + b - c)/uv(u + v) = 3\sigma$, then

$$\lim_{u, v \to 0} \frac{[\gamma \cdot (u + v)^3 - \alpha u^3 - \beta v^3\}}{uv(u + v)} = 0,$$

and it is readily seen (Exercise 2) that this implies

$$\lim_{u,\,v\to 0} \frac{a+b+c}{u+v} = \lim_{u,\,v\to 0} \frac{(a-b+c)}{u}$$

$$= \lim_{u,\,v\to 0} \frac{(-a+b+c)}{v}$$

$$= 2.$$

It then would follow that

$$k_M(p) = [2^3 \cdot 3\sigma]^{1/2} = k_H(p),$$

and the proof would be complete. It remains to establish that

$$\lim_{u,\,v\to 0} \frac{a+b-c}{uv(u+v)} = 3\sigma.$$

From the expressions for a, b, c given above,

$$a+b-c = \sigma \cdot [3uv(u+v)] + \alpha u^3 + \beta v^3 - \gamma \cdot (u+v)^3$$
$$< \sigma \cdot [3uv(u+v)] + \eta(u^3+v^3) - \gamma(u+v)^3,$$

and since, for $(u/2) \leq v \leq u$, it is easily shown that

$$\eta(u^3+v^3) - \gamma(u+v)^3 < 3\eta[3uv(u+v)],$$

it follows that

$$a+b-c < 3\sigma_2\,uv(u+v),$$

where $\sigma_2 = \sigma + 3\eta$ and $(u/2) \leq v \leq u$. In a similar way, it is seen that

$$3\sigma_1\,uv(u+v) < a+b-c,$$

where $\sigma_1 = \sigma - 3\eta$ and $(u/2) \leq v \leq u$. Therefore

(*) $\qquad 3(\sigma - 3\eta) < (a+b-c)/uv(u+v) < 3(\sigma + 3\eta),$

for all values of u, v satisfying $(u/2) \leq v \leq u$.

For any three points q, r, s of $A \cdot U(p; \delta)$ with $q \prec r \prec s$, assume the labeling so that $v \leq u$, and write $u = k \cdot v + t$, where k is a nonnegative integer, $0 < t \leq v$. Let $p_0, p_1, \ldots, p_k, p_{k+1}$ be points of the arc $A(q, s)$ such that $\ell(q, p_0) = t$, $\ell(p_i, p_{i+1}) = v$ $(i = 0, 1, 2, \ldots, k)$. It follows that $p_k = r$ and $p_{k+1} = s$. Writing $qp_i = d_i$ $(i = 0, 1, \ldots, k+1)$, and $p_i p_{i+1} = a_i$

$(i = 0, 1, \ldots, k)$, set

$$Q_i = \frac{[a_{i-1} + d_{i-1} - d_i]}{a_{i-1}}$$

$(i = 1, 2, \ldots, k + 1)$.

The ptolemaic inequality applied to the quadruple q, p_{i-1}, p_i, p_{i+1} gives

$$d_{i+1} a_{i-1} + d_{i-1} a_i \geqq d_i \cdot (p_{i-1} p_{i+1}),$$

from which it follows that

$$Q_{i+1} \leqq Q_i + \frac{d_i[a_i + a_{i-1} - p_{i-1} p_{i+1}]}{a_i a_{i-1}}$$

$(i = 1, 2, \ldots, k)$.

Since for the points p_{i-1}, p_i, p_{i+1} we have $u = v$, the inequalities (*) are valid. Moreover, $|a_{i-1} - v + \sigma v^3| < \eta v^3$, $|a_i - v + \sigma v^3| < \eta v^3$, and we conclude that

$$Q_{i+1} < Q_i + \frac{6\sigma_2 v(iv + t)}{(1 - \sigma_2 v^2)^2}$$

$(i = 1, 2, \ldots, k)$.

Direct computation of Q_1 gives

$$Q_1 < \frac{3\sigma_2 t(v + t)}{(1 - \sigma_2 v^2)^2} + \frac{10\eta v^2}{(1 - \sigma_2 v^2)},$$

and summation of Q_i $(i = 1, 2, \ldots, k)$ yields

$$\frac{(a + b - c)}{b} = Q_{k+1} < \frac{3\sigma_2 u(u + v)}{(1 - \sigma_2 v^2)^2} + \frac{10\eta v^2}{(1 - \sigma_2 v^2)}.$$

Since η is arbitrary and $\lim v/b = 1$, it follows that

$$\limsup_{u, v \to 0} \frac{(a + b - c)}{uv(u + v)} \leqq 3\sigma,$$

and so $\varepsilon' > 0$ implies the existence of a δ' $(0 < \delta' < \delta)$ such that if $q, r, s \in A \cdot U(p; \delta')$, then

(†)
$$\frac{(a + b - c)}{uv(u + v)} - 3\sigma < \varepsilon'.$$

It remains to find the limit inferior of the quotient $(a + b - c)/uv(u + v)$. To do this, let w denote a point of A, with $q \prec r \prec s \prec w$ $(q, r, s, w \in A \cdot U(p; \delta'))$, and $\ell[A(q, w)] = 2u + v$. Writing $f = qw$, $h = rw$, $g = sw$, and applying the ptolemaic inequality to the quadruple q, r, s, w, we obtain

$$qw \cdot rs + qr \cdot sw \geqq qs \cdot rw.$$

That is,

$$f \cdot b + a \cdot g \geqq c \cdot h,$$

and that inequality implies

$$b(a + h - f) - a(b + g - h) \leqq h(a + b - c).$$

From (∗) and (†) applied to the triples $\{q, r, w\}$ and $\{r, s, w\}$, respectively

$$a + h - f > 3(\sigma - 3\eta)u(u + v)(2u + v),$$
$$b + g - h < 3(\sigma + \varepsilon')uv(u + v),$$

and it follows that

$$a + b - c \geqq 3\sigma uv(u + v) - 3\sigma^2 uv^3(2u + v) - 9\eta uv(2u + v) - \varepsilon'u^2v.$$

Hence

$$\liminf_{u, v \to 0} \frac{a + b - c}{uv(u + v)} \geqq 3\sigma,$$

and consequently

$$\lim_{u, v \to 0} \frac{a + b - c}{uv(u + v)} = 3\sigma,$$

completing the proof of the theorem in the case that p is an interior point of A. Since $k_M(p)$ and $k_H(p)$ are continuous functions of p, the case of p an end point of A offers no difficulty, and we have established the following theorem.

THEOREM 10.3   *If p is any point of a metric ptolemaic arc, each of the curvatures $k_M(p)$, $k_H(p)$ exists provided one of them does, and then $k_M(p) = k_H(p)$.*

COROLLARY   *In any euclidean space, the curvatures $k_M(p)$ and $k_H(p)$ are equivalent.*

EXERCISES

1. Let A denote a rectifiable metric arc. If $p \in A$ such that $k_H(p)$ exists, show that

$$\lim_{q,r \to p} \frac{\ell[A(q, r)]}{qr} = 1,$$

where $q, r \in A$ and $\ell[A(q, r)]$ denotes the length of the subarc of A with end points q, r.

2. Verify the statement in the proof of Theorem 10.3 that refers to this exercise.

10.5   METRIC ARCS OF ZERO CURVATURE.
A METRIC DIFFERENTIAL
CHARACTERIZATION OF METRIC SEGMENT

Although in classical differential geometry an arc of euclidean space with zero curvature (in the classical sense) at each point is necessarily a segment, metric arcs exist with everywhere vanishing metric curvature (in either of the two senses defined above) which are *not* metric segments. In fact, a metric arc may even be locally a metric segment (and hence $k_M(p) = k_H(p) = 0$, at each point p) without being globally a metric segment. This is very simply exemplified by any arc of a circle (with shorter arc metric) greater than a semicircle. The question arises: in what spaces are "uncurved" arcs straight? An answer to this question is provided in this section.

THEOREM 10.4   *A ptolemaic metric continuum K is a metric segment if and only if $k_M(p) = 0$, $p \in K$.*

Proof.   Since for each three points q, r, s of a metric segment, $D(q, r, s) = 0$, it follows that $k_M(p)$ vanishes at each point p of the segment, and the necessity of the condition is established.

To prove the sufficiency, note that since $k_M(p) = 0$, $p \in K$, then $k_M(p)$ exists (finite) at each point of K, which has, consequently, properties Ω^* and Ω (see the proof of Theorem 10.2). Applying Theorems 9.5 and 10.1, we conclude that K is either (1) a rectifiable arc or (2) a rectifiable simple closed curve.

Case 1.   *K is a rectifiable arc.* Write $K = A(a, b)$, where a, b denote the end points of the arc.

Since $A(a, b)$ is a metric ptolemaic arc, it follows from Theorem 10.3 that $k_H(p) = k_M(p) = 0$, for each point p of $A(a, b)$, and hence there corresponds to each arbitrarily chosen $\varepsilon > 0$ a positive number δ such that for each two distinct points q, r of $A(a, b)$, with $qr < \delta$,

(*)
$$\frac{\ell(q, r) - qr}{\ell^3(q, r)} < \varepsilon,$$

where $\ell(q, r)$ denotes the length of the subarc of $A(a, b)$ with end points q, r.

Now let $a = p_0, p_1, p_2, \ldots, p_{n-1}, p_n = b$ denote $n + 1$ points of $A(a, b)$ with $\ell(p_i, p_j) = \alpha$, $|i - j| = 1$ $(i, j = 0, 1, \ldots, n)$, where $0 < \alpha < (\frac{1}{2})\delta$, and put $d_k = k\alpha - p_0 p_k = k\alpha - c_k$ $(k = 1, 2, \ldots, n)$. Note that d_k measures the difference between the length of an arc and the distance of its end points, and the proof of Case 1 will be complete when it is shown that $d_n = n\alpha - c_n = \ell[A(a, b)] - p_0 p_n = 0$, since a metric segment is the only metric arc whose length equals the distance of its end points (Exercise 1).

Setting $\ell[A(a, b)] = L$, we observe that $\ell(p_0, p_n) = n\alpha = L$, $c_k \le p_0 p_1 + p_1 p_2 + \cdots + p_{k-1} p_k \le k\alpha$, and so $d_k = k\alpha - c_k \ge 0$, $(k = 1, 2, \ldots, n)$. Further, since $p_0 p_1 \le \alpha < (\frac{1}{2})\delta$, $p_0 p_2 = c_2 \le 2\alpha < \delta$, application of (*) yields

(**)
$$d_1 = \alpha - c_1 < \varepsilon\alpha^3, \qquad d_2 = 2\alpha - c_2 < 8\varepsilon\alpha^3.$$

The ptolemaic inequality applied to the points $p_0, p_{k-1}, p_k, p_{k+1}$ yields

$$p_0 p_{k+1} \cdot p_{k-1} p_k + p_0 p_{k-1} \cdot p_k p_{k+1} \ge p_0 p_k \cdot p_{k-1} p_{k+1},$$

$(k = 2, 3, \ldots, n - 1)$, and since

$$p_{k-1} p_k \le \alpha, \; p_k p_{k+1} \le \alpha, \; p_{k-1} p_{k+1} > 2\alpha - 8\varepsilon\alpha^3,$$

$$p_0 p_{k-1} + p_0 p_{k+1} = (k - 1)\alpha - d_{k-1} + (k + 1)\alpha - d_{k+1}$$
$$= 2k\alpha - d_{k-1} - d_{k+1},$$

we readily obtain

$$(2k\alpha - d_{k-1} - d_{k+1})\alpha > (k\alpha - d_k)(2\alpha - 8\varepsilon\alpha^3).$$

From the above inequality we conclude

(†)
$$d_{k+1} - d_k < d_k - d_{k-1} + 8L\varepsilon\alpha^2$$

$(k = 2, 3, \ldots, n - 1)$.

Summing these inequalities for $k = 2, 3, \ldots, m$, $(m \leqq n - 1)$ yields

$$d_{m+1} - d_2 < d_m - d_1 + 8(m - 1)L\varepsilon\alpha^2.$$

Now $d_2 - d_1 = \alpha - (c_2 - c_1) \geqq \alpha - p_1 p_2 \geqq 0$, and (**) yields $d_2 - d_1 \leqq 8L\varepsilon\alpha^2$. Using $m\alpha \leqq (n - 1)\alpha < L$, we get

(††) $$d_{m+1} < d_m + 8L^2 \varepsilon\alpha.$$

Summation of inequalities (††) for $m = 2, 3, \ldots, n - 1$ gives

$$d_n < d_2 + 8(n - 2)L^2 \varepsilon\alpha,$$

and so

$$d_n < 8\varepsilon\alpha^3 + 8L^3\varepsilon - 16L^2\varepsilon\alpha < 8L^3\varepsilon.$$

Since the positive number ε is arbitrary and $d_n \geqq 0$, it follows that $d_n = n\alpha - c_n = L - ab = 0$, and consequently the rectifiable arc $K = A(a, b)$ is a metric segment.

Case 2.   K is a rectifiable simple closed curve.

Put $K = \Gamma$, a rectifiable simple closed curve. Since each arc of Γ is rectifiable, we conclude from Case 1 that each arc of Γ is a metric segment. It follows readily that Case 2 is impossible.

For select any point p_0 of Γ and let p^* denote any point of Γ such that $p_0 p_0^* = \max_{p \in \Gamma} p_0 p$ (Exercise 2). If $q \in \Gamma$, $p_0 \neq q \neq p_0^*$, the subarc of Γ with end points p_0, q which contains p_0^* is (according to the above remark) a metric segment, and since p_0^* is an interior point of that segment, then $p_0 q = p_0 p_0^* + p_0^* q > p_0 p_0^*$, contrary to the definition of p_0^*.

The proof of the theorem is complete.

EXERCISES

1. Show that a metric segment is characterized among all metric arcs by the property that its length equals the distance of its end points.

2. If Γ denotes a simple closed curve of a metric space, show that Γ contains for each of its points p at least one *diametral* point p^* (that is, $pp^* = \max_{q \in \Gamma} pq$).

10.6 CURVE THEORY IN CLASSICAL DIFFERENTIAL GEOMETRY.

EQUIVALENCE OF METRIC AND CLASSICAL CURVATURES

Since euclidean space is ptolemaic, Theorem 10.3 insures that if p is a point of any arc of $\mathbf{E}_3$, the existence of either $k_M(p)$ or $k_H(p)$ implies the existence of the other, and the two are equal. Hence for arcs of $\mathbf{E}_3$, we write $k(p) = k_M(p) = k_H(p)$, and the question arises: How does $k(p)$ compare with the classical curvature $K(p)$?

In the application of the differential calculus to the study of local properties of curves and surfaces of three-dimensional euclidean space that is commonly called *Differential Geometry*, the loci investigated are defined by equations that usually express the cartesian coordinates of their points in terms of functions of one or two variables. Thus, for example, a curve C is given (parametrically) by equations

$$C: \qquad x_i = x_i(t), \qquad (i = 1, 2, 3),$$

where (at least in the classical treatment of the subject) each of the functions $x_1(t)$, $x_2(t)$, $x_3(t)$ is real, single-valued, and analytic in an interval T of real numbers, and map that interval in a *one-to-one* manner onto the curve C (that is, to each $t \in T$ there corresponds exactly one point p, whose rectangular cartesian coordinates are $(x_1(t), x_2(t), x_3(t))$, while each point p of C is the correspondent of exactly one parametric value t, $(t \in T)$. In order that C should not degenerate into a point, it is assumed, moreover, that not all of the functions $x_i(t)$ $(i = 1, 2, 3)$ are constants. It follows that if t_0 is any arbitrary but fixed point of T, then for all values of τ less than a properly chosen positive constant, we may write

$$x_i(t_0 + \tau) = x_i(t_0) + x_i'(t_0)\tau + x_i''(t_0)\tau^2/2! + x_i'''(t_0) \cdot \tau^3/3! + \cdots,$$

where $x_i'(t_0)$, $x_i''(t_0)$, $\ldots$, denote the first, second, $\ldots$ derivatives of $x_i(t)$ with respect to t, evaluated at $t = t_0$ $(i = 1, 2, 3)$.

If, moreover, the first derivatives $x_1'(t)$, $x_2'(t)$, $x_3'(t)$ do not simultaneously vanish for $t \in T$, the curve C is called a *regular* analytic space curve and the parameter t is a regular parameter. Regular analytic space curves, parametrized by means of a regular parameter, are the curves whose theory is most extensively developed.

But even a regular space curve may contain *singular* points. They correspond to those values of t (if any) for which the rank of the matrix

$$\left\| \begin{matrix} x_1'(t) & x_2'(t) & x_3'(t) \\ x_1''(t) & x_2''(t) & x_3''(t) \end{matrix} \right\|$$

is 1. All other points of C are called regular points. The curve C is directed
by defining the positive direction along C as that of increasing parameter,
and the directed arc length s from a point $p_{t_0} = (x_1(t_0), x_2(t_0), x_3(t_0))$ of C
to a point p_t is easily shown to be expressible in the form

$$s = \int_{t_0}^{t} \sqrt{(x_1')^2 + (x_2')^2 + (x_3')^2} \, dt.$$

Let L denote a locus (line, plane, circle, or other) containing a point p of
curve C. If a point p' of C approaches p along C, the distance (p', L) of
p' from L, and the arc length $\ell(p, p')$ both approach zero. The infinitesimal
(p', L) has *order* $n + 1$ with respect to the infinitesimal $\ell(p, p')$ (n, a nonnegative integer) provided $\lim_{p' \to p} (p', L)/\ell^{n+1}(p, p')$ is finite, not zero. In
this case, the locus L is said to have *contact* of order n with C at p.

If L is a line and $n \geq 1$, then L is *tangent* to C at p; if L is a plane and $n \geq 2$,
then L is an *osculating plane* of C at p; and if L is a circle and $n \geq 2$, it is
an *osculating circle* of C at p. At a regular point p of C, $n = 1$ for the tangent
line and $n = 2$ for the (unique) osculating plane and osculating circle. It
follows easily from the definition that $(x_1'(t), x_2'(t), x_3'(t))$ are direction
numbers of the tangent line, and selecting the positive direction of the
tangent to coincide with the positive direction of C, the direction cosines
of the tangent are obtained by dividing each of the direction numbers by

$$\frac{ds}{dt} = [(x_1')^2 + (x_2')^2 + (x_3')^2]^{1/2}.$$

Hence the unit tangent vector α is given by

$$\alpha = \left(\frac{dx_1}{ds}, \frac{dx_2}{ds}, \frac{dx_3}{ds} \right).$$

The plane through p perpendicular to the tangent to C at p is called the
normal plane of C at p. It intersects the osculating plane of C at p in the
principal normal line, while the line through p perpendicular to the osculating
plane is the *binormal* line of C at p. These two lines carry the principal normal
unit vector β and the binormal unit vector γ, respectively, localized with α

at p and forming an orthogonal "right-handed" system with $\alpha = \beta \times \gamma$, $\beta = \gamma \times \alpha$, and $\gamma = \alpha \times \beta$, where $\times$ denotes the vector product.

The reciprocal $1/R$ of the radius R of the osculating circle of C at a regular point p is the (classical) curvature of C at p. It is expressed by

$$\frac{1}{R} = \left[\left(\frac{d^2 x_1}{ds^2} \right)^2 + \left(\frac{d^2 x_2}{ds^2} \right)^2 + \left(\frac{d^2 x_3}{ds^2} \right)^2 \right]^{1/2} \geq 0.$$

If $\not\!\times (\gamma_p, \gamma_{p'})$ denotes the angle between the unit binormal vectors at a regular point p of C and a neighboring point p', then the *torsion* $1/T$ of C at p is, within sign,

$$\lim_{p' \to p} \frac{\not\!\times (\gamma_p, \gamma_{p'})}{\ell(p, p')}.$$

This leads quickly to the expression

$$\left(\frac{1}{T} \right)^2 = \sum_{i=1}^{3} \left(\frac{d\gamma_i}{ds} \right)^2,$$

where $\gamma = (\gamma_1, \gamma_2, \gamma_3)$, and the sign of $1/T$ is chosen so that

$$\frac{d\gamma}{ds} = \frac{1}{T} \cdot \beta = \frac{\beta}{T}.$$

The so-called Frenet-Serret formulas

$$\frac{d\alpha}{ds} = \frac{\beta}{R}, \qquad \frac{d\beta}{ds} = -\frac{\alpha}{R} - \frac{\gamma}{T}, \qquad \frac{d\gamma}{ds} = \frac{\beta}{T}$$

are now easily derived. They play a predominant role in the classical differential geometry of curves. With their help, the vectors obtained by differentiating finitely often (with respect to s) the defining vector $x = (x_1(t), x_2(t), x_3(t))$ of the curve, are expressible as linear combinations of α, β, γ, with coefficients that are rational functions of $1/R$, $1/T$, and their derivatives with respect to s. This observation leads to the canonical representation of C in a neighborhood of a regular point p_0. Selecting the coordinate system so that p_0 is the origin, $\alpha = (1, 0, 0)$, $\beta = (0, 1, 0)$, and $\gamma = (0, 0, 1)$, the following parametrization of C (in terms of arc length s measured from the origin 0) is valid in a neighborhood of 0:

$$x_1(s) = s \qquad\qquad\qquad - (\tfrac{1}{6})(1/R_o)^2 s^3 + \cdots,$$

$$x_2(s) = (\tfrac{1}{2})(1/R_o)s^2 \qquad + (\tfrac{1}{6})\frac{d}{ds}(1/R)_o s^3 + \cdots,$$

$$x_3(s) = \qquad\qquad\qquad - (\tfrac{1}{6})(1/R_o)(1/T_o)s^3 + \cdots,$$

where $(1/R_0)$, $(1/T_0)$ denote the curvature and torsion, respectively, of C at the origin. The terms of these series that are not exhibited involve higher powers of s with coefficients that depend solely on the values of $(1/R)$, $(1/T)$ and their derivatives with respect to s, evaluated at the origin. Since the origin is a regular point of C, derivatives of all orders of $(1/R)$ and $(1/T)$ with respect to s exist there.

THEOREM 10.5 *Let C be a curve of E_3 that admits a parametrization $x_i = x_i(s)$, $(i = 1, 2, 3)$, in terms of arc length s, with each of the three functions $x_1(s)$, $x_2(s)$, $x_3(s)$ continuous, together with its first three derivatives, at each point of a neighborhood T of a point p of C. Then the metric curvatures $k_M(p)$ and $k_H(p)$ of C at p exist, and each equals the classical curvature $1/R_p$ of C at p.*

Proof. Select the coordinate system so that (1) p is the origin o (with s measured from o) and (2) the unit tangent, principal normal, and binormal vectors of C at o lie along the x_1, x_2, and x_3 axes, respectively. Let q, r denote any two distinct points of C, $q \neq p \neq r$, with $\ell(o, q) = u$, $\ell(q, r) = v$, and select the labeling of the points so that $v > 0$.

From the hypothesis, we may write

$$x_i(u + v) - x_i(u) = vx_i'(u) + (\tfrac{1}{2})v^2 x_i''(u) + (\tfrac{1}{6})v^3 x_i'''(u + \theta_i v),$$

$0 < \theta_i < 1$, $(i = 1, 2, 3)$.

Then

$$qr = \{\textstyle\sum [x_i(u + v) - x_i(u)]^2\}^{1/2}$$
$$= v\{\textstyle\sum [x_i'(u) + (\tfrac{1}{2})vx_i''(u) + (\tfrac{1}{6})v^2 x_i'''(u + \theta_i v)]^2\}^{1/2},$$

with all summations with respect to $i = 1, 2, 3$.

Hence

$$[\ell(q, r) - qr]/\ell^3(q, r) = [v - qr]/v^3$$
$$= [1 - \{\textstyle\sum [x_i'(u) + (\tfrac{1}{2})vx_i''(u)$$
$$+ (\tfrac{1}{6})v^2 x_i'''(u + \theta_i v)]^2\}^{1/2}]/v^2.$$

Rationalizing the numerator, the right side of the above equality becomes

$$1 - \textstyle\sum [x_i'(u) + (\tfrac{1}{2})vx_i''(u) + (\tfrac{1}{6})v^2 x_i'''(u + \theta_i v)]^2$$

divided by

$$v^2[1 + \{\sum [x_i'(u) + (\tfrac{1}{2})vx_i''(u) + (\tfrac{1}{6})v^2x_i'''(u + \theta_i v)]^2\}^{1/2}].$$

Denote this fraction by F.

The numerator of F can be written

$$1 - \sum [x_i'(u)]^2 - 2 \sum x_i'(u)[(\tfrac{1}{2})vx_i''(u) + (\tfrac{1}{6})v^2x_i'''(u + \theta_i v)]$$
$$- \sum [(\tfrac{1}{2})vx_i''(u) + (\tfrac{1}{6})v^2x_i'''(u + \theta_i v)]^2,$$

and since $\sum [x_i'(u)]^2 = 1$ and $\sum x_i'(u)x_i'' (u) = 0$, that expression equals

$$-v^2\{(\tfrac{1}{3}) \sum x_i'(u)x_i'''(u + \theta_i v) + \sum [(\tfrac{1}{2})x_i''(u) + (\tfrac{1}{6})vx_i'''(u + \theta_i v)]^2\}.$$

Dividing the numerator and denominator of F by v^2, and taking the limit as q, r approach o (that is, as u, $v \to 0$) gives, since the limit of the denominator is 2,

$$\lim_{q, r \to o} [\ell(q, r) - qr]/\ell^3(q, r) = -(\tfrac{1}{6}) \sum x_i'(0)x_i'''(0) - (\tfrac{1}{8}) \sum [x_i''(0)]^2.$$

Since the unit tangent vector of C at o lies along the x_1-axis, then $x_1'(0) = 1$, $x_2'(0) = 0 = x_3'(0)$. Use of the Frenet-Serret formulas readily yields $x_1''(0) = -(1/R_o)^2$ and $\sum [x_i''(0)]^2 = (1/R_o)^2$, where $1/R_o$ denotes the classical curvature of C at the origin.

Hence

$$\lim_{q, r \to o} [\ell(q, r) - qr]/\ell^3(q, r) = (\tfrac{1}{6})(1/R_o)^2 - (\tfrac{1}{8})(1/R_0)^2 = (\tfrac{1}{24})(1/R_o)^2,$$

and it follows that

$$k_H(o) = 1/R_o.$$

Since euclidean space is metric and ptolemaic, we conclude from a previous theorem that $k_M(o) = 1/R_o$, and the proof of the theorem is complete.

The proof of the following corollary is immediate.

COROLLARY   *At each regular point p of a regular analytic curve C of E_3, the metric curvatures $k_M(p)$ and $k_H(p)$ exist, and each equals the classical curvature of C at p.*

EXERCISES

1. Show that the plane curve $y = x^4 \sin(1/x)$, $x \neq 0$, $y = 0$, $x = 0$, does not have metric curvature at the origin.

2. If p is a regular point of a regular analytic curve C of E_3 show that

$$\text{(a)} \quad 1/R_p = 2 \lim_{q \to p} (\sin \theta)/pq,$$

where q is a neighboring point of p on C, $(q \neq p)$, and θ denotes the angle between the tangent to C at p and the secant determined by the points p and q; and

$$\text{(b)} \quad 1/R_p = 2 \lim_{q \to p} \lim_{r \to p} (\sin \sphericalangle \, p\!:\!q, r)/qr,$$

where q, r are neighboring points of p on C, $(q \neq p \neq r \neq q)$, and $\sphericalangle \, p\!:\!q, r$ denotes the angle at p of the triangle with vertices p, q, r.

10.7   METRIC TORSION τ_A

Since the radius of the circumscribed circle of three points is readily expressed in terms of the three mutual distances of those points, and the classical curvature of a curve C at a point p may be defined as the reciprocal of the radius of the osculating circle of C at p (which is itself definable at a regular point p of a regular analytic curve C as the limit of circles through points q, r, s of C as those points approach p along C) the metrization $k_M(p)$ of curvature (and the resulting extension of the notion to general metric continua) readily suggests itself.

No such "natural" approach to the metrization of the classical concept of torsion is available. To obtain such a metrization that is meaningful in any metric space, we consider first the situation in euclidean space $\mathbf{E}_3$, and prove the following theorem.

THEOREM 10.6   *If $\Delta\theta$ $(0 \leq \Delta\theta \leq \pi)$ denotes the angle between the osculating plane of C at a regular point p and the plane determined by the tangent to C at p and a point q of C $(q \neq p)$, then*

$$\left| \frac{1}{T_p} \right| = 3 \lim_{q \to p} \frac{\sin \Delta\theta}{pq}.$$

Proof. Selecting the coordinate system as in the proof of Theorem 10.5, the normal to the plane determined by $q = (x_1(s), x_2(s), x_3(s))$ and $\alpha = (1, 0, 0)$, the tangent to C at the origin, has direction numbers $(0, -x_3, x_2)$, and hence

$$\sin \Delta\theta = \frac{|x_3|}{[x_2^2 + x_3^2]^{1/2}}$$

Since $\lim_{q \to p} |s|/pq = 1$,

$$\lim_{q \to p} \frac{\sin \Delta\theta}{pq} = \lim_{q \to p} \frac{\sin \Delta\theta}{|s|} = \lim_{q \to p} \frac{|x_3|}{|s| \cdot [x_2^2 + x_3^2]^{1/2}},$$

and substituting for x_2 and x_3 the two series given in the canonical representation of C, we obtain after an easy computation the desired relation:

$$3 \lim_{q \to p} \frac{\sin \Delta\theta}{pq} = \left| \frac{1}{T_p} \right|.$$

Now let r_1, r_2 be points of C that determine with p and q the planes $\pi(p, r_1, r_2)$ and $\pi(q, r_1, r_2)$. Then

$$\frac{\sin \Delta\theta}{pq} = \lim_{r_1, r_2 \to p} \frac{\sin \angle [\pi(p, r_1, r_2), \pi(q, r_1, r_2)]}{pq},$$

where $\angle [\pi(p, r_1, r_2), \pi(q, r_1, r_2)]$ denotes the smaller of the dihedral angles made by the two planes, and so

$$\left| \frac{1}{T_p} \right| = 3 \lim_{p \to q} \lim_{r_1, r_2 \to p} \frac{\sin \angle [\pi(p, r_1, r_2), \pi(q, r_1, r_2)]}{pq}.$$

Hence a metrization of classical torsion will be obtained when the sine of a dihedral angle $\angle [\pi(p, r_1, r_2), \pi(q, r_1, r_2)]$ of a tetrahedron $T(p, q, r_1, r_2)$ is expressed in terms of the six mutual distances of the four points. We now direct ourselves to this task.

If, in the determinant

$$D(p, q, r_1, r_2) = \begin{vmatrix} 0 & 1 & 1 & 1 & 1 \\ 1 & 0 & pq^2 & pr_1^2 & pr_2^2 \\ 1 & pq^2 & 0 & qr_1^2 & qr_2^2 \\ 1 & pr_1^2 & qr_1^2 & 0 & r_1r_2^2 \\ 1 & pr_2^2 & qr_2^2 & r_1r_2^2 & 0 \end{vmatrix},$$

the fourth row (column) is subtracted from the second, third, and fifth rows (columns), application of the law of cosines gives

(*) $$D(p, q, r_1, r_2) = 8pr_1^2 \cdot qr_1^2 \cdot r_1r_2^2 \cdot \Delta,$$

where

$$\Delta = \begin{vmatrix} 1 & \cos \angle r_1 : p, q & \cos \angle r_1 : p, r_2 \\ \cos \angle r_1 : p, q & 1 & \cos \angle r_1 : q, r_2 \\ \cos \angle r_1 : p, r_2 & \cos \angle r_1 : q, r_2 & 1 \end{vmatrix},$$

with $\angle r_1 : x, y$ denoting the angle at the vertex r_1 of the triangle $T(r_1, x, y)$. Thus the three angles occurring in determinant Δ are the three face angles of the tetrahedron $T(p, q, r_1, r_2)$ at the vertex r_1.

Multiplying the last column of Δ by $\cos \angle r_1 : r_2, p$ and $\cos \angle r_1 : r_2, q$, and subtracting from the first and second columns, respectively, yields (upon application of the spherical law of cosines)

$$\Delta = \sin^2 \angle r_1 : p, r_2 \cdot \sin^2 \angle r_1 : q, r_2 \cdot \sin^2 \angle r_1, r_2 : p, q,$$

where $\angle r_1, r_2 : p, q$ denotes the dihedral angle made by the planes $\pi(p, r_1, r_2)$, $\pi(q, r_1, r_2)$ of the tetrahedron $T(p, q, r_1, r_2)$, with edge seg (r_1, r_2).

Since

$$-D(p, r_1, r_2) = 16A^2(p, r_1, r_2) = 4pr_1^2 \cdot r_1r_2^2 \cdot \sin \angle r_1 : p, r_2,$$

$$-D(q, r_1, r_2) = 16A^2(q, r_1, r_2) = 4qr_1^2 \cdot r_1r_2^2 \cdot \sin \angle r_1 : q, r_2,$$

substitution in (*) yields

$$\sin \angle r_1, r_2 : p, q = r_1r_2 \left[\frac{2D(p, q, r_1, r_2)}{D(p, r_1, r_2)D(q, r_1, r_2)} \right]^{1/2},$$

and hence a metrization of the absolute value of classical torsion is given by

(†) $$\left| \frac{1}{T_p} \right| = \lim_{q \to p} \lim_{r_1, r_2 \to p} \frac{r_1r_2}{pq} \cdot \sqrt{\frac{18D(p, q, r_1, r_2)}{D(p, r_1, r_2) \cdot D(q, r_1, r_2)}}.$$

Remarks. Since it was assumed that the point-triples p, r_1, r_2 and q, r_1, r_2 determine planes $\pi(p, r_1, r_2)$ and $\pi(q, r_1, r_2)$, respectively, neither of these triples is linear, and hence $D(p, r_1, r_2) < 0$ and $D(q, r_1, r_2) < 0$. Thus, the denominator of the radicand is positive, and since the points p, q, r_1, r_2 are in $\mathbf{E}_3$ (possibly in $\mathbf{E}_2$), $D(p, q, r_1, r_2) \geqq 0$ (Exercise 1). Hence the radicand in (†) is nonnegative for four points of $\mathbf{E}_3$ with two nonlinear triples.

If, now, p_1, p_2, p_3, p_4 is any *ordered metric quadruple*, the metric torsion τ_A of the ordered quadruple is defined by

$$\tau_A(p_1, p_2, p_3, p_4) = \frac{p_2\,p_3}{p_1\,p_4} \sqrt{\frac{18\,|D(p_1, p_2, p_3, p_4)|}{D(p_1, p_2, p_3) \cdot D(p_2, p_3, p_4)}},$$

provided neither of the determinants $D(p_1, p_2, p_3)$, $D(p_2, p_3, p_4)$ vanishes.

If p is an accumulation point of a metric space $\mathbf{M}$, the torsion $\tau_A(p)$ of $\mathbf{M}$ at p is defined by

$$\tau_A(p) = \lim_{s \to p}\ \lim_{q, r \to p} \tau_A(p, q, r, s)$$

provided the iterated limit exists and *is the same for each ordered quadruple obtained by keeping p fixed and permuting q, r, s.*

This metrization of torsion was given by the Hungarian mathematician George Alexits in 1938. It requires that

$$\lim_{s \to p}\ \lim_{q, r \to p} \tau_A(p, q, r, s) = \lim_{q \to p}\ \lim_{r, s \to p} \tau_A(p, r, s, q)$$

$$= \lim_{r \to p}\ \lim_{q, s \to p} \tau_A(p, s, q, r), \text{ etc.}$$

The use of an iterated limit and the lack of symmetry that necessitates the ordering of the quadruples (that is, the value of $\tau_A(p, q, r, s)$ depends on the *ordering* of the quadruple) must be regarded as disadvantages of the Alexits metrization of torsion. In the following section another metrization of torsion will be obtained that is free of those particular disadvantages and has the same value as $\tau_A(p)$ at a regular point p of a regular analytic curve of $\mathbf{E}_3$.

EXERCISES

1. Show that $D(p_1, p_2, p_3, p_4) \geqq 0$ for each four points p_1, p_2, p_3, p_4 of $\mathbf{E}_3$, and $D(p_1, p_2, p_3, p_4) = 0$ if and only if p_1, p_2, p_3, p_4 are in $\mathbf{E}_2$.

2. Obtain a "factorization" of $D(p_1, p_2, \ldots, p_{n+1})$ in terms of higher dimensional angles that extends formula (∗) of this section.

10.8   METRIC TORSION τ_B

Let p_1, p_2, p_3, p_4 be four points of $\mathbf{E}_3$, *without a linear triple*, and put

$$\tau_B^2(p_1, p_2, p_3, p_4)$$

$$= \frac{18D(p_1, p_2, p_3, p_4)}{[D(p_1, p_2, p_3) \cdot D(p_1, p_2, p_4) \cdot D(p_1, p_3, p_4) \cdot D(p_2, p_3, p_4)]^{1/2}}$$

Using the expansions of the D-determinants given in the preceding section, we readily obtain

$$\tau_B^2(p_1, p_2, p_3, p_4)$$

$$= \frac{9 \sin \not\prec p_1 : p_2, p_3 \cdot \sin \not\prec p_1 : p_2, p_4 \cdot \sin^2 \not\prec p_1, p_2 : p_3, p_4}{p_2 p_3 \cdot p_2 p_4 \cdot \sin \not\prec p_1 : p_3, p_4 \cdot \sin \not\prec p_2 : p_3, p_4}.$$

Upon multiplying numerator and denominator of the fraction by $4 \cdot p_3 p_4^2$,

$$\tau_B^2(p_1, p_2, p_3, p_4) = \frac{9.2 \sin \not\prec p_1 : p_2, p_3}{p_2 p_3} \cdot \frac{2 \sin \not\prec p_1 : p_2, p_4}{p_2 p_4}$$

$$\times \frac{p_3 p_4}{2 \sin \not\prec p_1 : p_3, p_4} \cdot \frac{p_3 p_4}{2 \sin \not\prec p_2 : p_3, p_4} \cdot \frac{\sin^2 p_1, p_2 : p_3, p_4}{p_3 p_4^2},$$

and so

(∗) $$\tau_B^2(p_1, p_2, p_3, p_4)$$

$$= \frac{9 k_M(p_1, p_2, p_3) k_M(p_1, p_2, p_4)}{k_M(p_1, p_3, p_4) k_M(p_2, p_3, p_4)} \cdot \frac{\sin^2 \not\prec p_1, p_2 : p_3, p_4}{p_3 p_4^2}.$$

If, now, p_1, p_2, p_3, p_4 are points of a euclidean continuum C and $k_M(p_1) \neq 0$, then clearly

$$\lim_{p_2, p_3 \to p_1} k_M(p_1, p_2, p_3) = \lim_{p_2, p_4 \to p_1} k_M(p_1, p_2, p_4) = \lim_{p_3, p_4 \to p_1} k_M(p_1, p_3, p_4)$$

$$= \lim_{p_2, p_3, p_4 \to p_1} k_M(p_2, p_3, p_4)$$

$$= k_M(p_1) = k_H(p_1) = \frac{1}{R_{p_1}} \neq 0,$$

by Theorem 10.5, and consequently (choosing the non-negative square root)

$$\lim_{p_2, p_3, p_4 \to p_1} \tau_B(p_1, p_2, p_3, p_4) = 3 \lim_{p_2, p_3, p_4 \to p_1} \frac{\sin \not\prec p_1, p_2 : p_3, p_4}{p_3 p_4}.$$

Defining the metric torsion $\tau_B(p)$ of a metric space $\mathbf{M}$ at an accumulation point p by

(†) $$\tau_B(p) = \lim_{q,r,s \to p} \frac{[18 \cdot |D(p, q, r, s)|]^{1/2}}{[D(p, q, r) \cdot D(p, q, s) \cdot D(p, r, s) \cdot D(q, r, s)]^{1/4}},$$

we have proved the following theorem.

THEOREM 10.7   *At a regular point p of a regular analytic curve C, $\tau_B(p) = |1/T_p| = \tau_A(p)$.*

Remarks.   The metrization $\tau_B(p)$ of torsion defined in (†) is free of the disadvantages of $\tau_A(p)$ mentioned in the preceding section, though (as we have just proved) both metrizations yield the same result (the absolute value of the classical torsion) when applied to a regular analytic curve of E_3 at a regular point. It was introduced by L. M. Blumenthal in 1939, together with its strengthened form $t_B^*(p)$ defined in the next section.

The metrizations of curvature and torsion permit a complete investigation of the differential geometry of curves by purely metric means—that is, without defining curves by means of functions, assumed differentiable to any desired order, expressing the various geometrical properties by means of analytical formulas, and pressing the whole development of the differential calculus into service.

EXERCISES

1. Show how the torsion of four points that is defined in formula (∗) of this section may be used to derive a law of sines for tetrahedra of E_3.

2. Is there a sine law for tetrahedra of E_3 in terms of the four trihedral angles of the tetrahedron?

3. For what class of tetrahedra are the sine laws the simplest (that is, direct extensions of the law of sines for triangles)?

10.9   A STRENGTHENED FORM OF τ_B

The metrization τ_B of torsion defined above may be strengthened in an obvious manner by defining

$$\tau_B^*(p) = \lim_{p_i \to p} \frac{[18 \cdot |D(p_1, p_2, p_3, p_4)|]^{1/2}}{[D(p_1, p_2, p_3)D(p_1, p_2, p_4)D(p_1, p_3, p_4)D(p_2, p_3, p_4)]^{1/4}},$$

for p an accumulation point of any metric space $\mathbf{M}$, where $p_1, p_2, p_3, p_4 \in \mathbf{M}$ that do not contain a linear triple. We define $\tau_B^*(p_1, p_2, p_3, p_4) = \tau_B(p_1, p_2, p_3, p_4)$.

Clearly, the existence of $\tau_B(p)$ does not imply the existence of $\tau_B^*(p)$; it is less obvious, perhaps, that $\tau_B^*(p)$ may exist without $\tau_B(p)$ existing (Exercise 2).

It has been shown in the Missouri doctoral dissertation of J. W. Sawyer that at a regular point p of a regular analytic curve of $\mathbf{E}_3$, the torsion $\tau_B^*(p)$ exists (and consequently equals the absolute value of the classical torsion). Even at a singular point p_0 of a regular analytic curve of $\mathbf{E}_3$, $\tau_B^*(p_0)$ exists (and equals the absolute value of the classical torsion at p_0) provided the curve has a nonnull tangent vector at p_0. From the formula (*) of the preceding section, it follows that if p is a point of any continuum C of $\mathbf{E}_3$ such that $k_M(p) \neq 0$, and $\tau_B^*(p)$ exists, then

$$\tau_B^*(p) = 3 \lim_{p_i \to p} \frac{\sin \measuredangle\, p_1, p_2 : p_3, p_4}{p_3\, p_4},$$

where $p_1, p_2, p_3, p_4 \in C$, with no triple linear, and $\measuredangle\, p_1, p_2 : p_3, p_4$ denotes the smaller of the dihedral angles made by the planes $\pi(p_1, p_2, p_3), \pi(p_1, p_2, p_4)$ determined by the nonlinear point-triples p_1, p_2, p_3 and p_1, p_2, p_4.

Defining a plane π to be the osculating plane, *in the strong sense*, of a euclidean continuum C at a point p of C provided $\pi(p_1, p_2, p_3) \to \pi$ as $p_i \to p$, $p_i \in C$ ($i = 1, 2, 3$), we establish the following theorem.

THEOREM 10.8   *If $\tau_B^*(p)$ exists and $k_M(p) \neq 0$, $p \in C$, a euclidean continuum, then C has at p an osculating plane in the strong sense.*

Proof.   From the existence of $k_M(p)$, we conclude that C is a rectifiable arc A in a neighborhood $U(p; \delta)$ of p (Theorem 10.1) and since $k_M(p) \neq 0$, then A is not a segment. Let $p_1, p_2, \ldots, p_6$ be points of A, with no three of the six points linear. We show that the planes $\pi(p_1, p_2, p_3)$, $\pi(p_4, p_5, p_6)$ approach the same plane as $p_i \to p$ ($i = 1, 2, \ldots, 6$).

Now

$$\tau_B^*(p) = 3 \lim_{\substack{p_i \to p \\ (i = 1, 2, 3, 5)}} \frac{\sin \measuredangle\, p_1, p_3 : p_2, p_5}{p_2\, p_5},$$

and consequently

$$\lim_{\substack{p_i \to p \\ (i = 1, 2, 3, 5)}} \sin \measuredangle\, p_1, p_3 : p_2, p_5 = 0.$$

It follows that the planes $\pi(p_1, p_2, p_3)$ and $\pi(p_1, p_3, p_5)$ approach coincidence.

Similarly, from

$$\lim_{\substack{p_i \to p \\ (i=1,3,4,5)}} \sin \measuredangle\, p_1, p_5 : p_3, p_4 = \lim_{\substack{p_i \to p \\ (i=1,4,5,6)}} \sin \measuredangle\, p_4, p_5 : p_1, p_6 = 0,$$

the planes $\pi(p_1, p_3, p_5)$, $\pi(p_1, p_4, p_5)$ approach coincidence, as do also the planes $\pi(p_1, p_4, p_5)$ and $\pi(p_4, p_5, p_6)$.

Hence $\pi(p_1, p_2, p_3)$ and $\pi(p_4, p_5, p_6)$ approach coincidence, and the theorem is proved.

EXERCISES

1. If $\tau_B^*(p)$ exists, $p \in C$, a euclidean continuum, show that

$$\lim_{p_i \to p} \frac{\sin \measuredangle\, p_1, p_2 : p_3, p_4}{p_3 p_4} = \lim_{p_i \to p} \frac{\sin \measuredangle\, p_2, p_3 : p_1, p_4,}{p_1 p_4}$$

$p_i \in C$ $(i = 1, 2, 3, 4)$.

2. Show that $\tau_B^*(p)$ is continuous wherever it exists. Is $\tau_B(p)$ continuous wherever it exists?

3. Give examples to show that either one of the torsions $\tau_B(p)$, $\tau_B^*(p)$ may exist without the existence of the other.

4. Give an example of a metric arc A such that (a) A is not congruently inbeddable in the euclidean plane, and (b) $k_M(p) \neq 0$, $\tau_B^*(p) = 0$, for every $p \in A$. *Hint.* Redefine distance for point-pairs of a planar arc A so that A is locally planar (not linear) in the new metric, and contains four points that are not congruently imbeddable in the plane.

5. Does an arc A of a metric ptolemaic space exist such that $k_M(p) \neq 0$, $\tau_B^*(p) = 0$, for every $p \in A$, but A is *not* congruently imbeddable in the euclidean plane?

6. Show that none of the torsions τ_A, τ_B, τ_B^* exists at a point of a metric segment.

10.10   LOCAL METRIC CHARACTERIZATION OF PLANE CONTINUA

If C denotes any continuum of the euclidean plane, then $\tau_B^*(p) = 0$ at each point p of C for which τ_B^* exists. This follows readily from the fact that $D(q, r, s, t) = 0$ for every four points q, r, s, t of the plane (Section 10.7, Exercise 1). This section is devoted to establishing a "restricted converse" of that remark. The proof is an adaptation of one due to G. Alexits.

THEOREM 10.9 *If at each point p of a continuum C of E_3, $k_M(p) > 0$ and $\tau_B^*(p) = 0$, then C is a plane continuum.*

Proof. Invoking Theorem 10.1, the existence of $k_M(p)$, $p \in C$, insures that C is either (1) a rectifiable arc or (2) a rectifiable simple closed curve, and since $k_M(p) > 0$ for each point p of C, it follows that each point p has a neighborhood on C that is free of linear triples.

Let q, r be two distinct points of C, and denote an arc of C with end points q, r by $A(q, r)$. (Clearly, $A(q, r)$ is unique in case C is itself an arc, and there are exactly two arcs of C with end points q, r, in case C is a simple closed curve.) Since $\tau_B^*(p) = 0$, $p \in A(q, r)$, to every $\varepsilon > 0$ there corresponds a $\delta > 0$ such that if x_1, x_2, x_3, $x_4 \in A(q, r)$ and $\text{diam}(x_1, x_2, x_3, x_4) < \delta$, then (Section 10.9, Exercise 1)

$$\frac{3 \sin \not\!\times\, x_2, x_3 : x_1, x_4}{x_1 x_4} < \varepsilon.$$

(Let the reader justify this.) Now select a natural number N so large that for every index $n > N > 4$, (1) each n-lattice $L_n = (p_0, p_1, p_2, \ldots, p_n)$ inscribed in $A(q, r)$ is unique ($p_0 = q$, $p_n = r$, $p_0 p_1 = p_1 p_2 = \cdots = p_{n-1} p_n$), (2) each four consecutive points of L_n has diameter less than δ and no three of the four points are linear (see Theorem 9.8 and Lemma 9.1), and $n \cdot \delta > \ell[A(q, r)]$. Then for each $i = 0, 1, 2, \ldots, n - 3$,

$$\sin \not\!\times\, p_{i+1}, p_{i+2} : p_i, p_{i+3} < \left(\frac{\varepsilon}{3}\right) \cdot p_i p_{i+3},$$

where $\not\!\times\, p_{i+1}, p_{i+2} : p_i, p_{i+3} = \not\!\times\, [\pi(p_i, p_{i+1}, p_{i+2}), \pi(p_{i+1}, p_{i+2}, p_{i+3})] \leqq \pi/2$.

Since for each angle x ($0 \leqq x \leqq \pi/2$), $x \leqq (\pi/2) \sin x$ (let the reader verify this), it follows that

$$\not\!\times\, p_{i+1}, p_{i+2} : p_i, p_{i+3} < \left(\frac{\pi}{6}\right) \varepsilon \cdot p_i p_{i+3} \qquad (i = 0, 1, \ldots, n - 3),$$

and hence

$$\sum_{i=1}^{n-4} \not\!\times\, p_{i+1}, p_{i+2} : p_i, p_{i+3} < \left(\frac{\pi}{6}\right) \varepsilon \cdot \sum_{i=1}^{n-4} p_i p_{i+3} < \left(\frac{\pi}{6}\right) \varepsilon \cdot 3\ell[A(q, r)].$$

That is,

$$(*) \qquad \sum_{i=1}^{n-4} \not\!\times\, p_{i+1}, p_{i+2} : p_i, p_{i+3} < \left(\frac{\pi}{2}\right) \varepsilon \cdot \ell[A(q, r)].$$

Assertion. *For* $n \geq 5,$

$$\measuredangle\,[\pi(p_1, p_2, p_3), \pi(p_{n-3}, p_{n-2}, p_{n-1})] \leq \sum_{i=1}^{n-4} \measuredangle\, p_{i+1}, p_{i+2} : p_i, p_{i+3}.$$

Observing that the Assertion is valid for $n = 5$, *assume its validity for* $n = k - 1$, and consider the angles

$$\measuredangle\,[\pi(p_1, p_2, p_3), \pi(p_{k-3}, p_{k-2}, p_{k-1})],$$
$$\measuredangle\,[\pi(p_1, p_2, p_3), \pi(p_{k-4}, p_{k-3}, p_{k-2})],$$
$$\measuredangle\,[\pi(p_{k-4}, p_{k-3}, p_{k-2}), \pi(p_{k-3}, p_{k-2}, p_{k-1})].$$

These angles are the *nonobtuse* angles formed by planes $\pi(p_1, p_2, p_3)$, $\pi(p_{k-3}, p_{k-2}, p_{k-1})$, $\pi(p_{k-4}, p_{k-3}, p_{k-2})$ of E_3, taken in pairs, and it is easily seen (Exercise 1) that they satisfy the triangle inequality.

Hence

$$\measuredangle\,[\pi(p_1, p_2, p_3), \pi(p_{k-3}, p_{k-2}, p_{k-1})]$$
$$\leq \measuredangle\,[\pi(p_1, p_2, p_3), \pi(p_{k-4}, p_{k-3}, p_{k-2})]$$
$$+ \measuredangle\,[\pi(p_{k-4}, p_{k-3}, p_{k-2}), \pi(p_{k-3}, p_{k-2}, p_{k-1})]$$
$$\leq \sum_{i=1}^{k-5} \measuredangle\, p_{i+1}, p_{i+2} : p_i, p_{i+3} + \measuredangle\, p_{k-2}, p_{k-3} : p_{k-4}, p_{k-1}$$
$$= \sum_{i=1}^{k-4} \measuredangle\, p_{i+1}, p_{i+2} : p_i, p_{i+3},$$

and so the Assertion is valid for $n = k$. This establishes the Assertion by complete induction.

Combining the Assertion with (∗) yields

(∗∗) $\measuredangle\,[\pi(p_1, p_2, p_3), \pi(p_{n-3}, p_{n-2}, p_{n-1})] < \left(\dfrac{\pi}{2}\right)\varepsilon \cdot \ell[A(q, r)].$

Now as $n \to \infty$, the points $p_1, p_2, p_3 \to q$, the points $p_{n-3}, p_{n-2}, p_{n-1} \to r$, and the planes $\pi(p_1, p_2, p_3)$, $\pi(p_{n-3}, p_{n-2}, p_{n-1}$, approach the osculating, planes π_q, π_r of C at q, r, respectively (each of which exists by virtue of Theorem 10.8). It follows from (∗∗) that $\measuredangle\,[\pi_q, \pi_r]$ is arbitrarily small and hence is zero.

Thus all of the osculating planes of C form a family of mutually parallel or identical planes, and so the continuum C lies in a plane, completing the proof of the theorem.

Extensions of the preceding theorem to continua of more general classes of spaces are highly desirable, but so far have not been obtained.

EXERCISES

1. If $\angle[\pi_1,\pi_2]$, $\angle[\pi_2,\pi_3]$, $\angle[\pi_1,\pi_3]$ denote the nonobtuse angles made by three planes π_1, π_2, π_3, of $\mathbf{E}_3$, taken in pairs, show that those angles satisfy the triangle inequality. *Hint:* Consider the three lines through a point that are perpendicular, respectively, to the three planes.

10.11   BIREGULAR ARCS OF $\mathbf{E}_3$

A metric arc is called *biregular* provided it has at each of its points p, a positive metric curvature $k_M(p)$ and a positive metric torsion $\tau_B^*(p)$. It follows that every biregular arc A of $\mathbf{E}_3$ is (1) rectifiable (Theorem 10.1), (2) has a tangent (in the strong sense) at each of its points (Section 10.1, Exercise 3), and (3) possesses an osculating plane (in the strong sense), and consequently a binormal, at every point (Theorem 10.8). Since A is rectifiable, we may suppose it parameterized $x = x(s)$, where the vector $x(s)$ is a function of the arc length s. Moreover, since at each point p of A, $k_M(p) = k_H(p)$ (Theorem 10.3, Corollary), then the ratio of arc length to chord approaches 1, as two of the parametric values approach a third, and consequently $\alpha = dx/ds$ is a unit vector along the tangent line.

Remark.   Let A be a biregular arc of $\mathbf{E}_3$ and π its osculating plane at one of its points p. The perpendicular projection on π of that part of A contained in the closure of a suitable neighborhood of p is an arc. Since the mapping defined by projection is continuous, it suffices to observe that it is also one-to-one in a neighborhood of p; for if this were not the case, then points p_1, p_2 of A, arbitrarily close to p, would exist that map into the same point of π, and the plane $\pi(p, p_1, p_2)$, being perpendicular to π, would not approach π as p_1, p_2 approach p, in contradiction to the assumption that π is the osculating plane of A at p.

THEOREM 10.10   *The curvature $k_M(p)$ of a biregular arc A of $\mathbf{E}_3$ at a point p equals the curvature at p of the planar arc obtained by projecting perpendicularly upon the osculating plane of A at p the common part A_1 of A and the closure of a neighborhood of p.*

Proof. By the preceding Remark, a $\delta > 0$ exists such that the perpendicular projection of $A_1 = A \cdot \bar{U}(p; \delta)$ on the osculating plane of A at p is an arc A_1'.

For every triple q, r, s of pairwise distinct points of A_1, each of which is sufficiently close to p, the plane $\pi(q, r, s)$ is arbitrarily close to the osculating plane of A_1 at p and we have $k_M(q, r, s) = 2(\sin \, \measuredangle \, q: r, s)/rs \neq 0$, and $k_M(q', r', s') = 2(\sin \, \measuredangle \, q': r', s')/r's' \neq 0$, where q', r', s' denote the projections on π of q, r, s, respectively. It is clear that the ratio $k_M(q, r, s)/k_M(q', r', s')$ can be made arbitrarily close to 1 by taking the points q, r, s sufficiently close to p (and hence $\pi(q, r, s)$ close to π), and it follows that the ratio of the curvatures of A_1 and A_1' at p is 1.

We note that the *regularity* of A (that is, $k_M(p) \neq 0$, $p \in A$) is all that is used in the proof.

We have remarked that the existence of the osculating plane at a point p of an arc implies the existence of the unit binormal vector of the arc at p, defined as the unit vector perpendicular at p to the osculating plane. But the binormal vector so defined is not necessarily a continuous function of p. The unit binormal vector defined as follows will be continuous wherever it exists.

Consider any three pairwise distinct points p_1, p_2, p_3 of A, encountered in the order of their subscripts as one traverses A from one end point to the other. If A is regular, the unit vector $\overrightarrow{p_1 p_2}/p_1 p_2$ approaches the unit tangent vector $\alpha(p)$ as $p_1, p_2 \to p$, $p \in A$ (Section 10.1, Exercise 3). Denote the unit vectors $\overrightarrow{p_1 p_2}/p_1 p_2$, $\overrightarrow{p_2 p_3}/p_2 p_3$ by u, v respectively. If, as p_1, p_2, p_3 approach p ($p \in A$), the limit of the *vector product* of the vectors u, v exists, the resulting vector is called the *binormal vector* of A at p, and normalizing it we obtain the unit binormal vector $\gamma(p)$ of A at p.

THEOREM 10.11   *At each point p of a biregular arc A of $\mathbf{E}_3$, the unit binormal vector $\gamma(p)$ exists and is a continuous function of p.*

Proof.   We may choose a positive δ so small that (1) neither the arc $A_1 = A \cdot \bar{U}(p; \delta)$ nor its projection A_1' on π, the osculating plane of A at p, contains any linear triple of pairwise distinct points, and (2) for each three pairwise distinct points q, r, s of A_1, the plane $\pi(q, r, s)$ makes an angle with π that is less than $15°$.

Now if $p_1, p_2, p_3, p_4 \in A_1$, occurring in the order of their subscripts, the directed plane $\pi(p_1, p_2, p_3)$ makes an acute angle with the directed plane

$\pi(p_2, p_3, p_4)$, for otherwise points p_1 and p_4 are separated by the plane through p_2, p_3 that is perpendicular to $\pi(p_1, p_2, p_3)$, and the polygon in π whose vertices are the ordered quadruple p'_1, p'_2, p'_3, p'_4 is not convex. But this implies that the arc A'_1 contains a linear triple of pairwise distinct points, contrary to the selection of δ. It follows from (2) that the planes $\pi(p_1, p_2, p_3)$, $\pi(p_2, p_3, p_4)$ make an angle less than 30° (Section 10.10, Exercise 1).

Let p_1, p_2, p_3 and q_1, q_2, q_3 be two triples of pairwise distinct points of A_1. Then $\{p_1, p_2, p_3\}, \{p_1, p_2, q_3\}, \{p_1, q_2, q_3\}, \{q_1, q_2, q_3\}$ are point-triples, each of which, properly ordered, determines planes successively making angles less than 30°. Hence $\pi(p_1, p_2, p_3)$ and $\pi(q_1, q_2, q_3)$ make an acute angle, and since these two directed planes make an acute angle, so do their unit normal vectors. But each of these planes approaches the osculating plane of A at p, as the points determining them approach p, and hence their unit normal vectors coincide in the limit, and define the binormal vector $\gamma(p)$ of A at p.

The reader is asked to prove the continuity of γ in an exercise.

From the existence and continuity of unit tangent and unit binormal vector, a continuous principal normal and unit vector $\beta(p)$ of A at p is defined to be the vector product of γ with α; that is, $\beta = \gamma \times \alpha$.

If $p_0, p_1, \ldots, p_n$ are $n + 1$ pairwise distinct points of $\mathbf{E}_3$ and v_i denotes the vector $\overrightarrow{p_{i-1} p_i}$ $(i = 1, 2, \ldots, n)$, put

$$q_i = v_i \times v_{i+1} \text{ (vector product)} \qquad (i = 1, 2, \ldots, n - 1),$$
$$\theta_i = \measuredangle(v_i, v_{i+1}) \qquad (i = 1, 2, \ldots, n),$$
$$\varphi_i = \measuredangle(q_i, q_{i+1}) \qquad (i = 1, 2, \ldots, n - 1).$$

Two polygons $P = (p_0, p_1, \ldots, p_n)$, $P' = (p'_0, p'_1, \ldots, p'_n)$ are *similarly oriented* provided

$$\text{sgn}[v_i, v_{i+1}, v_{i+2}] = \text{sgn}[v'_i, v'_{i+1}, v'_{i+2}] \qquad (i = 1, 2, \ldots, n - 2),$$

where the brackets indicate the *triple scalar product* (*determinant*) of the vectors enclosed.

An arc A of E_3 is *positively oriented* provided a positive δ exists such that for every polygon $P = (p_0, p_1, \ldots, p_n)$ inscribed in A, with $p_i p_{i+1} < \delta$ $(i = 0, 1, \ldots, n - 1)$, the triple scalar product $[v_i, v_{i+1}, v_{i+2}] > 0$ $(i = 1, 2, \ldots, n - 2)$. If each such product is negative, then the arc is *negatively oriented*.

THEOREM 10.12   *Each biregular arc of* $\mathbf{E}_3$ *is positively or negatively oriented.*

Proof. It evidently suffices to show that each point p of A is contained in a subarc A_1 of A with the property that each two of its ordered quadruples are similarly oriented.

Since the metric torsion of A at p is not zero, the subarc A_1 may be chosen so that no four of its points are planar. Let $P = \{p_1, p_2, p_3, p_4\}$ and $Q = \{q_1, q_2, q_3, q_4\}$ be two quadruples of A_1, encountered in the order of their subscripts when A_1 is traversed from one end point to the other.

If P_1 denotes a quadruple of A_1 obtained by replacing the point p_1 in P by any point of A_1 on the same side of p_2 as p_1, then P and P_1 are similarly oriented, for otherwise A_1 contains a point p_1' (on the same side of p_2 as p_1) such that p_1', p_2, p_3, p_4 are planar. In like manner, any quadruple of A_1 obtained by replacing p_2 in P by any point of A_1 between p_1 and p_3 is seen to have orientation similar to P; and so for the other points of P.

Applying the above procedure to the two quadruples P and Q, we see that we may pass from one quadruple to the other by a finite sequence of quadruples of A_1, each consecutive two of which are similarly oriented. Since that relation is clearly transitive, the theorem is established.

EXERCISES

1. Prove that the binormal vector $\gamma(p)$ is continuous wherever it exists.

2. Show that the vectors $\alpha(p)$, $\beta(p)$, $\gamma(p)$ are pairwise mutually orthogonal.

10.12 FUNDAMENTAL THEOREM OF CURVE THEORY

The primary importance of the concepts of arc length, curvature, and torsion in the differential geometry of curves of $\mathbf{E}_3$ is fully revealed by the so-called fundamental theorem of curve theory. As it is usually stated in classical differential geometry, that theorem asserts that a necessary and sufficient condition that two *regular analytic curves* (expressed parametrically in terms of a regular parameter) be congruent is the existence of a one-to-one correspondence between the points of the two curves that preserves arc length, curvature, and torsion. The usual proof of the theorem is based upon the Frenet-Serret formulas that exhibit the derivatives of the unit tangent, principal normal, and binormal vectors (with respect to arc length) as linear combinations of those vectors with curvature and torsion the scalar multipliers (Section 10.6).

The purpose of this section is to present a purely metric proof of the fundamental theorem for *biregular arcs*, based upon our metrizations of curvature and torsion. We first establish two lemmas.

LEMMA 10.1 *Let A denote any biregular arc of $\mathbf{E}_3$ and ε any positive number. There exists a positive integer N such that if $n > N$, each n-lattice $L_n = \{p_0, p_1, \ldots, p_n\}$ of A with side $\lambda(n)$ has the following properties:*

$$(1) \quad 1 - \varepsilon < \frac{\ell[A(p_i, p_{i+1})]}{\lambda(n)} < 1 + \varepsilon \qquad (i = 0, 1, \ldots, n-1),$$

$$(2) \quad 1 - \varepsilon < \frac{k_M(p_i)}{[\sin \theta_i]/\lambda(n)} < 1 + \varepsilon \qquad (i = 1, 2, \ldots, n-1),$$

$$(3) \quad 1 - \varepsilon < \frac{\tau_B^*(p_i)}{[\sin \varphi_i]/\lambda(n)} < 1 + \varepsilon \qquad (i = 1, 2, \ldots, n-2).$$

Proof. Property (1) is valid since the existence of $k_M(p)$, $p \in A$, implies that the limit of the arc-chord ratio is 1.

From the uniform continuity of $k_M(p)$ in A (Section 10.1, Exercise 3),

$$1 - \varepsilon < \frac{k_M(p_i)}{k_M(p_{i-1}, p_i, p_{i+1})} < 1 + \varepsilon$$

for n sufficiently large $(i = 1, 2, \ldots, n-1)$.

Since

$$k_M(p_{i-1}, p_i, p_{i+1}) = \frac{2[\sin(\pi - \theta_i)]}{p_{i-1}p_{i+1}}$$

$$= \frac{\sin \theta_i}{\lambda(n)} \cdot \frac{2\lambda(n)}{p_{i-1}p_{i+1}} \qquad (i = 1, 2, \ldots, n-1),$$

and for n sufficiently large, $2\lambda(n)/p_{i-1}p_{i+1}$ is arbitrarily close to 1, property (2) follows.

Similarly, the uniform continuity of $\tau_B^*(p)$ in A (Section 10.9, Exercise 2) yields

$$1 - \varepsilon < \frac{\tau_B^*(p_i)}{\tau_B^*(p_{i-1}, p_i, p_{i+1}, p_{i+2})} < 1 + \varepsilon,$$

and

$$1 - \varepsilon < \frac{\tau_B^*(p_{i-1}, p_i, p_{i+1}, p_{i+2})}{3[\sin \varphi_i]/p_{i-1}p_{i+2}} < 1 + \varepsilon \qquad (i = 1, 2, \ldots, n-2),$$

for n sufficiently large (Section 10.8, formula (*)).
    But

$$\frac{3 \sin \varphi_i}{p_{i-1}p_{i+2}} = \frac{\sin \varphi_i}{\lambda(n)} \cdot \frac{3\lambda(n)}{p_{i-1}p_{i+2}},$$

and since $3\lambda(n)/p_{i-1}p_{i+2} \to 1$ as $n \to \infty$, property (3) is valid.
    Hence for n sufficiently large, all three properties hold.

LEMMA 10.2   *Let A denote any biregular arc of $\mathbf{E}_3$, and ε any positive number. There exists a positive integer N such that for any $n > N$ and any n-lattice $L_n = \{p_0, p_1, \ldots, p_n\}$ inscribed in A, (1) $0 < \theta_i < \varepsilon$ $(i = 1, 2, \ldots, n)$, and (2) $0 < \varphi_i < \varepsilon$ $(i = 1, 2, \ldots, n-1)$.*

Proof.   Since A has a unique (continuous) tangent and binormal (in the strong sense) at each of its points, it follows that $\not\prec p_i : p_{i-1}, p_{i+1} \to \pi$, and $\not\prec p_i, p_{i+1} : p_{i-1}, p_{i+2} \to 0$, as $n \to \infty$. Hence for n sufficiently large, $0 < \theta_i < \varepsilon$ $(i = 1, 2, \ldots, n)$, and $0 < \varphi_i < \varepsilon$ $(i = 1, 2, \ldots, n)$, with the first "half" of each of the two inequalities being a consequence of the biregularity of A.

We are now in position to prove the fundamental theorem.

THEOREM 10.13   *A necessary and sufficient condition that two biregular analytic arcs of $\mathbf{E}_3$ be congruent is the existence of a one-to-one correspondence between their points that preserves arc length, metric curvature k_M, and metric torsion τ_B^*.*

Proof.   Since arc length, metric curvature, and torsion are obviously congruence invariants, the necessity of the condition is immediate. We turn to the sufficiency.
    Let A, A^* denote two biregular analytic arcs, and let $p^* = f(p)$ denote a one-to-one correspondence between their points that preserves arc length, curvature, and torsion, where $p \in A$ and p^* is the corresponding element of A^*.

If $L_n = \{p_0, p_1, \ldots, p_n\}$ and $L_n^* = \{p_0', p_1', \ldots, p_n'\}$ denote n-lattices inscribed in A and A^*, respectively, then for n sufficiently large, p_i' is arbitrarily close to $p_i^* = f(p_i)$ ($i = 0, 1, \ldots, n$), and hence for any $\varepsilon > 0$ there exists a positive integer N, such that if $n > N$,

$$1 - \varepsilon < \frac{k_M(p_i)}{k_M(p_i')} < 1 + \varepsilon,$$

$$1 - \varepsilon < \frac{\tau_B^*(p_i)}{\tau_B^*(p_i')} < 1 + \varepsilon$$

($i = 0, 1, \ldots, n$), for any n-lattices L_n, L_n^* of A, A^*, respectively. Since $\lim_{n \to \infty} n \cdot \lambda(n) = \ell[A] = \ell[A^*] = \lim_{n \to \infty} n \cdot \lambda^*(n)$, where $\lambda(n)$, $\lambda^*(n)$ denote the sides of L_n, L_n^*, respectively (Theorem 9.4), it follows that for n sufficiently large,

$$1 - \varepsilon < \frac{\lambda(n)}{\lambda^*(n)} < 1 + \varepsilon.$$

Also since

$$\frac{\sin \theta_i}{\sin \theta_i'} = \frac{(\sin \theta_i)/\lambda(n)}{(\sin \theta_i')/\lambda^*(n)} \cdot \frac{\lambda(n)}{\lambda^*(n)} = \frac{\dfrac{k_M(p_i)}{(\sin \theta_i)/\lambda(n)}}{\dfrac{k_M(p_i')}{(\sin \theta_i')/\lambda^*(n)}} \cdot \frac{\lambda(n)}{\lambda^*(n)} \cdot \frac{k_M(p_i')}{k_M(p_i)}$$

($i = 1, 2, \ldots, n$), use of Lemma 10.1 together with the foregoing inequalities shows that for all sufficiently large values of n,

$$1 - \varepsilon < \frac{\sin \theta_i}{\sin \theta_i'} < 1 + \varepsilon$$

($i = 1, 2, \ldots, n$).

Similarly, for all sufficiently large values of n,

$$1 - \varepsilon < \frac{\sin \varphi_i}{\sin \varphi_i'} < 1 + \varepsilon$$

($i = 1, 2, \ldots, n - 1$).

Now

$$\theta_i = \left[\frac{\theta_i}{\sin \theta_i} \right] \cdot \left[\frac{\sin \theta_i}{\lambda(n)} \right] \cdot \lambda(n)$$

($i = 1, 2, \ldots, n - 1$),

$$\varphi_i = \left[\frac{\varphi_i}{\sin \varphi_i} \right] \cdot \left[\frac{\sin \varphi_i}{\lambda(n)} \right] \cdot \lambda(n)$$

$(i = 1, 2, \ldots, n - 1)$, and it follows from Lemmas 10.2 that both $\theta_i/\sin \theta_i$ and $\varphi_i/\sin \varphi_i$ approach 1 as $n \to \infty$. Hence for all sufficiently large values of n,

$$1 - \varepsilon < \frac{\theta_i}{\theta_i'} < 1 + \varepsilon \qquad (i = 1, 2, \ldots, n),$$

and

$$1 - \varepsilon < \frac{\varphi_i}{\varphi_i'} < 1 + \varepsilon \qquad (i = 1, 2, \ldots, n - 1).$$

Finally, according to Theorem 10.12, for all sufficiently large values of n, $[v_{i-1}, v_i, v_{i+1}]$ has the same sign for $i = 1, 2, \ldots, n - 1$, as does the triple scalar product $[v_{i-1}', v_i', v_{i+1}']$.

Summarizing the preceding argument, there exists a positive integer N such that if $n > N$ and $L_n = \{p_0, p_1, \ldots, p_n\}$, $L_n^* = \{p_0', p_1', \ldots, p_n'\}$ are any n-lattices of A, A^*, with sides $\lambda(n)$, $\lambda^*(n)$, respectively, then

(a) we may suppose L_n and L_n^* similarly oriented; for if not, reflection of one of them, say L_n^*, in the origin yields a congruent lattice with orientation similar to that of L_n,

(b) $1 - \varepsilon < \lambda(n)/\lambda^*(n) < 1 + \varepsilon$,

(c) $1 - \varepsilon < \theta_i/\theta_i' < 1 + \varepsilon \qquad (i = 1, 2, \ldots, n)$,

(d) $1 - \varepsilon < \varphi_i/\varphi_i' < 1 + \varepsilon \qquad (i = 1, 2, \ldots, n - 1)$.

The proof of the theorem will be complete when it is shown that the distances $p_0 p_n$, $p_0' p_n'$ differ by a number that approaches zero with ε, and hence are equal; for the same argument can be applied to the subarcs of A and A^* that are determined by any two distinct points of A and their respective images, by the correspondence f, in A^*.

We assume that the similarly oriented lattices L_n, L_n^* have both p_0 and p_0' at the origin, that the vectors v_i and v_i' lie along the positive x-axis, and that the vectors q_1 and q_1' lie along the positive z-axis. If η_i denotes the angle between v_i and v_i', it is easily shown by induction that

$$\eta_i \le \sum_{j=1}^{i-1} |\theta_j - \theta_j'| + \sum_{j=1}^{i-2} |\varphi_j - \varphi_j'| < \varepsilon \cdot \sum_{j=1}^{i=1} \theta_j' + \varepsilon \cdot \sum_{j=1}^{i-2} \varphi_j',$$

where the last inequality follows from (c) and (d).

Since

$$\lim_{n \to \infty} \sum_{j=1}^{n} \theta_j' = \lim_{n \to \infty} \sum_{j=1}^{n} \left[\frac{\theta_j'}{\sin \theta_j'} \right] \cdot \left[\frac{\sin \theta_j'}{\lambda^*(n)} \right] \cdot \lambda^*(n)$$

$$= \text{the line integral of } k_M \text{ along } A^*,$$

it follows that $\sum_{j=1}^{n} \theta_j'$ is bounded. Similarly,

$$\lim_{n \to \infty} \sum_{j=1}^{n} \varphi_j' = \text{the line integral of } \tau_B^* \text{ along } A^*,$$

and so $\sum_{j=1}^{n} \varphi_j'$ is bounded. Denoting the values of these line integrals by K_{A^*} and T_{A^*}, respectively, we have, for every index i,

$$\eta_i < \varepsilon(K_{A^*} + T_{A^*}).$$

Now

$$|p_0 \, p_n - p_0' \, p_n'| \leq p_n \, p_n' = \left| \sum_{i=1}^{n} v_i - \sum_{i=1}^{n} v_i' \right| = \left| \sum_{i=1}^{n} v_i - v_i' \right|$$

$$\leq \sum_{i=1}^{n} |v_i - v_i'| = \sum_{i=1}^{n} [\lambda^2(n) + \lambda^{*2}(n) - 2\lambda(n)\lambda^*(n) \cdot \cos \eta_i]^{1/2}.$$

Since η_i becomes arbitrarily small with ε, it follows that

$$|p_0 \, p_n - p' \, p_n'| \leq \sum_{i=1}^{n} [(\lambda(n) - \lambda^*(n))^2]^{1/2} = \sum_{i=1}^{n} |\lambda(n) - \lambda^*(n)|$$

$$< \sum_{i=1}^{n} \varepsilon \cdot \lambda^*(n),$$

by use of (b).

Hence

$$|p_0 \, p_n - p_0' \, p_n'| < \varepsilon \cdot \sum_{i=1}^{n} \lambda^*(n) < \varepsilon \cdot \ell[A^*],$$

and since A^* is rectifiable, $|p_0 p_n - p_0' p_n'|$ is arbitrarily small with ε. Consequently, $p_0 p_n = p_0' p_n'$ and the theorem is proved.

EXERCISES

1. Let A denote the planar arc $x_1 = 0$, $x_2 = t$, $x_3 = t^3$ $(-1 \leq t \leq 1)$ and A^* the arc with equations $x_1 = 0$, $x_2 = t$, $x_3 = t^3$ $(0 \leq t \leq 1)$, $x_1 = t^3$, $x_2 = t$, $x_3 = 0$ $(-1 \leq t \leq 0)$. (The arc A^* may be obtained from arc A by rotating the lower half of the $x_2 x_3$-plane about the x_2-axis until it coincides with the positive half of the $x_1 x_2$-plane.) The arcs are not congruent, though there exists a one-to-one mapping $p^* = f(p)$ of A onto A^* that preserves arc length, $k_M(p) = k_M(p^*)$, $\tau_B^*(p) = \tau_B^*(p^*)$, $p \neq 0$. Show that the metric curvature of A^* at the origin does not exist, and that A^* has no osculating plane at the origin. What can be said about the torsion of A^* at the origin?

2. If p is a point of a biregular analytic arc A of E_3, show that $k_M(p) = |d\theta/ds|$, where θ is the angle between the unit tangent vectors at p and a neighboring point, and s denotes arc length.

3. Show that $\tau_B^*(p) = |d\varphi/ds|$ for $p \in A^*$, a biregular analytic arc of E_3.

REFERENCES

Section 10.1 The metrization of curvature $k_M(p)$ was introduced in Karl Menger, "Untersuchungen über allgemeine Metrik. Vierte Untersuchung. Zur Metrik der Kurven," *Math. Ann.*, **103**: 466–501 (1930).

The curvature $k_H(p)$ is an adaptation by J. Haantjes (*Proceedings, Akademie van Wetenschappen*, Amsterdam, **50**: 496–508 (1947)) of a definition due to P. Finsler (Ph.D. dissertation, Basle, 1918).

Section 10.2 J. Haantjes, *Sur la Géométrie Infinitésimale des Espaces Metriques*, Colloque de Géométrie Differentielle, Louvain, 1951, pp. 91–97.

Section 10.3 J. Haantjes, "Distance geometry. Curvature in abstract metric spaces," *Indagationes Math.*, **9**: 3–15 (1947).

Section 10.4 See the reference cited in Section 9.2.

Section 10.7 The metrization of torsion τ_A was introduced in G. Alexits, "La torsion des espaces distanciés," *Composito Math.*, **6**: 471–477 (1938–1939).

Sections 10.8 to 10.9 The metric torsions τ_B and τ_B^* are studied in L. M. Blumenthal, DG, pp. 84–87.

Sections 10.11 to 10.12 J. Gaddum, "Metric methods in integral and differential geometry," *Amer. J. Math.*, **75**: 30–42 (1953).

CHAPTER 11

Metrizations of Surface Curvature

11.1 INTRODUCTORY REMARKS

The problem of assigning a measure to the curvature of a curve that is in close agreement with our intuitive notion of that concept when applied to curves of euclidean space is quite satisfactorily solved by each of the formulas for curvature studied in the preceding chapter. The situation with respect to surface curvature is very different. None of the classical definitions of the curvature of a surface at one of its points can claim to yield a measure that conforms to intuition. A right circular cylinder, for example, is surely a curved surface, but by the measure of curvature developed by Gauss (the so-called *total* or Gauss curvature) it has *zero* curvature at each point. The surface obtained by revolving the catenary $y = a \cosh(x/a)$ about the x-axis is indeed a curved surface, but according to the measure of surface curvature introduced in 1831 by Sophie Germain (the *mean* curvature), that surface has zero (mean) curvature at each of its points. It is not unusual that when concepts which arose very early in our culture, such as "curve," "surface," "curvature," or "dimension," are finally made precise, something is lost in the process. The new, precise notions seldom agree completely with their intuitive counterparts.

After a very brief account of surface curvature in classical differential geometry, we present a metrization of Gauss curvature, obtained by A. Wald in 1936, which is the most important result yet established in the metric program for differential geometry. A modification of Wald's work, due to W. A. Kirk (1962), and the identification of his localization of a curvature notion introduced by W. Rinow in 1961 with Wald's curvature will conclude the chapter. Several lemmas of intrinsic interest will be proved.

11.2   DESCRIPTIVE REMARKS CONCERNING GAUSS CURVATURE

In seeking to extend the notion of curvature from curves to surfaces, it may be anticipated that the analogue for a surface S of the tangent line of a curve will play an important role. That analogue is the *tangent plane* of the surface at a point p, defined as the plane that contains the tangent lines of all curves on S that pass through p. By analogy, the curvature of S at p should measure the rate at which the surface leaves the tangent plane. But a difficulty arises which was not present before: although a curve can leave its tangent line in only one direction, a surface leaves its tangent plane in infinitely many directions, at a rate that, in general, clearly depends on the direction. This suggests that some kind of averaging procedure be applied.

Another, perhaps less obvious, difference between the two cases lies in the necessity of having surface curvature a signed quantity in order to distinguish between surfaces that are intrinsically different but have the same numerical "curvature," as, for example, a sphere and a pseudosphere. That necessity is not present in curve theory. The equation of a curve determines the square of its curvature (that is, $(1/R)^2 = \sum_{i=1}^3 [x_i''(t)]^2$), but not (uniquely) its curvature. We have taken $1/R \geq 0$, so that the center of the osculating circle at p lies on the positive half of the principal normal at distance R from p. It was, in part, the necessity of a *signed* surface curvature that created great difficulty in metrizing the notion.

The two new difficulties of the problem, brought about by an increase in the dimension, are resolved in classical differential geometry in the following manner.

Let a surface S have the tangent plane π at one of its points p. The normal to π at p is assigned a positive sense and a unit normal vector ζ is obtained. Consider now the sections of S made by the pencil of planes on the normal line to S at p. If one of these sections C with curvature $1/R$ at p opens in the direction of the vector ζ (that is, if its principal normal β coincides with ζ) we put $1/r = 1/R$, but if it opens in the opposite direction, we write $1/r = -1/R$. If the direction of C at p is λ, then $1/r$ is called the *normal curvature of S at p in the direction λ*.

For surfaces sufficiently smooth about p, the normal curvature varies continuously with the direction λ, and attains a maximum and a minimum value. These values, which, it may be shown, are assumed for mutually perpendicular directions, are called the *principal* normal curvatures of S at p. Their product is the Gauss or *total curvature* of S at p, and their arithmetic mean is the *mean* curvature of S at p. One of Gauss' most important contributions to differential geometry is the theorem that the total curvature of a surface at a point p is an *intrinsic* property of the surface; that is, it does not depend at all on the nature of the space containing the surface.

In introductory courses in differential geometry the foregoing discussion is formalized. The surface S is represented by the vector equation

$$x = x(u, v), \qquad a \leqq u \leqq b, \qquad c \leqq v \leqq d,$$

with all differentiability requirements assumed. The points of S have the *curvilinear* coordinates (u, v). A curve C on S is represented by equations $u = u(t)$, $v = v(t)$, and the element of arc length ds of C is given by

$$ds^2 = (dx \mid dx) = (x_u\, du + x_v\, dv \mid x_u\, du + x_v\, dv),$$

where $(z \mid w)$ denotes the scalar product of vectors z, w.

Hence $ds^2 = E\, du^2 + 2F\, du\, dv + G\, dv^2$, where $E = (x_u \mid x_u)$, $F = (x_u \mid x_v)$, and $G = (x_v \mid x_v)$. The quadratic differential form

$$E\, du^2 + 2F\, du\, dv + G\, dv^2$$

is called the *first fundamental form* of the surface S.

If p has the (curvilinear) coordinates (u_0, v_0), the curve $u = u_0$, $v = v$ is called the v-curve C_v through p, and the curve $u = u$, $v = v_0$ is the u-curve C_u through p.

The unit normal vector ζ to S at p is directed so that the triple scalar product or determinant $[x_u/\sqrt{E},\ x_v/\sqrt{G},\ \zeta] > 0$; that is, the unit tangent vector to C_u, the unit tangent vector to C_v, and the unit normal vector at p have, in the order given, the same disposition as the coordinate axes.

The negative of the scalar product of the vectors $dx = x_u\, du + x_v\, dv$ and $d\zeta = \zeta_u\, du + \zeta_v\, dv$ is a quadratic differential form (the *second* fundamental form),

$$e\, du^2 + 2f\, du\, dv + g\, dv^2,$$

where $e = -(x_u \mid \zeta_u), f = -(x_u \mid \zeta_v) = -(x_v \mid \zeta_u), g = -(x_v \mid \zeta_v)$.

The normal curvature of S at p in the direction $dv:du$ turns out to be

$$\frac{1}{r} = \frac{e\, du^2 + 2\, du\, dv + g\, dv^2}{E\, du^2 + 2F\, du\, dv + G\, dv^2},$$

where the coefficients e, f, g, E, F, G are evaluated at p. The principal normal curvatures $1/r_1$, $1/r_2$ are roots of the equation

$$(EG - F^2)x^2 - (Eg - 2Ff + gE)x + (eg - f^2) = 0,$$

and, denoting the Gauss curvature by K, it follows that

$$K = \frac{1}{r_1} \cdot \frac{1}{r_2} = \frac{eg - f^2}{EG - F^2}.$$

The presence in this formula of the coefficients e, f, g of the second fundamental form is somewhat misleading, since one might conclude from this that K depends on the imbedding space as well as on S itself. But such a conclusion would be in error, for Gauss expressed K entirely in terms of E, F, G, and their partial derivatives of the first and second orders. A well-known formula due to the German mathematician Frobenius (1849–1917) is

$$-4D^4K = 2D^3\left(\frac{\partial}{\partial u}\frac{G_u - F_v}{D} - \frac{\partial}{\partial v}\frac{F_u - E_v}{D}\right) + \begin{vmatrix} E & F & G \\ E_u & F_u & G_u \\ E_v & F_v & G_v \end{vmatrix},$$

where $D^2 = EG - F^2$.

11.3 WALD SURFACE CURVATURE

Anyone contemplating metrizing the Gauss curvature is likely to be intimidated and dismayed by a glance at Frobenius' formula for that notion. Nor will the dismay be much diminished by a knowledge of any of the simpler expressions for K that result from various choices of coordinates on S. In its very simplest aspect, the problem is that of metrizing a second partial derivative of a function which itself must be metrized! To express the Gauss curvature of a surface in E_3 at one of its points in a manner that would be meaningful when applied to a point in any metric space whatever would appear to be a grim task, indeed. Let us see how Abraham Wald (1902–1950) accomplished it.

The central idea of Wald's contribution (which solved the problem of attaching a sign to surface curvature) is what was subsequently called by L. M. Blumenthal the *imbedding curvature* of a metric quadruple $Q = \{p_1, p_2, p_3, p_4\}$. Such a quadruple has an imbedding curvature $K_W(Q) = 0$ provided the determinant $D(p_1, p_2, p_3, p_4) = 0$; it has an imbedding curvature $k > 0$ provided

$$k^{1/2} \cdot p_i p_j \leqq \pi, \text{ and } \det |\cos k^{1/2} \cdot p_i p_j| = 0 \qquad (i, j = 1, 2, 3, 4),$$

with all third-order principal minors nonnegative, and Q has an imbedding curvature $k < 0$ provided

$$\det |\cosh (-k)^{1/2} \cdot p_i p_j| = 0 \qquad (i, j = 1, 2, 3, 4).$$

Stated in geometrical terms, a metric quadruple Q has imbedding curvature *zero* provided Q is congruently contained in the euclidean plane E_2; Q has a *positive* imbedding curvature k $(k > 0)$ provided Q is congruently contained in a two-sphere $S_{2,k}$ of curvature k (that is, the surface of a sphere in E_3 of radius $1/\sqrt{k}$, with shortest arc metric), and Q has *negative* imbedding curvature k $(k < 0)$ provided Q is congruently contained in the hyperbolic plane $H_{2,k}$ of curvature k.*

Two important questions are raised by the foregoing definition. Does every metric quadruple have an imbedding curvature? Is an imbedding curvature $K_W(Q)$ unique, whenever it exists?

The answer to the first question is easily seen to be in the negative. The metric quadruple p_1, p_2, p_3, p_4 with $p_1 p_2 = p_1 p_3 = p_1 p_4 = 1$, and $p_2 p_3 = p_3 p_4 = p_2 p_4 = 2$ has p_1 metrically between each two of the three points p_2, p_3, p_4. It is clear that none of the surfaces $E_2, S_{2,k}, H_{2,k}$ contains such a quadruple, and so p_1, p_2, p_3, p_4 has no imbedding curvature.

The second question is also answered negatively by the trivial example of a *linear* quadruple Q; that is, Q is not only imbeddable in E_2 but also in $H_{2,k}$ for all (negative) values of k, and in $S_{2,k}$ for infinitely many (positive) values of k. But since linear quadruples will play no role in the theory,

---

* L. M. Blumenthal, *Theory and Applications of Distance Geometry*, The Clarendon Press, Oxford, 1953.

we are concerned with the number of imbedding curvatures of *nonlinear* metric quadruples. We shall prove that each such quadruple has *at most two* imbedding curvatures, and if a nonlinear quadruple contains a linear triple then it has *at most one* imbedding curvature.

DEFINITION OF WALD CURVATURE   A metric space **M** has at an accumulation point p a Wald curvature $K_W(p)$ provided

   (i) no neighborhood of p is linear,
   (ii) $\varepsilon > 0$ implies the existence of $\delta > 0$ such that each quadruple $Q = \{p_1, p_2, p_3, p_4\}$ of **M** with $pp_i < \delta$ ($i = 1, 2, 3, 4$) has an imbedding curvature $K_W(Q)$ such that $|K_W(p) - K_W(Q)| < \varepsilon$.

11.4   SOME PROPERTIES OF SPACES OF CONSTANT GAUSS CURVATURE

Let Σ_k denote the hyperbolic plane of curvature k, the euclidean plane, or the surface of the sphere of radius $1/\sqrt{k}$ (*with shortest arc metric*), according as k is negative, zero, or positive, respectively. The theorems of this section establish interesting inequalities between certain distances and angles of Σ_k and corresponding distances and angles of $\Sigma_{k'}$ ($k \neq k'$), one important consequence of which is *the uniqueness of the Wald curvature at a point of a convex metric space.*

   We observe that the lines of Σ_k, $k < 0$, are congruent with those of $\mathbf{E}_2$, while linear subsets of Σ_k, $k > 0$ (those subsets congruently contained in $\mathbf{E}_1$), are subsets of great *semicircles*. We denote the length of a great circle of Σ_k, $k > 0$, by $\gamma(k)$; that is, $\gamma(k) = 2\pi/\sqrt{k}$.

   If p, q, r are pairwise distinct points of Σ_k, let $\measuredangle q : p, r$ denote the smaller of the two angles formed by the two metric segments $\text{seg}[p, q]$, $\text{seg}[q, r]$. Clearly $0 \leqq \measuredangle q : p, r \leqq \pi$, and if p, q, r are linear, then $\measuredangle q : p, r = 0$ or $\measuredangle q : p, r = \pi$.

Property 11.1.   *If* $\measuredangle q : p, r = \pi$ $(p, q, r \in \Sigma_k, p \neq q \neq r \neq p)$, *then* pqr *holds* (*that is, q is metrically between p and r*) *for* $k \leqq 0$, *while for* $k > 0$ *either* pqr *subsists or* $pq + qr + pr = \gamma(k)$.

Proof.   The case $k \leqq 0$ is clear. If $k > 0$ and pqr does not hold, then p, q, r are not contained in a great semicircle. But since $\measuredangle q : p, r = \pi$, the three points are on a great circle, and so $pq + qr + pr = \gamma(k)$.

Property 11.2.   *If* p, a, b *is a nonlinear triple of* Σ_k, *and* c *is an interior point of* $\not\prec p: a, b$ *such that* $pc = pb$, *then* $ac < ab$.

Proof. If $k = 0$, the conclusion follows from the ordinary law of cosines. If $k < 0$, the result is given directly by the law of cosines in hyperbolic trigonometry, and the law of cosines in spherical trigonometry is utilized in case $k > 0$.

Property 11.3   *Let* a', b', c' *be a nonlinear point-triple of* $\Sigma_{k'}$, *and* a'', b'', c'', *a nonlinear point-triple of* $\Sigma_{k''}$, *with* $a'b' = a''b''$, $a'c' = a''c''$, *and* $\not\prec a': b', c' = \not\prec a'': b''c''$. *If* $k' < k''$, *then* $b'c' > b''c''$.

Proof.   Referring Σ_k to (geodesic) polar coordinates (r, θ), $r \geq 0$, we have

$$ds^2 = dr^2 + G_k^2(r)\, d\theta^2,$$

where $G_k(r) = [\sin r \sqrt{k}]/\sqrt{k}$, $k \neq 0$, $G_0 = \lim_{k \to 0}[\sin r \sqrt{k}]/\sqrt{k} = r$, and, for $k > 0$, $r < \frac{1}{2}\gamma(k)$. (Note that if $k < 0$, $[\sin r \sqrt{k}]/\sqrt{k} = [\sinh r \sqrt{-k}]/\sqrt{-k}$.) It is easily seen that for $k \neq 0$, the derivative $\dfrac{d}{dk} G_k(r)$ is negative, and so $G_k(r)$ is a decreasing function of k.

If $\not\prec a': b', c' = \pi$, then (since a', b', c' are not linear)

$$a'b' + b'c' + c'a' = \gamma(k') > \gamma(k'') = a''b'' + b''c'' + c''a'',$$

and so $b'c' > b''c''$.

To complete the proof of the property, it suffices to show that corresponding to each nonlinear triple a, b, c of Σ_k, with $\not\prec a: b, c \neq \pi$, there exists an $\varepsilon > 0$ such that if $k - \varepsilon < k^* < k + \varepsilon$, and $a^*, b^*, c^* \in \Sigma_{k^*}$ with $a^*b^* = ab$, $a^*c^* = ac$ and $\not\prec a^*: b^*, c^* = \not\prec a: b, c$, then $bc > b^*c^*$ if $k < k^*$, and $b^*c^* > bc$ if $k^* < k$.

Let a and a^* be the poles of systems of (geodesic) polar coordinates in Σ_k and Σ_{k^*}, respectively, so that b and c have the same polar coordinates in Σ_k as b^* and c^* have, respectively, in Σ_{k^*}. (Clearly, the hypotheses on the two point-triples a, b, c and a^*, b^*, c^* make this arrangement possible.) Since a, b, c are not linear, no two of those points are diametral (in case $k > 0$). Let $r = r(\theta)$ $(\theta_1 \leq \theta \leq \theta_2)$ be the equation of the segment (for $k \leq 0$) or the unique great circle arc (shorter) joining b and c in Σ_k.

For ε sufficiently small, there is an arc of Σ_{k*} joining b^* and c^* with the equation $r = r(\theta)$ $(\theta_1 \leq \theta \leq \theta_2)$. Since

$$ds^2 = dr^2 + G_k^2(r)\, d\theta^2, \qquad \text{for } r = r(\theta), \text{ in } \Sigma_k,$$

$$ds^2 = dr^2 + G_{k*}^2(r)\, d\theta^2, \qquad \text{for } r = r(\theta), \text{ in } \Sigma_{k*},$$

and, since $G_k^2(r)$ evidently decreases monotonically as k increases, then $bc > b^*c^*$ if $k < k^*$, and $b^*c^* > bc$ if $k^* < k$, completing the proof of the property.

We are now able to establish a very useful property that relates the angles of a geodesic triangle of Σ_k with the angles of a congruent geodesic triangle of Σ_{k*}, $k \neq k^*$.

Property 11.4 *If $a', b', c' \in \Sigma_{k'}$, $a'', b'', c'' \in \Sigma_{k''}$, with $a', b', c' \approx a'', b'', c''$ (that is, $a'b' = a''b''$, $b'c' = b''c''$, $a'c' = a''c''$) and a', b', c' not linear, then for $k' < k''$, each of the angles determined by the first point-triple is less than the corresponding angle determined by the second point-triple.*

Proof. Clearly, no angle of the first triple is π, for in the contrary case (since a', b', c' are not linear) it follows that $k' > 0$ and $a'b' + b'c' + a'c' = \gamma(k')$. But then the $\Sigma_{k''}$, $k'' > k'$ does not contain any three points congruent to a', b', c', since for each three points a'', b'', c'' of $\Sigma_{k''}$, with $k'' > k' > 0$, $a''b'' + b''c'' + a''c'' \leq \gamma(k'') < \gamma(k')$.

Now choose a point d'' of $\Sigma_{k''}$ so that $b''d'' = b''c''$ and $\not\!\angle\, b'': a'', d'' = \not\!\angle\, b': a', c'$. Property 11.3 applied to the triples a', b', c' and a'', b'', c'' gives $a'c' > a''d''$, and so $a''c'' > a''d''$. By Property 11.2, it follows that d'' is in the interior of $\not\!\angle\, b'': a'', c''$ and so $\not\!\angle\, b'': a'', c'' > \not\!\angle\, b'': a'', d'' = \not\!\angle\, b': a', c'$.

Similar arguments are applied for the two remaining pairs of corresponding angles.

Property 11.5. *Let $Q = \{a', b', c', d'\}$ denote a nonlinear quadruple of pairwise distinct points of $\Sigma_{k'}$, with $a'b' + b'c' = a'c'$, and let $Q'' = \{a'', b'', c'', d''\}$ denote a nonlinear quadruple of pairwise distinct points of $\Sigma_{k''}$, with $a'', b'', c'' \approx a', b', c'$, and $k' < k''$. Then*

(1) *if $a'd' = a''d''$ and $b'd' = b''d''$, then $c'd' > c''d''$,*
(2) *if $c'd' = c''d''$ and $b'd' = b''d''$, then $a'd' > a''d''$,*
(3) *if $a'd' = a''d''$ and $c'd' = c''d''$, then $b'd' < b''d''$.*

Proof. If a', b', d' are not linear, then by Property 11.4, $\not\!\times\, b': a', d' <$ $\not\!\times\, b'': a'', d''$, and so $\not\!\times\, b': c', d' > \not\!\times\, b'': c'', d''$. It follows easily that $c'd' >$ $c''d''$. (Why?)

If, on the other hand, a', b', d' are linear, then since Q' is not linear, $k' > 0$ and $b'a'd'$ holds (for since $a'b'c'$ subsists, if $a'b'd'$ does, then the quadruple Q' is linear, and linearity of Q' also results from $a'd'b'$). (Why?) Hence $\not\!\times\, a': c', d' = \not\!\times\, a': b', d' = \pi$ (since a', b', c', d' are on the great circle of $\Sigma_{k'}$ determined by a', b') and

$$a'c' + c'd' + a'd' = \gamma(k').$$

But clearly

$$a''c'' + c''d'' + a''d'' \leqq \gamma(k'') < \gamma(k') = a'c' + c'd' + a'd',$$

and since $a''c'' = a'c'$ and $a''d'' = a'd'$, then $c''d'' < c'd'$, and the proof of this case is complete.

The proof of (2) is entirely analogous to the foregoing.

Proceeding to the proof of (3), we assert that neither a', b', d' nor $a', c'\ d'$ are linear. For if a', b', d' are linear, then, as in the proof of (1), $k' > 0$ and $b'a'd'$ subsists, and we have

$$\gamma(k') = a'c' + c'd' + a'd' = a''c'' + c''d'' + a''d'' \leqq \gamma(k''),$$

which is impossible, since $\gamma(k') > \gamma(k'')$. If a', c', d' are linear, then the non-linearity of Q' implies $k' > 0$ and $a'd'c'$. But $a'b'c', a'd'c'$ and the nonlinearity of Q' require that a' and c' be a diametral point-pair. Hence $a'c' = \frac{1}{2}\gamma(k')$, and since $a''c'' = a'c'$, we have

$$\tfrac{1}{2}\gamma(k'') \geqq a''c'' = a'c' = \tfrac{1}{2}\gamma(k'),$$

and the same contradiction is encountered as before.

If a'', b'', d'' are linear, then $k'' > 0$, and (since $a'', c'', d'' \approx a', c', d'$ and by the above argument, a', c', d' are not linear) the triple a'', c'', d'' is not linear, and it follows that $b''a''d''$ must subsist; that is, $b''a'' + a''d'' = b''d''$. Since, as shown above, a', b', d' are not linear, then

$$b''d'' = b''a'' + a''d'' = b'a' + a'd' > b'd'.$$

If, on the other hand, a'', b'', d'' are not linear, let d^* denote a point of $\Sigma_{k'}$ such that $a'd^* = a''d''$ and $b'd^* = b''d''$. By (1), $c'd^* > c''d'' = c'd'$, and so by Property 11.2, $\not\!\times\, a': b', d^* > \not\!\times\, a': b', d'$. Consequently, $b'd' < b'd^* = b''d''$, and the proof of the property is complete.

The property just proved has as an immediate corollary the following statement, which is important in the theory of Wald curvature.

COROLLARY *A metric quadruple of pairwise distinct points has at most one imbedding curvature if three of the four points form a linear triple.*

DEFINITION If a, b, c, d are four pairwise distinct, nonlinear points of Σ_k, and $\measuredangle a: b, c + \measuredangle a: b, d + \measuredangle a: c, d = 2\pi$, point a is said to be *between* the points b, c, d, and the quadruple is called *triodic*. A quadruple of Σ_k is *atriodic* if no one of its points is between the other three points.

Property 11.6. *Let $Q = \{a, b, c, d\}$ denote a nonlinear quadruple of pairwise distinct points of Σ_k. If Q is triodic, then Q is not congruently imbeddable in any $\Sigma_{k''}$, for $k'' > k$. If Q is atriodic, then Q is not congruently imbeddable in any $\Sigma_{k'}$, for $k' < k$.*

Proof. If Q is triodic, suppose $a'', b'', c'', d'' \in \Sigma_{k''}$, $k'' > k$, such that $a, b, c, d \approx a'', b'', c'', d''$. Then Q contains no linear triple (Corollary, Property 11.5). Choosing the labeling so that a is between b, c, d, we have

$$\measuredangle a: b, c: \measuredangle a + b, d + \measuredangle a: c, d = 2\pi,$$

and by Property 11.4,

$$\measuredangle a'': b'', c'' + \measuredangle a'': b'', d'' + \measuredangle a'': c'', d'' > 2\pi,$$

which is impossible. Hence Q is not imbeddable in $\Sigma_{k''}$.

If Q is atriodic, assume $a', b', c', d' \in \Sigma_{k'}$, $k' < k$, such that $a, b, c, d \approx a', b', c', d'$. Then Q contains no linear triple (Corollary), and since no point of Q is between the remaining three points, the labeling may be chosen so that seg$[a, b]$ and seg$[c, d]$ have an interior point e in common. The seg$[a', b']$ of $\Sigma_{k'}$ is congruent to seg$[a, b]$ (since $ab = a'b'$) and hence contains a unique point e' such that $a'e' = ae$.

Application of (3) of Property 11.5 to the two quadruples c, a, e, b and c', a', e', b' yields $c'e' < ce$; and its application to d, a, e, b and d', a', e', b' gives $d'e' < de$. Hence $c'd' \leq c'e' + e'd' < ce + de = cd$, in contradiction with $a, b, c, d \approx a', b', c', d'$.

The following important corollary results.

COROLLARY   *A nonlinear metric quadruple of pairwise distinct points has at most two imbedding curvatures. If $k' < k''$, any quadruple Q'' of $\Sigma_{k''}$ that is congruently contained in $\Sigma_{k'}$ is triodic, and each quadruple Q' of $\Sigma_{k'}$ that is congruently contained in $\Sigma_{k''}$ is atriodic.*

We have given an example of a nonlinear metric quadruple of pairwise distinct points that has no imbedding curvature, and according to the Corollary of Property 11.5, any nonlinear quadruple Q of Σ_k has the unique imbedding curvature k if Q contains a linear triple. The following property establishes the existence of a class of nonlinear quadruples, each of which has two distinct imbedding curvatures, one of them being zero.

Property 11.7.   For each real number k, $k > 0$, each neighborhood of every point of Σ_k contains a nonlinear quadruple Q of pairwise distinct points, with imbedding curvature zero; that is, Q is congruently imbeddable in the euclidean plane E_2.

Proof.   Let p be any point of Σ_k, $k > 0$, and U any neighborhood of p.

Draw two great circles that intersect in a right angle at p, and denote the arcs of these circles contained in U by A_1 and A_2 (Figure 14). Select points q_1, q_2 on A_1 such that $pq_1 = pq_2 \neq 0$, so $pq_1 + pq_2 = q_1 q_2$, and let q be a point of A_2 such that $pq < \frac{1}{4}\gamma(k)$. Let q_1', q_2', q' be points of E_2 with $q_1', q_2', q' \approx q_1, q_2, q$, let h denote the altitude (length) of the isosceles triangle $T(q_1', q_2', q')$, from the vertex q', and let p' denote the midpoint of the base of the triangle.

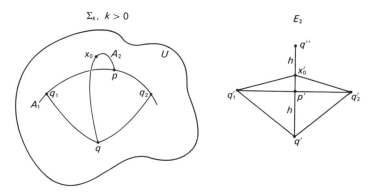

Figure 14

Applying part (3) of Property 11.5 to the quadruples q, q_1, q_2, p and q', q_1', q_2', p' yields $h < pq$. Now let x denote a variable point of A_2 between q and p, and let x' be the point of E_2 such that $q_1, q_2, x \approx q_1', q_2', x'$ and which is not on the same side of the line $G(q_1', q_2')$ that q' is. When $x = p$, $xq > x'q'$; when $x = q$, $x' = q''$ (see Fig. 14) and $xq = 0 < x'q' = 2h$. Accordingly, there exists a point x_0 of A_2, between p and q, such that $x_0 q = x_0' q'$. (Why?) Then $q_1, q_2, q, x_0 \approx q_1', q_2', q', x_0'$, with the "primed" points in E_2, and so the nonlinear quadruple q_1, q_2, q, x_0 of pairwise distinct points has an imbedding curvature zero, as well as the imbedding curvature k, $k > 0$.

Notice that the quadruple q_1, q_2, q, x_0 is triodic and the planar quadruple q_1', q_2', q', x_0', congruent with it is atriodic.

EXERCISES

1. Let a, b, c, d be four points equally spaced on the circumference of a small circle of Σ_k, $k > 0$. Show that these points do not satisfy the ptolemaic inequality, and hence are not congruently contained in any euclidean space. It follows that the sphere is not locally ptolemaic.

2. Show that the metric quadruple p_1, p_2, p_3, p_4 with $p_1 p_3 = p_1 p_4 = p_2 p_3 = p_2 p_4 = \pi$, $p_1 p_2 = p_3 p_4 = 3\pi/2$ has two positive imbedding curvatures, one of which is $\frac{1}{9}$, and the other k_0 is between $\frac{1}{4}$ and $\frac{4}{9}$. Verify that the quadruple is atriodic on $\Sigma_{1/9}$ and triodic on Σ_{k_0}.

3. Prove that hyperbolic three-dimensional space is ptolemaic. *Hint:* Apply Property 11.4 to show that the hyperbolic plane is ptolemaic.

11.5   PROPERTIES OF A REGULAR NEIGHBORHOOD OF A POINT

A detailed presentation of Wald's investigations that culminate in the theorem that identifies the curvature K_W with the Gauss curvature K_G would require more space than we can rightfully allot to it. (We avoid the tired phrase, "beyond the scope of this book.") But before stating the theorem, we shall pause to examine some of the stepping stones that enabled Wald to ascend to it. Due to the properties of Σ_k proved in the preceding section, we can easily establish the following basic result.

THEOREM 11.1   *A metrically convex metric space* **M** *has at each point* p *at most one curvature* $K_W(p)$.

Proof. In view of the corollary of Property 11.5, it evidently suffices to show that every spherical neighborhood $U(p; \delta)$ of $p, \delta > 0$, contains a nonlinear quadruple of pairwise distinct points, with one of its four point-triples linear.

We may assume that $U(p; \delta)$ contains a nonlinear triple of points a, b, c with each of the distances pa, pb, pc less than $\delta/2$. For if this is not the case, then the neighborhood $U(p; \delta/2)$ has all of its point-triples linear. This implies that $U(p; \delta/2)$ is itself linear (see DG, p. 117, Theorem 46.1), and hence $K_W(p)$ does not exist.

Let a, b, c denote a nonlinear point-triple of $U(p; \delta/2)$. Since M is metrically convex, a point d of $\mathbf{M}$ exists such that $b \neq d \neq c$ and $bd + dc = bc$ Since $bc \leq pb + pc < \delta$, then either $bd < \delta/2$ or $cd < \delta/2$. If $bd < \delta/2$, then $pd \leq pb + bd < \delta$, and $d \in U$. Similarly, $cd < \delta/2$ implies $d \in U$. Clearly, $d \neq a$ since b, c, d are linear, but a, b, c are not linear.

Hence for every $\delta > 0$, the neighborhood $U(p; \delta)$ of p contains a nonlinear quadruple a, b, c, d of pairwise distinct points, with the triple b, c, d linear.

DEFINITION A neighborhood $U(p; \rho)$, $\rho > 0$, of a point p is called *regular* provided every quadruple Q of pairwise distinct points, contained in $U(p; \rho)$, has an imbedding curvature $K_W(Q)$ satisfying the inequality $|K_W(Q)| < \pi^2/16\rho^2$.

Remark. If $K_W(p)$ exists, then each spherical neighborhood of p, with sufficiently small radius ρ, is regular.

We establish some properties of *a regular neighborhood $U(p; \rho)$ of a point p, of a compact and convex metric space* $\mathbf{M}$, *at which $K_W(p)$ exists.*

Property 11.8. *If four pairwise distinct points of $U(p; \rho)$ contain two linear triples, then the four points are linear.*

Proof. Let Q denote such a quadruple. Then $Q \approx Q'$, where Q' is a point-quadruple of Σ_k, with $|k| < \pi^2/16\rho^2$. If $k \leq 0$, then clearly Q' lies on a line. If $k > 0$, then Q' lies on half of a great circle of Σ_k (and hence is linear), for each two points of Q' have distance less than 2ρ, and $2\rho < \pi/2k^{1/2} = \pi r/2$, where r is the radius of Σ_k.

Property 11.9. *There exist two points x, y of $U(p; \rho)$ such that p, x, y are not linear.*

Proof. Suppose the contrary, and let a, b, c be pairwise distinct points of $U(p; \rho)$, each distinct from p; since $K_W(p)$ exists, $U(p; \rho)$ surely contains such points. Then, since p, a, b and p, a, c are linear triples, it follows from the preceding property that p, a, b, c are linear. Hence *every* three points of $U(p; \rho)$ are linear, and consequently $U(p; \rho)$ is linear. This contradicts the existence of $K_W(p)$.

Property 11.10.   A regular neighborhood $U(p; \rho)$ is completely convex; that is, if $x, z \in U(p; \rho)$ and $y \in$ M such that xyz subsists, then $y \in U(p; \rho)$.

Proof.   Let a, b denote any two distinct points of $U(p; \rho)$, and let c denote any point of M that is between a and b (that is, acb subsists). We show that $c \in U(p; \rho)$.

Case 1.   One of the points a, b is p. We may assume the labeling so that $p = a$. Then from pcb, we have $p \neq c \neq b$, and $pc + cb = pb$. Since $b \in U(p; \rho)$, then $pb < \rho$ and so $pc < \rho$. Hence $c \in U(p; \rho)$.

Case 2.   The points a, b, p are pairwise distinct and linear. If $c = p$ then $c \in U(p; \rho)$. If $c \neq p$, the quadruple a, b, c, p has its points pairwise distinct.

Now pab and acb imply pcb—that is, $pc < pb < \rho$, so $c \in U(p; \rho)$—and abp and acb imply pca—that is, $pc < pa < \rho$, and so $c \in U(p; \rho)$. Finally, if apb holds, suppose c exists with acb and $pc \geq \rho$. Since M is compact and convex, it contains a metric segment $\text{seg}[a, c]$, with end points a, c (Theorem 6.2), and since $pa < \rho$, $pb < \rho$, but $pc \geq \rho$, that segment contains a point y (distinct from a, b, p) such that $\max\{pa, pb\} < py < \rho$. Now ayc and acb imply ayb, so the quadruple a, p, y, b is contained in $U(p; \rho)$ and has two of its triples linear. We conclude (Property 11.8) that the four points are linear (in the order a, y, p, b or a, p, y, b) and so $py < \max\{pa, pb\}$, contradicting a previous inequality.

Hence the assumption that $pc \geq \rho$ is untenable, and the proof of this case is complete.

Case 3.   The points p, a, b are not linear. If $c \in$ M such that acb and $pc \geq \rho$, $\text{seg}[a, c]$, $\text{seg}[b, c]$ exist containing points c_1, c_2, respectively, such that $\max\{pa, pb\} < pc_1 = pc_2 < \rho$. From $ac_1 c$ and acb follows $ac_1 b$, and $ac_2 b$ follows from acb and $bc_2 c$. Then by Property 11.8, the quadruple a, c_1, c_2, b is linear, with $ac_1 c_2$ or $ac_2 c_1$ holding.

The quadruple $Q = \{p, a, c_1, c_2\} \subset U(p; \rho)$ and hence is congruent with $Q' = \{p', a', c_1', c_2'\} \subset \Sigma_k$, where $|k| < \pi^2/16\rho^2$. Since $a', c_1', c_2' \approx a, c_1, c_2$, then a', c_1', c_2' are linear, with $a'c_1'c_2'$ or $a'c_2'c_1'$ holding, and since $p'c_1' = pc_1 = pc_2 = p'c_2' > \max\{pa, pb\} \geq pa = p'a'$, it follows that k cannot be negative or zero. For $k > 0$, the inequality $p'c_1' > p'a'$ implies that the altitude of the isosceles spherical triangle $T(p', c_1', c_2')$ is at least $\pi/2\, k^{1/2}$. But since $p'c_1' = p'c_2' < \rho$ and $c_1'c_2' < 2\rho$, it follows that the altitude of that triangle is *less than* $2\rho < \pi/2\, k^{1/2}$.

The proof of the property is complete.

COROLLARY   *Each two distinct points of a regular neighborhood $U(p; \rho)$ are the end points of exactly one metric segment, and that segment lies wholly in $U(p; \rho)$.*

Proof. If $a, b \in U(p; \rho)$, then, since **M** is compact and convex, there exists in **M** at least one segment with end points a, b. If c is any point of any such segment, $a \neq c \neq b$, then acb holds, and by the property just proved, $c \in U(p; \rho)$. Hence $U(p; \rho)$ contains *all* of the segments with end points a, b. It follows easily from Property 11.8 that only one segment has a, b for end points.

DEFINITION   Let $U(p; \rho)$ be a regular neighborhood of a point p of a compact, convex, metric space **M**. Since every quadruple Q of $U(p; \rho)$ has an imbedding curvature between $-\pi^2/16\rho^2$ and $\pi^2/16\rho^2$, the set of numbers K such that each quadruple of $U(p; \rho)$ has an imbedding curvature $K_W(Q)$ with $|K_W(Q)| \leq K$ has a greatest lower bound K^* which is called the *curvature bound of $U(p; \rho)$.*

Let **M** have Wald curvature $K_W(p)$ at p $(p \in \mathbf{M})$, and let a, b be distinct points of a regular neighborhood $U(p; \rho)$ of p, with $a \neq p \neq b$ (Fig. 15). For each number x, $0 \leq x < \min\{pa, pb\}$, let $a(x), b(x)$ denote the points of seg$[p, a]$, seg$[p, b]$, respectively, with $pa(x) = x = pb(x)$, and put $d(x) = a(x)b(x)$.

Property 11.11.   *The limit $\lim_{x \to 0}[d(x)/x]$ exists.*

Proof. If p, a, b are linear, then $d(x) = 2x$ in case apb holds, and $d(x) = 0$ if pab or pba subsists. In the first case, $\lim_{x \to 0}[d(x)/x] = 2$, and in the second case $\lim_{x \to 0}[d(x)/x] = 0$.

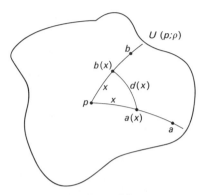

Figure 15

Suppose now that p, a, b are not linear. Choose two numbers k' and k'' such that $k' < -K^*$ and $k'' > K^*$, where K^* is the curvature bound of $U(p; \rho)$, and choose any number x_0 such that $0 < x_0 < \min[pa, pb]$. Denote by $a(x_0), b(x_0)$ the points on $\text{seg}[p, a]$, $\text{seg}[p, b]$, respectively, with distance x_0 from p.

Let $p', a'(x_0), b'(x_0)$ and $p'', a''(x_0), b''(x_0)$ be triples of $\Sigma_{k'}$ and $\Sigma_{k''}$, respectively, such that

$$p', a'(x_0), b'(x_0) \approx p, a(x_0), b(x_0) \approx p'', a''(x_0), b''(x_0)$$

(such triples surely exist for x_0 sufficiently small), and for any $x, 0 \leqq x \leqq x_0$ consider the points $a(x), a'(x), a''(x)$ on $\text{seg}[p, a(x_0)]$, $\text{seg}[p', a'(x_0)]$, $\text{seg}[p'', a''(x_0)]$, respectively, with distance x from p, p', p'', respectively. Let points $b(x), b'(x), b''(x)$ on $\text{seg}[p, b(x_0)]$, $\text{seg}[p', b'(x_0)]$, $\text{seg}[p'', b''(x_0)]$, respectively, be similarly defined. Clearly,

$$p', a'(x), a'(x_0) \approx p, a(x), a(x_0) \approx p'', a''(x), a''(x_0),$$

and

$$p', b'(x), b'(x_0) \approx p, b(x), b(x_0) \approx p'', b''(x), b''(x_0).$$

Now the quadruple $p, a(x_0), b(x_0), a(x)$ of $U(p; \rho)$ has an imbedding curvature $k^* > k'$; that is,

$$p, a(x_0), b(x_0), a(x) \approx p^*, a^*(x_0), b^*(x_0), a^*(x)$$

of Σ_{k^*}, and it follows from (3) of Property 10.5 that $b'(x_0)a'(x) < b^*(x_0)a^*(x) = b(x_0)a(x)$.

Similarly, the quadruple $p, b(x_0), b(x), a(x)$ of $U(p; \rho)$ has an imbedding curvature exceeding k'. Let $\bar{p}, \bar{b}(x_0), \bar{a}(x)$ be points of $\Sigma_{k'}$ with $\bar{p}, \bar{a}(x), \bar{b}(x_0) \approx$

p, $a(x)$, $b(x_0)$, and let $\bar{b}(x)$ denote that point of seg$[\bar{p}, \bar{b}(x_0)]$ such that

$$\bar{p}, \bar{b}(x), \bar{b}(x_0) \approx p, b(x), b(x_0).$$

Then by Property 11.5, $\bar{a}(x)\bar{b}(x) < a(x)b(x) = d(x)$, and since $b'(x_0)a'(x) < b(x_0)a(x) = \bar{b}(x_0)\bar{a}(x)$, it follows that $\nleftarrow \bar{p}: \bar{a}(x), \bar{b}(x) > \nleftarrow p': a'(x), b'(x)$, and hence

$$d'(x) = a'(x)b'(x) < \bar{a}(x)\bar{b}(x) < a(x)b(x) = d(x).$$

In a similar manner, it is shown that $d(x) < d''(x) = a''(x)b''(x)$, and so, for every x such that $0 \leq x < x_0$, we have $d'(x) < d(x) < d''(x)$. Consequently $\lim_{x\to 0}[d'(x)/x] \leq \lim_{x\to 0}[d(x)/x] \leq \lim_{x\to 0}[d''(x)/x]$, if the limits exist.

By trigonometry (spherical or hyperbolic)

$$\frac{\sin[\tfrac{1}{2}\sqrt{k'} \cdot d'(x)]}{\sin[\sqrt{k'} \cdot x]} = \frac{\sin[\tfrac{1}{2}\sqrt{k'} \cdot d'(x_0)]}{\sin[\sqrt{k'} \cdot x_0]} = L'(x_0) \qquad (x \leq x_0),$$

and hence $\lim_{x\to 0}[d'(x)/x] = 2L'(x_0)$.

A similar formula (replacing k' by k'') yields $\lim_{x\to 0}[d''(x)/x] = 2L''(x_0)$, and consequently all accumulation points of $d(x)/x$ lie between $2L'(x_0)$ and $2L''(x_0)$, which, in turn lie between 0 and 2. But

$$\lim_{x_0 \to 0} [L'(x_0) - L''(x_0)] = 0,$$

and it follows that $\lim_{x\to 0}[d(x)/x]$ exists.

The chief importance of the property just proved is that it permits the introduction of angles in $U(p; \rho)$. This is accomplished by means of the following definition.

DEFINITION   If $a, b \in U(p; \rho)$, $a \neq p \neq b$,

$$\nleftarrow p{:}a, b = 2\,\mathrm{Arcsin}\!\left(\tfrac{1}{2}\lim_{x\to 0}\frac{d(x)}{x}\right),$$

where Arcsin denotes the principal value of the function.

Remark 1.   For each two points a, b of $U(p; \rho)$, $a \neq p \neq b$, $0 \leq \nleftarrow p{:}\, a, b \leq \pi$, with the first equality sign holding if and only if pab or pba subsists, and the second one holds if and only if apb subsists. If $pa = pb$, then $\nleftarrow p{:}\, a, b = 0$ if and only if $a = b$.

Remark 2.   If $a, b \in U(p; \rho)$, $a \neq p \neq b$, and

$$p', a', b' \approx p, a, b \approx p'', a'', b'',$$

with the "primed" triple in $\Sigma_{k'}$, $k' < -K^*$, and the double "primed" points in $\Sigma_{k''}$, $k'' > K^*$, then

$$\nleftarrow p': a', b' \leqq \nleftarrow p: a, b \leqq \nleftarrow p'': a'', b''$$

(since $\lim_{x \to 0}[d'(x)/x] \leqq \lim_{x \to 0}[d(x)/x] \leqq \lim_{x \to 0}[d''(x)/x]$). If $pa = pa' = pa''$ and $pb = pb' = pb''$, and

$$\nleftarrow p: a, b = \nleftarrow p': a', b' = \nleftarrow p'': a'', b'',$$

then $a'b' > ab > a''b''$.

Remark 3.   Polar coordinates are introduced in $U(p; \rho)$ in the following manner. Let p_1, p_2 be points of $U(p; \rho)$ such that p, p_1, p_2 are not linear (Property 11.9). Associate with each point s of $U(p; \rho)$, $s \neq p$, the ordered pair of numbers $(r(s), \theta(s))$, where $r(s) = ps$, and $\theta(s) = \nleftarrow p: s, p_1$ or $2\pi - \nleftarrow p: s, p_1$ according as $|\nleftarrow p: p_1, p_2 - \nleftarrow p: s, p_1| = \nleftarrow p: s, p_2$ or $|\nleftarrow p: p_1, p_2 - \nleftarrow p: s, p_1| \neq \nleftarrow p: s, p_2$, respectively.

We are now in position to establish a very important property of $U(p; \rho)$ which is quite surprising: the existence of the Wald curvature at a point p of a compact and convex metric space **M** implies that a neighborhood of p is two-dimensional! (Compare this with Theorem 10.1.) The following remark is helpful in proving that property.

Remark 4.   Let $p_1, p_2, \ldots, p_n$ be n pairwise distinct points of $U(p; \rho)$, each distinct from p, and let $p_i(x)$ denote the unique point on the unique segment $\text{seg}[p, p_i]$ with distance x from p $(i = 1, 2, \ldots, n)$, for every x, $0 \leqq x \leqq \min[pp_1, pp_2, \ldots, pp_n]$, and put $p_0(x) = p$, for every x. Each four of the $n + 1$ points $p_0, p_1, \ldots, p_n$, say $p_{i_1}, p_{i_2}, p_{i_3}, p_{i_4}$ is congruent with a quadruple Q^* of points of Σ_{k^*}, where k^* depends on x and the indices i_1, i_2, i_3, i_4, and $k^* \to K_W(p)$ as $x \to 0$.

For x sufficiently small, the distance relation satisfied by the quadruple Q^* $(Q^* \subset \Sigma_{k^*})$, differs arbitrarily little from that satisfied by four points of E_2 (whose D-determinant vanishes), and the same observation is valid if the mutual distances of points of Q^* be divided by x. Putting $d_{ij} = \lim_{x \to 0}[p_i(x)p_j(x)]/x$ $(i, j = 0, 1, \ldots, n)$, then for each four indices i_1, i_2, i_3,

i_4, points p''_{i_1}, p''_{i_2}, p''_{i_3}, p''_{i_4} of E_2 exist such that

$$p''_i p''_j = d_{ij},$$

where $i, j = i_1, i_2, i_3, i_4$.

It follows that $n + 1$ points, $p''_0, p''_1, \ldots, p''_n$ of E_2 exist such that $p''_i p''_j = d_{ij}$ $(i, j = 0, 1, \ldots, n)$ (see DG, p. 118, Corollary; for $n = 4$, note that Σ_{k*} does not contain a pseudoplanar quintuple). It is clear, also that $\nmid p''_0 : p''_i, p''_j = \nmid p_0 : p_i, p_j$ $(i, j = 1, 2, \ldots, n)$.

Apply the foregoing to the points p_1 and p_2 of $U(p; \rho)$, which were used to define polar coordinates in $U(p; \rho)$, together with any other pair of distinct points p_3, p_4. Introduce in E_2 the polar coordinate system with pole p''_0, and initial ray, the half-line joining p''_0 to p''_1, oriented so that $\theta(p''_2)$, the second coordinate of p''_2, is less than π.

It follows that $\theta(p_i) = \theta(p''_i)$ $(i = 1, 2, 3, 4)$, and $\nmid p''_0 : p''_3, p''_4 = \nmid p_0 : p_3, p_4$.

Property 11.12. *A regular neighborhood $U(p; \rho)$ is homeomorphic with a subset U' of the euclidean plane.*

Proof.   Associate with p the pole p''_0 of a polar coordinate system in E_2, and if $s \in U(p; \rho)$, $s \neq p$, let its corresponding point of E_2 be the point s'' with polar coordinates $(r(s), \theta(s))$, where $(r(s), \theta(s))$ are the polar coordinates of s in the polar coordinate system defined above in $U(p; \rho)$. The mapping of $U(p; \rho)$ into E_2 so defined is a homeomorphism. (Let the reader apply Remark 4 to complete the proof.)

Our investigation of the properties of $U(p; \rho)$ is concluded with the following easily proved property.

Property 11.13.   *If $a, b \in U(p; \rho)$, $a \neq b$, and $c \in U(p; \rho)$ such that either pac or pbc subsists, then to each interior point q of seg$[a, b]$ there corresponds at least one point r of $U(p; \rho)$ such that pqr holds.*

Proof.   The property is obvious in case p, a, b are linear. If p, a, b are not linear and $c \in U(p; \rho)$ such that pbc holds, let q denote any interior point of seg$[a, b]$. There is a unique point r of seg$[a, c]$ such that $\nmid p : a, r = \nmid p : a, q$. It is easily shown that pqr holds.

A similar procedure is used in case pac holds.

11.6   IDENTIFICATION OF
GAUSS AND WALD CURVATURES

If **M** is a compact and metrically convex metric space that possesses Wald curvature $K_W(p)$ at each point p, then a positive number ρ_0 exists such that every $U(p; \rho)$, $p \in \mathbf{M}$, $\rho < \rho_0$, has the six properties established in the preceding section. Suppose that a continuous function $K_G(p)$ is defined at each p $(p \in \mathbf{M})$ such that if the continuously differentiable functions $r(t)$, $\theta(t)$ $(t_1 \leqq t \leqq t_2)$ define a curve in $U(p; \rho)$, its length is given by

$$\int_{t_1}^{t_2} \sqrt{dr^2 + G^2(r, \theta)\, d\theta^2},$$

where the function $G(r, \theta)$ is a solution of the differential equation

$$\frac{\delta^2}{\delta r^2} G(r, \theta) + K_G(r, \theta) \cdot G(r, \theta) = 0,$$

with $G(0, \theta) = 0$ and

$$\left[\frac{\partial}{\partial r} G(r, \theta) \right]_{r=0} = 1.$$

Then **M** is called a *Gauss surface*, and $K_G(p)$ the Gauss curvature of **M** at p. Wald proved the following two theorems.

THEOREM 11.2   *At each point p of a Gauss surface, the curvature $K_W(p)$ exists and equals the Gauss curvature $K_G(p)$.*

THEOREM 11.3   *Each compact and metrically convex metric space that has curvature $K_W(p)$ at each of its points is a Gauss surface, and at each point p, $K_G(p) = K_W(p)$.*

These theorems completely justify the assertion that $K_W(p)$ is a metrization of the Gauss curvature. It should be observed, moreover, that Theorem 11.3 provides a *metric characterization of Gauss surfaces among the class of metric spaces*, for it states wholly and explicitly in terms of "distance" those properties of a metric space by virtue of which it is a Gauss surface. Unfortunately, the proofs of the two theorems are too lengthy to give here.

One final observation about this matter is in order. The definition of $K_W(p)$ requires that in a sufficiently small neighborhood of p, *all* quadruples

Q of pairwise distinct points have an imbedding curvature $K_W(Q)$ that differs arbitrarily little from $K_W(p)$. Very soon the question arose whether "all quadruples Q" could be replaced by "all quadruples Q *containing a linear triple.*" The fact that such quadruples have at most *one* imbedding curvature gives added point to the query.

In his Missouri doctoral dissertation (1962), W. A. Kirk gave a thorough answer to that question. He observed that if Wald's curvature definition were modified in the suggested manner, then Property 11.12 is not valid (that is, regular neighborhoods are not necessarily homeomorphic with a subset of the plane) since, for example, any neighborhood of a point in E_3 has each quadruple Q (of pairwise distinct points) containing a linear triple planar, and so the modified definition gives $\mathbf{E}_3$ curvature zero at each point, but neighborhoods are not topologically planar. *But if the compact and convex metric space $\mathbf{M}$ be assumed locally homeomorphic with the plane,* Kirk showed that the rest of Wald's development is valid. This was accomplished by proving that in such a space, the modified definition implies the original one. From the standpoint of differential geometry, the assumption that the compact and convex metric space $\mathbf{M}$ is locally homeomorphic to the plane is quite acceptable.

11.8   RINOW CURVATURE

Let $\mathbf{M}$ denote a metric space that contains with each two of its distinct points p, q at least one *rectifiable* arc $A(p, q)$, with end points p, q. If we set $\gamma(p, q) = \inf \ell[A(p, q)]$ for all arcs $A(p, q)$, and $\gamma(p, p) = 0$, it is easily seen that

$$\gamma(p, q) \geqq 0, \ \gamma(p, q) = 0 \text{ if and only if } p = q, \ \gamma(p, q) = \gamma(q, p),$$

and for each three points p, q, r of $\mathbf{M}$, $\gamma(p, q) + \gamma(q, r) \geqq \gamma(p, r)$. Hence $\gamma(p, q)$ is a metric for the point set of $\mathbf{M}$. It is called the *geodesic* or *intrinsic* metric of $\mathbf{M}$. Clearly $\gamma(p, q) \geqq pq$ for each two points p, q of $\mathbf{M}$, since an assumption to the contrary implies the existence of an arc $A(p, q)$ with length $\ell[A(p, q)] < pq$. But $\ell[A(p, q)] \geqq pa + aq \geqq pq$, for each point a of $A(p, q)$. Let $\mathbf{M}_\gamma$ denote the space over $\mathbf{M}$, with metric γ.

In case $\mathbf{M}$ contains an arc $G(p, q)$ such that $\ell[G(p, q)] = \gamma(p, q)$, then $G(p, q)$ is a *geodesic* in $\mathbf{M}$ and a *metric segment* in $\mathbf{M}_\gamma$, since its length in $\mathbf{M}_\gamma$ (which is easily proved equal to $\ell[G(p, q)]$, its length in $\mathbf{M}$) equals the distance $\gamma(p, q)$ in $\mathbf{M}_\gamma$ of its end points. If each two distinct points of $\mathbf{M}$ are end points of a geodesic, then the space $\mathbf{M}_\gamma$ is metrically convex, though $\mathbf{M}$ itself may not be convex.

For example, let $\mathbf{M}$ denote the surface of a ball in E_3, with euclidean (chord) metric. Then $\mathbf{M}_\gamma$ is evidently that surface with shorter arc (geodesic) metric. It is a convex metric space, though $\mathbf{M}$ is not convex.

It is emphasized that the space $\mathbf{M}_\gamma$ is not necessarily convex—that is, two points of $\mathbf{M}_\gamma$ need not be joined by a metric segment in M_γ (a geodesic in $\mathbf{M}$).

For spaces $\mathbf{M}$ with intrinsic metric γ, W. Rinow recently introduced the notion of *region of bounded Riemann curvature*. (From now on, we denote distances in $\mathbf{M}_\gamma$ by pq instead of $\gamma(p,q)$, since the underlying metric (if any) plays no role.)

DEFINITION   A region R of a metric space $\mathbf{M}_\gamma$, with intrinsic metric, is a *region of Riemann curvature* $\leq k$ (k, a real number) provided R has the following properties.

I.   *Each two distinct points of R are joined by at least one metric segment.*

II.   *Each three points of R are congruently contained in Σ_k.*

III$_L$.   *If $p,q,r \in R$, $q \neq p \neq r$, let x,y denote points of* seg$[p,q]$, seg$[p,r]$, *respectively. If $p',q',r',x',y' \in \Sigma_k$ such that*

$$p,q,r \approx p',q',r',$$
$$p,q,x \approx p',q',x',$$
$$p,r,y \approx p',r',y',$$

then $xy \leq x'y'$.

If R has properties *I, II, III$_G$*, where *III$_G$* is obtained from *III$_L$* upon replacing "$xy \leq x'y'$" by "$xy \geq x'y'$," then R is a *region of Riemann curvature* $\geq k$.

The following localization of the notion of region of bounded Riemann curvature, suggested by Kirk, permits defining the Rinow curvature of a metric space *at a point*.

DEFINITION.   A space $\mathbf{M}$ (with intrinsic metric) has at an accumulation point p the *Rinow curvature* $K_R(p)$ provided (i) no neighborhood of p is linear, and (ii) corresponding to each $\varepsilon > 0$ there is a $\rho > 0$ such that the neighborhood $U(p;\rho)$ is a region of Riemann curvature less than or equal to $K_R(p) + \varepsilon$, and also a region of Riemann curvature greater than or equal to $K_R(p) - \varepsilon$.

EXERCISES

1. Prove that $(\mathbf{M}_\gamma)_\gamma = \mathbf{M}_\gamma$.

2. Is $\mathbf{M}_\gamma$ necessarily convex?

3. If $a, b, c \in \Sigma_k$, show that for each $k' < k$, a, b, c are congruently imbeddable in $\Sigma_{k'}$. What can be said when $k' > k$?

11.8 COMPARISON OF WALD AND RINOW CURVATURES

Let us compare K_W and K_R first in the context of a compact and convex metric space $\mathbf{M}$—the environment in which the curvature K_W was so successfully studied by Wald. Note that for such a space $\mathbf{M}$, the intrinsic metric is the same as the metric, since each two distinct points p, q of $\mathbf{M}$ are the end points of a metric segment, $\text{seg}[p, q]$ (not necessarily unique), and since $\ell[\text{seg}[p, q]] = pq$, then $\ell[\text{seg}[p, q]] \leqq \ell[A(p, q)]$, for each arc $A(p, q)$ of $\mathbf{M}$, with end points p, q. Hence $\gamma(p, q) = pq$.

THEOREM 11.4 *If the curvature $K_W(p)$ exists at a point p of a compact and convex metric space $\mathbf{M}$, then $K_R(p)$ exists, and $K_R(p) = K_W(p)$.*

Proof. Let $\varepsilon > 0$ be arbitrary, and let $\rho > 0$ be so chosen that (i) $U(p; \rho)$ is a regular neighborhood of p, (ii) every quadruple Q of pairwise distinct points of $U(p; \rho)$ has an imbedding curvature $K_W(Q)$ that satisfies the inequality $|K_W(p) - K_W(Q)| < \varepsilon$, and (iii) each three points of $U(p; \rho)$ are congruently imbeddable in $\Sigma_{k''}$, $k'' = K_W(p) + \varepsilon$.

The neighborhood $U(p; \rho)$ has the following properties.

I. *Each two distinct points of $U(p; \rho)$ are end points of a unique segment, and that segment is contained in $U(p; \rho)$.* The proof of Case 3, Property 11.10, may be applied to show that $U(p; \rho)$ contains every segment joining two of its points, the uniqueness of segments results from Property 11.8, and the existence of segments follows from Theorem 6.2.

II. *Each three points of $U(p; \rho)$ are congruent with three points of $\Sigma_{k'}$, where $k' = K_W(p) - \varepsilon$.* According to the choice of ε, every quadruple Q of pairwise distinct points of $U(p; \rho)$ has an imbedding curvature that exceeds

$K_W(p) - \varepsilon$; that is, Q is congruent with four points of a Σ_k, with $k > K_W(p)$ $- \varepsilon$. It follows that every *three* points of $U(p; \rho)$ are congruently contained in every Σ_{k^*}, with $k^* < k$, and hence, in particular, in $\Sigma_{k'}$, $k' = K_W(p) - \varepsilon$.

To prove that $U(p; \rho)$ is a region of Riemann curvature greater than or equal to $K_W(p) - \varepsilon$, it remains to show that $U(p; \rho)$ has property III_G with respect to $\Sigma_{k'}$.

Let a, b, c be points of $U(p; \rho)$, $b \neq a \neq c$, and let x, y denote points of seg$[a, b]$, seg$[a, c]$, respectively. Then by I, $x, y \in U(p; \rho)$. Now the points $a, x, b, c \approx a_1, x_1, b_1, c_1$, with the second quadruple in Σ_{k_1}, $k_1 > k' = K_W(p) - \varepsilon$. If $a', b', c' \in \Sigma_{k'}$, with $a', b', c' \approx a, b, c \approx a_1, b_1, c_1$, let x' denote the point of $\Sigma_{k'}$ such that $a', x', b' \approx a_1, x_1, b_1 \approx a, x, b$. Application of (3), Property 11.5, yields $c'x' < c_1 x_1 = cx$.

Now $\Sigma_{k'}$ contains a point x^* (on the same side of the geodesic $G(a', c')$ as x') such that $a', c' x^* \approx a_1, c_1, x_1 \approx a, c, x$. If x^* is an interior point of $\not\subset a' : x', c'$, then, since $a'x^* = ax = a'x'$, Property 11.2 yields $c'x' > c'x^* = cx$, in contradiction to the inequality established above. Hence x^* is not an interior point of $\not\subset a' : x', c'$, and consequently $\not\subset a' : x^*, c' \geq \not\subset a' : x', c'$. If, now, $y' \in \Sigma_{k'}$ such that $a', y', c' \approx a, y, c$, then $y'x^* \geq y'x'$.

The points $a, x, y, c \approx a_2, x_2, y_2, c_2$, with the second quadruple in Σ_{k_2}, $k_2 > k'$, and another application of Property 11.2 gives $y'x^* < y_2 x_2 = xy$. Hence $x'y' < xy$, and so $U(p; \rho)$ is a region of Riemann curvature $\geq K_W(p) - \varepsilon$.

A similar argument (which the reader is asked to supply) shows that $U(p; \rho)$ is a region of Riemann curvature $\leq K_W(p) + \varepsilon$. It follows from the definition of $K_R(p)$ that $K_R(p)$ exists at p and equals $K_W(p)$.

On the other hand, the existence of K_R at a point p of a compact, convex metric space does *not* imply the existence of K_W at p. At each point p of euclidean three-space E_3, it is easy to show that $K_R(p) = 0$, and $K_W(p)$ does not exist. The existence of $K_R(p)$ places no restriction on the dimensionality of the space in a neighborhood of p, in distinction to what is implied in that respect by the existence of $K_W(p)$. We shall see that the Rinow curvature is more closely related to the modified Wald curvature (referred to above and defined below) than it is to the Wald curvature itself. Of course, wherever the modified Wald curvature implies the Wald curvature—as, for example, in compact, convex metric spaces, locally homeomorphic with a subset of the plane—it clearly implies the Rinow curvature.

Curvature K_W'. A metric space **M** has *curvature* $K_W'(p)$ at a point p of **M** provided (i) no neighborhood of p is linear, and (ii) corresponding to each

$\varepsilon > 0$ there is a $\delta > 0$ such that if Q denotes any quadruple of pairwise distinct points of $U(p; \delta)$ and Q *contains a linear triple*, then Q has an imbedding curvature $K'_W(Q)$ such that $|K'_W(p) - K_W(Q)| < \varepsilon$.

Clearly the existence of $K_W(p)$ implies the existence of $K'_W(p)$, and the two curvatures are equal; but $K'_W(p) = 0$, $p \in E_3$, and $K_W(p)$ does not exist.

THEOREM 11.5 *If $K_R(p)$ exists at a point p of a metric space* **M** *with intrinsic metric, then so does $K'_W(p)$, and $K_R(p) = K'_W(p)$.*

Proof. Since $K_R(p)$ exists, no neighborhood of p is linear, and for each $\varepsilon > 0$ there is a neighborhood $U(p; \rho)$ of p which is a region of Riemann curvature less than or equal to $k'' = K_R(p) + \varepsilon$, and also a region of Riemann curvature greater than or equal to $k' = K_R(p) - \varepsilon$. If $k'' > 0$, choose $\rho < 1/(4\sqrt{k''})$. We wish to show that each quadruple Q of pairwise distinct points of $U(p; \rho)$, that contains a linear triple, has an imbedding curvature $K_W(Q)$ such that $|K_R(p) - K_W(Q)| < \varepsilon$.

Let p_1, p_2, p_3, p_4 be pairwise distinct points of $U(p; \rho)$, with $p_1 p_2 + p_2 p_3 = p_1 p_3$. Then points p'_1, p'_2, p'_3, p'_4 of $\Sigma_{k'}$ and points $p''_1, p''_2, p''_3, p''_4$ of $\Sigma_{k''}$ exist such that $p'_i p'_j = p_i p_j = p''_i p''_j$ for every pair of indices $i, j = 1, 2, 3, 4$, *except* possibly $i, j = 2, 4$ $(i \neq j)$. For those indices, we have

(†)
$$p'_2 p'_4 \leqq p_2 p_4 \leqq p''_2 p''_4.$$

For every value of k^* $(k' \leqq k^* \leqq k'')$, Σ_{k^*} contains points $p^*_1, p^*_2, p^*_3, p^*_4$ such that $p^*_i p^*_j = p_i p_j$, $i, j = 1, 2, 3, 4$, apart from the distances $p_2 p_4, p^*_2 p^*_4$. It is clear that the distance $p^*_2 p^*_4$ varies continuously with k^*, and hence, from (†), assumes the value $p_2 p_4$ for a value of k between k' and k''. Consequently, the quadruple p_1, p_2, p_3, p_4 of $U(p; \rho)$ has an imbedding curvature k, where $K_R(p) - \varepsilon \leqq k \leqq K_R(p) + \varepsilon$; that is, $|K_R(p) - K_W(Q)| < \varepsilon$ for every quadruple Q of pairwise distinct points of $U(p; \rho)$ that contains a linear triple. It follows that $K'_W(p) = K_R(p)$, and the theorem is proved.

THEOREM 11.6 *Let* **M** *be a space with intrinsic metric. If $K'_W(p)$ exists $p \in$* **M***, and if p has a neighborhood $U(p; \rho)$, each two points of which are joined by a metric segment, then $K_R(p)$ exists and $K_R(p) = K'_W(p)$.*

Proof. The proof is almost exactly the same as the proof of Theorem 11.4. Note that M is not necessarily locally homeomorphic with the plane.

Although Rinow introduced the notion of region of bounded Riemann curvature (from which Kirk derived the localization yielding the Rinow curvature $K_R(p)$ at a point p) in the context of a metric space with intrinsic metric, both the notion and its localization are meaningful in general metric spaces.

EXERCISES

1. Let a, b, c, x, y be points of Σ_k, and a', b', c', x', y' points of $\Sigma_{k'}$ with $a, b, c \approx a', b', c'$ nonlinear, $x \in \text{seg}[a, b]$ ($a \neq x \neq b$), $y \in \text{seg}[a, c]$ ($a \neq y \neq c$), $a, x, b \approx a', x', b'$ and $a, y, c \approx a', y', c'$. If $k < k'$, show that $xy < x'y'$.

2. Show that $K_R(p) = 0$, $p \in E_3$.

3. If Q is a metric quadruple of pairwise distinct points, containing a linear triple, find an equation whose single root is $K_W(Q) \neq 0$, in case a root exists. If $K_W(Q) = 0$, what relation is satisfied by the six distances determined by the four points of Q?

REFERENCES

Sections 11.3 to 11.6 A. Wald, "Begründung einer koordinatenlosen Differentialgeometrie der Flächen," *Ergebnisse eines math. Kolloquiums (Wien)*, **7**: 24–46 (1936).

Section 11.7 W. Rinow, *Die innere Geometrie der metrischen Raume*, Springer-Verlag, Berlin, 1961.

Section 11.8 W. A. Kirk, "On curvature of a metric space at a point." *Pacific J. Math.*, **14**: 195–198 (1964).

PART 4

CURVE THEORY

CHAPTER 12

Classical Definitions of Curves

12.1 THE PROBLEM

Strictly speaking, all material objects are three-dimensional, yet only such things as a metal sphere, a wooden block, or a rock are typical representations of solids. Paper, a piece of sheet-iron, and a membrane approximate what we mean when speaking of surfaces. Wire, threads, and streaks of chalk represent our idea of curves.

In elementary geometry, examples of solids (said to be three-dimensional) include cubes, spheres, and ellipsoids. Regions in the plane (bounded, for example, by squares or circles), the union of the six faces of a cube, the visible part of a sphere, and the like are called surfaces or two-dimensional. Circles, ellipses, lemniscates, and helices are called curves; along with segments of straight lines, polygons, the union of the eight edges of a cube, and other objects said to be one-dimensional, they are studied in the general theory of curves.

Everyone uses the words *solid*, *surface*, and *curve* in daily life and associates with them some intuitive ideas. But it is the task of geometry to formulate satisfactory definitions that make these notions precise.

When is a mathematical definition of a term that is used in everyday language satisfactory? *First* of all, the definition must be *appropriate* in the sense that it includes all objects that are always denoted by the term, and excludes all objects that are never so denoted. Thus an appropriate definition of the concept of curve would include circles, ellipses, lemniscates, and helices and would exclude a cube, the surface of a sphere, a square region, and dispersed sets (for example, sets containing only a finite number of points). *Second*, there are many objects unknown in daily life and not dealt with in ordinary language which are not covered by intuitive ideas. A satisfactory definition of a term should *extend* its daily use so as to include or exclude such unexplored or borderline objects. In the case of dimension, the ideal is to associate a number with *each* geometric object, even though in such a comprehensive definition a certain arbitrariness is inevitable. *Third*, a satisfactory mathematical definition must be *fruitful* in the sense that it yields many consequences—preferably theorems that are esthetically pleasing by their generality and simplicity. General theorems derived from the definition are the justification of its inevitable arbitrariness in borderline cases. (In the case of curves, no satisfactory general theorems could be obtained if straight segments and polygons were excluded.) Some of the general theorems extend statements from the restricted domain of the ordinary language to the enlarged realm covered by the definition. Others bring out interesting exceptions; often they even correct errors of our intuition. *Finally*, it is desirable that, by virtue of its definition, the concept be related to ideas treated in other theories or that consequences of the definition be applicable to other fields.

Various definitions of curves, attempted between 1870 and 1920, proved to be unsatisfactory as such but have resulted in concepts that will be discussed in this chapter because they are of great interest in their own right.

12.2   THE QUANTITATIVE APPROACH

Originally, mathematicians believed that curves and solids, surfaces and dispersed sets differed in *quantity*. Solids and surfaces seem to include *more* points than curves do; dispersed sets, *fewer* points. But this idea had to be abandoned in the light of four great discoveries by Georg Cantor in the 1870's.

I. All solids, surfaces, and curves as well as many dispersed sets include infinitely many points, and in the everyday language terms for quantitative distinctions are confined to finite sets. Hence, first of all, Cantor had to define the words *more*, *less*, and *equal* for infinite sets. Following him, two sets A and B, regardless of whether they are finite or infinite, are said to include *equally* many elements or to have the same power if, to each element of A, one can pair an element of B in such a way that each element of B is paired to exactly one element of A. Such a procedure is called a *one-to-one mapping* of A onto B. The set E of all even numbers and the set E^* of all odd numbers have the same power, as is seen, for example, by pairing to each even number the preceding odd number.

A statement valid for finite sets that cannot be upheld for infinite sets is the old saying that the whole is greater than any of its parts. The (whole) set N of all natural numbers $\{1, 2, \ldots, n, \ldots\}$ and its subset (or part) E have the same power, as is seen by pairing to each number n in N the even number $2n$ in E. But N also has the same power as the apparently much larger set P consisting of all pairs of natural numbers. If m and n are different numbers, there are two synonymous symbols for the set consisting of m and n—namely, $\{m, n\}$ and $\{n, m\}$. A one-to-one mapping of P onto N can be obtained by denoting all elements of P by those symbols $\{m, n\}$ for which $m < n$, and by pairing with $\{m, n\}$ the number $m + \frac{1}{2}n(n - 1)$ in N (see Exercise 1).

II. Since sets as unlike as N, E, and P contain equally many elements, one might suspect that *all* infinite sets have the same power. Calling sets having the same power as N *denumerable*, however, Cantor discovered infinite sets that are *not* denumerable. An example is the segment $[0, 1]$, the set of all real numbers such that $0 \leq x \leq 1$. That this set is not denumerable is demonstrated by proving that each *denumerable* subset S of $[0, 1]$ is a *proper* subset. Each number in $[0, 1]$ is equal to a nonterminating decimal fraction,

$$\sum_{n=1}^{\infty} \frac{a_n}{10^n} = .a_1 a_2 \ldots a_n \ldots,$$

where all a_k belong to the set $\{0, 1, \ldots, 9\}$. Visually different decimal fractions cannot be equal unless they end in 0's or 9's, as do $.500 \ldots 0 \ldots$ and $.499 \ldots 9 \ldots$, both of which are equal to $\frac{1}{2}$. Now suppose that S is any denumerable subset of $[0, 1]$ (in other words, that there is a one-to-one mapping of N onto S) and that this mapping pairs to n, say, the number

$$s_n = .s_{n1} s_{n2} \ldots s_{nn} \ldots.$$

Then the numbers in $[0, 1]$ not belonging to S include any number $t = .t_1 t_2 \ldots t_n \ldots$ such that (a) $t_n \neq s_{nn}$ for each n, which makes the fraction visually different from $.s_{n1} s_{n2} \ldots s_{nn} \ldots$ for each n, and (b) all t_n are $\neq 0$ and $\neq 9$, whence t and s_n are unequal for each n (see Exercise 4(a)).

III. Notwithstanding the existence of quantitative differences in the realm of infinite sets, Cantor found that *a segment on a line and a square region in the plane include equally many elements*. In the proof of this fact decimal fractions would not be convenient.

For many purposes, we shall make use of nonterminating g-ary fractions for some natural number $g > 1$,

$$\sum_{n=1}^{\infty} \frac{b_n}{g^n} = (.b_1 b_2 \ldots b_n \ldots)_g,$$

where all b_n belong to the set $\{0, 1, \ldots, g - 1\}$. For each g, the g-ary fractions share with the decimal fractions (obtained for $g = 10$) the inconvenience that one and the same number b may be equal to two visually different nonterminating fractions. This occurs if and only if b is what is called *g-ary rational*—that is, equal to a terminating g-ary fraction,

$$b = \frac{b_1}{g} + \frac{b_2}{g^2} + \cdots + \frac{b_k}{g^k},$$

where $b_k \neq 0$. In this case,

$$b = (.b_1 b_2 \ldots b_{k-1} b_k 00 \ldots 0 \ldots)_g$$
$$= (.b_1 b_2 \ldots b_{k-1}(b_k - 1)(g - 1)(g - 1) \ldots (g - 1) \ldots)_g.$$

Using binary fractions $(g = 2)$, F. Hartogs developed a one-to-one mapping of the set $]0, 1]$ of all numbers x such that $0 < x \leqslant 1$ onto the set N^{∞} of all infinite sequences of natural numbers. For each such x, consider that nonterminating binary fraction equal to x that contains infinitely many 1's; for example,

$(.1010 \ldots 10 \ldots)_2$ for $\frac{2}{3}$, $(.011 \ldots 1 \ldots)_2$ and *not* $(.100 \ldots 0 \ldots)_2$ for $\frac{1}{2}$,

and record the place numbers of the 1's in the fraction of x. A sequence of increasing natural numbers results; for example,

$(1, 3, 5, \ldots, 2n - 1, \ldots)$ for $\frac{2}{3}$, and $(2, 3, 4, \ldots, n, \ldots)$ for $\frac{1}{2}$.

The differences between consecutive numbers in this sequence constitute a sequence of natural numbers, which, written without commas, is called the *Hartogs sequence* of the number x; for example,

$(122 \ldots 2 \ldots)_H$ for $\frac{2}{3}$, and $(211 \ldots 1 \ldots)_H$ for $\frac{1}{2}$.

Clearly, in this way, (1) to each number x in $]0, 1]$ exactly one element of N^{∞} is paired, and (2) each element of N^{∞} is the Hartogs sequence of exactly one number. The sequence $(h_1 h_2 \ldots h_n \ldots)_H$ is the Hartogs sequence of the number

$$\frac{1}{2^{h_1}} + \frac{1}{2^{h_1 + h_2}} + \cdots + \frac{1}{2^{h_1 + h_2 + \cdots + h_n}} + \cdots.$$

Now consider the segment $]0, 1]$ and the Cartesian product set $S =]0, 1] \times]0, 1]$ of all points (x, y) in the Cartesian plane such that both x and y belong to $]0, 1]$. For each number t in $]0, 1]$, write its Hartogs sequence, $(t_1 t_2 t_3 t_4 \ldots t_{2n-1} t_n \ldots)_H$, and pair with t the point (x, y) of S for which x has the Hartogs sequence $(t_1 t_3 \ldots t_{2n-1} \ldots)_H$ and y has the Hartogs sequence $(t_2 t_4 \ldots t_{2n} \ldots)_H$. In this way each point of S is paired to exactly one point in $]0, 1]$; the point (x, y) whose coordinates have the Hartogs sequences $(x_1 x_2 \ldots x_n \ldots)_H$ and $(y_1 y_2 \ldots y_n \ldots)_H$ is paired to the point in $]0, 1]$ whose Hartogs sequence is $(x_1 y_1 x_2 y_2 \ldots x_n y_n \ldots)_H$. Thus the mapping of $]0, 1]$ onto S is one-to-one. Similarly, one can prove that the segment $]0, 1]$ and a cube have the same power.

IV. *There exist also dispersed sets in* $[0, 1]$ *that have the same power as the segment* $[0, 1]$, although, of course, some dispersed sets include fewer points than $[0, 1]$; for example, all finite sets or denumerable sets, such as the set of all points $1/n$ $(n = 1, 2, \ldots)$ or the set of all rational points (see Exercise 3). (In this chapter we use the term *dispersed* intuitively. In Chapter 13 we describe several ways to make it precise.)

Of particular importance among the dispersed sets having the same power as $[0, 1]$ is a set called the *discontinuum D** or Cantor's *middle-third set*. We give three ways of defining it.

A. *By deleting open segments.* First step: from $[0, 1]$ delete the open middle third, the interval $]\frac{1}{3}, \frac{2}{3}[$. Second step: from the two remaining closed segments, $[0, \frac{1}{3}]$ and $[\frac{2}{3}, 1]$, delete their open middle thirds, $]\frac{1}{9}, \frac{2}{9}[$ and $]\frac{7}{9}, \frac{8}{9}[$. At the end of the $(n-1)$st step, 2^{n-1} segments remain. The nth step consists in deleting each of their open middle thirds. The set D^* is then defined as the set of all points in $[0, 1]$ that are *not* deleted in *any* step of this procedure. The union of the deleted open segments is open in $[0, 1]$, and consequently D^* is closed. Clearly, each subsegment of $[0, 1]$, however short, contains a segment that has been deleted in some step of the procedure or is a part of such a deleted segment. Thus, according to any appropriate definition of the idea of "dispersed" set, the set D^* is dispersed.

B. *Arithmetically.* The deletion of middle thirds can conveniently be described in terms of ternary fractions $(g = 3)$. If a point $(.a_1 a_2 \ldots a_n \ldots)_3$ is deleted in the first step, then $a_1 = 1$. If it is deleted in the second step, then either $a_1 = 0$ and $a_2 = 1$ or $a_1 = 2$ and $a_2 = 1$; and so surely $a_2 = 1$. If it is deleted in the nth step, then $a_n = 1$. The points of D^* (that is, the points not deleted in any step) may be described as those *for which there exists a ternary fraction without any* 1's; that is to say, D^* consists of the numbers that are equal to a ternary fraction $(a_1 a_2 \ldots a_n \ldots)_3$ such that each a_n is either 0 or 2. (It should be noted, however, that such a number may also be equal to a ternary fraction that *does* include 1's; for example, $(.200 \ldots 0 \ldots)_3 = (.122 \ldots 2 \ldots)_3$ and $(.022 \ldots 2 \ldots)_3 = (.100 \ldots 0 \ldots)_3$. On the other hand, if a deleted number such as $\frac{4}{9}$ has two ternary fractions, then both contain 1's.)

C. *As the closure of a denumerable set.* Let D_1^* be the set of the end points of all deleted middle thirds. These end points (which are not deleted in any step and hence lie in D^*) are called the points of D^* *of the first kind.* Each of them possesses a terminating ternary fraction. Being a set of ternary rational points, D_1^* is denumerable (see Exercise 3). The other points of D^*, are called points *of the second kind.* For example, $\frac{1}{4}$ is ternary irrational but an accumulation point of D_1^*; it is the limit of the points

$$(.0200 \ldots 0 \ldots)_3, (.020200 \ldots 0 \ldots)_3, \ldots, (.0202 \ldots 0200 \ldots 0 \ldots)_3, \ldots.$$

Each point of the second kind equals only one ternary fraction, and this fraction contains infinitely many 0's as well as infinitely many 2's.

The subset D_2^* consisting of all points of the second kind of D^* has the same power as the set B of all binary irrational numbers in $[0, 1]$. Indeed, by pairing with each $(.a_1 a_2 \ldots a_n \ldots)_3$ of D_2^* (where $a_n = 0$ or 2) the number $(.b_1 b_2 \ldots b_n \ldots)_2$, where $b_n = \frac{1}{2} a_n$ for each n, one defines a one-to-one mapping of D_2^* on B. It readily follows that D^* itself has the same power as the segment $[0, 1]$ (see Exercise 8).

Not only some surfaces and solids but also some dispersed sets have the same power as a segment. Hence, whatever the difference between curves and noncurves may be, it certainly is not one of quantity.

EXERCISES

1. Show that the mapping of P onto N defined in this section is one-to-one by studying its inverse, which maps each n in N on that pair of natural numbers to which it has been paired. *Hint:* First map the numbers $1 + \frac{1}{2}k(k-1)$ on the pair $(1, k)$ for each k.

2. Establish one-to-one mappings onto N of (a) the set $N \times N$ of all sequences (m, n), where m and n belong to N; (b) the set of all sequences of unequal natural numbers. [We speak of the *sequence* (m, n) for, contrary to what one often reads, (m, n) is not an ordered pair of numbers. In fact, since m and n may be equal, (m, n) need not be either a pair or a set. A set either does or does not contain an element—it cannot contain it twice or with a higher multiplicity. The symbol (m, n) is an abbreviated notation for the set $\{(1, m), (2, n)\}$ or $\{(2, n), (1, m)\}$. In the abbreviation, the arguments of the sequence (that is, 1 and 2) are suppressed, and it is understood that m and n are written in the natural order of those arguments. Similarly, (n, m) is a brief symbol for the set $\{(1, n), (2, m)\}$ or $\{(2, m), (1, n)\}$. Elaborate this idea for infinite sequences such as those of the digits of a nonterminating decimal fraction.] (*Hint:* For part (a) of the exercise, pair the number $m + \frac{1}{2}(m + n - 1)(m + n - 2)$ of N with the element (m, n) of $N \times N$. What is the inverse mapping?)

3. Prove the denumerability of the sets of (a) all rational numbers; (b) all rational numbers in $[0, 1]$; (c) all ternary rational numbers in $[0, 1]$.

4. (a) Define numbers s_n and t equal to decimal fractions $.s_{n1} s_{n2} \ldots s_{nn} \ldots$ $(n = 1, 2, \ldots)$, and $.t_1 t_2 \ldots t_n \ldots$ such that $t_n \neq s_{nn}$ for each n and yet $t = s_1$. (b) Can binary fractions ($g = 2$) be used in proving that $[0, 1]$ is not denumerable? What about ternary fractions ($g = 3$)? (c) Use Hartogs sequences instead of decimal fractions in the proof. (d) Why can one not replace Hartogs sequences by decimal fractions in the proof given in this section that the segment and the square have the same power?

5. Prove that $]0, 1]$ has the same power as (a) a circle; (b) $[0, 1]$. *Hint:* The second task is more difficult. Decompose $[0, 1]$ into the set $\{0\}$ and infinitely many half-open segments. (c) Prove that $]0, 1]$ has the same power as $]0, 1[$.

6. Using the fact that $]0, 1[$ and $]0, 1[\times]0, 1[$ have the same power, prove that $[0, 1]$ and the full square $[0, 1] \times [0, 1]$ have the same power.

7. Prove that $]0, 1[$ and the cube consisting of all points (x, y, z) such that x, y, and z belong to $]0, 1[$ have the same power.

8. Using the facts that D_1^* is denumerable and that D_2^* has the same power as the set of all binary irrational numbers, prove that D^* has the same power as $[0, 1]$. Why can this not be proved by mapping any $(.a_1 a_2 \ldots a_n \ldots)_3$ with $a_n = 0$ or 2 on $(.a_1/2 \ a_2/2 \ldots a_n/2 \ldots)_2$? Use this mapping to prove that a certain *subset* of D^* has the same power as $[0, 1]$.

9. What is the combined length of all open segments deleted in constructing the discontinuum D^*? Answer the same question for

(a) The set of all points equal to a 5-ary fraction $(.a_1 a_2 \ldots a_n \ldots)_5$ without 2's, such that each a_i is 0, 1, 3 or 4. Construct this set by deleting open segments and as the closure of a denumerable set.

(b) The set obtained by deleting from [0, 1] the open middle fifth; then from each of the two remaining closed segments its open middle fifth; and so on. Without describing the remaining set arithmetically, define it as the closure of a denumerable set and prove that it has the same power as D^*.

10. Consider the Hartogs sequences $g = (g_1 g_2 \ldots g_n \ldots)_H$ and $h_m = (h_{m1} h_{m2} \ldots h_{mn} \ldots)_H$, where $m = 1, 2, \ldots$, and define a limit for Hartogs sequences by setting $\lim h_m = g$ if and only if for each number n there is a number m_n such that $h_{mi} = g_i$ for $i = 1, 2, \ldots, n$ and $m = 1, 2, \ldots, m_n$. If $|g|$ and $|h_m|$ denote the numbers in $]0, 1]$ with the Hartogs sequences g and h_m, respectively, prove (a) $\lim h_m = g$ implies $\lim |h_m| = |g|$; (b) $\lim |h_m| = |g|$ implies $\lim h_m = g$ if either $|g| = 1$ or g includes infinitely many numbers $\neq 1$; in other words, the mapping of the numbers in $]0, 1]$ on their Hartogs sequences is continuous for 1 and all binary irrational numbers; its inverse, which maps the sequences on the numbers, is continuous; (c) the mapping of the sequences on the numbers is discontinuous for all binary rational numbers in $]0, 1[$; (d) in terms of the sequences h_m and g, give necessary and sufficient conditions for $\lim |h_m| = |g|$ to be valid.

11. Note that two disjoint sets, A and B, are of equal power if and only if there exists a set P of pairs (not ordered pairs and not sequences of two elements!) such that (a) each pair in P consists of an element of A and an element of B, and (b) each element of A, as well as each element of B, is a member of exactly one pair in P. (This fact can be used in defining equal power purely in terms of sets, without such phrases as "one can pair or associate or map" and, moreover, in a way that is symmetric in A and B.) Apply this to the sets E and E' of all even and all odd numbers, respectively. Can it be applied to E and N? If A and B are nonidentical sets that have common elements, replace A by the set A', whose elements are the pairs $\{a, A\}$ for all elements a of A. This set A' is disjoint from B, and A and B are of the same power if and only if A' and B are.

12.3   CURVES AS TRAJECTORIES

Abandoning the quantitative approach, the French mathematician Camille Jordan (1838–1922) in the 1880's defined a curve as a continuous image of a segment or, in kinematic terminology, as the trajectory of a continuous motion. A *motion* of a particle can be described by specifying its position at any instant of some time interval. In the three-dimensional coordinate space, a motion is determined by three continuous functions f, g, h connecting the three coordinates of the particle with time in such a way that, when the clock reading is t, the particle is in the position $(f(t), g(t), h(t))$. In other words, if at the instant t the position is (x, y, z), then $x = f(t)$, $y = g(t)$, and $z = h(t)$.

Of course, one and the same set may be the trajectory of infinitely many different motions. The segment [0, 1] on the x-axis of the three-dimensional space can, for example, be traversed during the time interval [0, 1] uniformly, according to $f(t) = t$, $g(t) = 0$, $h(t) = 0$ for each t in [0, 1]; or in an accelerated motion, with $f(t) = t^2$; or in an oscillatory manner, say, according to $f(t) = \sin(3\pi t)$, whereby the point $(0, 0, 0)$ is traversed four times—namely, at the instants 0, $\frac{1}{3}$, $\frac{2}{3}$, and 1. Moreover, the segment can be traversed during another time interval. The trajectory of the motion with $f(t) = \cos t$, $g(t) = \sin t$, $h(t) = t$ for $0 \leq t \leq 2\pi$ is a cylindrical helix. The ellipse consisting of all points $(x, y, 0)$ such that

$$\frac{x^2}{a^2} + \frac{y^2}{b^2} = 1$$

is the trajectory of motions including, for example, those for which

$$f(t) = a \cos(2\pi t), \qquad g(t) = b \sin(2\pi t) \qquad \text{for all } t \text{ in } [0, 1],$$
$$f(t) = a \sin(2\pi t), \qquad g(t) = b \cos(2\pi t) \qquad \text{for all } t \text{ in } [0, 1],$$
$$f(t) = a \sin t, \qquad g(t) = b \cos t \qquad \text{for all } t \text{ in } [0, 2\pi].$$

In all cases, h is the constant function of value 0.

If all three functions, f, g, and h are constant, then the particle is at rest and the trajectory consists of a single point. No one would call a set consisting of a single point a curve. On the other hand, it is intuitively clear that the trajectory of a continuously moving particle that is not at rest cannot consist of exactly two points, and it will, in fact, be proved in the sequel that such a trajectory is an uncountable set of points—that is, neither finite nor denumerable.

Yet the definition of curves as nonconstant continuous images of a segment is not appropriate in the sense of Section 11.1. In 1890, the Italian mathematician Giuseppe Peano (1858–1932) made the unexpected discovery that this definition includes square regions and cubes, which certainly are among the objects that an appropriate definition of curves should exclude.

In continuously mapping a segment on a square the following concept is useful. Let S be a square in which one vertex is called the *initial* (and the opposite vertex, the *terminal*) point of S—briefly, *in S*, *tm S*. Divide S into 9 congruent squares, and arrange them in what will be called the *Peano chain* of S, shown in Figure 16, S_0, S_1, ..., S_8. In each S_i two opposite vertices are the tail and the head of a diagonal arrow. They will be called *in S_i* and

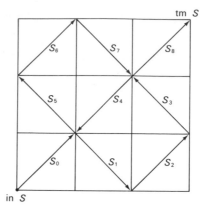

Figure 16

tm S_i, respectively. The chain satisfies the condition

(C_1) *in* $S = in\ S_0$, *tm* $S_i = in\ S_{i+1}$, *tm* $S_8 = tm\ S$ $(i = 0, 1, \ldots, 7)$.

Thus any two consecutive links of the chain have common points.

To map the segment $T = [0, 1]$ on the square $U = [0, 1] \times [0, 1]$, set *in* $U = (0, 0)$ and *tm* $U = (1, 1)$, and form the Peano chain $U_0, U_1, \ldots, U_8$ of U. For each i_1 in $\{0, 1, \ldots, 8\}$, let $U_{i_1 0}, U_{i_1 1}, \ldots, U_{i_1 8}$ be the Peano chain of U_{i_1}, and order the 9^2 squares $U_{i_1 i_2}$ lexicographically; that is, let $U_{i_1 i_2}$ precede $U_{j_1 j_2}$ if and only if either $i_1 < j_1$ or $i_1 = j_1$ and $i_2 < j_2$. One readily verifies that the 9^2 squares $U_{i_1 i_2}$ so arranged form a chain, any two consecutive links having common points. It will be called the *second-order Peano chain* of U. If the kth-order Peano chain of U has been defined, then for each of its links $U_{i_1 \ldots i_k}$, form the Peano chain $U_{i_1 \ldots i_k 0}, \ldots, U_{i_1 \ldots i_k 8}$, and order the 9^{k+1} squares $U_{i_1 \ldots i_k i_{k+1}}$ lexicographically. The result is a chain satisfying for each $i_1, \ldots, i_k$ the condition

(C_{k+1}) $\begin{aligned} &in\ U_{i_1 \ldots i_k} = in\ U_{i_1 \ldots i_k 0}, \qquad tm\ U_{i_1 \ldots i_k 8} = tm\ U_{i_1 \ldots i_k}, \\ &tm\ U_{i_1 \ldots i_k i_{k+1}} = in\ U_{i_1 \ldots i_k i_{k+1}+1} \qquad (i_{k+1} = 0, 1, \ldots, 7). \end{aligned}$

It will be called the $(k + 1)$st-order Peano chain of U.

In an analogous way, divide T into 9 congruent segments of length $1/9$, T_{i_1} beginning at the point $(.i_1)_9 = i_1/9$; and, for each k, into a chain of 9^k segments of length $1/9^k$, $T_{i_1 \ldots i_k}$ beginning at the point $(.i_1 \ldots i_k)_9$.

A continuous mapping of T on U can now be defined in various ways; for example, by the stipulation that all points of any segment $T_{i_1 \ldots i_k}$ should be mapped on points of the square $U_{i_1 \ldots i_k}$. (Kinematically speaking, this means that during the time interval $T_{i_1 \ldots i_k}$ the moving point should stay in the square $U_{i_1 \ldots i_k}$.) From this stipulation, it follows that the point $t = (.i_1 i_2 \ldots i_k \ldots)_9$ of T on which the segments $T_{i_1}, T_{i_1 i_2}, \ldots, T_{i_1 i_2 \ldots i_k}, \ldots$ close down must be mapped on the point of U on which the squares U_{i_1}, $U_{i_1 i_2}, \ldots, U_{i_1 i_2 \ldots i_k}, \ldots$ close down. If t is 9-ary irrational, then there is only one such sequence of segments closing down on t. If t is 9-ary rational, however, say $t = (.i_1 i_2 \ldots i_k)_9$, where $i_k \neq 0$, then t is equal to two non-terminating 9-ary fractions, namely $(.i_1 \ldots i_{k-1} i_k 00 \ldots 0 \ldots)_9$ and $(.i_1 \ldots i_{k-1}(i_k - 1)88 \ldots 8 \ldots)_9$, and both the segments $T_{i_1 \ldots i_{k-1} i_k 00 \ldots 0}$ and $T_{i_1 \ldots i_{k-1}(i_k-1)88 \ldots 8}$ close down on t. But then both the squares $U_{i_1 \ldots i_{k-1} i_k 00 \ldots 0}$ and $U_{i_1 \ldots i_{k-1}(i_k-1)88 \ldots 8}$ close down on one and the same point—the initial point of the former square and the terminal point of the latter square. (According to Condition (C_{k+1}) those points are identical.) This point of U is then the image of the point t of T. Thus each point of T has been mapped on exactly one point of U.

The mapping is uniformly continuous. For if t' and t'' are any two points of T having a distance $1/9^n$ from one another, then they lie either in one and the same segment $T_{i_1 \ldots i_n}$ or in two consecutive such segments. But then the images of t' and t'' lie either in the same or in two adjacent links of the nth-order Peano chain of U, and hence have a distance $< \sqrt{2}/3^n$ or $< 2\sqrt{2}/3^n$, thus certainly a distance $< 1/3^{n-1}$ from one another.

It remains to be shown that each point of U is the image of some point of T—in other words, that Peano's procedure maps T *onto* (and not only into) U. In kinematic terms, this means that each point u of U is traversed at some instant in the time interval $[0, 1]$. This follows from the fact that there are squares $U_{j_1}, U_{j_1 j_2}, \ldots, U_{j_1 j_2 \ldots j_k}, \ldots$ closing down on u (the kth square being a link of the kth-order Peano chain of U) and that the corresponding segments $T_{j_1}, T_{j_1 j_2}, \ldots, T_{j_1 \ldots j_k}, \ldots$ close down on a point t of T whose image obviously is u.

EXERCISES

1. Give examples of points of U which, in Peano's mapping, are the images of (a) exactly one point, (b) exactly two points, (c) exactly four points of T. Is any point of U the image of more than four points of T? Kinematically speaking, how often are the following points traversed?

$$(\tfrac{1}{9}, \tfrac{1}{9}), \quad (\tfrac{2}{3}, \tfrac{1}{3}), \quad (0, 0), \quad (\tfrac{1}{3}, \tfrac{1}{9}), \quad (\tfrac{2}{3}, 0), \quad (1, \tfrac{1}{3}).$$

2. Introduce auxiliary functions f^* and g^*, defined as follows.
  If $x = 0, 1, 2, 3, 4, 5, 6, 7, 8, 9$, then $f^*(x) = 0, 1, 2, 3, 2, 1, 0, 1, 2, 3$, and $g^*(x) = 0, 1, 0, 1, 2, 1, 2, 3, 2, 3$. By means of these functions, describe the co-ordinates of the initial points (a) of the 9 squares in Figure 16; (b) of the 9^2 squares of the second order Peano chain of U. (*Hint:* Introduce $d_i = f^*(i+1) - f^*(i)$ and $e_i = g^*(i+1) - g^*(i)$.) Let $f(t)$ and $g(t)$ be the co-ordinates of the image in U of the point t. Plot the $9 + 1$ points $(t, f^*(t))$ for $t = (.i_1)_9$, and join consecutive points by a segment. Then do the same for the $9^2 + 1$ points corresponding to $t = (.i_1 i_2)_9$. The polygons so obtained are the first and second approximations of the graph of f. Prove that f is continuous in $[0, 1]$, but nowhere differentiable. Then do the same for g. The first continuous and nowhere differentiable functions were discovered by Karl Weierstrass (1815–1897).

3. The German mathematician David Hilbert (1862–1943) mapped T on U, following Peano's general idea but using chains of 4^k (instead of 9^k) segments and squares. Define such a mapping.

4. Define a continuous mapping of $[0, 1]$ on the right triangle with the vertices $(0, 0)$, $(1, 0)$, and $(1, 1)$ by successive bisections of the segment and the triangle. (*Hint:* Let T be a right triangle in which the vertices of the hypotenuse are marked *in* T and *tm* T. Join the midpoint of the hypotenuse to the opposite vertex. Of the two right triangles into which T is split, denote the one con-taining *in* T by $\gamma_0 T$, and the other one by $\gamma_1 T$. Stipulate that *in* $\gamma_0 T = $ *in* T, *tm* $\gamma_1 T = $ *tm* T, and *tm* $\gamma_0 T = $ *in* $\gamma_1 T$. Represent the numbers in $[0,1]$ by nonterminating binary fractions.)

5. For every point $(.a_1 a_2 \ldots a_n \ldots)_2$ in $[0, 1]$, map the point $(.b_1 b_2 \ldots b_n \ldots)_3$ of D^*, where $b_n = 2a_n$ (see Exercise 8, Section 11.2) on the point $(.a_1 a_2 \ldots a_n)_3$ of $[0, 1]$. Show that this mapping of D into $[0, 1]$ is continuous but not one-to-one.

6. Show that the square U is a continuous image of the discontinuum D^*. (*Hint:* Use the result of the preceding Exercise. No arithmetical definition of the mapping is required.)

7. Show that the segment $[0, 1]$ is a continuous image of (a) the square U; (b) any square; (c) any cube.

8. Peano defined a continuous mapping of $[0, 1]$ on a cube using chains of 27^k segments and cubes. Define such a mapping using, instead, chains of 8^k segments and cubes. What is the kinematic meaning of such a mapping?

9. Show that a cube is a continuous image of a square, and that a square is a continuous image of a cube. (*Hint:* For the first part, use the results of the preceding Exercises.)

10. Map the ray R of all nonnegative numbers and the open segment $]0, 1[$ continuously onto one another and on $[0, 1]$.

11. Consider the following continuous mapping of the ray R_1 of all numbers ≥ 1 into a square of side 2. Let the image of t be the point (cos at, cos bt), where a and b differ by an irrational multiple of π. Kinematically speaking, this is a motion from 0 to eternity. Show that the trajectory is dense in U (that is, that U is the closure of the trajectory) but exhibits points of U not traversed in the course of this motion (called Lissajou curve).

12. Define a motion during the half-open time interval $[0, \pi/2[$ which has the same trajectory as the motion in Exercise 11. Can a motion during a closed time interval have that same trajectory?

12.4   CHARACTERIZATIONS OF TRAJECTORIES

The family of the continuous images of $[0, 1]$ (originally called *courbes de Jordan* and now sometimes referred to as Peano continua) not only includes noncurves but also excludes sets that deserve to be called curves. An important example is a subset S^* of the plane which, because of its relation to the sine curve, we shall call the *sinusoid*. This curve consists of all points $(x, \sin(1/x))$ for $0 < x \leq 1/\pi$, and all points $(0, y)$ for $-1 \leq y \leq 1$. It will be shown in part (a) of the proof of Theorem 12.1 that it is impossible to map $[0, 1]$ continuously on S^*.

Throughout Part 4, by a *set* whose elements are called points we mean a metric space—in many cases a subset of a more comprehensive metric space. For some sets, that space will be specified; for example, the sinusoid has been defined as a subset of the plane.

Although only overlapping with the family of curves, the family of continuous images of $[0, 1]$ is of considerable interest in its own right, and the question arises: by what intrinsic properties are its members characterized? A property of a set S is *intrinsic* if it can be defined in terms of points and subsets of S without any reference to sets more comprehensive than S (such as a space containing S) or to a set whose image S may be. The property of being closed is *not* intrinsic. For S is closed if S includes all points of the space that are accumulation points of S; and S may well be closed in one space without being closed in a more comprehensive space. For example, the subset P of the straight line consisting of all points with positive rational coordinates is closed if the space is the set of all rational points $\neq 0$; it is not closed if the space is the whole straight line or the set of all rational points. Similarly, compactness (as defined in Section 1.9) is not intrinsic.

On the other hand, the following important property is intrinsic. A set S (that is, a subset of a metric space) is called a *compactum* if, for each infinite subset T of S, there exists a point of S that is an accumulation point of T.

The reader can easily prove that a set S is a compactum if and only if S is both closed and compact. Thus the *intrinsic* property of being a compactum can also be defined in a *nonintrinsic* way as the conjunction of the properties of being closed and compact.

Another intrinsic property of a set is *connectedness* (introduced in Section 1.11), whose definition is due to the American mathematician N. J. Lennes (1874–1951) and to F. Hausdorff. We recall that a set S is connected if and only if S is not the union of two disjoint nonempty sets, each of which is closed in S (Theorem 1.5).

A *continuum* is a connected compactum containing more than one point. By Theorem 1.14, a continuous image C of a continuum is a continuum provided C contains more than one point. This section is devoted to intrinsic properties which characterize those continua that are continuous images of [0, 1] or, kinematically speaking, the trajectories of motions other than rest. Use will be made of some properties of connected sets.

LEMMA 12.1   *A necessary and sufficient condition for a set S to be connected is that each pair of points of S lie in a connected subset of S.*

If S is not connected, then S is the union of two nonempty disjoint sets, both closed in S—say S_1 containing a point p_1, and S_2 containing p_2. Then any subset T of S containing p_1 and p_2 is nonconnected because of the split $T = (S_1 \cap T) \cup (S_2 \cap T)$, and the sufficiency is established. The necessity is trivial.

LEMMA 12.2   *The union of two nondisjoint connected sets is connected.*

Suppose the union of two connected sets A and B, whose intersection $A \cap B$ is nonempty, is somehow split into two disjoint sets S_1 and S_2, both closed in $A \cup B$. We show that one of these sets is empty. Now A is the union of the disjoint sets $S_1 \cap A$ and $S_2 \cap A$, both closed in A. Since A is connected, one of them, say $S_2 \cap A$, must be empty. Since

$$A \cap B = (S_1 \cap A \cap B) \cup (S_2 \cap A \cap B) = S_1 \cap A \cap B$$

is nonempty, it follows that $S_1 \cap B$ is not empty. But then the connectedness of $B = (S_1 \cap B) \cup (S_2 \cap B)$ implies that $S_2 \cap B$ is empty. It follows that $(S_2 \cap A) \cup (S_2 \cap B)$ (that is, S_2) is empty.

An immediate consequence of these two lemmas is

LEMMA 12.3   *The union of any family of connected sets having a nonempty intersection is connected.*

Let S be a set, and p be a point of S. The union of all connected subsets of S containing p is called the *component of S containing p*. According to Lemma 12.3, this set is connected. By Lemma 12.2, the component of S containing p, and the component of S containing another point, q, are either identical or disjoint. Thus S is the union of a family of disjoint subsets that may be called simply the *components of S*. Each component of Cantor's discontinuum D^* consists of a single point. A connected set has only one component, the set itself.

LEMMA 12.4   *If S is a connected subset of T which, in turn, is a subset of cl S, then T is connected.*

If T is in some way split into two disjoint sets, T_1 and T_2, both closed in T, then the connected set S is the union of the disjoint sets $T_1 \cap S$ and $T_2 \cap S$, both closed in S, and so one of them, say $T_2 \cap S$, is empty. Hence $S = T_1 \cap S \subset$ cl S, and it follows that cl $T_1 =$ cl S. Since T_1 is closed in T it follows that $T_1 = T \cap$ cl $T_1 = T \cap$ cl $S = T$ and T_2 is empty.

LEMMA 12.5   *Each component of a set S is closed in S.*

If C is a component of S containing, say, the point p, then $C \subset S \cap$ cl $C \subset$ cl C. Hence, according to Lemma 12.4, $S \cap$ cl C is connected. But since C is the union of all connected subsets of S containing p, it follows that $S \cap$ cl $C \subset C$; that is to say, C is closed in S.

A finite sequence of points $(p = p_0, p_1, \ldots, p_{n-1}, p_n = p')$ of a metric space such that the distance between any two consecutive points is less than d is called a *d-chain* (more specifically, a *d*-chain *between the points p and p'* or *joining p and p'*).

LEMMA 12.6   *If S is a connected subset of a metric space, then for any two points p and p' of S and for any $d > 0$, S contains a d-chain between p and p'.*

Let P be the set of all points q of S such that S contains a d-chain between p and q. It is easily seen that P is both open and closed in S. If S is connected, then $S - P$ must be empty; that is to say, $S = P$, and p' belongs to P.

LEMMA 12.7 *If C is a continuum and U is an open set such that the sets $C \cap U$ and $C' = C \cap \text{cl com } U$ are nonempty, then each component of $C'' = C \cap \text{cl } U$ contains a point of C'.*

(It will be noted that in Part 4 a prime attached to the designation of a set does not in general denote the set of all accumulation points of that set.) Let K be a component of C''; p, a point of K; q, a point of C'. By Lemma 12.6, the continuum C contains for each n a $(1/n)$-chain

$$p = r_{n,1}, r_{n,2}, \ldots, r_n, k_n = q.$$

There is an index i_n such that $P_n = \{r_{n,1}, \ldots, r_{n,i_n}\} \subset C''$ and r_{n,i_n+1} belongs to C'. Let L be the set (called the *limit superior* of the sequence $(P_1, P_2, \ldots)$) of all points each of whose neighborhoods has a point in common with infinitely many sets P_n. Clearly, L is closed, contains p, and has a point in common with C'—namely, an accumulation point of the sequence $(r_{1,n_1}, r_{2,i_2}, \ldots, r_{n,i_n}, \ldots)$. The reader can easily prove that the assumption that L be nonconnected leads to a contradiction. Being a connected subset of C'' containing p, the set L is a subset of K. Since L contains a point of C' so does K.

The first intrinsic characterization of continua that are continuous images of $[0, 1]$ (by the property of local connectedness) was given independently by H. Hahn and the Polish mathematician Stefan Mazurkiewicz (1888–1946). We begin by defining a somewhat simpler and slightly stronger concept. A set S is *strongly locally connected at the point p* if and only if each neighborhood of p in S contains a connected neighborhood of p in S.

A segment (with or without its end points), a square domain, and the union of two disjoint segments are sets—the last one not connected—that are strongly locally connected at each of their points. The sinusoid S^* is easily seen to be strongly locally connected at each of its points (x, y) such that $x \neq 0$, but in none of the other points. Consider, for example, the point $(0, 0)$. The intersection, T, of S^* with the open circle of radius $\frac{1}{4}$ about $(0, 0)$ is a neighborhood of $(0, 0)$ in S^* that does not contain any connected neighborhood of $(0, 0)$ in S^*, since the component of T that contains $(0, 0)$

is a vertical segment each of whose points is an accumulation point of the wavy part of S^*.

A set S is *locally connected at the point p* if each neighborhood V of p in S contains a connected subset which (without necessarily being open in S) has a subset that is a neighborhood of p in S. If S is strongly locally connected at p, then S is, of course, locally connected at p. But the converse is not necessarily true. Consider the continuum S^{**} that is the union of the segments

L_0 of all points of $(x, 0)$ for $0 \leq x \leq 1$,

L_{mn} of all points of $\left(x, \ -\dfrac{1}{m}\left(x - \dfrac{1}{n}\right)\right)$ for $\dfrac{1}{n+1} \leq x \leq \dfrac{1}{n}$,

$$(n, m = 1, 2, \ldots).$$

Obviously S^{**} is strongly locally connected at all of its points (x, y) such that $y > 0$, as well as at $(1, 0)$. Indeed, the intersection of S^{**} with any open circle about $(1, 0)$ is a connected neighborhood of $(1, 0)$ in S^{**}. At the points $(x, 0)$ for $0 < x < 1$, S^{**} is not even locally connected. At $(0, 0)$, S^{**} is locally connected, for if V is any neighborhood of $(0, 0)$ in S^{**}, then for some natural number n_0, V contains the continuum S' consisting of all points (x, y) of S^{**} such that $x \leq 1/n_0$. Deleting the point $(1/n_0, 0)$ from S' one obtains a (nonconnected) neighborhood of $(0, 0)$ in S^{**}. Hence V contains a connected subset which, in turn, contains a neighborhood of $(0, 0)$ in S^{**}. But S^{**} is not strongly locally connected at $(0, 0)$, for if T is any neighborhood of $(0, 0)$ in S^{**} that is connected, then clearly T contains the point $(1/(n_0 + 1), 0)$ for some natural number n_0. Being open in S^{**}, the set T therefore also contains the points

$$\left(\frac{1}{n_0 + 1}, \ -\frac{1}{m_0}\left(\frac{1}{n_0 + 1} - \frac{1}{n_0}\right)\right)$$

for some natural number m_0 (in fact, for infinitely many m_0), and, being connected, T contains the entire segment $L_{m_0 n_0}$, including the point $(1/n_0, 0)$. By induction, it follows that T contains the point $(1, 0)$. Hence a neighborhood of $(0, 0)$ in S^{**} that does not contain $(1, 0)$ does not contain a connected neighborhood of $(0, 0)$ in S^{**}. The set S^{**}, which demonstrates the difference between local connectedness and strong local connectedness at a point (introduced by K. Menger), is due to N. Aronszajn.

The reader may verify that an example of a set that is not locally connected at each of its points is the continuum I^* obtained as follows. Let D^* be Cantor's discontinuum (see Section 12.2) on the segment $[0, 1]$ of the x-axis of a Cartesian plane. Then I^* is the union of the following semicircles that join two points of D^*:

1. in the upper half-plane, those about the point $(\tfrac{1}{2}, 0)$, for example, with radii $\tfrac{1}{2}$ and $\tfrac{1}{6}$;

2. in the lower half-plane, those about the point $(\tfrac{5}{6}, 0)$, for example, with the radii $\tfrac{1}{6}$ and $\tfrac{1}{18}$; those about the point $(\tfrac{5}{18}, 0)$, for example, with the radii $\tfrac{1}{18}$ and $\tfrac{1}{54}$; and, for any natural number k, those about $(5/(2 \cdot 3^k), 0)$.

A set S will be said to be (*strongly*) *locally connected* if and only if S is (strongly) locally connected at each of its points.

The continuous images of segments will now be characterized by various intrinsic properties.

THEOREM 12.1   *The six families of the following subsets of a metric space are identical.*

I. *Sets that are continuous images of a segment and contain more than one point.*

II. *Continua C with the following property: if V is any set that is open in C, then all components of V are open in C.*

III. *Strongly locally connected continua.*

IV. *Continua which, for every number $d > 0$, are unions of a finite family of continua with diameters $< d$.*

V. *Locally connected continua.*

VI. *Continua C that are uniformly locally connected in the sense that for each number $d > 0$ there exists a number $r_d > 0$ such that any two points of C with distance $< r_d$ lie in a subcontinuum of C with diameter $< d$.*

The theorem will be proved by showing that each family is contained in the next one and that Family VI is contained in Family I. (In particular, Families III and V are identical; that is to say, a continuum that is locally connected at *each* of its points is strongly locally connected at each of them.) The identifications of Family I with Families IV and II are due to W. Sierpiński and C. Kuratowski, respectively. The identity of Families I and V is the original discovery of Hahn and Mazurkiewicz.

(a) Family I $\subset$ Family II; that is, if C is a continuous image of [0, 1] and contains more than one point, then, if V is open in C, and W is a component of V, the set W is open in C. Assuming that W is not open in C, we derive a contradiction. According to the assumption, W contains an accumulation point p of $C - W$. Let $(p_1, p_2, \ldots)$ be a sequence of distinct points of $C - W$ converging to p. For each n, choose in [0, 1] a model of p_n, say r_n. Since the p_n are distinct, so are the r_n, whence the set R of all chosen r_n is infinite. The compactum [0, 1] therefore contains an accumulation point r of R. Let $(r_{i_1}, r_{i_2}, \ldots)$ be a sequence of points of R converging to r. The mapping of [0, 1] on C being continuous, the images p_{i_n} of the r_{i_n} (which converge to p) converge to the image of r, and consequently p is the image of r. Since V is open in C, the model of V is open in [0, 1], and therefore contains an *open* segment I including r. The image of I is a connected subset of V containing p, and hence a subset of W. But I includes points r_{i_n} (since they converge to r), whose images, p_{i_n}, lie in $C - W$, which is a contradiction.

(Since the intersection V of the sinusoid S^* with the open circle of radius $\frac{1}{2}$ about (0, 0) has a component that is not open in V (namely, a vertical segment), S^* cannot possibly be represented as a continuous image of [0, 1].)

(b) Family II $\subset$ Family III; that is, if C belongs to Family II, then C is strongly locally connected. Let p be any point of C, and V any neighborhood of p in C. Since C belongs to Family II, the component of V that contains p is open in C. Hence a connected neighborhood of p in C is contained in V; and C is strongly locally connected at p.

(c) Family III $\subset$ Family IV. Let C be a strongly locally connected continuum, and d a number >0. For each point p of C, the sphere about p of radius $d/2$ contains a connected neighborhood of p in C. By the Borel–Lebesgue property, a finite family of these sets which are open in C covers C. The closures of these sets are continua whose diameters are $\leqq d/2 < d$, and their union is C.

(d) Family IV $\subset$ Family V; that is, a continuum belonging to Family IV is locally connected. (That a strongly locally connected continuum—that is, a continuum in Family III—is locally connected would go without saying. But we wish to prove the local connectedness of a continuum which, for each $d > 0$, is the union of a finite family of continua whose diameters are $<d$.) Let p be any point of C, and V the intersection of C with any sphere about p, say, the sphere of radius r. By assumption, C is the union of a finite family of continua, $C_1, C_2, \ldots, C_n$ of diameters $<r/2$. The union of all

those C_i that contain p is a (connected) set contained in V and contains a set that is open in C, namely, the complement to C of the union of all those C_i that do not contain p. Hence C is locally connected at p.

(e) Family V $\subset$ Family VI; that is, a locally connected continuum C is uniformly locally connected. (The proof follows the pattern of the proof that every continuous function with a closed and compact domain is uniformly continuous.) If C is not uniformly locally connected, then, for some $d > 0$ and each natural number n, there exists a pair of points p'_n and p''_n of C with distance $< 1/n$ but not contained in a subcontinuum of C with diameter $< d$. Because C is a compactum this implies the existence of a point p_0 of C, every neighborhood of which contains a pair p'_m, p''_m. Since C is strongly locally connected at p_0, there exists a connected neighborhood V of p_0 in C with diameter $< d/2$. Then any two of the points p'_m, p''_m of V lie in a continuum with diameter $\leq d/2 < d$ (in the closure of V), which is a contradiction.

(f) Family VI $\subset$ Family I; that is, each uniformly locally connected continuum C is a continuous image of $[0, 1]$. In fact, any two distinct points, a and b, of C may be prescribed as the images of 0 and 1, respectively.

Remark 1.   The uniform local connectedness of C implies that for each natural number n there exists a number $d_n < (\frac{1}{2})^{n+1}$ such that any two points of C with distance less than d_n lie in a subcontinuum of C with diameter $< (\frac{1}{2})^{n+1}$. Hence any two points of C having distances $< d_n$ from one and the same point of C lie in a subcontinuum of C with diameter $< 1/2^n$.

Remark 2.   The continuum C contains a denumerable subset P that is dense in C; that is, $C = \text{cl } P$ (see Section 1.10). Serialize the set P and put $P_k = \{p_1, p_2, \ldots, p_k\}$ $(k = 1, 2, \ldots)$. It follows readily that for each n there exists an integer m_n such that each point of C has distance $< d_n$ from at least one of the points of P_{m_n}.

The following proof that Family VI $\subset$ Family I, which goes back to the American mathematician G. T. Whyburn, is based on the selection of a d_1-chain, Ch_1, of points of C joining a and b (the points chosen to be the images of 0 and 1, respectively), which contains P_{m_1}, and then on refining Ch_1, first to a d_2-chain Ch_2 containing P_{m_2}, and gradually to d_n-chains Ch_n containing P_{m_n}. Containing increasing sections of P, these chains get more and more dense in C, and they get finer, since d_n approaches 0. It is technically convenient to choose Ch_n such that the number of its points following the initial point, a, is a power of 2, say 2^{r_n}. Since a chain may contain the same point in several places, this choice can be effected by repeating a point if necessary—for example, by replacing a chain such as $(a = c_0, c_1, \ldots, c_5 = b)$ by the chain $(c_0, c_1, \ldots, c_5 = c_6 = c_7 = c_{2^3} = b)$. Then the binary rational number $i/2^3$ $(i = 0, 1, \ldots, 2^3)$ can be mapped on c_i, and the set of all binary rational numbers can be mapped on the union of all chains, which contains the entire set P (dense in C). Because of the increasing refinement of the chains, this mapping is uniformly continuous and hence can be extended to a continuous mapping of $[0, 1]$ on C.

More specifically, by Lemma 12.6, a and b can be connected by a d_1-chain

$$Ch_1 = (a = c_0^1, c_1^1, \ldots, c_{h_1}^1 = b) \qquad (h_1 = 2^{k_1})$$

that includes all points of P_{m_1}. (Why?) It may be assumed inductively that a d_n-chain is given,

$$Ch_n = (a = c_0^n, c_1^n, \ldots, c_{h_n}^n = b) \qquad (h_n = 2^{k_n}),$$

which includes all points of P_{m_n}. Then, by Remark 2, each point of C (in particular, each point of $P_{m_{n+1}}$) has a distance $<d_n$ from at least one point of Ch_n. Hence, if the set of all points of $P_{m_{n+1}}$ that have a distance $<d_n$ from c_{i-1}^n or c_i^n is denoted by $P'_{m_{n+1}}$ $(i = 1, \ldots, h_n)$, then the union of these h_n sets is $P_{m_{n+1}}$. By Remark 1, any two points with distances $<d_n$ from c_i^n (in particular, c_{i-1}^n and each point of $P'_{m_{n+1}}$) lie in a subcontinuum of C with diameter $<\frac{1}{2}^{n+1}$. Hence c_i^n and c_{i-1}^u can be joined by a d_{n+1}-chain

$$Ch_{n, i} = (c_{i-1}^n = c_{i, 0}^n, c_{i, 1}^n, \ldots, c_{i, h_n, i}^n)$$

including $P_{m_n}^i$. If g_n is a power of 2 exceeding the largest of the numbers $h_{n, i}$ $(i = 1, \ldots, h_n)$, then each $Ch_{n, i}$ may be assumed to include g_n points following c_{i-1}^n (including c_i^n). The total number of points following the point a in the h_n chains is $g_n \cdot h_n$ (a power of 2), which will be called $h_{n+1} = 2^{k_{n+1}}$. Uniting the h_n chains in this order into one chain, one may describe their points by the upper index $n + 1$ and lower indices from 0 to h_{n+1}. One has $a = c_0^{n+1}$. The other points of $Ch_{n, 1}$ get the lower indices from 0 to g_n in the order in which they appear in $Ch_{n, 1}$; the points of $Ch_{n, 2}$ in their proper order get the indices from g_n to $2g_n$; and so on. Finally, the points of Ch_{n, h_n} get the indices from $(g_n - 1) \cdot h_n$ to h_{n+1}. Thus one obtains a d_{n+1}-chain

$$Ch_{n+1} = (a = c_0^{n+1}, c_1^{n+1}, \ldots, c_{h_{n+1}}^{n+1} = b),$$

where $c_{i\cdot g_n}^{n+1} = c_i^n$ for $i = 0, 1, \ldots, h_n$, and, for $(i-1)\cdot g_n < i' < i\cdot g_n$, the point $c_{i'}^{n+1}$ has distance $<1/2^n$ from c_{i-1}^n as well as from c_i^n.

For the same reason, if $(i'-1)\cdot g_{n+1} < i'' < i'\cdot g_n$, the point $c_{i''}^{n+2}$ of Ch_{n+2} has distance $<1/2^{n+1}$ from $c_{i'-1}^{n+1}$ and $c_{i'}^{n+1}$, and hence a distance $<1/2^n + 1/2^{n+1}$ from c_{i-1}^n and c_i^n. By induction it is clear that any point between c_{i-1}^n and c_i^n in Ch_{n+r} is $<1/2^n + 1/2^{n+1} + \cdots + 1/2^{n+r-1}$ apart from those two points. Hence any point in the union of all the chains that lies between c_{i-1}^n and c_i^n is $<1/2^{n-1}$ apart from those points and so any two points between them have a distance $<1/2^{n-2}$.

One readily concludes that, however small the number $d > 0$, for sufficiently large n any two points in chains between c_i^n and either of its neighbors c_{i-1}^n or c_{i+1}^n have distance $<d$ from one another.

EXERCISES

1. Prove that (a) each finite subset of a metric space is a compactum; (b) a metric space S is a compactum if and only if S is compact; (c) for each subset T of a metric space there exists a space containing T in which T is closed; (d) a subset T of a euclidean space is a compactum if and only if T is closed and bounded (that is, contained in some sphere).

2. Show that the union of any family of connected sets, any two of which have a nonempty intersection, is connected.

3. Prove that in a metric space each connected subset containing more than one point is uncountable and, more specifically, has a subset that is a one-to-one image of the segment $[0, 1]$. (*Hint:* Consider the family of all spherical neighborhoods of a point of the set.)

4. Is the closure of a locally connected subset S of a metric space necessarily locally connected? What if S is open in a locally connected continuum?

5. Among all subsets of a straight line (including the nonconnected and/or noncompact sets), characterize the locally connected sets.

6. At which points, if any, are the following sets locally connected?
   (a) $[0, 1] \times D^*$ (that is, the set of all (x, y) such that x belongs to $[0, 1]$, and y belongs to Cantor's discontinuum D^*).
   (b) $[0, 1] \times D^* \cup D^* \times [0, 1]$.

7. Apply the proof of assertion (f) that $VI \subset I$ to the uniformly locally connected square U.

8. Using the identity of the Families I to VI, prove that a locally connected continuum is, for any number $d > 0$ the union of a finite family of *locally connected* continua of diameters $<d$.

9. Prove that the union of two locally connected continua, if connected, is a locally connected continuum. Show also that the unit square U is the union of two continua neither of which is locally connected.

10. Show that a segment, and hence a square, are continuous images of the sinusoid S^*.

11. On a straight line, let S be the set of all points $1/n$ $(n = 1, 2, \ldots)$ and $S_0 = S \cup \{0\}$. Show that S_0 cannot be continuously mapped on S, and define a continuous mapping of S on S_0.

12. Show that a continuous image of the ray R or of the interval $]0, 1[$ may be a continuum that is not locally connected. (*Hint:* The union of the sinusoid S^* and a semicircle joining the points $(0, -1)$ and $(1/\pi, 0)$ is such a continuum.)

13. Show that the continuum I^* (defined on p. 407) is not the union of two proper subcontinua. Continua with this property, first discovered by the Dutch mathematician L. E. J. Brouwer, are called *indecomposable*. Show that I^* contains subcontinua that are not indecomposable. The Polish mathematician Bronislaw Knaster constructed a continuum *each of whose subcontinua is indecomposable*.

14. Prove that in the plane the complement of any bounded set has exactly one unbounded component. (A subset of the plane is bounded if and only if it is contained in a circular disc.) Is this valid for euclidean spaces of every dimension?

15. Show that in the plane the set $D^* \times D^*$ of all points (x, y) such that both x and y have ternary fractions without any 1, does not contain any subcontinuum.

12.5 IRREDUCIBLE CONTINUA

Another approach to the idea of curves is due to the French mathematician Zoretti. A subset C of a metric space is said to be an *irreducible continuum between its points a and b* if C, but no proper subset of C, is a continuum containing a and b. The segment $[0, 1]$, for example, is an irreducible continuum between the points 0 and 1 but not between any other pair of its points. The sinusoid S^*, studied in Section 12.4, is an irreducible continuum between the point $(1/\pi, 0)$ and any point $(0, y)$ for $-1 \leq y \leq 1$, but not between any other pair of its points. A square domain and a cube are clearly not irreducible continua between any pair of points, and neither is any subset of the plane that contains a square domain. The family of irreducible continua thus includes some curves and excludes some objects that are not curves.

Nevertheless, the concept cannot be considered as an appropriate definition of curves. For, on the one hand, it excludes circles, lemniscates, and similar curves, because they are not irreducible continua between any pair of points, and on the other hand, in the three-dimensional space there exist irreducible continua that contain square domains as subsets.

Consider the set of the points

$$(\cos(a/t), \cos(b/t), 1/t) \quad \text{for all } t \geq 1,$$

where a and b differ by an irrational multiple of π. This set is the trajectory of a motion from 1 to eternity (see Exercise 11, Section 12.3). It is a connected set that comes arbitrarily close to each point of the square Q' consisting of all points $(x, y, 0)$ for $-1 \leq x \leq 1$, $-1 \leq y \leq 1$. The closure of this trajectory, which contains Q' as a subset, is an irreducible continuum between the point $(\cos a, \cos b, 1)$ and any point of Q'.

Irreducible continua are, however, of interest in their own right and also because of their relations to continua in general.

THEOREM 12.2.   *Let a, b denote any two distinct points of a continuum C. Then C contains a subset that is an irreducible continuum between a and b.*

We begin with a lemma that is important in itself.

LEMMA 12.8   *The intersection, K_0, of a decreasing sequence of continua $(K_1, K_2, \ldots)$ is connected.*

Note that since K_0 is a compactum, it will be a continuum if it contains at least two points.

Assume, on the contrary, that K_0 is the union of two disjoint nonempty sets K' and K'', both closed in K_0. This assumption leads to a contradiction: the sets K' and K'', being compact, have a distance $d > 0$ from one another, and the open sets V' and V'', consisting of all points having distances $< d/3$ from K' and K'', respectively, are disjoint. Each K_n has points in common with V' and V''—certainly the points of K' and K''—but, being connected, K_n is not a subset of $V' \cup V''$. Hence, for each n, the closed set L_n consisting of all points of K_n that do not belong to either V' or V'', is nonempty (see Exercise 3, Section 1.11). Consequently, the decreasing sequence $(L_1, L_2, \ldots)$ has a nonempty intersection which is a subset of K_0 as well as of the complement of $V' \cup V''$. But this is impossible, since K_0 is a subset of $V' \cup V''$, and the proof of the lemma is complete.

To prove Theorem 12.2, let $\Re$ be the family of all subcontinua of C that contain a and b. The family of spherical neighborhoods of radius 1 of points of C contains a finite subfamily $\mathfrak{S}_1$ covering C and consisting, say, of m_1 members. If K is a member of $\mathfrak{S}$, denote the number of spherical neighborhoods in $\mathfrak{S}_1$ that have common points with K by $f_1 K$. Clearly, $1 \leqq f_1 K \leqq m_1$, for each K. Let K_1 be one of the (perhaps many) members of $\Re$ for which $f_1 K$ assumes the smallest possible value (that is, such that $f_1 K_1 \leqq f_1 K$ for each K in $\Re$). Now suppose that K_{n-1} has already been defined. The family of spherical neighborhoods of radius $1/n$ of points of K_{n-1} contains a finite subfamily $\mathfrak{S}_n$ covering K_{n-1} and consisting, say, of m_n members. Denote the family of subsets of K_{n-1} belonging to the family $\Re$ by $\Re_n$ and the number of neighborhoods in $\mathfrak{S}_n$ that have points in common with a set K by $f_n K$. Clearly, $1 \leqq f_n K \leqq m_n$ for each K in $\Re_n$. Let K_n be a member of $\Re_n$ such that $f_n K_n \leqq f_n K$ for each K in $\Re_n$. The sequence $(K_1, K_2, \ldots)$ of subcontinua of C is decreasing. The intersection, J, of all K_n contains a and b, since each K_n does; and by the preceding lemma, J is a continuum. In order to prove that J is an irreducible continuum between a and b, let L be any closed proper subset of J. We will show that L does not belong to $\Re$.

Let p denote a point of $J - L$. Since L is closed, the point p has a positive distance from L, which, for some natural number n, is $> 1/n$. Since L is a subset of J, each member of $\Re_n$ that has common points with L has points in common with J. But there is a member of $\mathfrak{S}_n$ that contains the point p of J (and thus of K_n) that is disjoint from L. Hence if L belongs to $\Re$, then $f_n L < f_n K_n$, which contradicts the choice of K_n in $\Re_n$. Since no proper subset of J belongs to $\Re$, the subset J of C is an irreducible continuum between a and b.

Theorem 12.2 is a special case of the so-called Reduction Theorem (see Exercise 5), due to Brouwer.

EXERCISES

1. Let S be the set of all points $(x, \sin 1/(x - 1))$ for $1 < x \leqq 1 + 1/\pi$, all points $(x, \sin 1/x)$ for $0 < x \leqq 1$, and all points $(0, y)$ and $(1, y)$ for $-1 \leqq y \leqq 1$. Between which pairs of points is S an irreducible continuum?

2. Construct sets in the plane that are irreducible continua (a) between $(0, 0)$ and each point of the circle of radius 1 about $(0, 0)$, (b) between each point of the circle just mentioned and each point of the concentric circle of radius 2. At which points are the two continua locally connected?

3. Is the indecomposable continuum I^* an irreducible continuum between some pairs of points? (See p. 407.)

4. Prove that if, in the three-dimensional euclidean space, C is an irreducible continuum between two points, then C does not contain a cube as a subset. In the four-dimensional space such a continuum may contain a three-dimensional cube.

5. Let $\mathfrak{F}$ be a family of sets. A set F is said to be an *irreducible* member of $\mathfrak{F}$ if F, but no closed proper subset of F, belongs to $\mathfrak{F}$. A family $\mathfrak{F}$ is said to be *intersective* if, for each decreasing sequence $(F_1, F_2, \ldots)$ of closed sets in $\mathfrak{F}$, the intersection of all F_k belongs to $\mathfrak{F}$. According to Lemma 12.8, the family of all closed connected sets and the family $\mathfrak{K}$ are intersective. Give other examples of such families. Prove exactly as in the case of $\mathfrak{K}$ that *each compactum belonging to an intersective family $\mathfrak{F}$ contains a closed set that is an irreducible member of $\mathfrak{F}$* (Reduction Theorem).

6. Apply, if possible, the Reduction Theorem to the family of (a) all closed subsets of the space containing a certain set A as a subset; (b) all connected subsets of the space; (c) all subcontinua of the space; (d) all subcontinua containing the three points p, q, r; (e) all convex subsets of the plane containing a certain set S as subset. (A subset of the plane is *convex* if and only if it contains the segment joining any two of its points.)

7. Prove that if C is an irreducible continuum between a and b, and C' is homeomorphic to C, then C' is an irreducible continuum between the images of a and b. (Note that a continuum may be irreducibly connected between a and b or irreducible in some other sense between a and b. Hence one should refer to sets such as the sinusoid S^* as irreducible continua between two points rather than as continua that are irreducible between those points.)

12.6   ARCS

Since the family of one-to-one images, as well as the family of continuous images of segments, includes squares and cubes, mathematicians tried to define curves as images of $[0, 1]$ by mappings that are more restricted— namely, one-to-one *and* continuous. According to Theorem 1.15, since $[0, 1]$ is closed and compact, such a mapping has a continuous inverse and hence is *topological*. A topological image of $[0, 1]$ is called an *arc*; the images of 0 and 1 are referred to as the *end points* of the arc.

The sinusoid S^*, which is not even a continuous image of $[0, 1]$, is not an arc. Let A be an arc, and let g be the inverse of the topological mapping of $[0, 1]$ in A. Being continuous, g maps each connected subset of A on a connected subset of $[0, 1]$. But $[0, 1]$ has no connected proper subset containing both 0 and 1, and so A has no connected proper subset (and hence no proper subcontinuum) that contains the end points of A. It follows that *each arc is an irreducible continuum between its end points.* Hence circles and lemniscates as well as squares and cubes (none of which is an irreducible continuum between any two of its points) are excluded from the arcs. Nor can an arc, in contrast to some irreducible continua, contain a subset that is a square domain Q, for a connected set containing Q remains connected upon the deletion of any one point of Q, whereas $[0, 1]$ contains only two points x such that $[0, 1] - \{x\}$ is connected—namely, the points 0 and 1.

Thus, though it includes no noncurves, the family of arcs is too restricted for our purpose, since it excludes objects such as circles and lemniscates, which have always been regarded as curves. The family is wide enough, however, to include arcs with rather unexpected properties. If $(f(t), g(t))$ is the position of a point at time t in the course of a Peano motion traversing the unit square U during the time interval $[0, 1]$, associate with each t in $[0, 1]$ the point $(f(t), g(t), t)$ in the three-dimensional euclidean space. This mapping is one-to-one and continuous, and so the trajectory is an arc. But the projection of this arc into the XY-plane entirely fills the unit square!

Characterizations of arcs by intrinsic properties are due to Lennes, whose results are summarized in

THEOREM 12.3 *The five families of the following subsets of a metric space are identical.*

I. *Topological images of* $[0, 1]$.

II. *Continua containing two points between which they are irreducibly connected.* (A set S is *irreducibly connected* between the points p and q provided S, but no proper subset of S, is connected and contains both p and q. Such a set is, of course, an irreducible continuum between p and q. But the sinusoid S^* is an irreducible continuum between the points $(0, 0)$ and $(1/\pi, 0)$ without being irreducibly connected between them, since the (non-closed) set consisting of $(0, 0)$ and the points $(x, \sin 1/x)$, $0 < x \leq 1/\pi$, is a connected proper subset of S^* that contains those two points.)

III. *Irreducible continua between two points such that every other point they contain is a cut point.* (A point c is called a *cut point* of the continuum C provided $C - \{c\}$ is not connected.)

IV. *Continua C containing two points a and b, such that if $c \in C$ there exists an (a, c, b)-decomposition of C.* (An (a, c, b)-*decomposition* of C is given by two sets, A_c and B_c, containing a and b, respectively, with intersection c and union C. It will be shown that for each c, the sets A_c and B_c are connected, closed, and uniquely determined by the point c.)

The fifth family consists of sets with a certain order relation. By a *binary relation* in a set S we mean a subset R of $S \times S$. If the sequence (s_1, s_2) belongs to R, then s_1 is said to be in the relation R to s_2. A binary relation $R_{\prec}$ is called a linear order relation—briefly, an *order relation*—in S (and will be expressed by writing $s_1 \prec s_2$, if and only if (s_1, s_2) belongs to $R_{\prec}$) provided $R_{\prec}$ satisfies the following three conditions.

(1) *Of any two elements (p, q) and (q, p) of $S \times S$, at most one belongs to $R_{\prec}$; that is, if $p \prec q$, then $q \nprec p$.*

(2) *If $p \neq q$, then at least one of the elements (p, q), (q, p) of $S \times S$ belongs to $R_{\prec}$; that is, if $p \neq q$, then $p \prec q$ or $q \prec p$.*

(3) *If (p, q) and (q, r) belong to $R_{\prec}$, so does (p, r); that is, if $p \prec q$ and $q \prec r$, then $p \prec r$.*

(Properties (1), (2), (3) of $R_{\prec}$ are referred to as the *asymmetry*, the *completeness*, and the *transitivity* of the order relation, respectively.) We now describe the fifth family.

V. *Continua C for which there exists an order relation satisfying the following conditions.*

(4) *C contains two points, a and b such that $a \prec c \prec b$ for each point c of C that is distinct from a and b.*

(5) *For any two points d and e of C such that $d \prec e$, the set of all points p of C such that $d \prec p \prec e$ is nonempty and open in C. Its closure is obtained by adjoining the points d and e.*

The identity of the Families I to V will be proved by showing that each of them is contained in the next and that the last is contained in the first.

(a) Family I $\subset$ Family II. It has already been shown that an arc does not contain a connected proper subset including its end points.

(b) Family II $\subset$ Family III. It has also been shown that irreducible connectedness of a continuum C between a and b implies that C is an irreducible continuum between a and b. If the continuum C contains a point c that is not a cut point, then $C - \{c\}$ is a connected proper subset of C, and so C is not irreducibly connected between any two points of $C - \{c\}$. Consequently, if C is irreducibly connected between a and b, then each point c of C, $a \neq c \neq b$ is a cut point.

(c) Family III $\subset$ Family IV. If C is an irreducible continuum between a and b, and c is a cut point of C, $a \neq c \neq b$, then $C - \{c\}$ is the union of two disjoint nonempty sets V' and V'' that are open in C. The union of their closures, $A_c = V' \cup \{c\}$ and $B_c = V'' \cup \{c\}$, is C, and $A_c \cap B_c = \{c\}$. Since both V' and V'' are connected, by Lemma 12.4, A_c and B_c are connected.

These sets are subcontinua of C, neither of which contains both a and b, since C is an irreducible continuum between a and b. Let A_c be the set containing a, so that B_c contains b. If A and B are any two continua constituting an (a, c, b)-decomposition of C, then $A_c \cup B$, being the union of two continua containing c, is a continuum and contains a and b. It follows that $A_c \cup B = C$, and consequently, $(A_c \cup B) \cap B_c = C \cap B_c = B_c$. But

$$(A_c \cup B) \cap B_c = (A_c \cap B_c) \cup (B \cap B_c) = \{c\} \cup (B \cap B_c) = B \cap B_c.$$

Hence $B \cap B_c = B_c$; that is to say, B_c is a subset of B. In a similar way, one proves that B is a subset of B_c (and so $B = B_c$) and that $A = A_c$. The proof is completed by setting $A_a = \{a\}$, $B_a = C$ if $c = a$, and $A_b = C$, $B_b = \{b\}$ if $c = b$.

(d) Family IV $\subset$ Family V. The reader can readily show that if C belongs to Family IV, then C is an irreducible continuum between a and b. An order relation in C can be defined by setting $d \prec e$ if and only if A_d is a proper subset of A_e.

Condition (4) follows from the definitions of A_a and A_b; and (1) and (3) are immediate consequences of the fact that the proper-subset relation is asymmetric and transitive (though not complete !).

To prove Condition (2), we use the following

LEMMA 12.9   *If two closed sets, R and S, have exactly one common point, p, then each connected subset of $R \cup S$ that does not contain p is a subset of R or a subset of S* (see Exercise 3).

First we show that if, for two distinct points d and e of C, the point e lies in B_d, then $d \prec e$. It is clear that if e lies in B_d, then e is not contained in A_d. The connected set A_d, which does not contain e (the only point that is common to A_e and B_e), is, according to Lemma 12.9, a subset of one of them—clearly of A_e; and since A_d does not contain e, it is a proper subset of A_e, and so $d \prec e$. Similarly, one proves that if e lies in A_d, then B_d is a proper subset of B_e. Then A_e is a proper subset of A_d, and $e \prec d$. Since each point e distinct from d lies either in B_d or in A_d, condition (2) is satisfied. It is also clear that, for any two distinct points d and e, we have $d \prec e$ *if and only if e lies in B_d*.

The reader can easily verify that $d \prec p \prec e$ *if and only if $d \neq p \neq e$ and p belongs to $B_d \cap A_e$*. If $d \prec e$, then we denote the set of all points p such that $d \prec p \prec e$ by $]d, c[$. Being the complement of the closed set $A_d \cup B_e$ the set $]d, e[$ is open in C; and it is nonempty, for in the contrary case C would be the union of two disjoint continua, A_d and B_e. We call $]d, e[$ an *open interval*. Its only accumulation points not contained in $]d, e[$ are d and e. Thus Condition (5) holds. This concludes the proof that Family IV $\subset$ Family V.

We add a few remarks that are valid for any set C in Family V. Let $D = \{d_1, d_2, \ldots, d_m\}$ be a finite subset of C not including a or b. For a certain permutation $i_1, i_2, \ldots, i_m$ of $1, 2, \ldots, m$ (setting $a = d_{i_0}$ and $d_{i_{m+1}} = b$) we obviously have

$$d_{i_0} \prec d_{i_1} \prec \cdots \prec d_{i_j} \prec d_{i_{j+1}} \prec \cdots \prec d_{i_m} \prec d_{i_{m+1}}.$$

Thus D determines $m + 1$ open intervals $]d_{i_j}, d_{i_{j+1}}[$. As a corollary of Condition (5), the reader can easily prove that any continuum C in Family V satisfies the following condition.

(6) *If D is a finite subset of C, not including a or b, and C' is a subset of C that is dense in C, then each interval determined by D includes points of C'.*

(e) Family V $\subset$ Family I. Let C be a continuum with an order relation satisfying (4) and (5). It will be shown that C is homeomorphic to the continuum $T = [0, 1]$ in which an order relation satisfying (4) and (5) is defined by setting $x \prec y$ if and only if the number x is less than the number y; that is, $x \prec y$ if and only if $x < y$.

We construct a topological mapping of C onto $[0, 1]$. Being compacta, C and T contain countable dense subsets (Theorem 1.7). Let C' and T' be any two such sets not containing a, b and $0, 1$, respectively. The mapping to be constructed will map C' onto T'.

Let $(c_1, c_2, \ldots)$ and $(t_1, t_2, \ldots)$ be serializations of the elements of C' and T'. We rearrange them into sequences $(d_1, d_2, \ldots)$ and $(u_1, u_2, \ldots)$, respectively, as follows. Set $d_1 = c_1$ and $u_1 = t_1$, and assume that the sets

$$D_{n-1} = \{d_1, d_2, \ldots, d_{n-1}\} \quad \text{and} \quad U_{n-1} = \{u_1, u_2, \ldots, u_{n-1}\}, n > 1,$$

have been defined. If n is even, let d_n be the point c_k with the lowest index k that is not contained in D_{n-1}. Clearly, d_n lies in one of the n intervals $]d_{i_j}, d_{i_{j+1}}[$ determined by D_{n-1}. Let $]u_{i_j}, u_{i_{j+1}}[$ be the corresponding interval determined by U_{n-1} in $[0, 1]$. According to (6), this interval contains points of T'. The point t_k with the lowest index k that is in that interval is called u_n. If n is odd, let u_n be the point t_k with the lowest index that is not contained in U_{n-1} but in some interval determined by U_{n-1}, and let d_n be the point c_k with the lowest index k contained in the corresponding interval determined by D_{n-1}. Each point of C' and of T' occurs exactly once in the sequences $(d_1, d_2, \ldots)$ and $(u_1, u_2, \ldots)$, respectively. Indeed, each c_k is d_h for some

$h \leq 2(k-1)$; and each t_k is u_j for some $j \leq 2k-1$. By associating u_k with d_k for each k, one maps C' onto T' in a one-to-one way that is *order-preserving*; that is to say, if t and t' are the images of c and c', respectively, then $t < t'$ if and only if $c \prec c'$.

Now let d be any point of C. Call C_d' and C_d'' the sets of all points c' of C' such that $c' \prec d$ and $d \prec c'$, respectively. Their respective images T_1' and T_2'' in T' constitute a cut in T' or (if d belong to C' and hence its image u lies in T') a cut in $T' - \{u\}$—that is, a division of the set into two subsets such that each element of T_1' is less than each element of T_2. Such a cut determines an element, u, of $[0, 1]$, and this element we define as the image of d. The mapping of C onto T so defined is one-to-one, order-preserving, and, as the reader can easily verify on the basis of Condition (5), continuous. Since C is a compactum, the mapping is topological (Theorem 1.15). Thus C is homeomorphic to $[0, 1]$. This concludes the proof of Theorem 12.3.

EXERCISES

1. In the plane, give examples of topological mappings (a) of closed sets onto nonclosed sets, (b) of bounded sets onto unbounded sets.

2. Prove that the topological image of any locally connected continuum is a locally connected continuum.

3. Prove Lemma 12.8.

4. Prove Remark (6) of this section.

5. Show that each member of the Family IV defined in this section is an irreducible continuum between a and b.

6. In the plane, give an example of a continuum that is locally connected at some point p without containing an arc containing p.

7. In the plane, give an example of a connected set S having the following three properties: (a) S is irreducibly connected between two points; (b) any third point is a cut point of S; (c) there exists an order relation in S which satisfies Conditions (4) and (5), and *which is not an arc*.

12.7   ARCWISE CONNECTEDNESS

Any two points, a and b, of a continuum C are contained in a subset of C that is an irreducible continuum between a and b. But the sinusoid S^* contains points (for example, $(0, 0)$ and $(1/\pi, 0)$) that are not contained in one and the same subarc of S^*, though each of these points lies in some subarc of S^*. By constructing a chain of smaller and smaller continua homeomorphic to S^*, which converge to a point p_0, and by uniting those continua and $\{p_0\}$, the reader can easily define a continuum C containing a point (namely p_0) *that is not contained in any subarc of C.* There even exist in the plane continua (for example, Knaster's continuum, mentioned in Exercise 13, Section 12.4) *that do not contain any arc,* but such continua are difficult to construct.

A set S is said to be *arcwise connected* if any two points of S lie in an arc contained in S. Clearly, each arcwise connected set is connected; but it need not be either a compactum or locally connected. For example, each continuous image of $]0, 1[$ is readily seen to be arcwise connected; yet some of these images are not locally connected (see Exercise 12, Section 12.4), and some are not compacta. According to an important theorem due to the American mathematician R. L. Moore, however, local connectedness is sufficient for the arcwise connectedness of a continuum. In fact, Moore proved the following sharper assertion.

THEOREM 12.4   (MOORE'S THEOREM)   *If L is a locally connected continuum, then each connected subset of L that is open in L is arcwise connected.*

In view of the intrinsic characterizations of the continuous and topological images of $[0, 1]$, this theorem can be expressed in many ways; for example, *each continuous image of $[0, 1]$ (as well as certain of its subsets) contains for any two of its points a topological image of $[0, 1]$ joining them.* What will be proved is that *if L is a strongly locally connected continuum, M a connected subset that is open in L, and p, q are two distinct points of M, then M contains a continuum C containing p and q such that for each point r of C ($p \neq r \neq q$) there is a $(p, r, q) -$ decomposition of C.*

By a *chain of sets* we mean a finite sequence of sets $(F_1, F_2, \ldots, F_m)$ such that each F_i has a nonempty intersection with F_{i+1} $(i = 1, \ldots, m - 1)$. If $\mathfrak{F}$ is a family of sets, two points p and q are said to be *joined by a chain in $\mathfrak{F}$* provided $\mathfrak{F}$ contains a chain whose first member contains p, and the last member, q. If $\mathfrak{F}$ consists of sets that are open in a set S, and p is a point of S, then the set S_p of all points of S that are joined to p by a chain in $\mathfrak{F}$ is clearly open in S. If, furthermore, $\mathfrak{F}$ covers S (that is, if each point of S lies in at least one member of $\mathfrak{F}$), then S_p is also closed in S. (For if r is an accumulation point of S_p, then r lies in a member F of $\mathfrak{F}$ which, being open in S, contains a point q of S_p. If $(F_1, \ldots, F_m)$ is a chain in $\mathfrak{F}$ joining p and q, then $(F_1, \ldots, F_m, F)$ joins p and r.) If S is connected, then the only nonempty subset of S that is both open and closed in S is S, and so we have proved the following

Assertion.   If S is connected, and $\mathfrak{F}$ is a family of sets open in S that covers S, then any two points of S are joined by a chain in $\mathfrak{F}$.

A chain $(F_1, \ldots, F_m)$ between p and q is said to be *irreducible between p and q* if it does not contain a proper subchain between those points. *It is clear that each chain between p and q which is not irreducible between them contains a subchain that is.*

Now let L be a strongly locally connected continuum, M a connected subset (which will also be referred to as F_1^0) that is open in L and p, q two distinct points of M. Let $\mathfrak{F}^1$ be the family of all connected sets with diameter <1 that are open in L and whose closures are contained in $F_1^0 \ (= M)$. Since M is open in the strongly connected set L, the family $\mathfrak{F}^1$ covers M. Hence, according to the Assertion, p and q are joined by a chain $(F_1^1, \ldots, F_{m_1}^1)$ in $\mathfrak{F}^1$, which may be assumed to be irreducible between p and q.

Make the inductive assumption that an irreducible chain between p and q, $(F_1^k, \ldots, F_{n_k}^k)$, has been constructed, consisting of connected sets with diameters $<1/k$ that are open in L, such that the closure of each F_j^k is a subset of some set F_m^{k-1}. Then, for $i = 1, \ldots, n$, let $\mathfrak{F}_i^{k+1}$ be the family of all connected sets with diameters $<1/(k+1)$, that are open in L, and whose closures are subsets of F_i^k. Each $\mathfrak{F}_i^{k+1}$ covers F_i^k. Since p and any point r_1^k of the (nonempty) set $F_1^k \cap F_2^k$ lie in F_1^k, they are joined by a chain in $\mathfrak{F}_1^{k+1}$. Similarly, r_1^k and any point r_2^k in $F_2^k \cap F_3^k$ are joined by a chain in $\mathfrak{F}_2^{k+1}$; and so on. Finally, q and $r_{n_k-1}^k$ are joined by a chain in $\mathfrak{F}_{n_k-1}^{k+1}$. Uniting these chains in their proper order, one obtains a chain between p and q which contains a chain that is irreducible between p and q—say $(F_1^{k-1}, \ldots, F_{n_{k+1}}^{k+1})$. The inductive assumption is then satisfied for $k+1$. Let A^k be the union of the n_k members of the chain $(F_1^k, \ldots, F_{n_k}^k)$. Clearly, each A^k is a connected set containing p and q. Since the closure of A^{k+1} is a subset of A^k, the intersection C of all sets A^k (being equal to the intersection of their closures) is a continuum containing p and q. We show that C is an arc between p and q by proving that for each point r of C such that $p \neq r \neq q$ there exists a (p, r, q)-decomposition of C.

For each k, the point r lies either in exactly one member, $F_{j_k}^k$, of the kth chain or in two consecutive members and hence in $F_{j_k}^k \cup F_{j_k+1}^k$. In either case we define

$$P^k = F_1^k \cup \cdots \cup F_{j_k}^k \cup F_{j_k+1}^k \qquad \text{and} \qquad Q^k = F_{j_k}^k \cup F_{j_k+1}^k \cup \cdots \cup F_{n_k}^k.$$

Clearly, P^k and Q^k are continua containing r as well as p and q, respectively. Their union is A^k, their intersection is $F_{j_k}^k \cup F_{j_k+1}^k$, and hence has a diameter $<2/k$. The intersection C is thus the union of the intersection P of all P^k and the intersection Q of all Q^k. Since $(P^1, P^2, \ldots)$ is a decreasing sequence of continua containing p and r, their intersection P is a continuum containing p and r. Similarly, Q is a continuum containing r and q. The intersections $P^k \cap Q^k$, whose diameters converge to 0, close down on the point r. Hence $P \cap Q = \{r\}$. This completes the proof that $C = P \cup Q$ is a (p, r, q)-decomposition of C, and hence C is an arc.

The theorem leads to a sixth characterization of arcs.

THEOREM 12.5   *Families* I *to* V *described in Theorem* 12.3 *are identical with the following family.*
 VI. *Locally connected sets that are irreducible continua between two points.*

Indeed, each arc is locally connected and is an irreducible continuum between its end points. Conversely, if C is a locally connected continuum, then C contains an arc between any two points, a and b. If C is, moreover, an irreducible continuum between a and b, then this arc must be identical with C.

Thus the arcs are, so to speak, at the crossroads of the concepts of one-to-one images of $[0, 1]$, the continuous images of $[0, 1]$ (which are the locally connected continua), and the sets that are irreducible continua between two points. The arcs are the continuous one-to-one images of $[0, 1]$, the locally connected sets that are irreducible continua between two points, and the continua that are irreducibly connected between two points.

EXERCISE

1. Is the existence of an arc joining two points of a strongly locally connected continuum a consequence of the Reduction Theorem? (See Exercise 5, Section 11.5). Is the intersection of a decreasing sequence of locally connected continua necessarily locally connected?

12.8   UNIONS OF ARCS

Since the concept of arcs is much narrower than that of curves, in seeking a suitable definition of curves one might consider continua that are unions of families of arcs. But since the square and the cube are unions of families of segments, it is clear that the families have to be restricted in order to keep the concept from being too wide.

It will be shown in the sequel that the union of a *countable* family of arcs does not include square domains or even sets that include such domains as subsets (nor, in fact, any objects that should be excluded from the realm of curves). But, as the reader can easily prove, such unions exclude the continuum B^* in the plane (referred to as the *Cantor brush*) which is the union of the segments ("bristles") joining the point $(\frac{1}{2}, 1)$ to the points of Cantor's discontinuum D^* on the x-axis. They also exclude the (indecomposable) continuum I^* (Section 12.4) and even more important continua (to be discussed in the sequel) which should be included in the realm of curves. Thus the union of countable families of arcs is too narrow a concept for our purpose; nor do such unions appear to be of great intrinsic importance.

Similar objections may be raised to the unions of *finite* families of arcs, which do not even include the sinusoid S^*, though they do include, for example, the union of two planar arcs A' and A'' defined as follows: A' is the union of the discontinuum D^* and of all semicircles in the upper half-plane that have a deleted middle third as diameter; A'' is the reflection of A' in the X-axis. The intersection of A' and A'' is D^*.

A still narrower concept which, however, is of considerable interest, is that of *the union of a finite family of arcs any two of which have no common points other than possibly one or both of their end points*. The union G of such a family $\mathfrak{A}$ of arcs is called a *graph*. The circle, the lemniscate, and the letters T, P, and A are connected graphs. A point of G that is the end point of exactly k arcs of $\mathfrak{A}$ is said to be *of order k*. The points of G that are not endpoints of arcs of $\mathfrak{A}$ are said to be of order 2.

If A' and A'' are two arcs of $\mathfrak{A}$ that have a common end point p which is a point of order 2 of G, then $A' \cup A''$ is either an arc (of course not belonging to $\mathfrak{A}$) or a closed curve (that is, homeomorphic to a circle), called a *loop* of G, according as the other end points p' of A' and p'' of A'' are or are not distinct. If $p' = p'' = p^*$ is of order 2 in G, then the loop is a component of G. If p^* is of order >2, then p^* is the end point of at least one other arc of $\mathfrak{A}$. Being the only such point of the loop, p^* is called its *distinguished* point. In any case, replacement of A' and A'' by an arc or a loop eliminates at least one point of order 2 that is common to two arcs of $\mathfrak{A}$. Proceeding in this way one can eliminate, one by one, all these points and thus obtains

THEOREM 12.7   *Each graph G is the union of a finite family $\mathfrak{E}$ of what may be called the elements of G : that is, arcs and loops such that each common point of two elements E_1 and E_2 is (1) a point of order >2 of G, and (2) an end point or the distinguished point of E_i according as E_i is an arc or a loop ($i = 1, 2$).*

The elements of the letter P are an arc and a loop; those of a lemniscate are two loops; those of the letter A are four arcs (*not* a loop and two arcs !); those of the numeral 100 are an arc and two loops, each element being a component.

A point p of G which is (1) an end point of exactly s elements that are arcs and (2) the distinguished point of exactly t loops, where $s + t > 0$, is called a *node* of G and is said to be of *order $s + 2t$*. The nodes of order 1 are called the *end points* of G, those of an order >2, the *ramification points* of G. From Theorem 12.6, one readily deduces the following

COROLLARY   *If G is a graph without ramification point, then each component of G is either an arc or a closed curve : and the number of nodes (all of which are end points) of G is twice the number of components that are arcs. In particular, a graph without nodes is the union of a finite family of disjoint closed curves. A connected graph without ramification points is an arc or, if it contains no nodes, a closed curve.*

A graph that contains no subset that is a closed curve (in particular no loop) is called a *tree*. An example of a tree is, for any number m, the union of m arcs all of which have one common end point p, whereas the other m end points are distinct. Such a graph is called an *m-spoke* (in German: *m-Bein*) with the center p. The letter X is a 4-spoke. If $m \neq 2$, then each spoke is an element of the tree.

If p is a node of order m of the graph G, then G contains an m-spoke with the center p as a subset. Each point p that is not a node lies in the interior of a subarc of G, which is a 2-spoke with the center p. Hence such a point is said to be of *order 2*.

EXERCISES

1. Let P be the set of all nodes of the graph G. Prove that the elements of G are the closures of the components of $G - P$.

2. Construct a connected and a nonconnected graph having for each $k \neq 2$ the same number of points of order k.

3. Prove that a dumbbell-shaped curve and the letter θ are not homeomorphic even though both have exactly two nodes, which are of order 3.

4. Set up all types of (mutually nonhomeomorphic) graphs whose only nodes are (a) four points of order 3; (b) three points of order 4.

5. Prove that if the connected graph G has n nodes and m elements of which m' are arcs, then $m \geq m' \geq n - 1$. The graph is a closed curve if and only if $m = 1$ and $m' = n = 0$; it is a tree if and only if $m = m' = n - 1$.

6. Generalize the theorem in Exercise 5 so as to include also nonconnected graphs by considering (besides n, m, and m') the number c of the components of G.

12.9   ON n-CONNECTED GRAPHS

By a *graph of degree d* we mean the union G of (1) d arcs (called *the arcs of G*) any two of which are either disjoint or have exactly one common point, which is an end point of both arcs; (2) a finite (or empty) set F of points not in any of the arcs of G. The points belonging to F and the end points of the d arcs of G will be called *the points of G*, since the points in the interior of the arcs will not be considered. It will be noted that Condition (1) rules out pairs of arcs having both end points in common.

A *chain* in G is a sequence $(A_1, \ldots, A_m)$ of arcs of G such that any two consecutive arcs have a common point. If a_0 and a_m are the end points of A_1 and A_m that are not in A_2 and A_{m-1}, then the chain is said to *join a_0 and a_m* $(m \geq 1)$.

Throughout this section, H and J are two disjoint subgraphs of the graph G. A chain joining a point of H and a point of J is called an *HJ-chain*— more specifically, an *irreducible HJ*-chain if it has no points other than its end points in common with $H \cup J$. A set N of points of G is said to *separate* H and J if N has at least one point in common with each *HJ*-chain in G. It is not assumed that N is disjoint from $H \cup J$. In fact, H as well as J separates H and J in G.

We call *G* at least *n-connected between H and J*—briefly, *at least (HJ, n)-connected*—if each set separating *H* and *J* in *G* contains at least *n* points. If *G* contains a set of at most *n* points separating *H* and *J*, then *G* is said to be *at most (HJ, n)-connected*. If *G* is at least and at most *(HJ, n)*-connected, then *G* is said to be *(HJ, n)-connected*. In this case, if no proper subgraph *G'* of *G* is *n*-connected between *G' ∩ H* and *G' ∩ J*, the graph *G* is said to be irreducibly *(HJ, n)*-connected.

LEMMA *If G is irreducibly (HJ, n)-connected, then each point h of H is an end point of an irreducible HJ-chain in G and hence has a neighbor h' that is not in H.* (Two points of *G* are said to be *neighbors* if they are the end points of an arc in *G*.)

Indeed, if *G* contains no *HJ*-chain containing *h*, or if each such chain contains a point of *H* other than *h*, then let *G'* be the union of all subgraphs of *G* not containing *h*. For each set *N'* with less than *n* points, *G* contains an *HJ*-chain disjoint from *N'*. This chain contains an irreducible *HJ*-chain. Since the latter does not contain *h* it is a part of the subgraph *G'* which, consequently, is at least *n*-connected between *G' ∩ H* and *G' ∩ J* in contradiction to the assumption that *G* is irreducibly *n*-connected between *H* and *J*.

The following result, obtained by Menger in 1927, has since then been used in electrical engineering and in the theory of communications. The original proof has been elaborated by G. Nöbeling.

THEOREM 12.8 *If G is at least n-connected between the disjoint subgraphs H and J, then G contains n mutually disjoint HJ-chains.*

Since the assertion is trivial for graphs of degree 1 we assume its validity for all graphs of degree $<d$, and prove it for a graph *G* of degree *d*.

We first construct a set *N* of *n* points such that $N \not\subseteq H$, $N \not\subseteq J$ and that *N* separates *H* and *J* in *G*. Let *h* be a point of *H*. By the Lemma, *h* has a neighbor *h'* not belonging to *H*. Let *G'* be the union of all subgraphs of *G* in which *h* and *h'* are not neighbors. Then *h* and *h'* are not neighbors in *G'*, whence *G'* is of a degree $<d$ and, being a proper subgraph of *G*, at most $(n-1)$-connected between *G' ∩ H* and *G' ∩ J*. Consequently there exists a set *N'* of at most $n-1$ points separating *H* and *J* in *G'*. Setting $k = h'$ if $N' \subset H$, and $k = h$ otherwise, we claim that $N = N' \cup \{k\}$ has the desired properties. Because of the definition of *k*, we have $N \not\subseteq H$ and $N \not\subseteq J$. To

show that N separates H and J in G, let K be an HJ-chain in G not containing k and, therefore, not containing both h and h'. Consequently $K \subset G'$. Since K joins $G' \cap H$ and $G' \cap J$ the chain must have a point in common with N'. Thus N separates H and J in G. It follows that N contains at least n points. So do H and J, between which G is n-connected. Since N' contains at most $n - 1$ points, it follows that N and N' contain exactly n and $n - 1$ points, respectively.

We label the n points of N in such a way that

$$p_1, \ldots, p_h \text{ lie in } H \cap N; \qquad p_{h+1}, \ldots, p_{h+n} \text{ lie in } J \cap N;$$
$$p_{h+j+1}, \ldots, p_n \text{ lie in } N - H \cap N - J \cap N.$$

(Of course we may have $h = 0$ and/or $j = 0$ and/or $n - h - j = 0$.) Since neither H nor J is contained in N, the sets

$$H_1 = H - H \cap N \qquad \text{and} \qquad J_1 = J - J \cap N$$

are nonempty. Let p be any point of H_1. According to the Lemma, G contains an irreducible HJ-chain $(p, q_1, \ldots, q_g)$. Since N separates H and J in G, this chain has at least one point in common with N. Let q_i be the first of these points. Then $(p, q_1, \ldots, q_i)$ is irreducible between H_1 and $N_1 = N - H \cap N$ and disjoint from $H \cap N$. Let G_1 be the union of all chains in G which are irreducible between H_1 and N_1 and disjoint from $H \cap N$. Clearly, $H_1 \subset G_1$. We prove two properties of G_1.

1. G_1 has a degree $< d$. This is evident if $G_1 \cap J$ is empty. So let q be any point of $G_1 \cap J$. Belonging to G_1 it lies in a chain

$$K = (p, q_1, \ldots, q_{r-1}, q_r = q, q_{r+1}, \ldots, q_s)$$

which is irreducible between H_1 and N_1 and disjoint from $H \cap N$. The chain $(p, q_1, \ldots, q_{r-1}, q)$ is an HJ-chain and thus has a point in common with N and, because it is disjoint from $H \cap N$, with N_1. Since q_s is the only point of $K \cap N_1$ it follows that $q_s = q_r = q$. Thus q lies in N and $G_1 \cap J \subset N$. Since J_1 is nonempty, there is a point r of J that is not in G_1. According to the Lemma, G contains a neighbor r' of r. But since r is not in G_1, the points r and r' are not neighbors in G_1, whose degree thus is less than that of G.

2. G_1 is at least $(n - h)$-connected between H_1 and $N_1' = G_1 \cap N_1$. Let T separate H_1 and N_1 in G_1. Since $H \cap N$ consists of exactly h points it is sufficient to show that $T \cup (H \cap N)$ separates H and J in G—that is, contains a point of each HJ-chain. Let K be an HJ-chain in G that is disjoint from $H \cap N$. It contains an irreducible HJ-chain $(p_0, p_1, \ldots, p_r)$. If p_i is

its first point belonging to N and hence to N_1, then the chain $(p_0, \ldots, p_i)$ lies in G_1 and has a point in common with T, and so does K since K is disjoint from $H \cap N$.

Analogously, let G_2 be the union of all chains in G that are irreducible between J_1 and $N_2 = N - N \cap J$ and disjoint from $J \cap N$. Clearly, G_2 has the properties (1) and (2); that is, G_2 is at least $(n - j)$-connected between J_1 and $N_2' = G_2 \cap N_2$. We now prove:

3. $G_1 \cap G_2 \subset N$.

If r were a point of G in $G_1 \cap G_2 - N$, then, by the definitions of G_1 and G_2, r would lie in irreducible $H_1 N_2$- and $N_1 J_1$-chains. The first chain up to r, and the second from r on would, combined, constitute an HJ-chain disjoint from N. But such a chain does not exist, since N separates H and J in G_1.

Applying the inductive assumption to G_1 with the properties (1) and (2) we obtain $n - h$ disjoint chains K_i' $(i = h + 1, \ldots, h + j, h + j + 1, \ldots, n)$ such that K_i' joins H with a point t_1 of N_1 and has no other point in common with N since $G_1 \cap H \cap N$ is empty. Similarly we obtain $n - j$ disjoint chains K_i'' $(i = 1, \ldots, h, h + j + 1, \ldots, n)$ such that K_i'' joins J with a point t_i of N and has no other point in common with N. The joins K_i of the chains K_i' and K_i'' $(i = h + j + 1, \ldots, n)$ are HJ-chains. Altogether we obtain n HJ-chains

$$ K_1'', \ldots, K_h'', K_{h+1}', \ldots, K_{h+j}', K_{h+j+1}, \ldots, K_n. $$

Any two of these chains, say K^* and K^{**}, are disjoint. This is evident if either G_1 or G_2 contains both chains. If K^* is in G_1, and K^{**} in G_2, then because of (3) their intersection is a point of N. But N has only the points t^* and t^{**} in common with K^* and K^{**}, respectively, and so even in this case the two chains are disjoint. Thus the existence of n disjoint HJ-chains in G is demonstrated.

EXERCISE

1. Let H and J be disjoint closed subsets of the compactum S. We say that the set T *almost separates* H and J in S if $S - T = U \cup V$, where

$$ H - H \cap T \subset U, J - J \cap T \subset V, \text{ and } (U \cap \mathrm{cl}\, V) \cup (\mathrm{cl}\, U \cap V) \text{ is empty.} $$

We say that S is *strictly n-connected* between the disjoint finite sets H and J if each subset of T that almost separates H and J in S contains at least n points. Show that Theorem 12.8 has the following

COROLLARY   *Let G be a graph* (*in the sense of the union of a finite family of arcs such that the intersection of any two arcs is either empty or an end point of both arcs*). *If G is strictly n-connected between two disjoint finite subsets H and J, then G contains n pairwise disjoint arcs each of which joins a point of H and a point of J.*

12.10   CANTOR CURVES

A subset B of A that is not dense in A (that is, whose closure in A is a proper subset of A) may of course contain a subset that is dense in an open subset of A. The simplest example is a subset B consisting of one point of a set A that contains exactly two points. If, however, no subset of B is dense in any set that is open in A, then B is said to be *nowhere dense in A*. In this case, each nonempty set that is open in A contains a nonempty subset that is open in A and disjoint from B, and this condition is clearly also sufficient for B to be nowhere dense in A. Cantor's discontinuum D^* is nowhere dense in $[0, 1]$.

One readily sees that if B is nowhere dense in A, then so is the closure of B in A. Hence a set B that is closed in A is nowhere dense in A if and only if B does not contain a nonempty subset that is open in A.

One of the most important facts about nowhere-dense sets is a negative result due to the French mathematician René Baire. A special case of that result is expressed in

THEOREM 12.9   (BAIRE'S THEOREM)   *A compactum C is not the union of a countable family of sets that are nowhere dense in C.*

Let $(N_1, N_2, \ldots)$ be a (finite or denumerable) sequence of sets that are nowhere dense in C. There is a nonempty set V_1 that is open in C and disjoint from N_1. Let W_1 be a nonempty set that is open in C and whose closure, C_1, is contained in V_1 (and hence is disjoint from N_1). Now assume inductively that, for some natural number k, a nonempty set W_k is given that is open in C and whose closure C_k is disjoint from the union of $N_1, N_2, \ldots, N_k$. Since N_{k+1} is nowhere dense in C, the set W_k contains a nonempty subset V_{k+1} that is open in C and disjoint from N_{k+1}. Any nonempty subset W_{k+1} of V_{k+1} that is open in C and whose closure is contained in V_{k+1} satisfies the inductive assumption for $k + 1$. (Of course, the procedure breaks off if the sequence of the N_i is finite.) The decreasing sequence of closed nonempty subsets $(C_1, C_2, \ldots)$ of the compactum C has a nonempty intersection. Consequently, C contains a point not contained in any of the sets N_i.

A set that is nowhere dense in the space is said, briefly, to be *nowhere dense*. A closed subset S of the n-dimensional euclidean space is nowhere dense if and only if S does not contain an n-dimensional cube.

An appropriate concept of curves should exclude sets containing square domains as subsets. Hence, in the plane, only nowhere dense continua should fall under the concept. Cantor felt that, conversely, each nowhere dense subcontinuum of the plane warranted being called a curve. We shall call such curves *Cantor curves*; that is, *in the plane, Cantor curves are the nowhere dense continua*.

Within its scope (that is, in the realm of subsets of the plane) this concept of curve is perfectly appropriate. The exclusion, by definition, of undesirable sets keeps it from being too wide; and it is not too narrow, since it obviously includes all arcs in the plane, the sinusoid S^*, the Cantor brush B^*, and the continuum I^* mentioned in Sections 12.8 and 12.4.

As an immediate consequence of Baire's Theorem we obtain the following.

THEOREM 12.10 (SUM THEOREM FOR CANTOR CURVES) *If the union of a countable family of Cantor curves in a plane is a continuum, it is a Cantor curve.*

Otherwise, the union would contain a closed square domain Q, and the compactum Q would be the union of a countable family of nowhere dense subsets—namely, the intersections of Q with the curves of the family.

In particular, *each plane continuum that is the union of a countable family of arcs is a Cantor curve*.

An especially interesting Cantor curve that is not the union of a countable family of arcs is the continuum U_2^*, discovered by Sierpiński, which is in many ways analogous to Cantor's discontinuum. Like D^*, the curve U_2^* can be defined in three ways.

1. *By deleting open squares.* The first step consists in dividing the unit square U into 9 congruent squares of side length $\frac{1}{3}$ (as in Fig. 16) and deleting the open middle ninth. In the second step, the open middle ninth is deleted from each of the remaining 8 closed squares (forming a wreath-shaped continuum). At the end of the $(n-1)$st step, a continuum is left which is the union of 8^{n-1} closed squares of side length $1/3^{n-1}$. The nth step consists

in deleting the open middle ninths from each of those 8^{n-1} squares. The set U_2^* may be defined as the set of all those points of U that are not deleted in any step of this procedure. The union of all deleted open squares is open in U, and hence U_2^* is closed. Since it is the intersection of a decreasing sequence of continua, U_2^* is a continuum. Clearly, each subsquare of U, however small, contains a square that is deleted in some step of the procedure or is a part of such a deleted square. Thus U_2^* is nowhere dense in the plane and consequently is a Cantor curve.

2. *Arithmetically.* Let (a, b) be a point of U that is deleted in the construction of S, and let $(.a_1 a_2 \ldots a_n \ldots)_3$ and $(.b_1 b_2 \ldots b_n \ldots)_3$ be ternary fractions equal to a and b. If (a, b) is deleted in the first step of the construction, then $a_1 = b_1 = 1$; if in the second step, then certainly $a_2 = b_2 = 1$; if in the nth step, then $a_n = b_n = 1$. The points of U_2^* (that is, the points not deleted in any step) are those whose coordinates are expressible as ternary fractions that do not have 1's in the same places (see Exercise 4).

3. *As the closure of a denumerable set.* Let S_0 be the set of the vertices of all squares deleted from U. It is easy to see that U_2^* is the closure of S_0. Moreover, U_2^* is the closure of the union of the sides of all deleted squares. Still another definition of S can be obtained by defining the *grill* in a square Q as the union of the four sides of Q and the four segments dividing Q into 9 congruent squares. Now let S_1 be the union of the grill in U and, for each n, of the grills in the 8^{n-1} squares left at the $(n-1)$st step of the construction. U_2^* is the closure of S_1 (which contains S_0 as a subset).

The interest in U_2^* lies in the following important property (because of which U_2^* is called a *universal* curve in the euclidean plane E_2).

THEOREM 12.11 (ABOUT SIERPIŃSKI'S PLANE UNIVERSAL CURVE) *Each Cantor curve is homeomorphic to a subset of the curve U_2^*.*

We shall first illustrate Sierpiński's method by applying it to an analogous assertion about Cantor's discontinuum. *Each nowhere-dense compactum E of the straight line is homeomorphic to a subset of D^*.*

Instead of defining a topological mapping of E onto a subset of D^*, we construct a set D' containing E which is homeomorphic to D^*. Since E is compact, E lies in some interval $[a_0, a_1]$. We set $a_1 - a_0 = d$. Since E is nowhere dense, the middle third of $[a_0, a_1]$ contains an open interval that is disjoint from E. Denote its end points by $a_{1/3}$ and $a_{2/3}$. Clearly, E is contained in the union of the intervals $[a_0, a_{1/3}]$ and $[a_{2/3}, a_1]$, each of which has length $< \frac{2}{3}d$. The middle thirds of those intervals contain intervals

$[a_{1/9}, a_{2/9}]$ and $[a_{7/9}, a_{8/9}]$, which are disjoint from E; and each of the remaining 4 intervals whose union contains E has length $<(\frac{2}{3}d)^2$. Proceeding in this way, for each n one obtains 2^n disjoint intervals of lengths $<(\frac{2}{3}d)^n$ whose union contains E, and so E is contained in the intersection D' of those unions. A topological mapping of D' onto D^* can readily be obtained by stipulating that each point of D' contained in an interval $[a_{i/3^k}, a_{(i+1)/3^k}]$ should be mapped on a point of D^* in the interval $[i/3^k, (i+1)/3^k]$.

In a similar way, for each Cantor curve C, one can construct a set U' containing C and homeomorphic to U_2^*. Let Q be a square containing C, and let d denote the length of its diagonal. All rectangles considered in the following outline of the construction have sides parallel to those of Q.

The middle ninth of Q contains a square Q_1 that is disjoint from C. Extending the sides of Q_1 divides Q into a wreath of 8 rectangles surrounding Q_1 whose diagonals have lengths $<\frac{2}{3}d$. The middle ninth of each of these 8 rectangles contains a square that is disjoint from C. We trim these 8 squares in such a way that the triples of squares along either horizontal side of Q have the same Y-projection, and the triples along either vertical side of Q have the same X-projection (see Exercise 6). By extending the sides of the trimmed squares one decomposes each of the 8 rectangles of the first wreath in such a way that, besides the middle ninth (which is disjoint from C), there is a wreath of 8 rectangles whose diagonals have lengths $<(\frac{2}{3}d)^2$. Clearly, C is contained in the union of the 64 rectangles constituting the 8 wreaths. In the middle ninth of each of these rectangles, there is a square that is disjoint from C. The 64 squares can be trimmed in such a way that all squares in a horizontal (vertical) row have the same Y-projection (X-projection).

In the nth step of the construction one obtains 8^n rectangles whose union contains C. So does the intersection of all these unions, which is the set U', homeomorphic to U_2^*. The definition of a topological mapping of U' on U_2^* (utilizing the trimming of the squares disjoint from C) is left to the reader.

Though perfect within its scope, Cantor's definition has two shortcomings. First, it is not in terms of an intrinsic property of the continua to be called curves, but is based on a property that is relative to the surrounding plane—their nowhere denseness. Second, its scope is limited. On the surface of a sphere or a torus, it still would make good sense to define curves as the nowhere-dense continua. But the definition cannot be extended to higher-dimensional spaces. In the three-dimensional space, the entire surface of a sphere is a nowhere-dense continuum without, of course, being a curve. In space, all that can be said is that the nowhere-dense continua are not solids. But this would not discriminate between curves and surfaces.

EXERCISES

1. Using Baire's Theorem, prove that $[0, 1]$ as well as Cantor's discontinuum D^* is uncountable.

2. Assuming that for every compactum that is nowhere dense in the line, D^* contains a homeomorphic subset, prove the same for every nowhere dense subset of the line.

3. Show that for every nowhere dense subset of the line, the set D_2^* of all points of D^* of the second kind contains a homeomorphic subset.

4. Denote a periodic ternary fraction by dots above the periodic digits; for example, $(.20\dot{1})_3 = (.20101 \ldots 01 \ldots)_3$. Which of the following points belong to the Sierpiński curve U_2^*?

    (a) $((.\dot{1}\dot{2})_3, (.\dot{0}\dot{1})_3)$,   (b) $((.1\dot{2}\dot{0})_3, (.0\dot{1}\dot{2})_3)$,

    (c) $((.\dot{1}\dot{2})_3, (.0\dot{1}\dot{2})_3)$,   (d) $((.0\dot{2})_3, (.\dot{1}\dot{2})_3)$.

5. Assuming that for any Cantor curve C there is a homeomorphic subset of U_2^*, prove the same for any nowhere dense subset of the plane.

6. Perform the trimming mentioned in the construction of U_2^*. (*Hint:* Start with a square Q_1 in the rectangle in the lower left corner of the wreath. Then keep its right neighbor Q_2 between the horizontal lines that bound Q_1, and the right neighbor Q_3 of Q_2 between the horizontal lines that bound Q_2. Then trim Q_1 and Q_2 according to Q_3, and proceed vertically to Q_4).

7. Prove that the set $D^* \times D^*$ of all points (x, y) in the plane such that x and y belong to D^* is homeomorphic to D^*.

8. What is the relation between the plane universal curve U_2^* and the set $D^* \times [0, 1] \cup [0, 1] \times D^*$?

9. Prove that the curve U_2^* is not the union of a countable family of arcs.

CHAPTER 13

Introduction to
Modern Curve Theory

13.1   THE DEFINITION OF CURVES

To find out what distinguishes curves from noncurves we go back to the physical objects mentioned at the beginning of the preceding chapter: blocks of wood (which are typical solids), paper models of surfaces, and objects made of wire (which approximate the idea of curves). The characteristic differences between the various types of objects become apparent in a simple experiment; namely, by removing from an object a point and its surroundings. In performing such an experiment it is seen that objects that are of unlike types require the application of unlike tools. To remove a piece of a wooden solid a saw is needed to sever the piece from the rest of the solid by cutting along surfaces. For the model of a surface a pair of scissors is sufficient, and the paper has to be cut along curves. For the curvelike object only pincers are needed to sever the wire at separate points.

This description is intrinsic (that is, entirely in terms of the objects described and their parts). But there is a second way to discriminate between solids, surfaces, and curves, which makes reference to the space containing those objects. If p is a point of a *solid* and V is any small neighborhood of p (such as a sphere, an ellipsoid, or a cube), then the solid penetrates the frontier of V in a surface. For example, if p is a corner of a cube, and V is a small sphere about p, then the solid penetrates an octant of the frontier (that is, the surface) of V. A point of a *surface* lies in small neighborhoods whose frontiers are penetrated by the surface only along curves. For example, a corner of a cubic surface lies in small spheres with whose frontiers the surface has the union of three circular arcs in common. A point of a *curve* is contained in small neighborhoods, whose frontiers the curve pierces only in dispersed points. For example, a corner of the union C of the twelve edges of a cube lies in small spheres whose frontiers have only three points in common with C.

All that can be expected of a curve—*this must be strongly emphasized*—is that, for each point p of C there should exist *some* neighborhoods (yet neighborhoods as small as may be desired !) whose frontiers have dispersed intersections with C. It cannot be expected that C has dispersed intersections with *all* small neighborhoods of p. This would not even be the case for a segment in the plane, as is demonstrated by the following simple example, which will be referred to repeatedly in the sequel.

Let C be the segment $[0, 1]$ on the X-axis, and let p denote the point $(\frac{1}{4}, 0)$. Denote by Z^* the snail-shaped neighborhood of p which is the union of (1) the (open) segment $]0, \frac{1}{2}[$ on the X-axis; (2) the open semicircular domain, in the upper half-plane, about p of radius $\frac{1}{4}$; (3) the open semicircular domain, in the lower half-plane, about the point $(\frac{1}{2}, 0)$ of radius $\frac{1}{2}$. The frontier of Z^* consists of the two semicircles and the segment $[\frac{1}{2}, 1]$ on the X-axis, and thus has this entire segment and the point $(0, 0)$ in common with C. In fact, p lies in such snail-shaped neighborhoods that are arbitrarily small, and the frontier of each of them has a segment and an isolated point in common with C. But of course there *also* exist some arbitrarily small neighborhoods whose frontiers have only two points in common with C (namely, the small circular domains about p), and their existence suffices to guarantee that the segment is a curve in the sense of the following tentative definition.

A curve is a continuum C each point of which lies in arbitrarily small neighborhoods whose frontiers intersect C in dispersed sets.

To make this definition perfectly explicit and operative, three points must be clarified: (1) what is meant by *arbitrarily small*? (2) what are *frontiers*? (3) what is the meaning of *dispersed*?

1. If $\mathfrak{S}$ is a family of sets, then a point is said to lie in arbitrarily small members of $\mathfrak{S}$ if each neighborhood of p has a subset that is a member of $\mathfrak{S}$ containing p. For example, each point of the three-dimensional space lies in arbitrarily small closed cubes, as well as in arbitrarily small cubic neighborhoods. The points at which a subset S of the space is strongly locally connected are the points that lie in arbitrarily small neighborhoods whose intersections with S are connected.

2. The frontier of an open set V (symbolized fr V) is the set of all accumulation points of V that do not lie in V; that is, fr $V = \text{cl } V \cap \text{com } V$. Applied to regions (that is, connected open sets) in the plane, this definition leads to the traditional definition of the frontiers (also called boundaries) of the domains studied in the theory of complex functions. Since both cl V and com V are closed sets, *the frontier of every open set is closed.*

3. The idea of dispersed sets can be made precise in various ways. A set is called *discontinuous* if it does not contain any subcontinuum. A set S is said to be *disconnected* if every connected subset of S consists of just one point—in other words, if each component of S consists of a single point. Cantor's discontinuum is disconnected and hence, a fortiori, discontinuous. A third approach to the idea of dispersed sets is through the basic physical experiment which, by the use of various tools, has led from the (three-dimensional) solids to surfaces (which are two-dimensional); from surfaces to curves (which are one-dimensional); and from curves to dispersed sets, which, therefore, it is natural to call 0-dimensional sets. If one applies to such a set the experiment of removing a point and all points near by, then no tool whatsoever is required since there is nothing to be dissected. In other words, from 0-dimensional sets the experiment leads to the empty set (which, therefore, may be conveniently considered as (-1)-*dimensional*).

We thus arrive at

DEFINITION 13.1 (OF 0-DIMENSIONALITY)   A set S is called 0-*dimensional* if S is nonempty and each of its points is contained in arbitrarily small neighborhoods whose frontiers have empty intersections with S (that is, are disjoint from S).

It will be shown in Section 13.6 that, in the class of compacta, the families of the discontinuous, the disconnected, and the 0-dimensional sets are identical, whereas among the noncompact sets each of the three properties describes a stronger degree of dispersedness than the preceding. Since, by the definition given below, curves are compacta, and since the frontiers of neighborhoods are closed sets, their intersections with curves are compacta. Hence in our definition of curves it does not make any difference whether we speak of discontinuous, disconnected or 0-dimensional intersections with frontiers. Summarizing this (nonintrinsic) approach we thus arrive at the following definition, *applicable to metric and topological spaces.*

DEFINITION 13.2 (OF CURVES IN SPACES)   A subset C of a space is a *curve* if and only if C is a continuum, each point of which lies in arbitrarily small neighborhoods whose frontiers have discontinuous intersections with C—that is, if and only if, for each point p of C, each neighborhood V of p contains a neighborhood W of p such that $C \cap \mathrm{fr}\, W$ does not contain a subcontinuum.

If C is the space itself, this definition implies the following intrinsic one.

DEFINITION 13.3 (OF CURVES AS SPACES)   A space is a *curve* if and only if it is a continuum, each of whose points lies in arbitrarily small neighborhoods with discontinuous frontiers.

The connection with the approach suggested by the physical experiment is obvious. The intersections of C with the frontiers of the neighborhoods of p (or those frontiers themselves if C is the space) are the sets that have to be cut in order to remove p and the points near p.

What requires clarification is the relation between the two definitions of curves (in spaces and as spaces). Each subset C of a space S is itself a space. Is this space C a curve if C is a curve in S? The answer is affirmative; but the point warrants several remarks.

Remark 1.   If each point of the subset B of a space lies in arbitrarily small neighborhoods with whose frontiers B has continua in common and if B is considered as a space, then its points do not necessarily lie in neighborhoods whose frontiers contain continua.

This is instanced by a segment C in the plane, whose points lie in arbitrarily small snail-shaped neighborhoods such as Z^*, the frontiers of which have continua in common with C, whereas if C is considered as a space, the frontier of each open set is discontinuous.

If V is an open subset of the space, then $V \cap C$ is open in C. By the frontier of $V \cap C$ in C—briefly, $\mathrm{fr}_C(V \cap C)$—is meant the set of all accumulation points of $V \cap C$ in $C - V \cap C$.

Remark 2.   If C is considered as a space, then $\mathrm{fr}_C(V \cap C)$ is the frontier of the open set $V \cap C$, and $\mathrm{fr}_C(V \cap C) \subset C \cap \mathrm{fr}(V \cap C)$. But $\mathrm{fr}_C(V \cap C)$ may be a proper subset of $C \cap \mathrm{fr}(V \cap C)$.

This is again shown by the snail-shaped neighborhood Z^* of p considered above, where $C = [0, 1]$. The neighborhood Z^* of the point $(\frac{1}{4}, 0)$, however, contains neighborhoods W of $(\frac{1}{4}, 0)$ whose frontiers have only points of $\mathrm{fr}_C Z^*$ in common with C—for example, the circular domain W' and the square domain W'' having $[0, \frac{1}{2}]$ as a diameter and as a diagonal, respectively. The closures (and the frontiers) of both W' and W'' of course contain the two points of $\mathrm{fr}_C Z$ in the complement of Z^*. But cl W' (as well as fr W') contains, besides, a semicircle belonging to fr Z^*, whereas cl W'' (and fr W''), except for those two points, is contained in the open set Z^*. The important fact that there always exist open sets of the type of W'' is expressed in

Remark 3.   If C is any subset of a metric space, Z is open in C, and V is an open set such that $C \cap V = Z$, then there exist open sets W such that $C \cap W = Z$, $C \cap \mathrm{fr}\, W = \mathrm{fr}_C Z$, and cl $W - \mathrm{fr}_C Z$ is a subset of V.

The reader can easily verify that the set W of all points of V that are closer to Z than to the complement of V satisfies Remark 3.

Now let $\mathfrak{F}$ be a family of sets including, for each of its members F, all subsets of F as members. We shall call such a family of sets *hereditary*. If S is a set and p lies in arbitrarily small neighborhoods V whose frontiers have intersections with S that belong to the family $\mathfrak{F}$, then p lies in arbitrarily small neighborhoods in S (that is, sets containing p that are open in S) whose frontiers in S are, by Remark 2, subsets of the sets $S \cap \mathrm{fr}\, V$ and hence belong to $\mathfrak{F}$. From Remark 3 it follows that, conversely, if p lies in arbitrarily small neighborhoods Z in S whose frontiers belong to $\mathfrak{F}$, then p

lies in arbitrarily small neighborhoods such that the intersections of their frontiers with S are the sets $\mathrm{fr}_S Z$ and hence belong to $\mathfrak{F}$.

A set S is called $\mathfrak{F}$-*regular at one of its points* p (and p will be called an $\mathfrak{F}$-*regular point of* S) if and only if p lies in arbitrarily small neighborhoods in S whose frontiers belong to $\mathfrak{F}$. From what has just been proved one obtains an equivalent formulation of this intrinsic definition.

Nonintrinsic condition for $\mathfrak{F}$-regularity.   If $\mathfrak{F}$ is a hereditary family of sets, then S is $\mathfrak{F}$-regular at the point p if and only if p lies in arbitrarily small neighborhoods such that the intersections of their frontiers with S belong to $\mathfrak{F}$.

With regard to a nonhereditary family of sets, such as the family of all sets that are not discontinuous, the nonintrinsic condition is not in general sufficient, as Remark 1 demonstrates. The family of all discontinuous sets, however, is hereditary; whence the nonintrinsic condition (which is satisfied at each point of a continuum C if and only if C is a curve) is sufficient as well as necessary. Consequently, we have

DEFINITION 13.4 (INTRINSIC DEFINITION OF CURVES IN SPACES)

A continuum C is a *curve* if and only if each point of C lies in arbitrarily small neighborhoods in C whose frontiers in C are discontinuous; in other words, if for each point p of C, each neighborhood of p in C contains a neighborhood of p in C whose frontier in C is discontinuous.

It should be noted that even the intrinsic definition of a curve C merely requires that each point lie in *some* arbitrarily small neighborhoods in C whose frontiers in C are discontinuous. The following curve (which, in the sequel, will be referred to as the *circle union* K^*) demonstrates that a point of a curve may *also* lie in arbitrarily small sets that are open *in the curve* and whose frontiers *in the curve* are continua.

In the plane, let C_m, for each natural number m, be the circle about $(0, 0)$ of radius $1/2^m$, and $C_{m,n}$, for any two natural numbers m and n, the concentric circle of radius $(1/2^m) - (1/2^{m+n})$. Let K^* be the union of all these circles and the segment $[0, 1]$ on the X-axis. The reader can easily verify that K^* is a curve. In particular, $(0, 0)$ lies in arbitrarily small neighborhoods in K^* whose frontiers in K^* consist of single points (for example, the sets of all points of K^* at a distance $< 1/3^m$ from $(0, 0)$). Yet $(0, 0)$ has also arbitrarily small neighborhoods in K^* (namely, the sets of all points of K^* whose distance from $(0, 0)$ is $< 1/2^m$) whose frontiers in K^* are the circles C_m.

EXERCISES

1. Prove that the following sets are curves: (a) each graph (see Section 12.8), (b) the sinusoid S^*, (c) the continuum I^* (p. 407), (d) the Cantor brush (p. 425), (e) the circle union K^* (p. 441), (f) the plane universal curve U_2^*.

2. Show that each point of the vertical segment of the sinusoid S^* is contained in arbitrarily small neighborhoods in S^* whose frontiers in S^* contain continua.

13.2   RAMIFICATION ORDER

Before drawing inferences from the (equivalent) definitions of curves or proving the propriety of the curve concept so defined, we demonstrate the great fertility of the underlying principle by applying it to define various local properties and important types of curves and other sets.

A set S is said to be $\mathfrak{F}$-regular if and only if S is $\mathfrak{F}$-regular at each point of S. First of all, one may study $\mathfrak{F}$-regularity with regard to the hereditary families $\mathfrak{F}$ of all countable and of all finite sets. A point p of a set S will be said to be a *rational* or a *regular* point of S if and only if p is contained in arbitrarily small neighborhoods such that the intersections of S with their frontiers are countable or finite, respectively. (It goes without saying that these and all following similar definitions might also be formulated intrinsically in terms of neighborhoods in S and their frontiers in S.) We call a compactum *rational* or *regular* if each of its points is rational or regular, respectively. A segment is a regular curve. The sinusoid is rational but not regular, and so is the circle union K^* (p. 441). The Cantor brush (p. 425) and the plane universal curve U_2^* are irrational (that is, not rational). Clearly, p is an *irrational* or *irregular* point of the set S if and only if p has a neighborhood V such that S and the frontier of each neighborhood of p contained in V have uncountable or infinite intersections, respectively.

If, for some natural number n, the point p lies in arbitrarily small neighborhoods whose frontiers have at most n points in common with S, then S is said to have at p a *ramification order* $\leq n$—briefly, an *order* $\leq n$. The following definition stipulates that S is of order n at p if S is of order $\leq n$, but not of order $\leq n-1$, at p.

DEFINITION 13.5 (OF RAMIFICATION ORDER) The point p is said to be a point *of order* n of the set S if p lies in arbitrarily small neighborhoods whose frontiers have at most n points in common with S and if there exists a neighborhood V of p such that the frontier of each neighborhood of p contained in V has at least n points in common with S.

The reader can readily prove that each point of an arc is of order 2, except the two end points, which are of order 1. Points of order 1 of any set S will be called *end points* of S—an extension of this term from the theory of graphs to general sets. The concept of order is another such extension. For a point p of a graph is of order $\leq n$ (in the sense of $\mathfrak{F}$-regularity with regard to the family $\mathfrak{F}$ of all sets including at most n points) if and only if p is of order $\leq n$ in the sense of Section 12.8.

Clearly, for each n, a point of order n of S is a regular point of S. But the converse is not generally true. In the plane, for each natural number n, let S_n be the segment consisting of the points $(x, x/n)$ for $0 \leq x \leq 1/n$. The union C of all the S_n is regular; in particular, the frontiers of all circular neighborhoods of $(0, 0)$ have finite intersections with C. But for no natural number n is $(0, 0)$ a point of order n of C. A regular point of a set S that is not of any finite order is said to be a point of order ω. The set C will be referred to as an ω-*spoke* with the center $(0, 0)$.

In the other direction, the definition of order can be extended from natural numbers n to 0. A point p of S is said to be of *order* 0 if p lies in aribtrarily small neighborhoods whose frontiers have no common point with S. At such a point, S is also said to be 0-*dimensional*, since the sets that are of order 0 at each of their points are those that have been called 0-dimensional. A point p of S that is not of order 0 has a neighborhood V with the property that the frontier of each neighborhood of p contained in V has points in common with S. At such a point, S is said to be *of positive order*, as well as *of positive dimension*.

The general concepts of curve and of ramification order were developed independently by K. Menger and the Russian mathematician Paul Urysohn (1898–1924) in 1921 and the following years. Presentations of the two theories are contained in Urysohn's *Mémoire sur les multiplicités Cantoriennes II*, Verhandelingen Akad. Wetensk. Amsterdam 13, 1927, and in the papers by Menger in *Mathematische Annalen* 95, 1926, and *Fundamenta Mathematicae* 10, 1927, and are summarized in the book *Kurventheorie* by Karl Menger, edited in collaboration with Georg Nöbeling, Leipzig and Berlin, 1932, reprinted by Chelsea Publ. Co., Bronx, N.Y., 1967. Most of the theorems proved in the subsequent sections are due to both founders of the theory. Hence only the authorship of results obtained by merely one of them, or by other mathematicians, will be specified. The particular form of the definitions expounded in the present section, as well as the physical experiment motivating and illustrating them, are due to Menger.

EXERCISES

1. Prove in detail that an *n*-spoke with the center *p* is of order *n* at *p*.

2. Give an example of a family of four arcs each of which has the point *p* as an end point, and whose union is of order 3 at *p*. Then give two examples where the order at *p* of the union is 2 and 1, respectively. Can it be more than 4?

3. Prove that the union of two arcs is regular and contains no point of an order >4.

4. Show that each point of a segment *S* lies in arbitrarily small neighborhoods in *S* whose frontiers in *S* are uncountable but discontinuous.

5. Determine the orders of the circle union K^* (p. 441) at its various points. Which points besides $(0, 0)$, if any, lie in arbitrarily small neighborhoods in K^* whose frontiers in K^* contain continua?

6. Prove that if a subset *S* of a metric space is of positive dimension at the point *p*, then the intersection of *S* with each neighborhood of *p* is uncountable.

7. Prove that a connected set is of positive dimension at each of its points. (A 0-dimensional set is, therefore, disconnected and hence discontinuous.)

8. Prove that if *C* is an irreducible continuum between *a* and *b*, then no point *c* of *C*, distinct from *a*, *b*, is an end point.

9. Show that if *S* is 0-dimensional at the point *p* of *S*, then *p* is contained in arbitrarily small sets that are both closed and open in *S*.

10. $\mathfrak{F}$-regularity of a set *S* can also be defined at a point of cl $S - S$. For instance, the open interval $]0, 1[$ may be said to be of order 1 at 0 and at 1. A connected set *S* having at each of its points the order 2 can have various orders at a point of cl $S - S$. Give three different examples.

13.3 IMMEDIATE CONSEQUENCES OF THE DEFINITIONS

We begin with the following theorem, whose obvious proof is left to the reader.

THEOREM 13.1 *Each subcontinuum of a curve, a rational curve, and a regular curve is a curve, a rational curve, and a regular curve, respectively. If p is a rational point, a regular point, or a point of an order $\leq n$ of a set S,* *then each subset of S whose closure contains p is rational, regular, or of an order $\leq n$ at p, respectively.*

For the second part of the theorem, see Exercise 10 in Section 13.2.

Let a subset S of a metric space be mapped by a homeomorphism on a subset S' of the same or of another metric space, each point p on a point p'. Clearly the sets V open in S and the sets V' open in S' correspond to each other, and the image of $\operatorname{fr}_S V$ is $\operatorname{fr}_{S'} V'$. Let $\mathfrak{H}$ be a family of sets open in S, and $\mathfrak{H}'$ the family of their images. If p lies in arbitrarily small neighborhoods belonging to $\mathfrak{H}$, then clearly p' lies in arbitrarily small neighborhoods belonging to $\mathfrak{H}'$. The homeomorphic images of discontinuous, countable, and finite sets and of sets containing at most n points are discontinuous, countable, finite, and contain at most n points, respectively. Using the intrinsic definition of curves and orders, one therefore obtains the following invariance theorem.

THEOREM 13.2   (TOPOLOGICAL INVARIANCE OF CURVES AND ORDER) *If C is a curve and C' is homeomorphic to C, then C' is a curve. If p is a point of an order $\leq n$, or a regular point, or a rational point of a set S, and S is topologically mapped on the set S', then the image of p is of an order $\leq n$, or a regular point, or a rational point of S', respectively.*

By the *kernel* of a family $\mathfrak{H}$ of open sets is meant the set of all points that lie in arbitrarily small neighborhoods belonging to $\mathfrak{H}$. The *kernel of $\mathfrak{H}$ in a set S* is the intersection of S with the kernel of $\mathfrak{H}$. For each natural number n, the set K_n of all points that lie in a member of $\mathfrak{H}$ with diameter $< 1/n$ is open. Clearly, the kernel of $\mathfrak{H}$ is the intersection of all these sets K_n, and we are led to

THEOREM 13.3   (ABOUT KERNELS) *The kernel of each family of open sets is a G_δ, that is, the intersection of a countable family of open sets. Its complement is an F_σ, that is, the union of a countable family of closed sets.*

Each closed and each open subset of a metric space is both a G_δ and an F_σ. Every countable set is an F_σ, since each set consisting of a single point is closed. For example, the set of all points of $[0, 1]$ with a rational coordinate is an F_σ. But it is not a G_δ; for if a set S without isolated points is a G_δ, then

S contains a subset that is in one-to-one correspondence with Cantor's discontinuum (see Exercise 5). It follows that the set of all points in [0, 1] with an irrational coordinate is not an F_σ.

The intersection of a G_δ (or F_σ) with a set S is called a G_δ (or F_σ) in S.

Applying the theorem about kernels to the family of all connected sets that are open in a set S, one obtains the following

COROLLARY   *The set of all points at which a set S is strongly locally connected is a G_δ in S.*

If $\mathfrak{F}$ is a family of sets, then the set of all $\mathfrak{F}$-regular ($\mathfrak{F}$-irregular) points of any set S will be called the $\mathfrak{F}$-*regularity set* (the $\mathfrak{F}$-*irregularity set*) of S. If $\mathfrak{F}$ is hereditary, then the $\mathfrak{F}$-regularity set is the kernel of the family of all open sets whose frontiers have intersections with S that belong to $\mathfrak{F}$, and Theorem 13.3 yields

THEOREM 13.4   (ABOUT $\mathfrak{F}$-REGULARITY SETS)   *For each set S and each family $\mathfrak{F}$ of sets, the $\mathfrak{F}$-regularity set of S is a G_δ in S, and the $\mathfrak{F}$-irregularity set is an F_σ in S. In particular, for each natural number n, the set of all points of an order $\leq n$ of S, the set of all regular points of S, and the set of all rational points of S are G_δ-sets in S. The sets of all irrational points, of all irregular points and of all points of an order $>n$ of S are F_σ-sets in S.*

Note that for $n > 1$ the set of all points of order n is the difference between two G_δ-sets as well as the difference between two F_σ-sets, but such sets are not necessarily either G_δ or F_σ. The theorem yields, however, the following

COROLLARY   *The set of all points at which a set S is of order 0 or 0-dimensional is a G_δ in S. The set of all irrational points of a set S is an F_σ in S.*

EXERCISES

1. If S is (strongly) locally connected at p, and S' is homeomorphic to S, is S' (strongly) locally connected at the image of p?

2. Is the set of all points at which a set S is locally connected a G_δ in S?

3. Prove that a G_δ in a set S is the intersection of a countable family of sets that are open in S. State the corresponding theorem for an F_σ in S.

4. In the euclidean plane, let S be the graph of a function—that is, a set not containing two points with equal abscissas and unequal ordinates. Let $\mathfrak{H}$ be the family of all open sets V with the following property: if V contains the point (x_0, y_0) of S, then V also contains, for some positive number d (depending on the point), all points (x, y) of S such that $|x - x_0| < d$. What is the $\mathfrak{H}$-kernel of S, and what follows from the theorem about kernels?

5. In a compactum, let the subset S be a G_δ without isolated points (so that each point of S is an accumulation point of S; a set with that property is called *dense-in-itself*). Suppose S is the intersection of the sequence of open sets $(V_1, V_2, \ldots)$. Choose two open subsets W_0 and W_1 of V_1 which have nonempty intersections with S and whose closures are disjoint. Then choose open subsets W_{00} and W_{01} of W_0, and W_{10} and W_{11} of W_1 with disjoint closures, which are also contained in V_2 and have nonempty intersections with S. Proceeding in this way, construct a subset of S that is homeomorphic to the discontinuum D^*. The hypothesis of a compactum may be replaced by the weaker assumption that S be a subset of a metric space that is *complete* —that is, contains an accumulation point of each Cauchy sequence of points $p_1, p_2, \ldots$ (in which for each $d > 0$ there exists a point p_m having distances $< d$ from $p_{m+1}, p_{m+2}, \ldots$). That in a complete metric space each G_δ without isolated points has a subset homeomorphic to D^* was first proved by the British mathematician W. H. Young.

13.4   THE FRONTIERS OF OPEN SETS

Because of the role that frontiers of neighborhoods play in the particular form of the curve definition expounded here, a study of frontiers is important.

If Z is an open set, fr Z can be defined as cl $Z - Z$, or equivalently as cl $Z \cap$ com Z. Hence, if V and W are two open sets, then, using the De Morgan formulas, one obtains

$$\begin{aligned}
\mathrm{fr}(V \cup W) &= \mathrm{cl}(V \cup W) \cap \mathrm{com}(V \cup W) \\
&= (\mathrm{cl}\, V \cup \mathrm{cl}\, W) \cap (\mathrm{com}\, V \cap \mathrm{com}\, W) \\
&= (\mathrm{cl}\, V \cap \mathrm{com}\, V \cap \mathrm{com}\, W) \cup (\mathrm{cl}\, W \cap \mathrm{com}\, V \cap \mathrm{com}\, W) \\
&= (\mathrm{fr}\, V \cap \mathrm{com}\, W) \cup (\mathrm{fr}\, W \cap \mathrm{com}\, V).
\end{aligned}$$

Since fr V and fr W are subsets of com V and com W, respectively, we have

$$\begin{aligned}
\mathrm{fr}(V \cup W) &= (\mathrm{fr}\, V \cap \mathrm{com}\, V \cap \mathrm{com}\, W) \cup (\mathrm{fr}\, W \cap \mathrm{com}\, V \cap \mathrm{com}\, W) \\
&= (\mathrm{fr}\, V \cup \mathrm{fr}\, W) \cap \mathrm{com}\, V \cap \mathrm{com}\, W \\
&= (\mathrm{fr}\, V \cup \mathrm{fr}\, W) \cap \mathrm{com}(V \cup W).
\end{aligned}$$

We thus have established

LAW I   (FRONTIERS OF UNIONS) *For any two open sets, V and W,*

$$fr(V \cup W) = (fr\ V \cap com\ W) \cup (fr\ W \cap com\ V)$$
$$= (fr\ V \cup fr W) \cap com(V \cup W),$$

and

$$fr(V_1 \cup V_2 \cup \cdots \cup V_n)$$
$$= (fr\ V_1 \cup fr\ V_2 \cup \cdots \cup fr\ V_n) \cap com(V_1 \cup V_2 \cup \cdots \cup V_n)$$

for any finite family of open sets.

(The second half follows from the first by induction.)

Let W be an open set. From $W \subset cl\ W$ it follows that $com\ cl\ W \subset com\ W$, which implies $cl\ com\ cl\ W \subset cl\ com\ W = com\ W$ (since W is open). This justifies the following

Remark.   For each open set W, $com\ cl\ W \subset cl\ com\ cl\ W \subset com\ W$.

Now let V and W be two open sets and put $Z = V - cl\ W = V \cap com\ cl\ W$. Then

$$fr\ Z = cl(V \cap com\ cl\ W) \cap com(V \cap com\ cl\ W).$$

Since $cl(X \cap Y) \subset cl\ X \cap cl\ Y$, it follows that

$$fr\ Z \subset (cl\ V \cap cl\ com\ cl\ W) \cap (com\ V \cup cl\ W).$$

By the preceding remark, this set is a subset of $(cl\ V \cap com\ W) \cap (com\ V \cup cl\ W)$, which, by the distributive law, is

$$(cl\ V \cap com\ W \cap com\ V) \cup (cl\ V \cap com\ W \cap cl\ W)$$
$$= (fr\ V \cap com\ W) \cup (fr\ W \cap cl\ V).$$

We thus have established

LAW II   (FRONTIERS OF DIFFERENCES) *For any two open sets, V and W,*

$$fr(V - cl\ W) = fr(V \cap com\ cl\ W) \subset (fr\ V \cap com\ W) \cup (fr\ W \cap cl\ V).$$

If V is the space, then cl $V = V$ and fr V is empty, whence the following

COROLLARY (FRONTIERS OF COMPLEMENTS) *For any open set W,*

$$\text{fr(com cl } W) \subset \text{fr } W.$$

Consider, next, the intersection of two open sets, V and W.

$$\begin{aligned}
\text{fr}(V \cap W) &= \text{cl}(V \cap W) \cap \text{com}(V \cap W) \\
&\subset (\text{cl } V \cap \text{cl } W) \cap (\text{com } V \cup \text{com } W) \\
&= (\text{cl } V \cap \text{cl } W \cap \text{com } V) \cup (\text{cl } V \cap \text{cl } W \cap \text{com } W) \\
&= (\text{fr } V \cap \text{cl } W) \cup (\text{fr } W \cap \text{cl } V).
\end{aligned}$$

LAW III (FRONTIERS OF INTERSECTIONS) *For any two open sets, V and W,*

$$\text{fr}(V \cap W) \subset (\text{fr } V \cap \text{cl } W) \cup (\text{fr } W \cap \text{cl } V).$$

Laws I, II, III yield the following important

COROLLARY *For any two open sets, V and W, the sets $\text{fr}(V \cup W)$, $\text{fr}(V - \text{cl } W) = \text{fr}(V \cap \text{com cl } W)$, and $\text{fr}(V \cap W)$ are subsets of $\text{fr } V \cup \text{fr } W$.*

Law I, concerning the unions of finite families, has no direct extension to unions of denumerable families. If, in the three-dimensional euclidean space, V_n is the sphere about the point $(0, 0, 0)$ of radius $1 - 1/2^n$ for $n = 1, 2, \ldots$, then the union of all V_n is the concentric sphere of radius 1 whose frontier is disjoint from the union of the frontiers of all V_n. The following extension, however, is valid.

LAW IV (FRONTIERS OF DENUMERABLE UNIONS) *Let $\mathfrak{H}$ be a denumerable family of open sets. Denote their union by V, the union of their frontiers by F, and let M denote the set of all points p with the following property: each neighborhood of p contains points of infinitely many sets of the family $\mathfrak{H}$. Then $\text{fr } V \subset (M \cup F) \cap \text{com } V$.*

To establish this relation, note that cl V is the union of M and the closures of all sets in $\mathfrak{H}$. The intersection of the latter with com V is equal to the union of the frontiers of all sets in $\mathfrak{H}$ intersected with com V.

EXERCISES

1. In the plane, give examples of open sets W for which fr(com cl W) is a proper subset of fr W.

2. Under the assumptions of Law IV, give an example of a family H such that fr V is a proper subset of $(M \cup F) \cap \text{com } V$.

3. If A is closed and B is a closed subset of A, let V_1, V_2, ... be a sequence of open sets whose diameters approach 0, and whose union V contains $A - B$. Show that fr $V \subset B + F$, where F is the union of fr V_1, fr V_2, ...

4. Does the application of Law IV to a finite family H yield Law I?

13.5   THE SET OF END POINTS AND ITS COMPLEMENT

The proofs of many theorems contain references to open sets whose closures are subsets of other open sets. It is convenient to have at one's disposal a symbol for this relation. If V_1 and V_2 are open sets, we shall write

$$V_1 \Subset V_2 \qquad \text{if and only if} \qquad \text{cl } V_1 \subset V_2.$$

THEOREM 13.5   (ABOUT END POINTS AND POINTS OF ORDER >1)   *In each curve C the set C_1 of all points of order >1 is connected and dense in C, and any two points of C_1 lie in a subcontinuum of C_1. The set of all end points of C is a 0-dimensional G_δ.*

Since, according to Theorem 13.4, the set of all points of orders ≤ 1 is a G_δ and a continuum is free of points of order 0, in each curve C the set $C - C_1$ of all its end points is a G_δ. We next prove that this set is 0-dimensional. Let p be an end point of C, and Z any neighborhood of p such that there is a point p' of C not in cl Z. There exists a neighborhood V of p such that $V \Subset Z$ and that fr V has exactly one point q in common with C. We assume that q is an end point of C and derive a contradiction.

The assumption implies the existence of a neighborhood W of q whose frontier has exactly one point, say r, in common with C and may be assumed to be so small that its closure does not contain p and is contained in Z—that is, $W \Subset Z$. The point r cannot lie in fr V (which has only q in common with C). Hence r lies either in V or in com cl V.

In the first case, set $Y_1 = V \cup W$. By the law concerning the frontiers of unions,

$$\text{fr } Y_1 = (\text{fr } V \cup \text{fr } W) \cap \text{com}(V \cup W).$$

Hence $C \cap \text{fr } Y_1$ is the intersection of $\text{com}(V \cup W)$ with $C \cap (\text{fr } V \cup \text{fr } W)$. But the latter set consists of the two points q and r, neither of which lies in $\text{com}(V \cup W)$, since q lies in W, and p in V. Hence $C \cap \text{fr } Y_1$ is empty, which, in view of the connectedness of C, contradicts the assumption that a point p' of C lies outside of $\text{cl } Z$.

In the second case, set $Y_2 = V - \text{cl } W$. By the law concerning the frontiers of differences,

$$\text{fr } Y_2 \subset (\text{fr } V \cap \text{com } W) \cup (\text{fr } W \cap \text{cl } V).$$

Hence $C \cap \text{fr } Y_2$ is contained in the union of the intersection of $C \cap \text{fr } V$ with $\text{com } W$, and the intersection of $C \cap \text{fr } W$ with $\text{cl } V$. But $C \cap \text{fr } V$ consists of the point q, which lies in W; and $C \cap \text{fr } W$ consists of the point r, which lies in $\text{com cl } V$. Hence $C \cap \text{fr } Y_2$ is empty, which again leads to a contradiction. It thus follows that q cannot be an end point of C.

Hence Z contains a neighborhood of p whose frontier is disjoint from the set of all end points, and consequently $C - C_1$ is 0-dimensional at p. It further follows that each neighborhood of the end point p contains a point of C which is not an end point, and so C_1 is dense in C.

Since C is a continuum, by Theorem 12.2, any two points q and r of C lie in a subset K of C that is an irreducible continuum between q and r. This set K does not contain any end point other than possibly q and r (see Exercise 8, Section 13.2). If q and r are points of C_1, then K is a subset of C_1. Hence any two points of C_1 lie in a subcontinuum of C_1. According to Lemma 12.1, the set C_1 thus is connected. This completes the proof of the theorem.

The proof of the 0-dimensionality of the set of all end points is the simplest example of the *method of modification of open sets near their frontiers*, developed by Menger, which is very useful in curve theory and in dimension theory. The open set V has been modified near its frontier, by either adding or subtracting W in order to rid the frontier of the point q.

The theorem itself bears out our intuition about end points and points of higher order. In a graph (see Section 12.8), the end points (being finite in number) are altogether exceptional points. Even in a curve in the general sense, the end points (constituting a 0-dimensional set) are in a way more "thinly" spread than the other points, which constitute a dense, connected set, and therefore seem to form the bulk of the curve. The following example, however, will refine and, to some extent, correct our intuition by demonstrating that in the general sense the end points of a curve may also constitute a dense set, which, in a certain way, is even richer than its complement.

Example of a curve in which the end points are dense.   There exists a curve T* consisting only of points of orders 3, 2, and 1, in which the set of all end points is dense. Whereas the set T_1^* of the points of order >1 is the union of denumerably many segments each of which is nowhere dense in T*, the set $T^* - T_1^*$ of all end points is not the union of a countable family of sets that are nowhere dense in $T^* - T_1^*$.

Let K be any horizontal or vertical closed segment of length d in the plane. Closed segments perpendicular to K at the binary rational points of K will now be defined—segments pointing upward if K is horizontal, and to the right if K is vertical. For all natural numbers n and all numbers $m = 0, 1, \ldots,$ $2^{n-1} - 1$, let $\alpha_{n,m} K$ be the segment of this kind that begins at the point of K at distance $(2m + 1)d/2^n$ from an end point of K and whose length is $d/2^{n+1}$. Let βK be the union of all these segments. Clearly, βK lies inside the square with the diagonal K, and the squares with the diagonals $\alpha_{n,m} K$ ($n = 1, 2, \ldots$; $m = 0, 1, \ldots, 2^{n-1} - 1$) can easily be shown to be mutually disjoint.

Begin the construction of T^* with a horizontal segment H. Then βH is the union of the denumerable family of vertical segments $V_{n,m} = \alpha_{n,m} H$. For each $V_{n,m}$, the set $\beta V_{n,m}$ is the union of a denumerable family of horizontal segments. Since they lie in the mutually disjoint squares with diagonals $V_{n,m}$, the various sets $\beta V_{n,m}$ are mutually disjoint. Denote their union by $\beta^2 H$. Proceeding in this way one obtains, alternatingly, unions $\beta^{2k} H$ of horizontal segments and unions $\beta^{2k+1} H$ of vertical segments. Let N be the union of H and all the unions $\beta^k H$. Clearly, N is connected; but since N is the union of a denumerable family of segments each of which is nowhere dense in N, by Baire's Theorem 12.7, N is not a compactum and hence not closed. Because of the relation between the segments making up N and the squares mentioned in the construction, it is easy to prove that N contains only points of orders 3, 2, and 1. The points of order 3 are among the end points of the segments that constitute N, and so the set of these points is denumerable, and dense in N. So is the set of the other end points of those segments, which are points of order 1 in N. All other points of N are easily seen to be of order 2.

The closure of N is a continuum T^*. Each point of $T^* - N$ lies in arbitrarily small squares, mentioned in the construction, whose frontiers have two common points with N—a point of order 3 and a point of order 1. But each such square can be slightly enlarged to a quadrangle whose frontier has only the point of order 3 in common with T^*. Hence each point of $T^* - N$ is an end point of T^*. The set of all end points of T^* is dense in T^* and has no isolated points. Being a G_δ, it is not the union of countably many sets that are nowhere dense in $T^* - T_1^*$ (see Exercise 3).

EXERCISES

1. Construct a curve containing only points of order 4, 2, and 1, in which the set of all end points is dense.

2. Prove that if the set of all end points is dense in a curve then it is uncountable.

3. Following R. Baire one says that a set S is of *first* or *second category* according as S is or is not the union of a countable family of sets that are nowhere dense in S. Prove Baire's Theorem (of which Theorem 12.9 is a special case) that *each dense G_δ subset of a compactum (or of a complete metric space, see Exercise 5, Section 13.3) is of second category.*

4. Why would it be insufficient, in the proof of Theorem 13.5, to assume only $V \subset Z$ (rather than $V \Subset Z$) or $W \subset Z$?

13.6   THE IRREGULARITY SET

A family $\mathfrak{F}$ of compacta is said to be *additive* if the union of any finite family of members of $\mathfrak{F}$ belongs to $\mathfrak{F}$.

THEOREM 13.6 *If $\mathfrak{F}$ is a hereditary and additive family of compacta, then each point (if any) of the $\mathfrak{F}$-irregularity set of a compactum S lies in a subcontinuum of the $\mathfrak{F}$-irregularity set of S.*

A neighborhood W of a point or of a set will be called an $\mathfrak{F}$-neighborhood if $S \cap$ fr W belongs to the family $\mathfrak{F}$. Now assume that p is an $\mathfrak{F}$-irregular point of S, and define C_p as the set of all points q of S such that the closure of each $\mathfrak{F}$-neighborhood of q contains p. Clearly, C_p contains p and is a subset of S (in fact, of the $\mathfrak{F}$-irregularity set of S), since any $\mathfrak{F}$-regular point r of S lies in arbitrarily small $\mathfrak{F}$-neighborhoods whose closures do not contain

p, and so r does not lie in C_p. The set C_p will now be proved to be a continuum.

First of all, C_p is a compactum. Indeed, if r is a point of S not belonging to C_p, then, by definition of C_p, there exists an $\mathfrak{F}$-neighborhood W_r of r whose closure does not contain p. Since no point of C_p can lie in W_r, it follows that r is not an accumulation point of C_p. Hence C_p is closed in the compactum S and thus is a compactum. It remains to be shown that C_p is connected and contains more than one point. This will be deduced from

LEMMA 13.1   *If V is an open set whose closure does not contain p and whose frontier is disjoint from C_p, then there exists an $\mathfrak{F}$-neighborhood of V whose closure does not contain p.*

Since S is a compactum, so is the set $R = S \cap \text{fr } V$. If r is any point of R, then, since r does not belong to C_p, one can associate with r an $\mathfrak{F}$-neighborhood W_r whose closure does not contain p. A finite family of such associated neighborhoods, say, $\{W_{r_1}, W_{r_2}, \ldots, W_{r_n}\}$ covers the compactum R. Their union W and the set $X = V \cup W$ are open sets whose closures do not contain p, since none of the sets cl V and cl W_{r_1} do. By the law concerning the frontiers of unions,

$$\text{fr } X \subset (\text{fr } V \cap \text{com } W) \cup (\text{fr } W \cap \text{com } V).$$

$S \cap \text{fr } V$, being a subset of W, is disjoint from com W, whence $S \cap \text{fr } X$ is a subset of $S \cap \text{fr } W$, which in turn is a subset of the union of the n compacta $S \cap \text{fr } W_{r_i}$. Since they belong to $\mathfrak{F}$ and the family $\mathfrak{F}$ is additive and hereditary, $S \cap \text{fr } W$, and hence $S \cap \text{fr } X$, belong to $\mathfrak{F}$. It follows that X is an $\mathfrak{F}$-neighborhood of V whose closure does not contain p, and the Lemma is proved.

From the Lemma it follows that C_p is connected. For otherwise C_p would be the union of two nonempty, disjoint closed sets C_1 and C_2. Let C_1 be the set containing p, and let q be a point of C_2. The set V of all points that are closer to C_2 than to C_1 is then a neighborhood of C_2 and of q, whose closure is disjoint from C_1. Hence fr V is disjoint from C_1, as well as from C_2, which is a subset of V. Consequently fr V is disjoint from C_p. According to the Lemma, V (and hence q) is contained in an $\mathfrak{F}$-neighborhood whose closure does not contain p. This is impossible, since q is a point of C_p.

The Lemma further implies that p is not the only point of C_p. For if it were, then each neighborhood Z of p would contain a neighborhood Y of p whose frontier would be disjoint from C_p. Set $V = \text{com cl } Y$. By the Corollary about the frontiers of complements, fr $V \subset$ fr Y, whence the frontier of V would be disjoint from C_p. The Lemma implies the existence of a neighborhood X of V whose closure does not contain p. If $Z =$ com cl X (containing p), then, by the corollary just quoted, fr $Z \subset$ fr X; whence also Z is an $\mathfrak{F}$-neighborhood of p. But the existence of arbitrarily small $\mathfrak{F}$-neighborhoods of p contradicts the assumption that p is $\mathfrak{F}$-irregular.

This completes the proof of the theorem that p lies in a subcontinuum of the $\mathfrak{F}$-irregularity set of S, which in this form is due to the Polish-born mathematician Witold Hurewicz (1904–1956).

We apply the theorem first to the (hereditary and additive) family $\mathfrak{F}$ consisting of the empty set. The $\mathfrak{F}$-irregularity set of a set S is the set S^+ of all points of S at which S is of positive dimension (or positive order). From Theorem 13.6, it follows that if C is a compactum, then each point of C^+ lies in a subcontinuum of C^+. Consequently, if C is discontinuous, then C^+ is empty; in other words, a nonempty discontinuous compactum is 0-dimensional. Combining this result with the theorem that each 0-dimensional set is disconnected, and hence also discontinuous (see Exercise 7, Section 13.2) one obtains the following result.

THEOREM 13.7   (ABOUT DISPERSED COMPACTA)   *In the class of the nonempty compacta, 0-dimensionality, disconnectedness, and discontinuity are equivalent; that is to say, each compactum having one of these three properties has the other two also.*

Applying Theorem 13.6 to the families of all finite and all countable sets (which are hereditary and additive) one obtains

THEOREM 13.8   (ABOUT REGULARITY AND RATIONALITY SETS OF CURVES)   *The irregularity set and the irrationality set of a compactum contain, for each of their points, a subcontinuum containing that point.*

EXERCISES

1. Prove that each point at which a compactum C is of positive dimension lies in a subcontinuum of C consisting of such points. Is the fact that each point of C of order >1 lies in a subcontinuum of C consisting of points of order >1 a consequence of Theorem 13.6?

2. Prove that each open set containing a point of a continuum C contains a subcontinuum of C. (It seems plausible that even each open set containing a point of a *connected* set S contains a *connected* subset of S with more than one point. But this statement is not in general valid. It fails, for example, for a set S, constructed by Knaster and Kuratowski, which is obtained by deleting from Cantor's brush (see Section 12.8) all points with irrational ordinates from the bristles joining $p = (\frac{1}{2}, 1)$ with points of D_2^* (see p. 396) and all points with rational ordinates <1 from the bristles leading to points of D_1^*. It is easy to see that after the deletion of the point p the set S becomes disconnected, whence no open set that does not contain p contains a connected subset of S with more than one point. Harder to prove is the astounding fact that S is connected.)

3. Each closed discontinuous subset of a continuum C is nowhere dense in C. (*Hint:* Use the result of Exercise 2.)

4. A compactum of positive dimension is not the union of a countable family of closed 0-dimensional sets. (*Hint:* Use the result of Exercise 3.)

5. Show that the segment [0, 1] is the union of two 0-dimensional sets, and that a square is the union of three 0-dimensional sets. (*Hint:* For $r = 0, 1, 2$, denote by $S_{r, 2-r}$ the subset of the plane consisting of all points with r rational and $2 - r$ irrational coordinates. Each point of $S_{r, 2-r}$ lies in very simple neighborhoods whose frontiers are disjoint from $S_{r, 2-r}$.) *Note:* A cube is the union of four 0-dimensional sets, for example, of the analogous sets $S_{r, 3-r}$ for $r = 0, 1, 2, 3$. For $r = 0, 2, 3$ each point of $S_{r, 3-r}$ lies in arbitrarily small cubes whose frontiers are disjoint from $S_{r, 3-r}$. But the 0-dimensionality of $S_{1, 2}$ is much harder to prove, since this set has points in common with every surface consisting of the points $(x, y\ F(x, y))$ for all (x, y) in a square domain, where F is any function that is continuous in that square. Similarly, $S_{1, 2}$ has common points with each analogous set of points $(x, F(x, z), z)$ and $(F(y, z), y, z)$. This fact was noticed by the Austrian mathematician Otto Schreier.

13.7   THE IRRATIONALITY SET

The set of all irregular points of a continuum and even its closure are not necessarily irregular; for example, the irregularity set of a sinusoid is a closed segment. Nor is, as a more subtle example will demonstrate, the set of all irrational points of a continuum necessarily irrational; but its closure is. This follows from a general theorem that has also other applications.

The theorem deals with $\mathfrak{F}$-irregularity sets with regard to hereditary families $\mathfrak{F}$ of compacta that are not only additive but *σ-additive* in the sense *that each compactum that is the union of a countable family of members of $\mathfrak{F}$ belongs to $\mathfrak{F}$*. In any space, the family of all countable compacta is σ-additive.

The additive family of all finite subsets of a space containing an accumulation point is not σ-additive. The study of regularity with regard to hereditary and σ-additive families of sets was initiated by Hurewicz. The following theorem has not been heretofore proved in its full generality.

THEOREM 13.9   (ABOUT THE CLOSURE OF $\mathfrak{F}$-IRREGULARITY SETS)   *If $\mathfrak{F}$ is a hereditary, σ-additive family of compacta, C a compactum and C^* the $\mathfrak{F}$-irregularity set of C, then the closure of C^* is $\mathfrak{F}$-irregular at each point of C^*.*

The proof (by the method of modifying neighborhoods near their frontiers) is based on the following.

COVERING THEOREM FOR DIFFERENCES OF COMPACTA   *Let A be a compactum; B, a closed subset of A; Z, an open set containing A; and $\mathfrak{W}$, a family of open sets whose kernel (see Section 13.3) contains $A - B$. Then $\mathfrak{W}$ includes a countable family $\mathfrak{B}$ of sets such that $A - B$ is contained in their union and that B contains all points p with the property that every neighborhood of p has points with infinitely many sets of the family $\mathfrak{B}$ in common.*

For each natural number n, let Y_n be the set of all points of A having a distance $<1/n$ from B. Clearly, the set $X_n = A \cap (\text{cl } Y_n - Y_{n+1})$ is a compactum $\subset A - B$. With each point of X_n one can associate a neighborhood belonging to $\mathfrak{W}$ with diameter $<1/(n + 1)$. Each point of such a neighborhood has a distance $<1/n + 1/(n + 1) < 2/n$ from B. A finite family $\mathfrak{B}_n$ of such open sets covers the compactum X_n. The union $\mathfrak{B}$ of the families $\mathfrak{B}_1, \mathfrak{B}_2, \ldots, \mathfrak{B}_n, \ldots$ covers $A - B$. Let p be a point not belonging to B, and let d be the positive distance between p and B. If $2/n < d$, the sphere about p of radius d is disjoint from all the open sets belonging to $\mathfrak{B}_n, \mathfrak{B}_{n+1}, \ldots$. The only sets belonging to $\mathfrak{B}$ with which this sphere may have common points belong to $\mathfrak{B}_1, \mathfrak{B}_2, \ldots, \mathfrak{B}_{n-1}$. This completes the proof of the Covering Theorem.

Turning to the proof of Theorem 13.9 we show that if p is a point of C at which cl C^* is $\mathfrak{F}$-regular, then also C is $\mathfrak{F}$-regular at p (so that p does not belong to C^*). Let Z be any neighborhood of p. We must demonstrate the existence of a neighborhood W of p such that $W \subset Z$ and that $C \cap \text{fr } W$ belongs to $\mathfrak{F}$.

Since cl C^* is $\mathfrak{F}$-regular at p there certainly exists a neighborhood X of p such that $X \in Z$ and that cl $C^* \cap \text{fr } X$ belongs to $\mathfrak{F}$.

Each point of $(C - \operatorname{cl} C^*) \cap \operatorname{fr} X$, being an $\mathfrak{F}$-regular point of C, lies in arbitrarily small neighborhoods $\subset Z$ whose frontiers have intersections with C that belong to $\mathfrak{F}$. By virtue of the Covering Theorem for Differences of Compacta, there exists a countable family $\mathfrak{B} = \{V_1, V_2, \ldots\}$ of such open sets covering $(C - \operatorname{cl} C^*) \cap \operatorname{fr} X$ and such that each point every neighborhood of which has common points with infinitely many sets belonging to $\mathfrak{B}$ lies in $\operatorname{cl} C^* \cap \operatorname{fr} W$. Let V be the union of all V_n. By virtue of the law about the frontiers of denumerable unions in Section 13.4, $\operatorname{fr} V$ is a subset of the union of $\operatorname{cl} C^*$ and all $\operatorname{fr} V_n$. Since $\operatorname{cl} C^* \cap \operatorname{fr} W$ as well as all sets $C \cap \operatorname{fr} V_n$ belong to the hereditary and σ-additive family $\mathfrak{F}$, so does the compactum $C \cap \operatorname{fr} V$. We set $X \cup V = W$ and prove that W has the desired properties. Clearly, W contains p and $W \subset Z$. According to the law about the frontiers of unions,

$$\operatorname{fr} W = \operatorname{fr}(X \cup V) = (\operatorname{fr} X \cap \operatorname{com} V) \cup (\operatorname{fr} V \cap \operatorname{com} X).$$

Hence $C \cap \operatorname{fr} W = (C \cap \operatorname{fr} X \cap \operatorname{com} V) \cup (C \cap \operatorname{fr} V \cap \operatorname{com} X)$. But

$$(C - \operatorname{cl} C^*) \cap \operatorname{fr} X \subset V.$$

Consequently, $C \cap \operatorname{fr} X \cap \operatorname{com} V \subset \operatorname{cl} C^*$. Since the family $\mathfrak{F}$ is hereditary the compactum $C \cap \operatorname{fr} X \cap \operatorname{com} V$, being a subset of the member $\operatorname{cl} C^*$ of $\mathfrak{F}$, belongs itself to $\mathfrak{F}$, as does $C \cap \operatorname{fr} V$. It follows that $C \cap \operatorname{fr} W$ belongs to $\mathfrak{F}$. This concludes the proof of the theorem.

The application to the family of all countable compacta yields

THEOREM 13.10 (about the closure of irrationality sets)   *The closure of the irrationality set C^* of a compactum C is irrational at each point of $\operatorname{cl} C^*$.*

It should be noted, however, that $\operatorname{cl} C^*$ may be rational and even regular at points of $\operatorname{cl} C^* - C^*$ (see Exercise 1).

Let $\mathfrak{D}$ be the family of all discontinuous (or 0-dimensional) compacta. The $\mathfrak{D}$-regular continua are the curves. More generally, according as p is a $\mathfrak{D}$-regular or $\mathfrak{D}$-irregular point of the compactum C we say that C is *at most 1-dimensional* or *higherdimensional* at p. A set is said to be *higherdimensional* if it contains at least one point at which it is higherdimensional. The family $\mathfrak{D}$ is hereditary and σ-additive (see Exercise 4, Section 13.6). Hence Theorem 13.9 has the following

COROLLARY   *If C is a higherdimensional compactum, then the closure of the set C* of all points at which C is higherdimensional is higherdimensional at each point of C*.*

From a general dimensiontheoretical result of Menger it follows that even the set *C** *itself* is higherdimensional at each point of *C**. This assertion can be based on the fact that each at most 1-dimensional set is contained in an at most 1-dimensional G_δ (see Exercise 2). A countable set, however, is not necessarily contained in a countable G_δ; and this is why irrational curves can exist whose irrationality set is rational even though its closure is irrational. Such compacta are called *weakly irrational*. The following example was discovered by Knaster and Mazurkiewicz.

Example of a weakly irrational curve.   There exists in the plane an irrational curve *C* whose irrationality set *C** is the union of a denumerable family of straight segments, and for which *C − C** is the set of the end points of *C*.

Being the union of a countable family of segments, *C** is rational (see Exercise 11, Section 13.2). It is of course easy to construct a sequence of segments whose union, *C**, has an irrational closure. But it is difficult to construct the sequence in such a way that *C** is the entire irrationality set of *C* = cl *C**—in other words, that no point of *C − C** is an irrational point of *C*. In the following example, all points of *C − C** are end points of *C*.

The basic idea of the definition of *C* is the iteration of a process that may be described as *attaching two symmetric fans to the end points of a segment*. Let *D* be the segment (of length *d*) between the points *a'* and *a''*. The first fan consists of a sequence of segments $D'_1, D'_2, \ldots$ of length $\frac{2}{3}d$ issuing from *a'*, all of them on one and the same side of *D*, and such that if D'_1 and *D* form the angle *α*, then D'_m and *D* form the angle $\alpha/2^{m-1}$. The second fan is symmetric to the first with regard to the midpoint of *D* and consists of segments $D''_1, D''_2, \ldots$ issuing from *a''* on the other side of *D*. It will be convenient to unite the two sequences in one by setting

$$D'_m = D_{2m-1} \quad \text{and} \quad D''_m = D_{2m} \quad \text{for } m = 1, 2, \ldots.$$

In iterating this procedure (that is, in attaching pairs of fans to all the segments D_n) we want to keep the various 2nd order fans mutually disjoint. In order to achieve this we attach to the end points of *D* fans whose blades are *rhombuses* $R_1, R_2, \ldots$ (that is, closed sets consisting of open parallelograms and their four sides of equal length) such that D_n is the longer diagonal of R_n and that $R_1, R_2, \ldots$ are mutually disjoint except for the point a' in the R_{2m-1} and the point a'' in the R_{2m}. Then the choice of the 2nd order fans attached to D_n inside of R_n guarantees their mutual disjointness except for the points a' and a''. It should be clear, however, that the rhombuses play only an auxiliary role and that their sides will not be subsets of the curve to be defined.

For the sake of a smooth definition of the iterative process we shall assume that the given segment D itself is the longer diagonal of a rhombus R. Let ω be the angle between D and a side of R. We then choose the angle α between D'_1 and D less than $\omega/2$ and so small that the segment D'_1 of length $\frac{2}{3}d$ is contained in R. (Such a choice is obviously possible.) We then set $D_m = \delta_m(R)$ and denote by $R_m = \rho_m(R)$ a rhombus whose longer diagonal is D_m and whose sides form with D_m the angle $\alpha/2^{m+1}$. Since the diagonals D_m of R_m and D_{m+1} of R_{m+1} form an angle of $\alpha/2^m$ it is clear that R_m and R_{m+1} have no common point other than the end point of D to which they are attached. Use will be made also of the following simple

Remark 1.   The union of D and the rhombuses $R_m = \rho_m(R)$ is a closed set.

The starting point of the definition of C consists of a square, R_0, and one of its diagonals, D_0, whose length is 1. In the first step of the construction, we obtain the auxiliary rhombuses $\rho_{n_1}(R_0) = R_{n_1}$ for $n_1 = 1, 2, \ldots$, and their diagonals $\delta_{n_1}(R_0) = D_{n_1}$ of length $\frac{2}{3}$. In the second step we construct, for each R_{n_1}, the auxiliary rhombuses $\rho_{n_2}(R_{n_1}) = R_{n_1, n_2}$ for $n_2 = 1, 2, \ldots$ and their diagonals $\delta_{n_2}(R_{n_1}) = D_{n_1 n_2}$. At the end of the kth step, the rhombuses $R_{n_1, n_2, \ldots, n_k}$ and their diagonals $D_{n_1, n_2, \ldots, n_k}$ of length $(2/3)^k$ are present. In the $(k + 1)$th step, we construct for each of these rhombuses the rhombuses $\rho_{n_{k+1}}(R_{n_1, n_2, \ldots, n_k}) = R_{n_1, \ldots, n_k, n_{k+1}}$ for $n_{k+1} = 1, 2, \ldots$, and their diagonals $D_{n_1, \ldots, n_k, n_{k+1}}$ of length $(2/3)^{k+1}$.

For each natural number k, let D'_k be the union of D_0 and all segments $D_{n_1, n_2, \ldots, n_k}$ with at most k indices. From Remark 1 it readily follows that D'_k is a closed set for each k. The union of all D'_k (that is, the union of all diagonals obtained in *all* steps) is the set C' whose closure is a continuum C that will be proved to be weakly irrational.

Another way to obtain C is as follows. Let R'_1 be the union of D_0 and all rhombuses R_{n_1} with one index; and, more generally, let R'_k be the union of D'_{k-1} and all rhombuses $R_{n_1, \ldots, n_k}$ with k indices. From Remark 1 it follows that the set R'_k is closed for each k. So is, therefore, the intersection C of all R'_k. From the definition of R'_{k-1}, R'_k, and D'_{k-1} it follows that $R'_{k-1} \supset R'_k \supset D'_{k-1}$ for each k. In particular, $R'_k \supset D'_k$, whence

$R'_{k-1} \supset D'_k$. One readily concludes that, more generally, $R'_k \supset D'_h$ for each k and h. Consequently, $C \supset D'$.

There actually exist points of $C - D'$. Consider, for example, the point p, which is the intersection of the rhombuses:

$$R_1, R_{1,2}, R_{1,2,1}, R_{1,2,1,2}, \ldots.$$

Clearly, p lies in C. But p does not lie in D_0, since p lies in $R_{1,2}$; nor in D'_1, since p lies in $R_{1,2,1}$; nor in D'_2, since p lies in $R_{1,2,1,2}$; and so on. Not lying in any D'_k, the point p does not lie in their union D'.

Let p be any point of $C - D'$. Then p is an accumulation point of D'. Indeed, p lies in rhombuses with arbitrarily many indices, and the length of any diagonal $D_{n_1, \ldots, n_k}$ is $(2/3)^k$. Hence $C = \text{cl } D'$. It is furthermore clear that p lies in the interior of those rhombuses, since the only points on their frontiers belonging to C are the two end points of the longer diagonals; and these two points belong to D'. Thus p lies in arbitrarily small open parallelograms whose frontiers have exactly two common points with C. Thus p is a point of order ≤ 2.

But it is easy to show that p is an end point of C. For the rhombus $R_{n_1, \ldots, n_{k-1}, n_k}$ can be replaced by a slightly larger quadrangle whose frontier has only one point in common with C—namely, the point where the rhombus is attached to $D_{n_1, \ldots, n_k}$.

It remains to be shown that C is irrational. Actually, C' is the irrationality set of C; but it is sufficient to prove that one point of C', for example the end point a' of D, is irrational. Now for each small neighborhood V of a' there exist two disjoint rhombuses of the first step, R_{i_1} and R_{j_1}, which have points in common with V as well as with com cl V and hence with fr V. We denote these rhombuses by T_0 and T_1. Each of them contains two disjoint rhombuses of the second step that have common points with fr V. We denote them by T_{00}, T_{01} and T_{10}, T_{11}. Continuing in this way, for each n we obtain 2^n disjoint rhombuses having common points with fr V. Let S_n be the union of these rhombuses. The set $S_1 \cap S_2 \cap \ldots \cap S_n \cap \ldots$ clearly is a subset of $C \cap \text{fr } V$ that is homeomorphic to Cantor's discontinuum and hence uncountable. It follows that a' is an irrational point of C.

EXERCISES

1. Give a simple example of a continuum C for which the closure of the irrationality set C' is regular at a point of cl $C' - C'$. (The weakly irrational curve is a complicated example of a more complex phenomenon.)

2. Prove that each 0-dimensional set is contained in a 0-dimensional G_δ.

3. Each F_σ subset S of a compactum has the following property: If S is contained in the kernel of a family $\mathfrak{W}$ of open sets, then for each $d > 0$, S is contained in the union of a sequence of sets of $\mathfrak{W}$ whose diameters are $< d$ and converge to 0. Give an example of a G_δ set S without this property.

13.8   THE UNION OF CURVES

The results of Exercises 3 and 4 in Section 13.6 are capable of broad generalizations.

LEMMA 13.2   *Let $\mathfrak{F}$ be a hereditary family. If C is a compactum in which the $\mathfrak{F}$-irregularity set of C is dense, then each $\mathfrak{F}$-regular closed subset of C is nowhere dense in C, and C is not the union of a countable family of $\mathfrak{F}$-regular closed subsets.*

The second assertion follows from the first because of Baire's Theorem 12.7. To prove the first assertion let B be an $\mathfrak{F}$-regular closed subset of C, and Z any open set that is not disjoint from C. We must prove that Z has an open subset Z' that is disjoint from B but not from C. Since, by assumption, the $\mathfrak{F}$-irregularity set C' is dense in C, there exists in Z a point p of C'. Being $\mathfrak{F}$-irregular, p has a neighborhood $W \subset Z$ such that for no neighborhood $V \subset W$ of p does the set $C \cap \mathrm{fr}\ V$ belong to $\mathfrak{F}$. If p does not belong to the closed set B, then some neighborhood of p is disjoint from B and satisfies the conditions concerning Z'. If p does belong to the $\mathfrak{F}$-regular set B, then there exists a neighborhood $V \subset W$ of p such that $B \cap \mathrm{fr}\ V$ belongs to $\mathfrak{F}$. But since $V \subset W$, the set $C \cap \mathrm{fr}\ V$ does not belong to the hereditary family $\mathfrak{F}$ and consequently is not a subset of B. It contains therefore a point q that does not belong to B. But then some neighborhood Z' of q is disjoint from the closed set B.

If $\mathfrak{F}$ is a hereditary and σ-additive family of compacta, and K is an $\mathfrak{F}$-irregular compactum, then, according to Theorem 13.9, the closure of the $\mathfrak{F}$-irregularity part K' of K is $\mathfrak{F}$-irregular in each point of K'. Consequently, C is a compactum in which the $\mathfrak{F}$-irregularity set of C is dense. According to Lemma 13.2, C is not the union of any countable family of $\mathfrak{F}$-regular sets. Since $C \subset K$ neither is K such a union, and we have the following result.

THEOREM 13.11   *If $\mathfrak{F}$ is a hereditary, σ-additive family of sets, then an $\mathfrak{F}$-irregular compactum is not the union of a countable family of $\mathfrak{F}$-regular compacta; in other words, if the union of a countable family of $\mathfrak{F}$-regular compacta is a compactum, then it is $\mathfrak{F}$-regular. In particular, the union of every finite family of $\mathfrak{F}$-regular compacta is $\mathfrak{F}$-regular.*

Applying this theorem to the families of all discontinuous and all countable compacta we obtain the following

COROLLARY *If the union S of a countable family of 1-dimensional compacta is a compactum, then S is 1-dimensional. In particular, if S is a continuum, then S is a curve. The curve S is rational if all members of the family are rational. If the union of a finite family of curves (of rational curves) is connected, then it is a curve (a rational curve).*

The sinusoid is the union of a countable family of regular curves and yet irregular. The question arises, however, *whether a curve that is the union of a finite family of regular curves is necessarily regular?* Urysohn found that the answer to this question is negative.

Example of two regular curves whose union is irregular.   Let S be the union of the following families of segments H_n and $K_{n,m}$.

The horizontal segment	is the set of all points
H_1	$(x, \frac{1}{2})$ for $-\frac{1}{2} \leq x \leq \frac{1}{2}$,
H_2	$(x, \frac{1}{4})$ for $-\frac{3}{4} \leq x \leq \frac{3}{4}$,
$\cdots$	$\cdots$
H_n	$\left(x, \dfrac{1}{2^n}\right)$ for $-1 + \dfrac{1}{2^n} \leq x \leq 1 - \dfrac{1}{2^n}$ $(n = 1, 2, \ldots)$
H_∞	$(x, 0)$ for $-1 \leq x \leq 1$

The vertical segment	begins at the point of H_∞ with the abscissa	is of length
K_0	0	1
$K_{1,-1}$ $\quad$ $K_{1,1}$	$-\frac{1}{2},\ \frac{1}{2}$	$\frac{1}{2}$
$K_{2,-2},\ K_{2,-1},\ K_{2,1},\ K_{2,2}$	$-\frac{3}{4},\ -\frac{1}{4},\ \frac{1}{4},\ \frac{3}{4}$	$\frac{1}{4}$
$\cdots$	$\cdots$	$\cdots$
$K_{n,-2^n-1},\ K_{n,-2^{n-1}+1},\ \ldots,\ K_{n,2^n-1}$	$-1 + \dfrac{1}{2^{n-1}},\ -1 + \dfrac{3}{2^{n-1}},\ \ldots, 1 - \dfrac{1}{2^{n-1}}$	$\dfrac{1}{2^n}$

Thus $K_{n,m}$ is the set of all points

$$\left(\frac{2m-1}{2^n}, y\right)$$

for $0 \le y \le 1/2^n$ for $m = 1, 2, \ldots, 2^{n-1}; 0, -1, -2, \ldots, -2^{n-1}-1$, and $n = 1, 2, \ldots$.

Clearly S is regular at each point not on H_∞. But each point p of H_∞ is an irregular point of S, since the frontier of every sufficiently small neighborhood of p intersects infinitely many segments H_n. Even though S is irregular, S is the union of two regular curves R_1 and R_2, which will now be defined in terms of the segments H_n and $K_{n,m}$.

The upper end points of $K_{1,-1}$ and $K_{1,1}$ coincide with the end points of H_1. But for $n > 1$, the upper end points of the 2^n segments $K_{n,m}$ divide H_n into $2^n - 1$ segments. For example, the end points of the segments $K_{2,m}$ divide H_2 into the three segments for which

$$-\tfrac{3}{4} \le x \le -\tfrac{1}{4}, \quad -\tfrac{1}{4} \le x \le \tfrac{1}{4}, \quad \tfrac{1}{4} \le x \le \tfrac{3}{4}.$$

They will be denoted by $H_{2,-1}$, $H_{2,0}$, and $H_{2,1}$, respectively. Similarly, denote the $2^n - 1$ segments into which the upper end points of the segments $K_{n,m}$ divide H_n by

$$H_{n,-1-2^{n-1}}, \ldots, H_{n,-1}, H_{n,0}, H_{n,1}, \ldots, H_{n,2^{n-1}-1}.$$

Let Q_1 (which is shaped like a family tree, if each member of the family has exactly two children) be the union of the following segments:

1. the upper halves of all the vertical segments $K_{n,m}$;
2. the segments $H_{n,m}$ for all odd integers m and all n.

We define R_1 as the closure of Q_1, which is $Q_1 \cup H_\infty$.

Let Q_2 (which somewhat resembles the front view of a cemetery with crosses behind one another) be the union of the lower halves of all $K_{n,m}$ and the segments $H_{n,m}$ for all even integers m and all n.

R_2 is defined as the closure of Q_2, which is $Q_2 \cup H_\infty$.

Clearly, $R_1 \cup R_2 = S$, and R_1 and R_2 are regular at all points outside H_∞. But R_1 and R_2 (in contrast to S) are also regular at the points of H_∞. Each point of H_∞ lies in arbitrarily small circular neighborhoods each of which has its center at some point $(k/2^n, 0)$ and a radius $1/2^{n+k}$ ($k > 2$). The frontier of such a circle has four points in common with R_1 (two of them with H_∞). Each point of H_∞ lies also in arbitrarily small rectangular neighborhoods two of whose sides are vertical segments with abscissae

$$\frac{k}{2^n} \pm \frac{1}{2^{n+1}} \mp \frac{1}{2^{n+2}} \pm \mp \cdots = \frac{1}{2^n}(k + \tfrac{1}{3})$$

for some integer k between -2^{n-1} and 2^{n-1}, and whose frontiers have finite intersections with R_2. This completes the proof that R_1 and R_2 are regular curves.

We are now going to sharpen the assertion of Theorem 13.11 concerning the $\mathfrak{F}$-regularity of the union of a countable family of $\mathfrak{F}$-regular compacta by dropping the restriction that the union be a compactum.

THEOREM 13.12 (about the union of $\mathfrak{F}$-regular compacta) *If $\mathfrak{F}$ is a hereditary and σ-additive family of compacta, then the union of any countable family of $\mathfrak{F}$-regular compacta is $\mathfrak{F}$-regular.*

We illustrate the method of the proof first in the simplest case, where $\mathfrak{F}$ consists of the empty set alone, and the nonempty $\mathfrak{F}$-regular sets are 0-dimensional. Let $(N_1, N_2, \ldots)$ be a sequence of 0-dimensional compacta and let N denote their union. We shall show that, for any point p of N, each neighborhood Z_0 of p contains a neighborhood of p whose frontier is disjoint from N. It is clear that p has a neighborhood $V_1 \Subset Z_0$ whose frontier is disjoint from N_1. There exists (see Exercise 6) an open set Z_1 such that $V_1 \Subset Z_1 \Subset Z_0$ and that the set N_1 is disjoint from the frontier of every open set W *between* V_1 and Z_1 (that is, such that $V_1 \subset W \subset Z_1$).

Assume inductively the existence of two open sets V_k and Z_k such that $V_k \Subset Z_k \Subset Z_{k-1}$ and that the frontier of each open set between V_k and Z_k is disjoint from $S_k = N_1 \cup N_2 \cup \ldots \cup N_k$. There exists (see Exercise 5) an open set V_{k+1} such that $V_k \Subset V_{k+1} \Subset Z_k$ and fr V_{k+1} is disjoint from N_{k+1} and hence from S_{k+1}. In the same way as Z_1 has been obtained, one obtains an open set Z_{k+1} which, together with V_{k+1}, satisfies the inductive assumption for $k + 1$.

Now let W denote the union of all open sets of the increasing sequence $(V_1, V_2, \ldots)$. Clearly, $V_k \Subset W \Subset Z_k \subset Z_0$ for each k, and so fr W is disjoint from each S_k and hence from N. This completes the proof of the fact that *the union of any sequence of* 0-*dimensional compacta is* 0-*dimensional.*

Turning to the proof of Theorem 13.12 we consider a sequence of $\mathfrak{F}$-regular compacta $(C_1, C_2, \ldots)$. For each k we set $S_k = C_1 \cup C_2 \cup \ldots \cup C_k$ and denote by S the union of all C_i. Let p be a point of S, and Z_0 a neighborhood of p. We must prove the existence of a neighborhood $W \subset Z_0$ of p such that $S \cap \operatorname{fr} W$ belongs to $\mathfrak{F}$. There certainly exists a neighborhood $V_1 \Subset Z_0$ such that the set $C_1 \cap \operatorname{fr} V_1$, which we shall call A_1, belongs to $\mathfrak{F}$. Except for the case where $\mathfrak{F}$ consists of the empty set alone, however, there does not necessarily exist an open set Z_1 such that $V_1 \Subset Z_1 \Subset Z_0$ and that $C_1 \cap \operatorname{fr} W$ belongs to $\mathfrak{F}$ for every W between V_1 and Z_1. But let Z_1 be the (open) set of all points of Z_0 that are closer to V_1 than to $(C_1 \cap \operatorname{com} V_1) \cup \operatorname{com} Z_0$. Then it is easy to see that $\operatorname{cl} V_1 - A_1 \subset Z_1$ and that $C_1 \cap \operatorname{fr} W$ belongs to $\mathfrak{F}$ for any open set W between V_1 and Z_1. Indeed, each point p of $\operatorname{fr} W$ is at most as far from $C_1 \cap \operatorname{com} V_1$ as from V_1. If p belongs to C_1 it has the distance 0 from $C_1 \cap \operatorname{fr} V_1 = A_1$. Hence $C_1 \cap \operatorname{fr} W$ is a subset of A_1 and thus belongs to $\mathfrak{F}$.

The set $A_2 = C_2 \cap \operatorname{fr} V_1$ of course need not belong to $\mathfrak{F}$. But by modifying V_1 near its frontier we shall obtain a set $V_2 \supset V_1$ such that (1) $\operatorname{cl} V_2 - A_1 \subset Z_0$, whence $C_1 \cap \operatorname{fr} V_2$ belongs to $\mathfrak{F}$, and (2) $C_2 \cap \operatorname{fr} V_2$ belongs to $\mathfrak{F}$. Consequently, $(C_1 \cup C_2) \cap \operatorname{fr} V_2$ will belong to $\mathfrak{F}$. In order to construct V_2 we set $C_2 \cap C_1 \cap \operatorname{fr} V_1 = B_2$. Clearly, B_2 belongs to $\mathfrak{F}$. Each point of $A_2 - B_2$ lies in arbitrarily small neighborhoods (in particular, in neighborhoods $\Subset Z_1$) whose frontiers intersect C_2 in sets belonging to $\mathfrak{F}$.

By the covering theorem for differences of compacta in Section 13.7, $A_2 - B_2$ can be covered with a sequence $(Y_1, Y_2, \ldots)$ of such neighborhoods in such a way that each point, every neighborhood of which has points in common with infinitely many Y_k, lies in B_2. Let Y be the union of all Y_k. By Law IV in Section 13.4 concerning the frontiers of denumerable unions, $\operatorname{fr} Y$ is a subset of the union of B_2 and the sets $\operatorname{fr} Y_k$. The compactum $C_2 \cap \operatorname{fr} Y$, therefore, is a subset of the union of B_2 and the sets $C_2 \cap \operatorname{fr} Y_k$, all of which are compacta belonging to the hereditary and σ-additive family $\mathfrak{F}$. Consequently, $C_2 \cap \operatorname{fr} Y$ belongs to $\mathfrak{F}$. Since $Y_k \Subset Z_1$ for each k, we have $\operatorname{cl} Y - B_2 \subset Z_1$.

Now set $V_2 = V_1 \cup Y$. Then $V_1 \subset V_2 \subset Z_1$ and $\operatorname{cl} V_2 - A_2 \subset Z_1$. Just as in the proof of Theorem 13.9, we conclude that

$$C_2 \cap \operatorname{fr} V_2 = (C_2 \cap \operatorname{fr} V_1 \cap \operatorname{com} Y) \cup (C_2 \cap \operatorname{fr} Y \cap \operatorname{com} V_1).$$

Both compacta whose union is $C_2 \cap \operatorname{fr} V_2$ belong to $\mathfrak{F}$: the first, because it is part of $A_2 \cap \operatorname{com} Y$ and (since $A_2 - B_2 \subset Y$) also part of B_2, which belongs to $\mathfrak{F}$; and the second, because it is part of $C_2 \cap \operatorname{fr} Y$, which belongs to $\mathfrak{F}$. Consequently, $C_2 \cap \operatorname{fr} V_2$ belongs to $\mathfrak{F}$. Since V_2 is between V_1 and Z_1 we see that also $C_1 \cap \operatorname{fr} V_2$ belongs to $\mathfrak{F}$. So does, consequently, $S_2 \cap \operatorname{fr} V_2$ which we call A_2.

In the same way as we obtained Z_1 from V_1 we obtain an open set $Z_2 \subset Z_1$ such that cl $V_2 - A_2 \subset Z_1$ and that $S_2 \cap$ fr W belongs to $\mathfrak{F}$ for each open set W between V_2 and Z_2.

Inductively, we assume that two open sets V_k and $Z_k \subset Z_{k-1}$ be given such that cl $V_k - A_k \subset Z_k$ (where $A_k = S_k \cap$ fr V_k) and that $S_k \cap$ fr W belongs to $\mathfrak{F}$ for each open set between V_k and Z_k. Just as V_2 and Z_2 have been obtained from V_1 and Z_1 one obtains V_{k+1} and Z_{k+1} from V_k and Z_k; that is to say, V_{k+1} is obtained by modifying V_k near its frontier; and Z_{k+1} is constructed in such a way that cl $V_{k+1} - A_{k+1} \subset Z_{k+1}$ and that $S_{k+1} \cap$ fr W belongs to $\mathfrak{F}$ for each open set W between V_{k+1} and Z_{k+1}.

Now let W be the union of all V_k. For each k, we have $V_k \subset W \subset Z_k$, and so $S_k \cap$ fr W belongs to $\mathfrak{F}$ for each k. But $S \cap$ fr W is the union of all the compacta $S_k \cap$ fr W and hence also belongs to $\mathfrak{F}$. Thus W is the neighborhood of p with the desired property. This concludes the proof of the theorem.

Remark. In dimension theory, Theorem 13.12 is further generalized: If the subset S of a compactum is the union of a sequence of one-dimensional (or 0-dimensional) sets that are *closed in* S (without necessarily being compacta), then S is one-dimensional (or 0-dimensional).

EXERCISES

1. Prove that *the union of two regular curves having an* 0-*dimensional intersection is a regular curve.*

2. In order that a nonempty set S that is the union of a countable family of compacta be discontinuous either of the following conditions is both necessary and sufficient: (1) S is 0-dimensional; and (2) each member of the family is discontinuous.

3. *If the compactum C_0 is 0-dimensional at the point p, and the compacta $C_1, C_2, \ldots$ are 0-dimensional, then the union of the entire countable family is 0-dimensional at the point p.*

4. *Let $\mathfrak{F}$ be a hereditary and $\mathfrak{F}$-additive family of compacta. If the compactum C_0 is $\mathfrak{F}$-regular at the point p, then the union of C and any countable family of $\mathfrak{F}$-regular compacta is $\mathfrak{F}$-regular at p.*

5. *If C is a 0-dimensional (a one-dimensional) compactum and U and W are open sets such that $U \Subset W$, then there exists an open set between U and W whose frontier is disjoint from C (has a 0-dimensional or empty intersection*

with C). By examples demonstrate that (1) the assumption that C is a compactum cannot be dispensed with; (2) even if C is a compactum but has the dimension 0 or 1 only at the point p, then between two neighborhoods U and W of p there does not necessarily exist an open set of the desired kind.

6. *If the compactum C is disjoint from the frontier of the open set U, then there exists an open set W such that $U \Subset W$ and that C is disjoint from the frontier of every open set between U and W.*

7. *If C is a one-dimensional compactum, and N is a closed 0-dimensional subset of C, then each point of C is contained in arbitrarily small neighborhoods whose frontiers have 0-dimensional or empty intersections with C and are disjoint from N.*

13.9   DECOMPOSITION PROPERTIES OF CURVES

If $\mathfrak{B} = (V_1, V_2, \ldots)$ is a sequence of open sets, we set

$$V_1' = V_1 \text{ and } V_k' = V_k \cap \operatorname{com} \operatorname{cl}(V_1 \cup V_2 \cup \ldots \cup V_{k-1}) \text{ for } k > 1$$

and call $\mathfrak{B}' = (V_2', V_2', \ldots)$ the sequence of sets derived from $\mathfrak{B}$ or, briefly, the *derived sequence*. The derived sets have the following properties:

(1) Each V_k' is an open subset of V_k.

(2) If $i \neq j$ then V_i' and V_j' are disjoint, whence

$$\operatorname{cl} V_i' \cap \operatorname{cl} V_j' \subset \operatorname{fr} V_i' \cap \operatorname{fr} V_j'.$$

(3) $\operatorname{cl} V_1' \cup \cdots \cup \operatorname{cl} V_m' = \operatorname{cl} V_1 \cup \cdots \cup \operatorname{cl} V_m$

(4) $\operatorname{fr} V_k' \subset \operatorname{fr} V_1 \cup \operatorname{fr} V_2 \cup \cdots \cup \operatorname{fr} V_{k-1} \cup \operatorname{fr} V_k \ (k = 2, \ldots, m)$.

Property (4) readily follows from the definition of V_k' in view of the laws about the frontiers of unions and differences.

THEOREM 13.13 (CHARACTERIZATION OF $\mathfrak{F}$-REGULAR COMPACTA)   *If $\mathfrak{F}$ is an additive and hereditary family of compacta in a metric space, then a compactum C is $\mathfrak{F}$-regular if and only if, for each $d > 0$, C is the union of a finite family of closed sets with diameters $<d$ such that the intersection of any two of them belongs to $\mathfrak{F}$.* The closed sets may be assumed to be *pieces* of C, that is, closures of sets that are open in C.

Necessity. For each $d > 0$, each point of an $\mathfrak{F}$-regular compactum C lies in an open set having a closure $<d$ and a frontier whose intersection

with C belongs to $\mathfrak{F}$. A finite sequence $\mathfrak{B} = (V_1, \ldots, V_m)$ of such open sets covers the compactum C. We set $C_i = C \cap \text{cl } V_i'$, where $(V_1', \ldots, V_m')$ is the sequence derived from $\mathfrak{B}$. From Property (1) of the derived sets it follows that the C_i are pieces of C with diameters $<d$; from (3), that their union is C; from (2), that $C_i \cap C_j \subset C \cap \text{fr } V_i' \cap \text{fr } V_j'$; and from (4), that the latter set is a subcompactum of the union of the m sets $C \cap \text{fr } V_i$. Since each $C \cap \text{fr } V_i$ belongs to the additive hereditary family $\mathfrak{F}$, so does their union and each subcompactum $C_i \cap C_j$ of that union.

Sufficiency. For some $d > 0$, let the compactum C satisfy the following condition:

(α_d) C is the union of a finite family $(C_1, \ldots, C_m)$ of closed sets whose diameters are $<d/2$ and whose intersections belong to $\mathfrak{F}$.

Being hereditary, $\mathfrak{F}$ includes also the union S of all sets $C_i \cap C_j$. Let p be any point of C, and denote by B_p the union of all C_i that do not contain p. Then the set $V_p = C \cap \text{com } B_p$ contains p and is open in C. Since each point p of C whose distance from p is $\geq d/2$ lies in B_p, each point of V_p has a distance $<d/2$ from p, whence the diameter of V_p is $<d$. Clearly, $\text{fr}_C V_p$ is a subset of the set $B_p \cap \text{cl } V_p$ each point of which lies in some C_i contained in B_p as well as in some C_j contained in cl V_p and, consequently, is S. Since $\mathfrak{F}$ is hereditary the compactum $\text{fr}_C V_p$ belongs to $\mathfrak{F}$. Thus from the condition (α_d) it follows that p has a neighborhood in C whose diameter is $<d$ and whose frontier in C belongs to $\mathfrak{F}$. If (α_d) holds for *each* $d > 0$, then C is $\mathfrak{F}$-regular.

Applying the theorem to the family $\mathfrak{F}$ consisting of the empty set alone one obtains

THEOREM 13.14 (CHARACTERIZATION OF 0-DIMENSIONAL COMPACTA) *In order that a nonempty compactum N be 0-dimensional (or discontinuous) it is necessary and sufficient that, for each $d > 0$, N be the union of a finite family of mutually disjoint closed sets with diameter $<d$.*

Compacta of dimension one have decomposition properties beyond those that follow from Theorem 13.13.

THEOREM 13.15 (DECOMPOSITION PROPERTIES OF CURVES, RATIONAL AND REGULAR CURVES) *A one-dimensional, rational, or*

regular compactum is, for every d > 0, the union of a finite family of closed sets <d such that any three sets are disjoint and any two have, respectively, 0-dimensional or empty, denumerable, and finite intersections.

If $\mathfrak{F}$ is the family of all compacta that are 0-dimensional or empty, denumerable or finite, then by Theorem 13.13 a one-dimensional, rational or regular compactum C is for each $d > 0$ the union of a finite family $C_1, \ldots, C_n$ of closed sets $<d$ such that for $i \neq j$ the set $C_i \cap C_j$ belongs to $\mathfrak{F}$. The union B of all these intersections is a 0-dimensional compactum. Hence, by Exercise 7, Section 13.8, each point of B is contained in a neighborhood $<d$ whose frontier is disjoint from B and has an intersection with C that belongs to $\mathfrak{F}$. The compactum B can be covered with a finite family of such open sets $V_1, \ldots, V_m$ whose closures are disjoint. We set $B_i = C \cap \mathrm{cl}\, V_i$. Let V be the union of the V_i and set $C'_j = C_j - C_j \cap \mathrm{cl}\, V$. If $j \neq k$ then C'_j and C'_k are disjoint since $C_j \cap C_k \subset V$. One readily sees that the sets C'_j and B_i have the union C, and that the intersection of any two of them belongs to $\mathfrak{F}$. If A_1, A_2, A_3 are three of these $m + n$ sets, then at least two of them belong either to the B_i or to the C'_j and therefore are disjoint.

The decomposition properties described in Theorem 13.15 will now be shown to be characteristic for curves among the continua.

THEOREM 13.16   (CHARACTERIZATION OF CURVES)   *In order that a continuum C be a curve each of the following conditions is sufficient:*

(1) for each d > 0, C is the union of a finite family of closed sets <d such that the intersection of any two of them is discontinuous (that is, 0-dimensional or empty),

(2) for each d > 0, C is the union of a finite family of closed sets <d such that the intersection of any three of them is empty.

According to Theorem 13.15, each of these sufficient conditions is also necessary. The sufficiency of Condition (1) follows from Theorem 12.13 in the case where $\mathfrak{F}$ is the family of the discontinuous compacta. The sufficiency of Condition (2) is a result of Urysohn. We sketch the proof, which somewhat resembles that of Theorem 13.12. Let C be a compactum which for each $d > 0$ is the union of a finite family of closed sets $<d$ no three of which have a common point. If p is a point of C, and Z_0 is a neighborhood of p we have to exhibit a neighborhood $V \subset Z_0$ of p such that $C \cap \mathrm{fr}\, V$ is discontinuous.

Let U_0 and W_0 be neighborhoods of p such that $U_0 \Subset W_0 \Subset Z_0$. One can decompose C into closed sets with diameters <1 no three of which have a common point and which are so small that none of them has points in common with both cl U_0 and com W_0. It is easy to obtain an open set U_0 such that $U_0 \subset U_1 \Subset W_0$ and that U_1 has the following

Property (P_1). The set $C \cap \mathrm{fr}\ U_1$ is the union of a finite family of mutually disjoint closed sets with diameters <1.

One then readily constructs an open set W_1 such that $U_1 \Subset W_1 \subset W_0$ and that each open set between U_1 and W_1 has the Property (P_1). Next, by decomposing C into closed sets with diameters $<\frac{1}{2}$ one obtains an open set U_2 such that $U_1 \subset U_2 \Subset W_1$ and that U_2 has the Property (P_2) that $C \cap \mathrm{fr}\ U_2$ is the union of a finite family of mutually disjoint closed sets $<\frac{1}{2}$. One then constructs an open set W_2 such that $U_2 \Subset W_2 \subset W_1$ and that each open set between U_2 and W_2 has the Property (P_2). Continuing in this way one obtains a sequence of pairs of open sets U_n and W_n $(U_n \Subset W_n)$ such that each open set X between U_n and W_n is the union of a finite family of mutually disjoint closed sets with diameters $<1/n$. Then it is clear that the union V of all U_n is a neighborhood $\subset Z_0$ of p which lies between U_n and W_n for each n, whence $C \cap \mathrm{fr}\ V$ is, for each n, the union of a finite family of mutually disjoint closed sets with diameters $<1/n$. By Theorem 13.14, $C \cap \mathrm{fr}\ V$ is 0-dimensional or empty.

13.10   ON REGULAR CURVES

A continuum C is said to be *hereditarily locally connected* if C and each subcontinuum of C is locally connected.

THEOREM 13.17   *Each regular curve is hereditarily locally connected.*

Since each subcontinuum of a regular curve C is a regular curve it is sufficient to prove that C is locally connected. According to Theorem 13.15, C is, for each $d > 0$, the union of a finite family $\{C_1, \ldots, C_m\}$ of closed sets $<d$ any two of which have finite intersections. According to Lemma 12.7, the number of components of such a set C_i cannot exceed the number of points that C_i has in common with the union of the $m-1$ other C_j. Hence the family of all components of C_i is finite and so is the family of all components of all m sets $C_1, \ldots, C_m$. Consequently, for each $d > 0$, C is the union of a finite family of continua $<d$ and, by Theorem 12.1, C is locally connected.

On the other hand, there are examples, due to H. M. Gehman and Knaster, of *hereditarily connected continua that are not regular curves*; for instance, in the plane the union of (1) the segment $0 \leqslant x \leqslant 1$, $y = 0$; (2) all upper semicircles with centers at $((2k - 1)/2^n, 0)$ and radius $1/2^n$ for $k = 1, 2, \ldots,$ 2^{n-1}; (3) all lower semicircles with centers at $((2k - 1)/(2 \cdot 3^n), 0)$ and radius $1/(2 \cdot 3^n)$ for $k = 1, 2, \ldots, 3^n$ and all natural numbers n. It can be shown that C is hereditarily locally connected and irregular at each point of the segment.

THEOREM 13.18 (CHARACTERIZATION OF GRAPHS) *In order that a continuum C be a graph it is necessary and sufficient that the set of all points of orders $\neq 2$ be finite.*

The necessity of the condition is evident. To prove its sufficiency, let C be a continuum such that the set P of all points of orders $\neq 2$ is finite. An irregular continuum contains infinitely many (irregular) points which are of order $\neq 2$. Hence C is regular and thus hereditarily locally connected. Let K be a component of $C - P$. Since $C - P$ is open in a locally connected continuum, it follows from Theorem 12.1 that K is open in C. There are two possibilities.

Case I. The closure of K contains at least two points of P, say p' and p''. Being a subcontinuum of a regular curve, cl K is locally connected and hence contains an arc A joining p' and p''. Let B be the set obtained by deleting p' and p'' from A. Clearly, $B \subset K$. We prove $B = K$ by deriving a contradiction from the assumption that there exists a point q of $K - B$. Let p be any point of B. Being open in a locally connected continuum, K contains an arc B' joining p and q. Moving on B' from q to p one meets a first point, r, belonging to A, which is in B. (Why is r distinct from p' and p''?) But then r is the center of a 3-spoke (consisting of B' and the two arcs into which B is divided by r) and hence of order > 2 contrary to the assumption that r lies in $C - P$. From $K = B$ it follows that cl $K = A$.

Case II. The closure of K contains at most one point of P. Then choose two points, p and q, in cl K, where p is the point of $P \cap$ cl K if this set is nonempty. The regular curve C contains an arc A between p and q. Since the end point q of A is of order 2 in K there exists a point, r, in $K - A$. The locally connected continuum cl K contains arcs A' and A'' joining r to p and q, respectively. Again it is easy to derive a contradiction from the assumption that cl K contains a point outside of $A \cup A' \cup A''$, and one readily shows that any two of these three arcs have exactly one point in common, namely, a common end point. Hence cl K is a closed curve which is identical with C or has only p in common with $C - K$ according as $P \cap$ cl K is or is not empty.

Thus the closure of each component of $C - P$ is an arc or a closed curve. Unless P is empty and C a closed curve, by Lemma 12.7 the closure of each component of $C - P$ contains a point of P. If p is a point of order m of C, then p cannot belong to the closure of more than m components of $C - P$. Thus the number of all components of $C - P$ is finite and C is a graph whose elements are the closures of the components of $C - P$.

An immediate consequence of the proof is the following

COROLLARY (CHARACTERIZATION OF ARCS AND CLOSED CURVES) *In order that a continuum C be an arc or a closed curve it is necessary and sufficient that C contain no point of order >2; it is a closed curve if and only if all its points are of order 2.*

EXERCISES

1. Show that a continuum in which the set of all points of order $\neq 2$ is finite is not necessarily a graph.

2. Show that a continuum containing at most one point of an order $\neq 2$ is a closed curve.

13.11   THE POINTS OF FINITE ORDER

The circle contains only points of order 2. Are there continua all of whose points are of order m for any number $m \neq 2$? A negative answer to this question is provided by the following result of Urysohn.

THEOREM 13.19 (ABOUT POINTS OF FINITE ORDER) *If for a natural number m all points of a compactum C have ramification orders $\leq m$ then C contains points of order $\leq m/2 + 1$, and the set of all these points is dense in C.*

An illustration of the case $m = 3$ (the case $m = 2$ is trivial) is the continuum T^* in Section 13.5, which contains only points of orders ≤ 3 and in which, indeed, the sets of all points of orders 1 and 2 are dense. The proof of the theorem is based on three lemmas.

LEMMA 13.3 *If* (1) X *and* Y *are open sets,* (2) C *is a set having a finite intersection, say at most m common points, with* fr Y, (3) q *is a point of* $C \cap Y \cap$ fr X, (4) q *is the only point that* cl Y *has in common with* $C \cap$ fr X, (5) $Y_1 = X \cap Y$ *and* $Y_2 = Y - $ cl Y_1, *then the frontier of at least one of the sets* Y_1, Y_2 *has at most $m/2 + 1$ points in common with* C.

(See Fig. 17, where $m = 3$ and only those points of C are marked that belong to fr X or fr Y.)

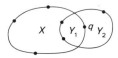

Figure 17

By (4), $C \cap$ fr Y is disjoint from fr X. Hence

$$C \cap \text{fr } Y \cap \text{cl } X = C \cap \text{fr } Y \cap X$$

and

$$C \cap \text{fr } Y \cap \text{com } X = C \cap \text{fr } Y \cap \text{com cl } X.$$

The set $C \cap$ fr Y, which by (2) contains at most m points, is thus the union of the two disjoint sets $C \cap$ fr $Y \cap X$ and $C \cap$ fr $Y \cap$ com X, at least one of which, therefore, contains at most $m/2$ points.

Because of (5), the laws concerning the frontiers of intersections and differences in Section 13.4 imply

$$\text{fr } Y_1 \subset (\text{fr } Y \cap \text{cl } X) \cup (\text{fr } X \cap \text{cl } Y),$$
$$\text{fr } Y_2 \subset (\text{fr } Y \cap \text{com } X) \cup (\text{fr } X \cap \text{cl } Y).$$

Taking assumption (4) into account we obtain

$$C \cap \text{fr } Y_1 \subset (C \cap \text{fr } Y \cap \text{cl } X) \cup \{q\}$$
$$C \cap \text{fr } Y_2 \subset (C \cap \text{fr } Y \cap \text{com } X) \cup \{q\}.$$

Consequently, at least one of the sets $C \cap$ fr Y_1 and $C \cap$ fr Y_2 contains at most $m/2 + 1$ points.

LEMMA 13.4 *In addition to the Assumptions* (1)–(5) *of the preceding lemma, assume that* (6) $C \cap \text{fr } X$ *is a finite set consisting, say, of n points, and* (7) $C \cap X$ *contains a point p with the following property: If V is any neighborhood of p such that $V \subset X \cap Y$, then $C \cap \text{fr } V$ contains at least n points. Then neither of the sets $C \cap Y_1$ and $C \cap Y_2$ is empty.*

Assuming that either $C \cap Y_1$ or $C \cap Y_2$ is empty, one arrives at a contradiction. Let S be a sphere with center q whose closure does not contain p and is a subset of Y. Define X^* as $X \cup S$ if $C \cap Y_2$ is empty, and as $X - \text{cl } S$ if $C \cap Y_1$ is empty. In either case, X^* is a neighborhood of p contained in $X \cup Y$, and $\text{fr } X^* \subset \text{fr } X \cup \text{fr } S$. In view of $\text{cl } S \subset Y$ it follows from (4) that $C \cap \text{fr } S$ is empty. Hence $C \cap \text{fr } X^* \subset C \cap \text{fr } X$. By (6), the latter set consists of n points, one of which (namely q) does not belong to $\text{fr } X$. Hence $C \cap \text{fr } X^*$ contains at most $n - 1$ points in contradiction to (7).

A combination of the preceding lemmas yields

LEMMA 13.5 *Under the assumptions* (1)–(7), Y *contains an open subset (namely, Y_1 or Y_2) whose intersection with C is not empty and whose frontier has at most $m/2 + 1$ common points with C.*

Proceeding now to the proof of the theorem, let C be a compactum each of whose points is of order $\leq m$, and let Z be an open set containing a point p_1 of C. It must be shown that Z contains a point of C of an order $\leq m/2 + 1$.

Let n_1 ($\leq m$) be the order of p_1. Then there exists a neighborhood Z_1 of p_1 with diameter <1, contained in Z and having the following property: the frontier of each neighborhood of p that is part of Z_1 contains at least n_1 points. Furthermore, there exists a neighborhood X_1 of p_1 with $X_1 \Subset Z_1$ and such that $C \cap \text{fr } X_1$ contains exactly n_1 points. Let q be one of them. Since q is of an order $\leq m$, the sphere of radius $\frac{1}{2}$ about q contains a neighborhood $Y \Subset Z_1$ such that $\text{cl } Y$ does not contain any of the other $n_1 - 1$ points of $C \cap \text{fr } X_1$ and that $\text{fr } Y$ has at most m common points with C. The open sets X_1 and Y satisfy the conditions (1)–(7). By Lemma 13.5, Y has an open subset Z_2 which contains a point p_2 of C and whose frontier has at most $m/2 + 1$ common points with C. Clearly, $Z_2 \Subset Z_1$.

Suppose, inductively, that open sets $Z_1, Z_2, \ldots, Z_k$ have been defined such that (a) $Z_{i+1} \Subset Z_i$, (b) Z_i has a diameter $<1/i$, (c) $C \cap \text{fr } Z_i$ contains at most $m/2 + 1$ points (for $i = 1, 2, \ldots, k$), and (d) Z_k contains a point

p_k of C. If p_k is of order n_k and is treated in the same way as p_1 of order n_1 has been, then one obtains an open set $Z_{k+1} \in Z_k$ satisfying the inductive assumption for $k + 1$. The sequence cl Z_1, cl Z_2, ..., cl Z_k... closes down on a point of the compactum C in Z that is obviously of order $m/2 + 1$. This completes the proof of the theorem.

In a graph, all points except a finite number of end points and ramification points are of order 2. In a curve in the general sense, the set of all points of orders $\neq 2$ may, as the curve T^* in Section 12.5 demonstrates, be dense. Still, in that example, also the set of all points of order 2 is dense. In 1915, however, Sierpiński constructed what amounts to an

Example of a regular curve without any points of order 2.   There exists a curve Δ^* all of whose points are of the orders 3 and 4.

The curve is a slight modification of a plane curve Δ, called Sierpiński's *triangle curve*, which besides points of order 3 and 4 contains three points of order 2. It can be obtained by applying to a triangle Cantor's method of defining his discontinuum.

Let T be a disk bounded by an equilateral triangle of side length 1 and vertices p_1, p_2, p_3. Divide T into four congruent triangles, denote by T_1, T_2, T_3 the triangles containing p_1, p_2, p_3, respectively, and delete the interior of the fourth triangle, T_4. In each of the triangles T_1, T_2, T_3 repeat the construction; that is, divide T_i into four congruent triangles, denote by T_{i1}, T_{i2}, T_{i3} the triangles that contain the vertices of T_i (by T_{ij} the triangle closest to p_j), and delete the interior of the fourth triangle, T_{i4}. Iterating this construction ad infinitum we obtain for each sequence of numbers $(i_1, i_2, \ldots, i_n)$ from $\{1, 2, 3\}$ four triangles $T_{i_1 i_2 \ldots i_n j}$ and delete the interior of the middle $T_{i_1 i_2 \ldots i_n 4}$. The set of all points of T that are *not* deleted in any step of this construction is the continuum Δ. Clearly, Δ may also be obtained as the closure of the union of the sides of T and those of all deleted triangles $T_{i_1, i_2, \ldots, i_n, 4}$.

Each point p that is common to two triangles $T_{i_1, i_2, \ldots, i_n, i}$ and $T_{i_1, i_2, \ldots, i_n, j}$ ($i \neq j$) is of order 4. For if V is the circular domain centered at p, whose radius is $1/2^m$ for some m, then fr V has at most 4 common points with Δ, and so p is of order ≤ 4. But since p lies in two sides of each of the two triangles, p is the center of a 4-spoke contained in Δ, and consequently of order ≥ 4. In a similar way, the reader can easily prove that p_1, p_2, p_3 are of order 2.

All other points of Δ are of order 3. For let q be a point $\neq p_1, p_2, p_3$ lying for each n in exactly one triangle $T_{i_1, i_2, \ldots, i_n}$, which we denote, briefly, by S_n. The frontier of the circular domain that circumscribes S_n has exactly 3 points in common with Δ; namely, the vertices of S_n. Hence q is of order ≤ 3. We now prove that q is of order ≥ 3 by constructing a 3-spoke with center q. For each n, the triangles S_n and S_{n+1} have exactly one common vertex, $c_n = c_{n+1}$. Let a_n and b_n be the other two vertices of S_n, and denote by a_{n+1} and b_{n+1} the vertices of S_{n+1} which are the midpoints of c_n and a_n, and c_n and b_n, respectively. Finally, let A_n and B_n denote the segments joining a_n to a_{n+1}, and b_n to b_{n+1}, respectively. (Each of these segments is one half of a side of S_n.) Then the reader can readily verify that the union of $\{q\}$ and the segments A_n and B_n for all n is a 3-spoke with center q. For these segments fit together to form three arcs with the end point q, but without other common points.

If in each of four alternating triangular faces of an octahedron one inscribes such a continuum Δ, then any two of these four Δ's have exactly one common point, namely a vertex; and the union Δ^* of the four Δ's is a continuum containing only points of order 4 and 3.

From Theorem 13.19 it follows that even in continua all of whose points have orders ≤ 5, the sets of points of orders ≤ 3 are dense. Similarly, in continua all of whose points have orders ≤ 6 or even ≤ 7, the sets of all points of order 4 is dense; and so on. The theorem thus yields the following

COROLLARY *If all points of a continuum C have orders $\geq n$, then C contains points of orders $\geq 2n - 2$; and the set of all these points is dense in C.*

Generalizing Sierpiński's curve Δ^*, Urysohn constructed, for each n, a continuum consisting only of points of the orders n and $2n - 2$ (for example, curves consisting only of points of the orders 4 and 6), as well as curves all of whose points are of order ω. G. Beer in Vienna constructed two curves consisting only of points of the orders 3 and 6, and 3 and 8, respectively.

G. T. Whyburn and W. L. Ayres proved that *for each $n > 2$ the set of all points of order n of a curve is discontinuous and even 0-dimensional*, as is the set of all end points. The only points of a curve that can fill continua are thus the points of order 2. This seems to bear out our intuition. Ayres has proved, however, that only if a curve C contains a *free arc*, that is an arc whose interior is open in C, is the set of all points of order 2 in C positive-dimensional. Thus when one thinks of the points of order 2 as the *ordinary* points of curves it is because one thinks primarily of graphs or the like. In curves that are not somehow made up of free arcs, the points of order 2 need not be exceptionally plentiful. In a curve without free arcs, the set of all points of any order n (including $n = 2$) is 0-dimensional or empty. For the proofs the reader is referred to Chapter III (especially Section 5) of the author's book *Kurventheorie*, quoted on p. 443.

The fundamental property of ramification order is described in the following result of Menger about regular curves, which G. Nöbeling extended to all locally connected continua and so as to include also the points of order ω.

THEOREM 13.20 (THE n-SPOKE THEOREM) *A locally connected continuum contains for any point p of order n an n-spoke, and for any point of order ω an ω-spoke, with the center p.*

Since no continuum C can contain an $(n + 1)$-spoke whose center is a point of order n in C, the theorem has the following

COROLLARY *In each locally connected continuum C, the ramification order of each regular point p is the largest number n such that C contains an n-spoke with the center p.*

Since each locally connected continuum C is arcwise connected, each point p of C is the center of a 1-spoke, that is, an end point of an arc in C. One thus might expect that the theorem could be proved by induction. If in C a k-spoke S with the center p (where $1 \leqq k < n$) has been constructed, then one might try to extract from the union of p and $C - S$ a $(k + 1)$th arc ending in p and otherwise disjoint from S. In general, however, this turns out to be impossible. In a plane with polar coordinates, let C be the union of the segment $[0, 1]$ on the polar axis and the spirals $(r = 1/\theta, \theta)$ and $(r = 2/\theta, \theta)$ for $\theta > \pi/2$. The curve C is of order 2 at the pole p since p lies in arbitrarily small circular neighborhoods whose frontiers have exactly 2 points in common with C; and the union of p and the two spirals is a 2-spoke with the center p. But if one first selects the segment $[0, 1]$, then the union of p and $C - [0, 1]$ does not contain any second arc with the end point p.

Neither does the Reduction Theorem seem to be of any avail in proving the theorem. For let p be a point of order n of C, and U a neighborhood of p such that $C \cap \text{fr } U$ consists of n points, and that $C \cap \text{fr } V$ contains at least n points for each neighborhood $V \subset U$ of p. If C contains an n-spoke N with the center p and end points in $C \cap \text{fr } U$, then N is irreducible with regard to the property of being a subcontinuum of C that is of order n at p and contains the n points of $C \cap \mathrm{fr}\ U$. But the intersection of a sequence of continua with this property does not necessarily have this property.

A point of order n in a set C is contained in a sequence of neighborhoods $U_1, U_2, \ldots (U_{k+1} \Subset U_k)$ such that $C \cap \mathrm{fr}\ U_k$ consists of n points, and that for no neighborhood $V \Subset U_1$ of p the set $C \cap \mathrm{fr}\ V$ contains less than n points. If C is a locally connected continuum, then each ring $R_k = C \cap (\mathrm{cl}\ U_k - U_{k+1})$ is a locally connected compactum and contains two disjoint n-tuples of points, $P = \{p_1, \ldots, p_n\}$ on $\mathrm{fr}\ U_k$, and $Q = \{q_1, \ldots, q_n\}$ on $\mathrm{fr}\ U_{k+1}$, which cannot be separated in R_k by less than n points; that is to say, for no set M containing less than n points is $R_k - M$ the union of two disjoint sets that are open in R_k and of which one contains P, and the other, Q. By a rather complicated procedure, for the details of which the reader is referred to Chapter VII of the book *Kurventheorie* by Menger, one can extract from R_k a graph G that is strictly n-connected between P and Q (see the exercise in Section 12.9). By the Corollary of Theorem 12.7, G, and consequently R_k, contains n mutually disjoint arcs each of which joins a point in P and a point in Q. The n-tuples of arcs in consecutive rings R_k and R_{k+1} fit together and the union of p and all these arcs is an n-spoke with the center p.

EXERCISES

1. Carry out in detail the proof that each point of order 3 in the curve Δ^* is the center of a 3-spoke.

2. In the example of the curve T^* (see Section 13.5) verify the statement that in a curve without free arcs the set of all points of order 2 is 0-dimensional. Prove that a curve in which either the set of all end points or the set of all ramification points is dense does not contain any free arc.

REFERENCES

Besides the material covered in this chapter and references to the literature, the author's book *Kurventheorie*, Bronx, N.Y., 1968, contains further curve-theoretical concepts and theorems as well as an extensive treatment of cut points, trees, and cyclically connected continua. For all these matters, the reader is also referred to C. Kuratowski's book *Topologie*, Warsaw 1948, especially vol. I, Sections 20 to 22, and vol. II, Sections 41 to 47, and to G. T. Whyburn's book *Analytic Topology*, Amer. Math. Soc. Coll. Lectures 28, 1942, especially Chapters 4 and 5.

CHAPTER 14

The Propriety of the Modern Curve Concept

Chapter 13 has demonstrated the fertility of the new curve definition and of the concept of ramification order. These ideas partly confirm, partly contradict, the intuition developed from experience with the simple curves that are familiar to everyone. They extend some of the elementary laws in a formerly undreamed-of generality and reveal totally unexpected exceptions to others. The new concept has also been shown to fulfill the first condition that an appropriate curve concept must satisfy: it includes the objects usually called curves such as arcs, graphs, and the sinusoid.

What still must be verified is the second condition: that the concept excludes such objects as a cube, the closure of a square region, and sets containing such squares as subsets—objects found among the one-to-one images of a segment, the trajectories of continuous motions and the irreducible continua. This problem and related questions will be studied in the present chapter.

14.1   EACH PLANE CURVE IS A CANTOR CURVE

By squares and triangles we shall in this chapter mean the closures of square and triangular regions in the plane. We begin by formulating one of the main results of this chapter.

THEOREM 14.1   *A set that contains a subset homeomorphic to a square is not one-dimensional (nor, of course, 0-dimensional). In particular, a continuum containing a subset homeomorphic to a square is not a curve.*

Since a plane continuum not containing a square is called a Cantor curve, an immediate consequence of the theorem is the following

COROLLARY   *In the plane, each curve is a Cantor curve.*

Since, by Theorems 13.1 and 13.2, each subset S of a set of dimension 1 or 0 (as well as each set homeomorphic to S), is 1- or 0-dimensional, Theorem 14.1 can be proved by showing that *a square is not of dimension* 1 *or* 0. Now according to Theorem 13.15, a compactum that is 1-dimensional or 0-dimensional is the union of finite families of arbitrarily small closed sets no three of which have a common point. Hence Theorem 14.1 will be established by demonstrating that a square lacks this property.

THEOREM 14.2   *If a square is decomposed into a finite family of sufficiently small closed sets, then some point of the square belongs to at least three of the sets.*

This theorem is due to the French mathematician Henri Lebesgue (1875–1941), but his own proof of it was inadequate. Brouwer gave another rather involved proof, which he later had to correct. In 1928, Hurewicz and the German mathematician Emanuel Sperner independently reduced the assertion to perfectly transparent combinatoric statements. Sperner decomposes a triangle rather than a square.

To begin with, let S be a segment with the end points a_0 and a_1, and let $\{S_1, S_2, \ldots, S_r\}$ be a *segmentation* of S; that is, a finite family of segments having the union S and such that the intersection of any two segments

S_i and S_j is either empty or a single point (which clearly is an end point or, as we shall say, a *vertex* of both S_i and S_j). Suppose, further, that for this segmentation a (*linear*) *marking function* has been defined; that is to say, that with each vertex p of S_i a number $m(p)$ is somehow associated, subject only to the following conditions:

1'. $m(p) = 0$ or 1 for each vertex p.

2'. $m(a_0) = 0$ and $m(a_1) = 1$.

Let $n(S_i)$ denote the number of vertices of S_i that are marked 0 $(i = 1, \ldots, r)$. Clearly, $n(S_i)$ is 2 or 0 or 1 according as the two vertices of S_i have the marks $0, 0$ or $1, 1$ or $0, 1$. Let s_2, s_0, s_1 be the numbers of all segments having the marks 2, 0, and 1, respectively. The s_1 segments of the third type will be called *distinguished segments*. Clearly,

$$n(S_1) + \cdots + n(S_r) = 2s_2 + 0s_0 + 1s_1 = 2s_2 + s_1.$$

Let v_0 denote the number of all vertices marked 0. Each of them except a_0 belongs to exactly two segments S_i, whence

$$n(S_1) + \cdots + n(S_r) = 1 + 2(v_0 - 1) = 2v_0 - 1.$$

Equating the two expressions for the sum of the $n(S_i)$ one obtains

$$2s_2 + s_1 = 2v_0 - 1 \qquad \text{or} \qquad s_1 = 2(v_0 - s_2) - 1.$$

Hence s_1 is an odd number. This establishes

LEMMA 14.1   *For each segmentation of a segment and for each marking function satisfying conditions 1' and 2', the number of distinguished segments is odd.*

Now let T be a triangle with the vertices a_0, a_1, a_2. The union of the three segments $[a_0, a_1]$, $[a_0, a_2]$, and $[a_1, a_2]$ will be called the *rim* of T. Let $\{T_1, \ldots, T_s\}$ denote a *triangulation* of T—that is, a finite family of triangles having the union T and such that the intersection of any two T_i and T_j is either empty, or a vertex of both T_i and T_j, or an edge (that is, a side) of both T_i and T_j.

Suppose, further, that for this triangulation a (*planar*) *marking function* has been defined; that is, with each vertex of each T_j a number is associated subject only to the following conditions:

1''. $m(p) = 0$ or 1 or 2 for each vertex p.

2". On the rim of T, the mark of each vertex on $[a_0, a_1]$ is 0 or 1; on $[a_0, a_2]$, 0 or 2; and on $[a_1, a_2]$, 1 or 2.

Since a_0 lies on $[a_0, a_1]$ and on $[a_0, a_2]$, it follows that $m(a_0) = 0$; similarly, $m(a_1) = 1$.

If one vertex of an edge of a triangle T_j is marked 0 and the other is marked 1, then the segment is called a *distinguished edge*. Condition 2" immediately yields

Remark 1. On the rim of T, only $[a_0, a_1]$ can contain distinguished edges.

Clearly the edges of the triangles contained in $[a_0, a_1]$ constitute a segmentation of that segment and the marks of the vertices define a linear marking function on it, since $m(a_0) = 0$ and $m(a_1) = 1$. The distinguished segments of this segmentation are the distinguished edges of the triangulation that are contained in the segment $[a_0, a_1]$. Applying Lemma 14.1 one sees that the number of distinguished edges on $[a_0, a_1]$ is odd. Hence Remark 1 yields

Remark 2. The number of distinguished edges on the rim of T is odd.

Let $n(T_j)$ be the number of distinguished edges of T_j $(j = 1, \ldots, s)$. Clearly,

$n(T_j)$ is	if and only if the vertices of T_j have the marks
2	0, 0, 1 or 0, 1, 1
0	0, 0, 0 or 0, 2, 2 or 2, 2, 2 or 1, 1, 1 or 1, 1, 2 or 1, 2, 2
1	0, 1, 2

Let n_2, n_0, n_1 be the numbers of all T_j in the triangulation for which $n(T_j) = 2, 0, 1$, respectively. Then

$$n(T_1) + \cdots + n(T_s) = 2n_2 + 0n_0 + 1n_1 = 2n_2 + n_1.$$

If an edge E is contained in the rim of T, then E lies in exactly one triangle T_j. Every other edge belongs to exactly two triangles. This holds, in particular, for the distinguished edges. Let e be the number of *all* distinguished edges, and d the number of the distinguished edges *on the rim*, which according to Remark 2 is odd. Then

$$n(T_1) + \cdots + n(T_s) = d + 2(e - d) = 2e - d.$$

Equating the two expressions for the sum of the $n(T_j)$ one obtains

$$2n_2 + n_1 = 2e - d \qquad \text{or} \qquad n_1 = 2(e - n_2) - d.$$

Since d is an odd number, so is n_1. Hence $n_1 > 0$, and we have established

LEMMA 14.2 *For each triangulation of a triangle T, and each marking function satisfying conditions $1''$ and $2''$, there exists at least one distinguished triangle—that is, one whose vertices display all three marks 0, 1, 2.*

Now let T, as before, be a triangle with the vertices a_0, a_1, a_2, and let $\{R_1, \ldots, R_m\}$ be a family of closed sets having the union T and such that none of the R_i has points in common with each of the three segments $[a_0, a_1]$, $[a_0, a_2]$, and $[a_1, a_2]$. Since each of the three vertices of T lies in two of these segments, it follows that any set R_i containing one of these vertices is disjoint from the third side, which does not contain that vertex.

Denote by U_0 the union of all those R_i which are disjoint from $[a_1, a_2]$; by U_1, the union of all R_i that are not contained in R_0 and are disjoint from $[a_0, a_2]$; and by U_2, the union of the remaining R_i. Clearly, U_0, U_1, U_2 contain the points a_0, a_1, a_2, respectively. The segment $[a_0, a_1]$ is contained in the union of U_0 and U_1. It is furthermore clear that any point common to U_0, U_1, U_2 is contained in at least three of the sets R_i.

Now consider a triangulation $\{T_1, \ldots, T_s\}$ of T, and associate the mark 0 with each vertex contained in U_0; the mark 1, with each of the remaining vertices that is contained in U_1; and the mark 2 with each remaining vertex. In this way a marking function is defined, since a_0, a_1, a_2 obtain the marks 0, 1, and 2, respectively; the vertices on $[a_0, a_1]$, which is disjoint from U_2, have the marks 0 and 1, and those of $[a_0, a_2]$ and $[a_1, a_2]$ have the marks 0 or 2 and 1 or 2, respectively. By Lemma 14.2, there exists at least one distinguished triangle. Its vertices lie in U_0, U_1, and U_2 and hence in at least three of the sets R_i.

For each natural number n, there is a triangulation of T whose triangles have diameters $< 1/n$. If T^n is a distinguished triangle of the nth triangulation and p^n one of the vertices of T^n, then T contains an accumulation point p of the sequence of these p^n. Each neighborhood of p contains points belonging to at least three of the sets R_i, and since these sets are closed, p itself lies in at least three of these sets. Thus the following refinement of Theorem 14.2 has been demonstrated. *If a triangle T is the union of a finite family of closed sets, none of which has points in common with all three sides of T, then there exists at least one point of T that lies in at least three of the sets.*

This concludes the proof of Theorems 14.2 and 14.1.

EXERCISES

1. Using Lemma 14.1 show that any two nonempty closed sets whose union is the segment $[a_0, a_1]$ have at least one common point; in other words, prove in this way that the segment is connected.

2. Give an example of a decomposition of T into a finite family of triangles which is not a triangulation, even though any two triangles are either disjoint or have exactly one point or a segment in common.

3. Using the refined Theorem 14.2, prove that a triangle is not homeomorphic to a segment.

4. Note the analogy between the Lemmas 14.1 and 14.2, and formulate a theorem about decompositions of a tetrahedron in tetrahedra, with proper assumptions concerning their intersections and a properly defined marking function. Prove the theorem about a tetrahedron, applying the result concerning triangles to one of its faces. (*Hint:* Use must be made of the fact that the number of distinguished triangles in a triangulation is odd.)

5. Using the theorem obtained in Exercise 4, prove Lüroth's Theorem: *A tetrahedron and a square are not homeomorphic.*

6. By induction, extend Lemma 14.2 and its proof to all dimensions. From the extended Lemma, derive the extension of Theorem 14.2. An immediate consequence is Brouwer's theorem that *an m-dimensional and an n-dimensional simplex ($m \neq n$) are not homeomorphic* ("Topological invariance of dimension").

14.2 A SECOND PROOF

We shall now outline a second proof of the fact that a square Q is not a curve—a proof that is based directly on the definition of curves. If Q were a curve, then each point of Q would be contained in arbitrarily small neighborhoods with discontinuous frontiers. But this is not the case. *In the plane, the frontier of every bounded open set contains a continuum.* This fact, demonstrated by the Swedish mathematician Phragmén in 1883, is readily seen to be a corollary of

THEOREM 14.3   *In the plane, the complement of a discontinuous compactum is connected.*

We first prove that in all arcwise connected spaces, each set with a certain covering property has a connected complement.

THEOREM 14.4   *In an arcwise connected metric space, A, let S be a set with the following property Π: For every $d > 0$, S can be covered by a finite family of open sets $< d$ with mutually disjoint closures and connected frontiers. Then $A - S$ is connected. (In fact, any two points of $A - S$ lie in a continuum $\subset A - S$.)*

If p and q are any two points of $A - S$, then a subcontinuum of $A - S$ containing p and q will be exhibited. By assumption, A contains an arc C joining p and q. If C is disjoint from S, then C is the desired continuum. If $C \cap S$ is nonempty, then let d be a positive number less than half the distances between p and S, and q and S. Since S has the property Π, there exists a finite family $\{V_1, \ldots, V_m\}$ of open sets $< d$ that cover S, whose frontiers are connected and whose closures are disjoint. Since S lies in the union of the open sets V_i and their closures are disjoint, the frontiers of $V_1, \ldots, V_m$ are disjoint from S. Neither p nor q belongs to the closure of any V_i because of the smallness of their diameters. But moving on C from p to q, one encounters a point of S and hence a first point, q_1, belonging to the closure of some V_i, say to cl V_{i_1}. Clearly, q_1 lies in fr V_{i_1}. So does the last point on C belonging to cl V_{i_1}; call it p_1. Since fr V_{i_1} is connected and disjoint from S, the union of fr V_{i_1} and the arc C_{pq_1} of C joining p and q_1 is a continuum that is disjoint from S and contains p and p_1. If the arc $C_{p_1 q}$ is disjoint from S, then its union with the above mentioned continuum is the desired set. If $C_{p_1 q} \cap S$ is nonempty, then there is a first point on C belonging to the closure of some V_i, say to cl V_{i_2}. A repetition of this procedure yields a continuum that is disjoint from S and contains p and the last point, p_2, of C in fr V_{i_2}. After a finite number of steps, one obtains a continuum in $A - S$ joining p to a point p_k such that the arc $C_{p_k q}$ is disjoint from S, and thus the desired continuum that is disjoint from S and contains p and q.

In each metric space, a discontinuous compactum S can be covered with a family $\{V_1, \ldots, V_m\}$ of open sets $< d$ with disjoint closures. In claiming that in some spaces the set S has the property Π, the emphasis therefore lies on the connectedness of the frontiers of those open sets. For example, in the straight line bounded open sets with connected frontiers (and consequently sets with Property Π) don't exist. In the n-dimensional Euclidean spaces with $n \geq 2$, however, all discontinuous compacta do have that property. But in what follows, we shall prove only the case $n = 2$ of this statement, since this is all that is needed in deriving Theorem 14.3 from 14.4.

THEOREM 14.5   *In the plane, each discontinuous compactum S has the property Π.*

In the plane, the sets $V_1, \ldots, V_m$ will be replaced by open sets $X_1, \ldots, X_n$ ($n \leq m$) covering S which have diameters $<d$, disjoint closures, and connected frontiers. This will be achieved in three steps. *First*, each V_i will be replaced by a polygonal domain W_i of diameter $<d$ such that $V_i \subset W_i$ and that cl W_i and cl W_j are disjoint for $i \neq j$. *Second*, $W_1, \ldots, W_m$ will yield the desired sets $X_1, \ldots, X_n$ by a construction based on the assumption that in the plane the complement of a closed polygon is not connected. This fact will be proved in the *third* and final step.

First Step. In constructing the polygonal domains we cover the plane with a net of hexagons rather than with a net of squares since the use of hexagons facilitates the second step. If $\mathfrak{H}$ is a hexagonal tesselation (that is, a family of congruent regular hexagons covering the plane), then each element of $\mathfrak{H}$ is called an $\mathfrak{H}$-*cell*; each segment common to two $\mathfrak{H}$-cells, an $\mathfrak{H}$-*edge*; each point common to three $\mathfrak{H}$-cells (and to three $\mathfrak{H}$-edges), an $\mathfrak{H}$-*vertex*. A set that is the union of a family (a finite family) of $\mathfrak{H}$-cells is called an $\mathfrak{H}$-*system* (a *finite* $\mathfrak{H}$-*system*). The union of all $\mathfrak{H}$-cells not belonging to an $\mathfrak{H}$-system K will be denoted by K^*. The union of all $\mathfrak{H}$-edges that belong to both K and K^* will be called the *boundary* of K as well as of K^*—briefly, bd K and bd K^*. One readily obtains

Remark 1.   If K is an $\mathfrak{H}$-system, then $K -$ bd K is an open set whose closure is K and whose frontier is bd K.

Returning to the family $\{V_1, \ldots, V_m\}$ of open sets let $\mathfrak{H}$ be a hexagonal tesselation whose cells are so small that (1) for each V_i, the union V_i' of V_i and all $\mathfrak{H}$-cells having nonempty intersections with V_i has a diameter $<d$; and (2) any two $\mathfrak{H}$-systems V_i' and V_j' are disjoint ($i \neq j$). If for each V_i one sets $W_i = V_i' -$ bd V_i', then Remark 1 yields a family $\{W_1, \ldots, W_m\}$ of polygonal domains with diameters $<d$ covering S and having mutually disjoint closures.

Second Step. If p is an $\mathfrak{H}$-vertex belonging to the boundary of an $\mathfrak{H}$-system K, then two of the three $\mathfrak{H}$-cells meeting in p belong to one of the two systems K, K^* (denote it by K_1), and the third belongs to the other one. Of the three $\mathfrak{H}$-edges meeting in p, one lies in K_1 alone, and the two others belong to both K and K^*. Consequently each $\mathfrak{H}$-vertex in bd K lies in exactly two $\mathfrak{H}$-edges belonging to bd K. The Corollary to Theorem 12.6 thus yields the important

Remark 2.   Each component of the boundary of a finite $\mathfrak{H}$-system is a closed polygon.

(In nets of squares rather than hexagons, no simple analogue of Remark 2 is valid. This is why the use of hexagonal tesselations is preferable.)
Use will also be made of

Remark 3.   If L is a connected $\mathfrak{H}$-system, then any two points r' and r'' of L can be joined by a polygon which has no points, except possibly r' and/or r'', in common with bd L.

Clearly, the cells H' and H'' containing r' and r'', respectively, can be joined by a chain $(H', H_1, \ldots, H_n, H'')$ of $\mathfrak{H}$-cells in L such that any two consecutive cells have an $\mathfrak{H}$-edge in common. If c', c_i, c'' are the centers of the cells H', H_i, H'', then the polygon with the vertices

$$(r', c', c_1, \ldots, c_n, c'', r'')$$

lies in L and has no point except possibly r' and/or r'' in common with bd L.
Moreover, one readily proves

Remark 4.   If K is a finite $\mathfrak{H}$-system, then the union, K^+, of K and the bounded components of K^* is a finite $\mathfrak{H}$-system that contains K, has the same diameter as K, and such that both K^+ and $(K^+)^*$ are connected.

The following Remark 5 is by no means obvious. If the plane is modified by deleting the interior of two disjoint hexagonal cells and joining the two holes by a prismatic hollow handle, then in the resulting surface the handle K is the closure of an open set with the following property: both K and the closure K^* of the complement of K (that is, the plane with two holes) are connected, whereas the boundary of K (being the union of the rims of the hexagons) is not connected. Hence in Remark 5 and its consequences it is essential that the underlying space is the (unmodified) plane.

Remark 5.   In the plane, if the finite $\mathfrak{H}$-system K as well as K^* is connected, then bd K is connected and hence (by Remark 2) is a closed polygon.

Calling a set S *polygonally connected* if any two points of S lie in a polygon contained in S one can describe the property of the plane that accounts for the validity of Remark 5 as follows.

THEOREM 14.6   *In the plane, the complement of a closed polygon is not polygonally connected.*

The proof of this theorem will be given later. At this point, its validity will be assumed in deriving Remark 5. To this end, let C be any component of bd K. According to Remark 2, C is a closed polygon. Remark 5 will be proved by showing that if p is any point of bd K, then p belongs to C. Lying on bd K, the point p belongs to an $\mathfrak{H}$-edge E that is a side of two hexagons, $H' \subseteq K$ and $H'' \subseteq K^*$. Let q' and q'' be the centers of H' and H''! According to Theorem 14.6, there exist two points, r' and r'', that cannot be connected by a polygon in the complement of the closed polygon C. By Remark 3, r' and r'' cannot both lie in either K or K^*. Suppose r' lies in K, and r'' in K^*. Then, again by Remark 3, K contains a polygon Q joining r' and q' which, with the possible exception of r', is disjoint from bd K and thus from C; and K^* contains a polygon Q^* joining q' and r'' which, with the possible exception of r'', is disjoint from C. The union of Q, the segment $[q', q'']$, and Q^* is a polygon joining r' and r'', and hence must have a point in common with C. Since $(Q \cup Q^*) \cap C$ is empty $[q', q''] \cap C$ is nonempty; and clearly this intersection is the midpoint of the $\mathfrak{H}$-edge E, which is common to H and H'. But then this entire edge including the point p belongs to C, which concludes the proof of Remark 5.

It is now possible to complete the *second* of the previously announced steps. Let $W_1, \ldots, W_m$ be mutually disjoint $\mathfrak{H}$-systems of diameters $<d$ which cover the discontinuous compactum S. Let W_i' be the union of W_i and all bounded components of W_i^*. By Remarks 4 and 5, each W_i' is an $\mathfrak{H}$-system containing W_i, having a diameter $<d$ and a connected boundary. But $W_1', \ldots, W_m'$ need not be mutually disjoint; for example, W_2' may be contained in a bounded component of W_1 and hence in W_1'. But setting $W_1'' = W_1'$; denoting by W_2'' one of the sets W_i' that is disjoint from W_1''; by W_3'',

one of the W_i' that is disjoint from W_1'' and W_2''; and so on, one finally obtains a family $\{W_1'', \ldots, W_n''\}$ $(n \leq m)$ of mutually disjoint $\mathfrak{H}$-systems $<d$ that cover S and whose frontiers are connected. Then $X_i = W_i'' - \text{bd } W_i''$, $(i = 1, \ldots, n)$ are the desired open sets.

Third Step. It remains to prove Theorem 14.6—that is, to derive it from simpler properties of the plane. As such one may use the analogue of the fact that a straight line is divided by each of its points into two components (called *open rays;* their closures, *closed rays*).

Property Λ.   A plane is divided by each of its straight lines L into two components (called *open half-planes; their closures, closed half-planes*). *If two points, q, q', lie in the same open half-plane, then the latter contains the entire closed segment $[q, q']$ joining them; whence $[q, q']$ is disjoint from L. If q and q' lie in opposite half-planes, then the segment $[q, q']$ has exactly one point in common with L.*

This property of the plane, which we take for granted, readily follows from Pasch's even more elementary axioms of plane geometry (see Section 5.12). One of the consequences of Property Λ is

LEMMA 14.3   *In the plane, the number of points that a closed polygon P has in common with a straight line L that does not contain any vertex of P is even (possibly 0).*

Call the vertices of P that lie in one of the open half-planes complementary to L *positive,* those in the other open half-plane, *negative.* (None of the vertices lies *on* L.) If all vertices of P are positive or all are negative, then P and L are disjoint (that is, have 0 common points). If the sequence $S = (p_0, p_1, \ldots, p_{n-1}, p_n = p_0)$ of the vertices of P includes points of both kinds, then we may assume that the first k_1 points are positive, the next k_2 points negative, the next k_3 positive, and so on. Since p_n $(=p_0)$ is positive, the last k_{2r-1} points are positive. Here, $r > 1$. The segment $[p_k, p_{k+1}]$ is disjoint from L if p_k and p_{k+1} are either both positive or both negative; it has exactly one point in common with L if p_k and p_{k+1} have opposite signs. Consequently P and L have as many common points as there are changes of sign in the sequence S; that is, $2(r - 1)$, which proves the assertion.

Property Λ is capable of the following refinement.

Property Λ^*. *The plane is divided into two components by the union V of any two rays with a common vertex.* (The components are called *open sectors,* their closures, *closed sectors.*) *If q and q′ are in the same sector, then* $[q, q′]$ *is either contained in that sector and thus disjoint from V or it has exactly two points in common with V—one with each ray.*

Each line being the union of two rays with a common vertex, property Λ^* is a generalization of property Λ and can likewise be derived from Pasch's axioms concerning the plane. Correspondingly, Lemma 14.3 can be extended to

LEMMA 14.4   *In Lemma* 14.3, *the line L may be replaced by the union V of any two rays with a common vertex v provided that V does not contain any vertex of the polygon P, and P does not contain v.*

Indeed, if all vertices of P in one open sector are called *positive,* and those in the other sector, *negative,* then one demonstrates Lemma 14.4 by repeating the proof of Lemma 14.3 with one single modification: a segment joining two consecutive vertices of P with the same sign has either 0 or 2 points in common with V. But this obviously does not change the fact that the total number of common points is even.

To prove Theorem 14.6 it suffices to show that *the complement of any closed polygon P contains two points, q and r, such that each polygon joining q and r contains a point of P.*

Let K be a circular domain containing P in its interior, and L a line containing no vertex but at least one point of P. Let q be one of the two points of $L \cap$ fr K. Clearly, q *is the end point of a ray that is disjoint from P.* Moving along L from q to the other point of $L \cap$ fr K one meets a second point of P, say $p′$. It will be proved that any point r of L between p and $p′$ has the desired property that each polygon joining r and q contains a point of P. This will be demonstrated by showing that if s is any point joined to r by a polygon that is disjoint from P,

$$(r = r_1, r_2, \ldots, r_{m-1}, r_m = s),$$

then no ray with the end point s is disjoint from P, whence $s \neq q$.

For the sake of brevity, a ray that contains an odd number of points of P but no vertex of P will be called an *odd ray.* One first observes that r_1 is the end point of an odd ray, namely the ray containing the single point p of P. Inductively, it may be assumed that for some $k \geq 1$ the point r_k is the end point of an odd ray M_k. Let H_k be the ray issuing from r_k and containing r_{k+1}. Clearly, $H_k \neq M_k$. Two cases are possible.

1. H_k is free of vertices of P. Then by Lemma 14.4, P and $V_k = H_k \cup M_k$ have an even number of common points. Since M_k is an odd ray, so is H_k and, since $[r_k, r_{k+1}] \cap P$ is empty, so is the ray M_{k+1} contained in H_k and issuing from r_{k+1}.

2. H_k contains at least one vertex of P. Being disjoint from P, $[r_k, r_{k+1}]$ has a neighborhood that is disjoint from P. Since the number of straight lines through r_k and r_{k+1} which contain vertices of P is finite, there clearly exists (see Fig. 18) a point r'_{k+1} such that (a) the triangle $r_k r'_{k+1} r_{k+k}$ and its interior are disjoint from P, and (b) the rays H'_k and H'_{k+1} issuing from r_k and r_{k+1}, respectively, and containing r'_{k+1}, are free of vertices of P. By Lemma 14.4, the number of points in $P \cap (M_k \cup H'_k)$ is even. Since M_k is odd, so is H'_k and (since $[r_k, r'_{k+1}]$ is disjoint from P) so is the ray H''_k contained in H'_k and issuing from r_{k+1}. Again by Lemma 14.4, the union of H''_k and the ray H''_{k+1} contained in H'_{k+1} and issuing at r'_{k+1} has an even number of points in common with P. Consequently H''_{k+1} is an odd ray and so is the ray M_{k+1} issuing from r_{k+1} and containing r'_{k+1}.

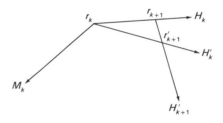

Figure 18

By induction, one arrives at an odd ray M_m issuing at $r_m = s$. Now let H be any other ray with the end point s. Either H contains a vertex of P or, by Lemma 14.4, $H \cup M_m$ contains an even number of points of P, whence H is odd and thus contains at least one point of P. This completes the proof of Theorem 14.6 and, thereby, the last step of the proof of Theorem 14.5.

Theorem 14.6, asserting that the complement of a closed polygon has at least two components, can be elaborated in two ways. (1) It can be shown that that complement has at most two components; consequently *each closed polygon P divides the plane in exactly two connected open sets.* Exactly one of those two sets, called the *interior* of P, is bounded. (2) Statement (1) can be extended from closed polygons to closed curves; that is to say, *each topological image of a circle divides the plane into two connected sets exactly one of which is bounded.* In this form, the theorem was first announced and proved by C. Jordan in 1893. The first correct proof of Jordan's Theorem for closed polygons was given by Hans Hahn. In what precedes, we have reproduced his proof of that half of the theorem which is used in demonstrating Theorems 14.3 and 14.5.

EXERCISES

1. Give examples of connected subsets of the plane that are not polygonally connected. Prove that the complement of a closed polygon is not even connected.

2. Generalizing Remark 4, prove that in any euclidean space the union of a bounded open set and the bounded components of its complement is connected.

3. In the proof of Theorem 14.6, two points q and r have been defined which cannot be joined by a polygon that is disjoint from the closed polygon P. Prove that each point of the complement of P can be joined either with q or with r by a polygon that is disjoint from P. (This is the second half of Jordan's Theorem for polygons.)

14.3   EACH CANTOR CURVE IS A CURVE

We now prove the converse of the Corollary to Theorem 14.1.

THEOREM 14.7   *Each Cantor curve is a curve in the general sense.*

This assertion is an immediate consequence of the following four statements:

1. Each Cantor curve is homeomorphic to a subcontinuum of the plane universal curve U_2^* (Theorem 12.9).

2. U_2^* is a curve (Exercise 1, Section 13.1).

3. Each subcontinuum of a curve is a curve (Theorem 13.1).

4. Each set that is homeomorphic to a curve is a curve (Theorem 13.2).

Theorem 14.7 has the following

COROLLARY *If S is a plane continuum not containing a square, then each point of S lies in arbitrarily small neighborhoods whose frontiers have discontinuous (0-dimensional or empty) intersections with S.*

This corollary can be sharpened in two ways: (1) by dispensing with the assumption that S is a connected compactum; (2) by showing that among the neighborhoods mentioned in the corollary there are some of very special types: circles, if S is a compactum, and convex neighborhoods for any set S. Use will be made of the following

Remark 1. If F is the frontier of a convex open set in the plane, then a nonempty subset G of F is 0-dimensional (as well as discontinuous) if and only if $F - G$ is dense in G (see Exercise 1).

THEOREM 14.8 *If S is any plane set not containing a square, then each point of S lies in arbitrarily small convex neighborhoods whose frontiers have discontinuous intersections with S.*

Let $(p_0, p_1, \ldots, p_{n-1}, p_n = p_0)$ be the sequence of vertices of a polygon that is *strictly convex*; that is, convex without any triple of collinear vertices and such that the sum of any two consecutive angles exceeds 180° (which implies $n \geq 5$). In the line containing the segment $[p_{i-2}, p_{i-1}]$, let R_{i-1}^+ be the ray issuing from p_{i-1} that does not contain the point p_{i-2}. In the line containing $[p_i, p_{i+1}]$, let R_i^- be the ray issuing from p_i that does not contain p_{i+1}. Set $R_n^+ = R_0^+$ and $R_n^- = R_0^-$. By the triangle *adjacent* to the side $[p_i, p_{i+1}]$ of P is meant the open set Q_i bounded by that segment and parts of the rays R_{i+1}^+ and R_i^-. (If the two rays are parallel, then Q_i is half of a strip.) The reader can easily prove

Remark 2. If q is any point in Q_i, then $(p_0, p_1, \ldots, p_{i-1}, q, p_i, \ldots, p_{n-1}, p_n = p_0)$ are the vertices of a strictly convex polygon.

This remark is used in demonstrating

LEMMA 14.5 *If the set T is dense in the strictly convex domain Z, and $(s_0, s_1, \ldots, s_{m-1}, s_m = s_0)$ is the sequence of vertices of a strictly convex* *polygon in Z, then $T \cap Z$ contains points $t_1, t_2, \ldots, t_m$ satisfying the following conditions:*

1. The points $(s_0, t_1, s_1, t_2, s_2, \ldots, s_{m-1}, t_m, s_m = s_0)$ are the vertices of a strictly convex polygon.

2. Each of the points t_i has distances from s_{i-1} and s_i that exceed half the distance betwen s_{i-1} and s_i.

Since Z is strictly convex one can choose a point t_1 of $T \cap Z$ in the domain adjacent to the side $[s_0, s_1]$ of P. According to Remark 2, $(s_0, t_1, s_1, s_2, \ldots, s_{m-1}, s_m)$ are the vertices of a strictly convex polygon. Applying Remark 2 to the side $[s_1, s_2]$ of *this* polygon (that is to say, using t_1 rather than s_0 as the point preceding s_1) one obtains a point t_2 of T. Proceeding in this way one finally obtains t_m (using s_m and t_1 rather than s_1 in defining R_m^+). If each t_k is chosen close to the line bisecting $[s_{k-1}, s_k]$, then also condition (2) is satisfied and the Lemma is proved.

Now let S be a set whose complement T is dense in the plane, p, a point of S; and Z, a circle about p. One can choose three noncollinear points p_0, p_1, p_2 in $T \cap Z$ such that the triangle with the vertices $(p_0, p_1, p_2, p_3 = p_0)$ contains p in its interior. From this triangle one readily obtains a strictly convex hexagon inside Z whose vertices include those of the triangle and belong to T. Applying to this hexagon the lemma, and iterating this procedure, one obtains a sequence of strictly convex polygons whose vertices belong to T, each including the vertices of the preceding polygon. The union V of all the neighborhoods of p bounded by these polygons is easily seen to be a convex neighborhood of p inside Z; and since the approximating polygons satisfy Condition (2), the set of their vertices is easily seen to be dense in fr V. Hence $T \cap$ fr V is dense in fr V. From Remark 1 it follows that $S \cap$ fr V is discontinuous and 0-dimensional or empty, which concludes the proof of Theorem 14.8.

Much more can be asserted if S is a closed set.

THEOREM 14.9   *If C is a closed subset of the plane that does not contain any square, then each point of C is the center of arbitrarily small circles whose intersections with C are discontinuous.* More precisely: *If S is any closed set in the plane, and p a point of S such that, for some number $r_0 > 0$, each circle about p with a radius $\leq r_0$ has an arc in common with S, then p lies in the closure of a set S' that is open (in the plane) and contained in S.* (As a point p on the boundary of a triangle S demonstrates, p is not necessarily a point of S'.)

The first half of the theorem immediately follows from the second. To prove the latter, let p be the pole of a polar coordinate system and denote the circle of radius r about p by C_r. Let $[c, d]$ be any segment contained in $]0, r_0]$. For each r in $[c, d]$, let the arc consisting of the points (r, φ) for all φ in $[a_r, a_r + m_r]$ be a subset of $S \cap C_r$. For each natural number k, let R_k be the set of all numbers r in $[c, d]$ such that $m_r \geqq 2\pi/k$. For example, R_1 is the (possibly empty) set of those radii r for which the entire circle C_r is a subset of S. Clearly, $R_k \subset R_{k+1}$. Each R_k is a closed subset of $[c, d]$. For let r be the limit of a sequence $(r_1, r_2, \ldots)$ of points belonging to R_k. We have to show that r belongs to R_k. The intersection of C_{r_i} with S contains the arc A_{r_i} consisting of the points (r_i, φ) for all φ in $[a_{r_i}, a_{r_i} + 2\pi/k]$. Since any bounded part of S is a compactum, there exists an accumulation point (r, a^*) of the set $\{(r_1, a_{r_1}), (r_2, a_{r_2}), \ldots\}$; and S obviously contains the arc consisting of the points (r, φ) for all φ in $[a^*, a^* + 2\pi/k]$. Each m_r is $\geqq 2\pi/k$ for some k. Hence $[c, d]$ is the union of all R_k. By Baire's Theorem 12.7, at least one of the sets R_k, say R_{k_0}, is dense in some subsegment $[c', d']$ of $[c, d]$. Being closed, R_{k_0} contains $[c', d']$. For each r in $[c', d']$, the circle C_r and S have in common an arc that subtends an angle $\geqq 2\pi/k_0$. Now divide each circle C_r whose radius belongs to $[c', d']$ into $2k_0$ equal sectors, $C_{r, 1}, C_{r, 2}, \ldots, C_{r, 2k_0}$, where $C_{r, j}$ consists of the points (r, φ) for all φ in $[2(j - 1)\pi/2k_0, 2j\pi/2k_0]$. Clearly, for each r in $[c', d']$ the circle C_r and S have at least one of the $2k_0$ arcs $C_{r, j}$ in common. For each of the numbers $j = 1, 2, \ldots, 2k_0$, let Q_j be the set of all numbers r in $[c', d']$ for which C_r and S have $C_{r, j}$ in common. Each of the $2k_0$ sets Q_j is easily seen to be closed and since each r in $[c', d']$ belongs to at least one set Q_j, their union is $[c', d']$. Again by Baire's Theorem, one concludes that at least one of the sets Q_j, say Q_{j_0}, is dense in some subsegment $[c'', d'']$ of $[c', d']$ and, being closed, contains $[c'', d'']$. But then the ring sector consisting of all points (r, φ) such that r is in $[c'', d'']$ and φ in $[2(j_0 - 1)\pi/2k_0, 2j_0\pi/2k_0]$ is a subset of S. Since one can choose d as close to 0 as one pleases it is clear that p is an accumulation point of an open subset of S.

EXERCISES

1. Prove Remark 1 by utilizing the fact that in the plane the frontier of each convex open set is homeomorphic to a circle or a straight line.

2. Show by an example that without the use of Condition (2) in the construction demonstrating Theorem 14.8, the resulting convex set V might have a frontier

in which the complement of S is not dense. (*Hint:* V might not be strictly convex.)

3. Prove that for a closed subsets of a square Q to contain some square either of the following conditions is sufficient. S has a segment in common (1) with each segment in Q that is parallel to, and equally long as, one side of Q (the same side for all segments); (2) with the frontier of each square domain that is concentric with Q and has sides parallel to Q. Generalize the latter condition so as to include other domains.

4. Give two examples of one-dimensional subsets of the plane respectively containing (1) every straight line with the exception of at most two points; (2) every polygon with the exception of a finite set of points. (*Hint:* Obtain the first set as the complement of a denumerable set P which is dense in the plane and does not include three collinear points. Construct P inductively. In choosing the nth point avoid the lines joining any two of the $n - 1$ points already chosen.)

5. Give an example of a one-dimensional subset of the plane which has a continuum in common with every algebraic curve. (*Hint:* By how many points is an algebraic curve $F(x, y) = 0$ determined if $F(x, y)$ is a polynomial expression including only powers of x and y not exceeding n?)

6. Under the assumptions of Theorem 14.8, let C and C' be two circles centered at the point p (mentioned in the proof) and contained in Z. Modify the proof in such a way that the frontier of V lies between C and C'. In other words, show that *each point of a one-dimensional subset S of the plane lies in arbitrarily small strictly convex neighborhoods whose frontiers have discontinuous intersections with S and which are as close to circles as may be desired.*

14.4   REMARKS ABOUT ANALYTIC COMPLEX FUNCTIONS

Recently, Theorem 14.7 and other results of curve theory have been used in a characterization of complete analytic functions as introduced by Weierstrass. Such a function is a set of ordered pairs of complex numbers defined in terms of the following three concepts.

A *power series p* of (positive) radius r_p about the complex number z_0 is a function p with the following property: There exists a ("determining") sequence of complex numbers

$$(z_0; w_0, a_1, a_2, \ldots) \text{ with lim inf} |a_n|^{-1/n} = r_p > 0$$

such that for each z within the set C_p of all z satisfying $|z - z_0| < r_p$ ("the circle of convergence of p") the function p assumes the value

$$p(z) = w_0 + a_1(z - z_0) + a_2(z - z_0)^2 + \cdots$$

The function p is the set of the pairs $(z, p(z))$ for all z in C_p.

A *direct continuation* of p is a power series p' with a determining sequence $(z_0'; w_0', a_1', a_2', \ldots)$ such that $|z_0' - z_0| < r_p$ and that for each z belonging to both circles C_p and $C_{p'}$ (that is, satisfying $|z - z_0| < r_p$ and $|z - z_0'| < r_{p'}$) the functions p and p' assume equal values:

$$w_0 + a_1(z - z_0) + a_2(z - z_0)^2 + \cdots = w_0' + a_1'(z - z_0') + a_2'(z - z_0')^2 + \cdots$$

A *continuation* of p is any power series p^* such that there exists a finite chain $(p, p_1, \ldots, p_n, p^*)$ in which each power series is a direct continuation of its predecessor.

A *complete analytic function* is a set F consisting, for some power series p, of all pairs of numbers (z, w) belonging to p or some continuation of p. Being a set of pairs of complex numbers F is a subset of the Cartesian product $Z \times W$ of two Argand planes. The question arises: How are the complete analytic functions characterized among the subsets of $Z \times W$?

By a *discontinuity configuration* we mean a sequence

$$(a_0, b_0); (a_1, b_1), (a_2, b_2), \ldots, (a_n, b_n), \ldots$$

of elements of $Z \times W$ such that the a_n converge to a_0 whereas the b_n do not converge to b_0. Now let C be a subset of $Z \times W$ with the following properties:

1. *Each point of C has a neighborhood that does not contain any discontinuity configuration contained in C.*

2. *C is connected.*
Then it can be proved that each point of C has a neighborhood N in $Z \times W$ such that $C \cap \mathrm{cl}\, N$ is homeomorphic to a subset of Z. A countable family $\{N_1, N_2, \ldots\}$ of such neighborhoods covers C. Now assume further

3. *C is not 1-dimensional.* (Being connected, C is not 0-dimensional.) Then at least one of the sets $C \cap \mathrm{cl}\, N_i$—call it N^*—is neither 1- nor 0-dimensional since by an extension (proved in dimension theory) of Theorem 13.12 the union S of any countable family of 1- or 0-dimensional sets that are *closed in S* (even if they are not compacta) is 1- or 0-dimensional.

Let M^* be the subset of Z that is homeomorphic to N^*. By Theorem 13.2, N^* is neither 1- nor 0-dimensional. Hence, according to Theorem 14.7, M^* contains a circular domain in the Argand plane Z. Consequently, N^* contains a disc (that is, a set homeomorphic to a circular domain) and this set is open in C. Finally assume:

4. *C is weakly homogeneous* in the sense that any two points of C have disjoint homeomorphic neighborhoods in C.

Then it follows that each point of C has a neighborhood in C that is a disc and a countable family of such neighborhoods covers C. A set with the properties (1)–(4) will be called a *patch*. It thus has been established that *each patch is what is called a 2-dimensional manifold.* (It will be noted that neither local connectedness nor even local compactness or completeness has been assumed.)

In the definition of the patch C, assumptions (3) and (4) can be replaced by the following assumption.

3*. *For each point of C, the frontiers of all sufficiently small neighborhoods of that point in $Z \times W$ have intersections with C that are nonlinear* (that is, not homeomorphic to a subset of the straight line.).

A necessary and sufficient condition for a patch C to be a subset of a complete analytic function is the following property of C.

5. *C is smooth,* that is, for each point (z, w) of C and each sequence (z_n, w_n) of points of C converging to (z, w) such that $z_n \neq z$, the quotients $(w_n - w)/(z_n - z)$ have a finite limit.

It then can be shown that a subset F of $Z \times W$ is a complete analytic function if and only if

A. *F is smoothly connected,* that is, any two points of F lie in a smooth patch contained in F;

B. *F is the union of a countable family of smooth patches;*

C. *F is saturated with regard to Properties A and B,* that is, $Z \times W$ does not contain a subset with the properties A and B containing F as a proper subset.

EXERCISES

1. Show that the set of all (z, w) such that $(w - z)^2 - z^3 = 0$ is not an analytic function but becomes one upon deletion of the pair $(0, 0)$.

2. Show that the set of all $(z, w) \neq (0, 0)$ such that $w^2 - z(z - 1)^2 = 0$ is a complete analytic function but not a patch.

REFERENCES

For the characterization of analytic functions cf. Menger, *Mathematische Annalen*, 167 (1966), 177–194 and *Comptes Rendus*, Paris Acad. Sci., 261 (1965), 4968. See vol. 267 (1968), 11 for a related characterization of the algebraic functions by T. Taner and interesting problems formulated by J. Leray and C. Arf. The simplification of the original characterization of analytical functions by introducing discontinuity configurations is due to H. I. Whitlock.

CHAPTER 15

The Universal Curve

15.1 THE DEFINITION OF U*

The plane universal curve U_2^* contains topological images of all curves in the Euclidean plane E_2. We are now going to construct in the three-dimensional Euclidean E_3 a curve U_3^* that contains topological images of all curves in E_3. But U_3^* is universal in an even much wider sense. According to a theorem of Menger, each curve contained in any higherdimensional Euclidean space, in the Hilbert space, and in fact in any metric space, is homeomorphic to a curve contained in E_3. Consequently, U_3^* is universal in an absolute sense—a fact whose proof requires methods quite different from those used in proving Theorem 12.10. Because of its universality, U_3^* will be denoted simply by U^*.

The construction of U^* is, just as that of U_2^*, a generalization of Cantor's construction of the discontinuum D^*. Like D^* and U_2^*, the universal curve can be obtained in three ways.

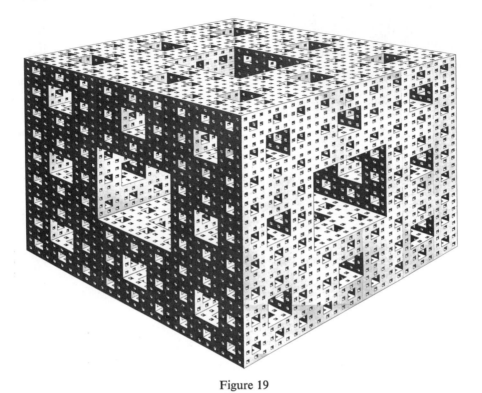

Figure 19

1. *By deleting cubes.* Divide a unit cube K_0 into 27 cubes of side $\frac{1}{3}$. Let K' be the union of the innermost cube and the six cubes with each of which it has a face in common. Delete this three-dimensional cross K' with the exception of the faces that are common to K' and the other cubes. Clearly, the remaining set K_1 is the union of those 20 closed cubes which contain points of the edges of K_0. One may also visualize K_1 as the result of drilling three mutually perpendicular passages through K_0, each joining the center squares of two opposite faces of K_0, the three passages having the interior of the innermost of the 27 cubes in common.

If this procedure is repeated in each of the 20 cubes of K_1 one arrives at a set K_2, which is the union of 20^2 closed cubes with sides of length $1/3^2$. Proceeding in this way, at the nth step one obtains the union K_n of 20^n closed cubes with the sides of length $1/3^n$. The universal curve U^* is defined as the set of the points in all the sets $K_1, K_2, \ldots, K_n, \ldots$. Being the intersection of a decreasing sequence of continua, U^* is a *continuum*. For each n, U^* is the union of a finite family of continua $< 1/n$ (for example, of the intersections of U^* with the 20^n cubes in K_n). Hence U^* is *locally connected*.

2. *As the closure of a union of segments*, namely the 12 edges of K_0, the edges of the 20 cubes in K_1 and, for each n, the edges of the 20^n cubes in K_n. The union F_n of all edges of these 20^n cubes is called the *nth framework* of U^*. Clearly, $F_n \subset F_{n+1}$. The set $F_{n+1} - F_n$ can easily be shown to be connected. The points of $F_n \cap \mathrm{cl}(F_{n+1} - F_n)$ are called the *nodes* of F_n. The union F of all F_n is called the *framework* of $U^* = \mathrm{cl}\, F$.

3. *Arithmetically.* The reader can easily verify that, in a properly chosen coordinate system, U^* is the set of all points (x, y, z) such that at least two of the three numbers x, y, z have ternary fractions without any 1.

Let c be any ternary irrational number whose ternary fraction contains 1 in at least one place. For each point common to U^* and the plane consisting of all points (c, y, z), both y and z have ternary fractions without any 1. The set of all these points is discontinuous (see Exercise 15, Section 11.4). Similarly, the planes consisting of all points (x, c, z) and (x, y, c) have 0-dimensional intersections with U^*. Clearly, each point of U^* is contained in arbitrarily small cubes whose six faces lie in planes of these three types and whose frontiers have consequently 0-dimensional intersections with U^*. Hence U^* is one-dimensional. The preceding results can be summarized as follows.

THEOREM 15.1   *The continuum U^* is a locally connected curve.*

15.2   HOMOLOGOUS RAMIFICATION SYSTEMS

All sets considered in this section are *closed subsets of compact metric spaces.* Two families, $\mathfrak{S}$ and $\mathfrak{S}'$, of sets in the same or in different spaces are said to be *homologous* if there is a one-to-one correspondence between their sets satisfying the following conditions:

(a) If A and B are sets in $\mathfrak{S}$ such that $A \subset B$, then $A' \subset B'$ for the corresponding sets in $\mathfrak{S}'$, and conversely.

(b) If $A_1, \ldots, A_r$ are sets in $\mathfrak{S}$ with an empty intersection, then the corresponding sets in $\mathfrak{S}'$ have an empty intersection, and conversely.

By a *ramification system* in a (compact) space we mean a family $\mathfrak{R}$ of (closed) sets of the space inductively defined as follows:

1. $\mathfrak{R}$ contains a finite family of sets, $R_1, \ldots, R_m$, called the sets *of order 1* in $\mathfrak{R}$.

2. If $R_{i_1,\ldots,i_n}$ (for some natural numbers n, $i_1,\ldots,i_n$) is a set of order n in $\mathfrak{R}$, then $\mathfrak{R}$ contains, for some natural number k (depending on $i_1,\ldots,i_n$) a finite family of subsets of $R_{i_1,\ldots,i_n}$,

$$R_{i_1,\ldots,i_n,1}, \ldots, R_{i_1,\ldots,i_n,k},$$

called sets *of order $n+1$* in $\mathfrak{R}$.

3. If d_n is the largest diameter of a set of order n, then $\lim_{n\to\infty} d_n = 0$.

For each n, let S_n be the union of all sets of order n in $\mathfrak{R}$. Each set S_n is closed and $S_{n+1} \subset S_n$. The intersection S of all sets S_n is called the set *generated* by the ramification system $\mathfrak{R}$. Clearly, S is a compactum. For each point p in S, let $S_n(p)$ be the union of all sets of order n containing p. The closed set $S_n(p)$ has a diameter $\leq 2d_n$, and $S_{n+1}(p) \subset S_n(p)$.

Let $\mathfrak{R}'$ be a ramification system in the same space as $\mathfrak{R}$ or in another space, and d_n' the largest diameter of a set of order n in $\mathfrak{R}'$. If $\mathfrak{R}$ and $\mathfrak{R}'$ are homologous, then the simplest way to describe this relation is by denoting corresponding sets by $R_{i_1,\ldots,i_n}$ and $R'_{i_1,\ldots,i_n}$.

THEOREM 15.3 *The sets generated by homologous ramification systems are homeomorphic.*

If S and S' are the compacta generated by $\mathfrak{R}$ and $\mathfrak{R}'$, one can define a mapping of S into S' as follows. For each point p of S, let $S_n'(p)$ be the union of all sets in $\mathfrak{R}'$ that correspond to the sets in $\mathfrak{R}$ containing p. Since the latter sets have a common point (namely p), so do because of (b) the sets in $\mathfrak{R}'$ with the union $S_n'(p)$. Hence $S_n'(p)$ has a diameter $\leq 2d_n'$. From (a) it follows that $S_{n+1}'(p) \subset S_n'(p)$. The diameters of the sets $S_1'(p), \ldots, S_n'(p), \ldots$ converge to 0, whence these sets have at most one point in common. Since the sets are closed in a compact space their intersection is not empty. Hence this intersection consists of a single point, p'. Associating p' with p one establishes, as is easily seen, a one-to-one correspondence between S and S' which is continuous, whence S and S' are homeomorphic.

EXERCISES

1. Let $\mathfrak{S}$ consist of four points in a plane and the lines joining them. Define a homologous family of subsets of the set $\{1, 2, 3, 4\}$ depending upon whether no triple or exactly one triple or all four points are collinear.

2. For each of the following sets define two ramification systems generating the set: (a) a disc (a circular domain); (b) the discontinuum D^*; (c) the plane universal curve U_2^*.

15.3   THE UNIVERSALITY OF U*

THEOREM 14.2   *Each curve in any metric space is homeomorphic to a subset of U^*.*

The proof will be outlined by sketching, for any curve C, the construction of a ramification system generating C and of a homologous ramification system of polyhedra in E_3 generating a subset of U^*. A polyhedron is the union of a finite family of tetrahedra that is homeomorphic to a sphere (which excludes, for example, the union of two tetrahedra having only one vertex or one edge in common). The boundary of the polyhedron P is denoted by bd P.

The construction of the ramification systems is based on the following elementary

Remark.   If G is a graph which, except for a finite set of points, is contained in the interior of the polyhedron P, and if p' and p'' are two points of P, then P contains a polygon joining p' and p'' which has no points except possibly p' and/or p'' in common with $G \cup$ bd P. If G is contained in the framework F of U^*, and p' and p'' are nodes of F, then F contains a polygon joining p' and p'' that has no points except possibly p' and/or p'' in common with G.

One chooses two sequences of numbers, $d_1, d_2, \ldots$ and $d_1', d_2', \ldots$ both having the limit 0. The diameters of the sets of order n in the ramification systems generating C and the subset of U^* will be $< d_n$ and $< d_n'$, respectively.

It is not difficult to construct the sets of order 1 of the two systems. By Theorem 13.15, C is the union of closed sets $< d_1$, say $C_1, \ldots, C_m$, such that all intersections $C_i \cap C_j$ and $C_i \cap C_j \cap C_k$ are discontinuous and empty, respectively. Using the first half of the preceding Remark one can fairly easily construct a homologous family of polyhedra $< d_1'$, say $P_1, \ldots, P_m$, such that $P_i \cap P_j$ is a polygonal domain $Q_{ij} \subset$ bd $P_i \cap$ bd P_j if $C_i \cap C_j$ is not empty, and that P_i and P_j are disjoint if C_i and C_j are disjoint. By shrinking the P_i, if necessary, one may assume the Q_{ij} to be mutually disjoint and of diameter $< d_2$.

The construction of the sets of order 2 is more complicated for two reasons. First of all, in constructing the sets C_{ij} with first index i one must take into account all the mutually disjoint sets $D_{ij} = C_i \cap C_j$ that are nonempty, and the corresponding polygonal domains Q_{ij} in bd P_i. Let their union be Q_i. One can represent C_i as the union of closed sets $<d_2$, say $C_{i1}, \ldots, C_{in}$, such that each point of a D_{ij} lies in exactly one of the n sets C_{ik}. Again from the first half of the preceding Remark one can infer that P_i contains a homologous family of polyhedra, $P_{i1}, \ldots, P_{in}$, such that $P_{ij} \cap$ bd P_i is empty or a polyhedral domain $\subset Q_{ij}$ according as $C_i \cap D_{ij}$ is empty or nonempty.

The second difficulty lies in the fact that the polyhedra P_{ij} just mentioned are not necessarily $<d_2'$. But if, say, $P_{i1} = P'$ has a diameter $\geq d_2'$, then one can replace P' by a chain of polyhedra $P_1', \ldots, P_k'$ of diameters $<d_2$ such that (1) $P_i' \cap P_{i+1}'$ is a polygonal domain $\subset$ bd $P_i' \cap$ bd P_{i+1}' whereas P_i' and P_j' are disjoint for $|i - j| > 1$. (2) Each polyhedral domain $P' \cap Q_{ij} (j = 2, \ldots, n)$ is a part of exactly one of the sets $P_1', \ldots, P_k'$. (3) $P' \cap$ bd $P' \subset Q_1$ for $i = 1, \ldots, k$. One then can replace C_1 by $C_1', \ldots, C_k'$ such that $C' \cap C_{i+1}'$ is discontinuous and C_i and C_j are disjoint for $|i - j| > 1$. In cases where the discontinuous set $C_i' \cap C_{i+1}'$ is empty one replaces P_i' and P_{i+1}' or one of them by slightly smaller polyhedra which are disjoint. These resulting polyhedra are homologous to $C_1', \ldots, C_k'$. If this procedure is followed for all P_{ij} that are $\geq d_2'$ one obtains two homologous families of closed sets $C_{i_1 i_2}$ and polyhedra $P_{i_1 i_2}$, which are the sets of order 2 of two homologous ramification systems.

Using the second half of the introductory Remark one can show that all the polyhedra used in these constructions may be assumed to be unions of cubes used in defining U^*, the intersections of any two of the polyhedra being squares that are common to two of those cubes. In this way, the set in E_3 generated by the ramification system of polyhedra, which are homeomorphic to C, is a subset of U^*.

These brief indications may serve as a guide to the study of the details of the proof, contained in Chapter XII of the book *Kurventheorie*.

INDEX